VIIth International Colloquium on Amphipoda

Developments in Hydrobiology 70

Series editor
H. J. Dumont

VIIth International Colloquium
on Amphipoda

Proceedings of the VIIth International Colloquium on Amphipoda held in Walpole,
Maine, USA, 14—16 September 1990

Edited by
L. Watling

Reprinted from Hydrobiologia, vol. 223 (1991)

Springer-Science+Business Media, B.V.

Library of Congress Cataloging-in-Publication Data

```
International Colloquium on Amphipoda (7th : 1990 : Walpole, Me.)
    VIIth International Colloquium on Amphipoda : proceedings of the
VIIth International Colloquium on Amphipoda, held in Walpole, Maine,
USA, September 14-16, 1990 / edited by L. Watling.
        p.    cm. -- (Developments in hydrobiology ; 70)
    Includes bibliographical references and index.
    ISBN 978-94-010-5568-0      ISBN 978-94-011-3542-9 (eBook)
    DOI 10.1007/978-94-011-3542-9
    1. Amphipoda--Congresses.    I. Watling, Les.   II. Title.
III. Title: 7th International Colloquium on Amphipoda.   IV. Series.
QL444.M315I58   1990
595.3'71--dc20                                         91-31549
```

ISBN 978-94-010-5568-0

Printed on acid-free paper

Contents

vi

Hydrobiologia **223**: vii, 1991.
L. Watling (ed.), VIIth International Colloquium on Amphipoda.

Preface

This volume contains papers given in association with the VIIth International Colloquium on Amphipoda held at the University of Maine's Darling Center in Walpole, South Bristol, Maine, U.S.A., on 14–16 September 1990. Four additional papers by persons unable to come to Maine are also included.

The amphipod colloquia began in Verona in 1979, originally as a meeting of specialists interested in the systematics of *Gammarus* and *Niphargus*. After a second colloquium, held in Lyon in 1973, the third was convened in conjunction with the 1st International Symposium on Groundwater Ecology in Schlitz, Germany, in 1975. These two groups met together in Blacksburg (1978) and Lodz, Poland (1980 and 1981). The VIth Colloquium for the amphipod side was held in Ambleteuse, France. More complete histories of the early meetings are given in Crustaceana, Supplement 4, pp. 1–2, and Supplement 6, pp. 1–3, and Polskie Archiwum Hydrobiologii, volume 29, no. 2, p. 1. Since Ambleteuse, the meeting has been more generally concerned with amphipods of all kinds.

Most of the contributions given at the colloquia have been published in the following volumes:

I.   Verona, 1969. Memorie del Museo civico di Storia naturale di Verona, vol. 5 (1972).
II. Lyon, 1973. Crustaceana, vols. 27(3), 28(1), 28(2), 29(1) (1974–75).
III. Schlitz, 1975. Crustaceana, Suppl. 4 (1977).
IV. Blacksburg, Virginia, U.S.A., 1978. Crustaceana, Suppl. 6 (1980).
V.   Lodz, Poland, 1980 and 1981. Polskie Archiwum Hydrobiologii, Vol. 29, No. 2 (1982).
VI. Ambleteuse, France, 1985. Crustaceana, Suppl. 13 (1988).
VII. Walpole, Maine, U.S.A., 1990. This volume.

There were 42 attendees, from the following countries: Australia, 5; Brazil, 1; Canada, 10; Chile, 1; Germany, 3; Japan, 2; The Netherlands, 4; Poland, 2; U.S.A., 13; and U.S.S.R., 1. All functions were held on the grounds of the Darling Marine Center, including meals, which were taken under a large canopy–the weather still being quite pleasant. The conference ended with a trip to Camden, one of Maine's most picturesque harbor villages, on Friday evening. On Saturday there was an excursion by boat to Monhegan Island, located 15 km offshore, followed by an old-fashioned Maine lobster bake later that evening. The fine food, and good conversation conducted in a variety of accents, is still very fondly remembered by the local hosts, the Damariscotta Region Chamber of Commerce.

I would like to give special thanks to Madolyn Musick, Irene Leeman, and the rest of the staff and students of the Darling Marine Center who worked very hard to make the meeting a success. Also, the warm welcome extended by the Damariscotta Region Chamber of Commerce and the owners of the Pemaquid Hotel to our overseas guests was greatly appreciated. The University of Maine's Center for Marine Studies helped financially, ultimately making the attendance of some participants possible. For this, I would like to thank Dr. Robert Wall, the Center's Director, who appreciates the value of scientific dialogue.

L. WATLING
University of Maine

This special volume on the biology of Amphipoda is dedicated to

J. Laurens Barnard
1928–1991

who died suddenly on 16 August 1991, while working in Florida. Jerry was unquestionably the most important contributor to our knowledge of the Amphipoda since the early years of Sars and Stebbing, and his death leaves an enormous void. He described many hundreds of species and was responsible for several syntheses, the most important of which were the two volumes published privately on the freshwater Amphipoda of the world and the just published world monograph, an update of his 1969 treatment of the families and genera of marine gammaridean amphipods.

To many of us, though, Jerry was more than a source of descriptions and details. Jerry's work, and personality, influenced the lives and careers of many people. His devotion to amphipods and birds, his appreciation of the natural world, and his sense of humor and generosity were an inspiration to those who knew him. He always encouraged us to continue our taxonomic studies, recognizing that for many of us it would mean only part-time devotion to this field. He was also a sympathetic sounding board, always willing to try out new ideas and approaches.

We are, sadly, in a period where taxonomic studies do not seem to have the value they once had, where new taxonomists are not being trained, or, if trained, are unemployable. Jerry lamented this decline in systematics, but continued to advance amphipod taxonomy until the day he died. We can only hope to have the opportunity to carry on his work, and to follow his example in stimulating a new generation of scientists to take up the science of taxonomy.

LES WATLING & JIM THOMAS

Hydrobiologia **223**: 1–9, 1991.
L. Watling (ed.), VIIth International Colloquium on Amphipoda.
© 1991 *Kluwer Academic Publishers.*

Comparative fore-gut morphology of Antarctic Amphipoda (Crustacea) adapted to different food sources

Charles Oliver Coleman
Universität Oldenburg, Fachbereich 7, Arbeitsgruppe Zoomorphologie, Postfach 2503, D-2900 Oldenburg, Germany

Key words: Antarctic, Crustacea, Amphipoda, fore-gut morphology, food preference

Abstract

The fore-gut morphology of ten species of Antarctic amphipods utilizing different food sources was investigated. There are considerable differences in shape, relative lengths of the stomachs and their structures. Relative lengths of the stomachs range from more than 30% to 2% compared to the total body lengths. The relative length of the anterior rough filter corresponds in general with the relative stomach length. Stomachs with long rough filter share in general a small fine filter area.

Interspecific differences of stomach lateralia might be used for phylogenetic analysis, but are apparently not related to different food sources.

Different speculative selective pressures that might have had influence on the evolution of fore-guts are discussed.

Introduction

Most Antarctic amphipods are considered to be mainly omnivorous or necrophagus (Arnaud, 1970, 1977). The exact food preference of more than 500 species from Antarctic and Subantarctic waters (Lowry & Bullock, 1976) is unknown.

Very few investigations of fore-gut and mid-gut contents of Antarctic amphipods have been carried out (Bone, 1972; Rakusa-Suszczewki, 1972; Richardson & Whitaker, 1979; Stockton, 1982; Bregazzi, 1972, 1973; Slattery & Oliver, 1986; Coleman, 1989a; Coleman, 1990a).

One might assume that Antarctic amphipods have food preferences similar to related and better investigated species from boreal regions. On the other hand, many species are endemic for the Antarctic (Knox & Lowry, 1974).

One of the major problems in investigating fore-gut and midgut contents of Antarctic amphipods is that in most cases a small number of specimens is available for dissection and thus statistics cannot be used for supporting the results. Alternatively, functional morphology of the mouthparts can be used to support the results of the gut content examinations. Food specialists may have special adaptations to their food source. For example the mandible of *Gnathiphimedia mandibularis* K.H. Barnard 1930 is adapted for crushing zooids of bryozoans rather than for biting (Coleman, 1989b). The sharp dentated incisor of the mandible of *Maxilliphimedia longipes* is used for cutting soft mucous cnidarian tissue (Coleman, 1989a). Through the use of its distal maxillipedal endites *Anchiphimedia dorsalis* possibly pushes detritus together which then might be transported to the mouth opening by the maxillae (Coleman, in press).

In the following study this morphological approach has been extended to the fore-gut structures. Stomachs of species utilizing different food sources have been compared to check for differences that might show adaptations to their food preference.

The morphology and the function of the fore-gut is relatively well understood. For example Kanneworff & Nicolaisen (1969) worked out the functional morphology of the fore-gut of *Bathyporeia sarsi* and Martin (1964) that of *Marinogammarus obtusatus*. Icely & Nott (1984) published a detailed study on the anatomy of *Corophium volutator*. The structures, muscles and functions of the fore-gut of the isopod *Asellus aquaticus* were described in detail by Scheloske (1976). Kanneworff & Nicolaisen (1969) and Thiem (1942) made in vivo observations of small specimens of *Bathyporeia* and *Synurella* respectively. They obtained data of the movement of the stomach structures, which helped to understand the function of the stomach elements.

Food is sucked through the esophagus and is pushed into the stomach by the lateralia, which are lateral invaginations of the anterior stomach region (Fig. 1A, B). Food is digested in the storage cavity. Breakdown products of fine consistency pass into the mid-gut gland for absorption. The remains enter the mid-gut and leave after having passed through the hind-gut and rectum.

A cross section through the anterior stomach region reveals three compartments: First a dorsal cavity, which is never filled with food, is separated by rows of setae: the superolateralia (Fig. 1B, C). Enzymes are probably transported in this cavity anteriorly and are discharged into a second compartment: the storage cavity, where the food is digested (Kanneworff & Nicolaisen, 1969). Ventrally, there is a third compartment separated by lateral invaginations (inferolateralia anteriores) with rows of setae (clatri setarum anteriores). An elevation of the ventral stomach wall (inferomedianum anterius) divides this part into two channels (Fig. 1C). The digested food is filtered through these setal rows. Digestive fluids and small particles are pressed through the anterior rough filter unit, and move posteriorly in the channels. They unite posteriorly and then are divided again by a second medial ventral elevation (inferomedianum posterius). This ridge-like structure bears two grates comprising rows of setulated setae laterally (clatri setarum posteriores). They cover small channels which are connected with the mid-gut gland (Fig. 1D). This unit works as a fine filter. Only microscopic food particles can pass through it and are processed in the digestive gland.

Dorsally, the inferomedianum posterius is covered by infoldings of the lateral stomach walls (inferolateralia posteriores). These folds partly overlap each other medially. They prevent penetration of any food particles from the dorsal storage room into the ventral filter unit.

The food thus is filtered twice and only fluid and very fine food particles enter the mid-gut gland. A thin lamella (valvula postero-ventralis) is located posteroventrally to the stomach and extends into the atrium of the mid-gut gland. Its function is not known.

Material and methods

The material was collected during the Antarctic expedition ANT VI/2 (1987) of RV 'Polarstern'. The benthic species were obtained with a commercial fishery bottom-trawl near Elephant Island (South Shetland Islands), the planktonic species with a rectangular midwater trawl at station 12/208, situated in Bransfield Strait (63° 05,8′ S, 61° 54,3′ W), depth 1164–1310 m. The animals were preserved in 4% Formalin. Ten to fifteen specimens of each species were dissected. Stomach and midgut contents were determined using light microscopy.

A stomach from each species was heated for 2 hours in concentrated potassium hydroxide solution to remove tissue. The unstained stomachs were transferred into glycerol. All structures can easily be seen through the transparent cuticle. Small pieces of glass were used to stabilize the stomachs in the watchglass. Drawings were made with a Wild M5 dissecting

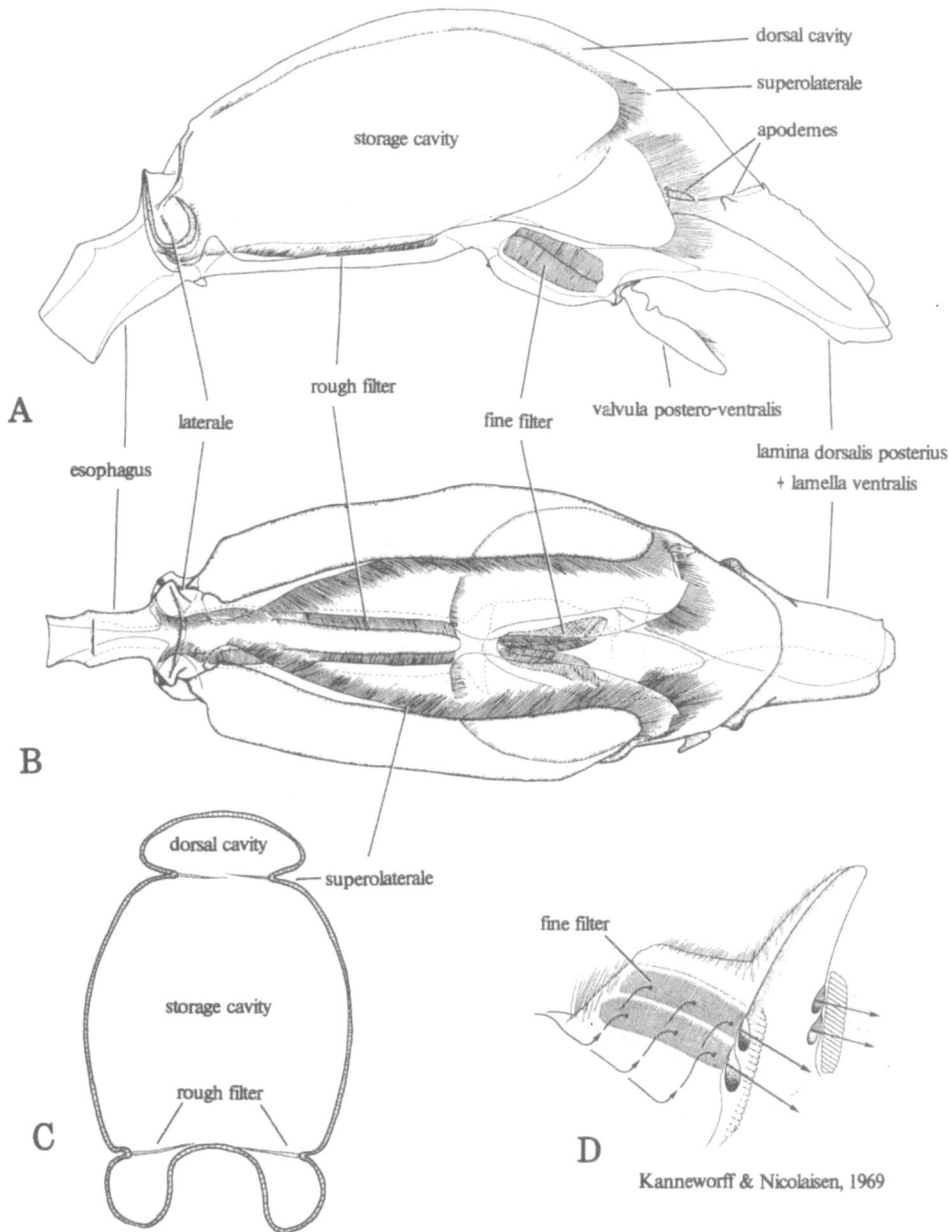

Fig. 1 A–D. Fore-guts of amphipods. A: Lateral view of the fore-gut of *Epimeria georgiana*. B: Dorsal view of the fore-gut of *Epimeria georgiana*. C: Schematic cross-section through the anterior stomach region. D: Fine filter unit (drawing from Kanneworff & Nicolaisen, 1969).

microscope and a Leitz 20 EB microscope with a camera lucida.

The lengths of the specimens was measured from the tip of the rostrum to the end of the telson. The length of each stomach was determined from the anterior margin of the lateralia to the posterior tip of the inferomedianum posterius. Measurements were also made of the lengths of the rough filters. Using an image analyser the area of the fine filter was determined. In order to avoid comparing an area to a length because of the different dimensions, the square root of the areas was taken and put in relation to the stomach lengths.

Results

Diversity of fore-guts

The shape of the fore-guts of the examined species is surprisingly diverse. Intraspecific differences in shape, relative lengths of structures and armature

of the lateralia in *Eusirus perdentatus* appear to be minimal (Fig. 2A).

The following stomach morphologies show some extreme examples of shape and structure.

The stomach of *Waldeckia obesa* is elongated and takes up a large part of the pereon. In *Cyphocaris richardi* the stomach is very much shorter and laterally expanded. It reaches only into the first pereomere. The lamina dorsalis posterius and lamella ventralis are conspicuously short. In *Paraceradocus stenepimerus*, however, these structures are elongated. The caprellid *Aeginoides*

Table 1. List of the examined species.

Species	Food preference (# of specimens examined)	Reference
Lysianassidae: *Waldeckia obesa* (Chevreux, 1905)	necrophagus	Arnaud, 1970
Stegoecephalidae: *Parandania boecki* (Stebbing, 1888)	medusae (12)	Moore & Rainbow (1989), Coleman (1990b)
Iphimediiae: *Maxilliphimedia longipes* (Walker, 1906)	Cnidaria (10)	Coleman (1989)
Paramphithoidae: *Epimeria georgiana* Schellenberg, 1931	carnivorous (15)	Coleman (unpublished)
Gammaridae: *Paraceradocus stenepimerus* Andres, 1984	detritus (13)	Coleman (1989c)
Ampeliscidae: *Byblis subantarctica* Schellenberg, 1931	detritus (10)	Coleman (unpublished)
Iphimediidae: *Anchiphimedia dorsalis* K. H. Barnard 1930	detritus (10)	Coleman (in press)
Eusiridae: *Eusirus perdentatus* Chevreux 1912	carnivorous (11)	Coleman (unpublished)
Lysianassidae: *Cyphocaris richardi* Chevreux, 1905	carnivorous (10)	Hopkins, 1985 Coleman (unpublished)
Caprellidae: *Aeginoides gaussi* Schellenberg, 1926	carnivorous? (direct observation of living specimens)	Coleman (unpublished)

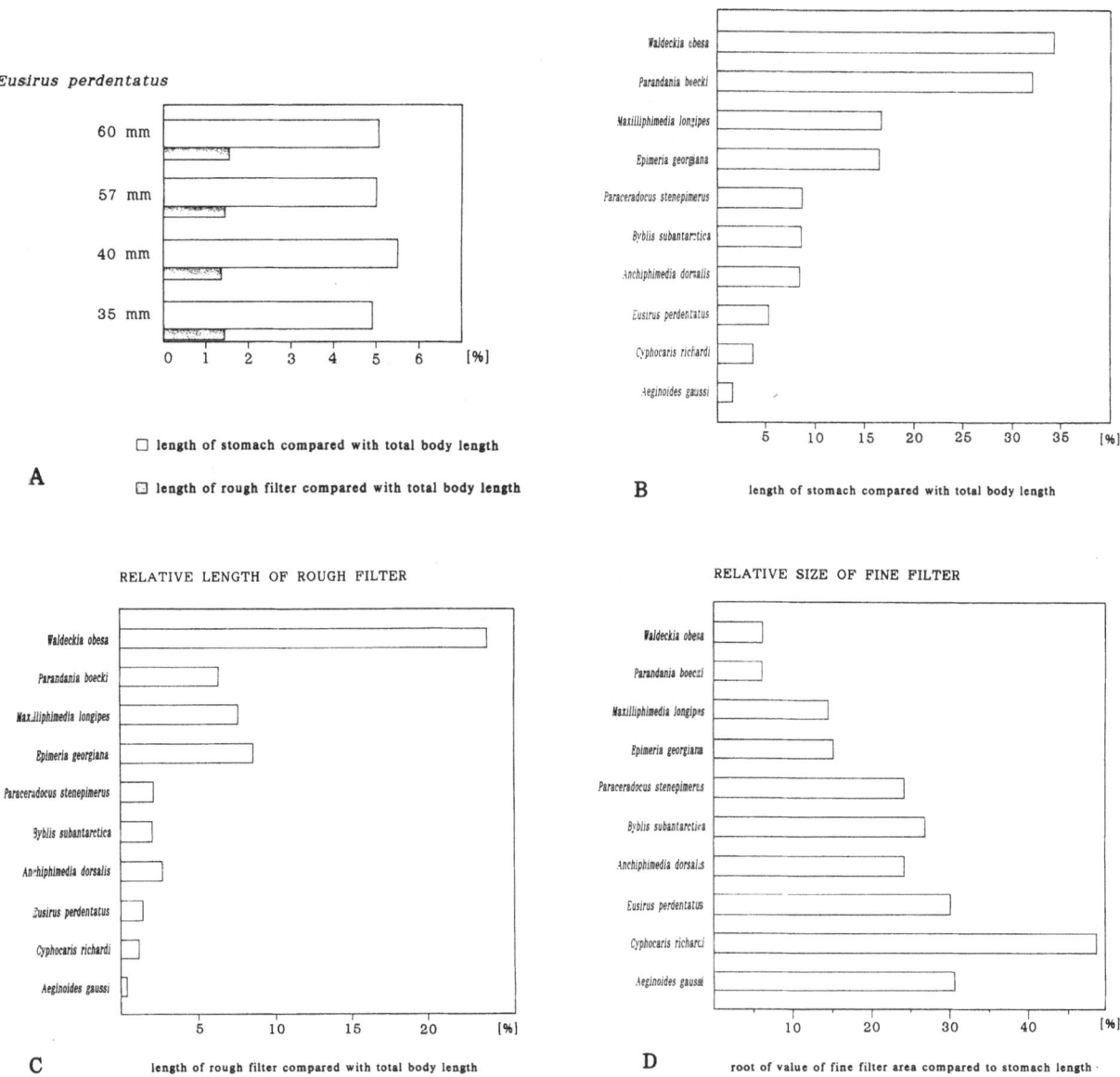

Fig. 2 A–D. Relative lengths of amphipod stomachs and their structures. A: Comparison of relative stomach and rough filter lengths of four specimens of *Eusirus perdentatus* showing very little intraspecific difference. B: Relative stomach lengths of ten species of Antarctic amphipods arranged according to their length. C: Relative length of rough filters to body length. D: Relative size of fine-filters.

gaussi has the smallest stomach, relatively and absolutely. It is only half a millimeter long.

Parandania boecki feeds on medusae and has a most remarkable fore-gut, unique among amphipods. The lumen of the esophagus is reduced to a V-like slit which may be enlarged by contraction of its muscles. The esophagus is extremely restricted near the entrance to the stomach by folds that resemble the doors of a gate. The superolateralia are not situated dorsally, as in all other known amphipod stomachs, but anterolaterally. The inferomedianum anterius is wider anteriorly than posteriorly. Thus the ventral channels are not parallel as in the other stomachs. The stomach is a very large sack. The posterior stomach region consists of very small cuticular folds which might provide extra elasticity. There are none of the typical lamellae posteriorly, which project into the mid-gut in other amphipods, but just a small, ventral opening which is connected to the mid-gut gland. There is apparently no connection between the stomach and the mid-gut, the latter beginning blindly. It is conceivable that the function of the midgut has changed into an additional excretion organ (Coleman, 1990b). If there are any undigestable remains, they must be reurgitated. Digested tissue can only be absorbed in the mid-gut glands. Certain compounds are demobilized in the cells of the midgut glands in form of crystals (Moore & Rainbow, 1989).

The cuticle of the stomachs shows interspecific differences. Stomachs of the genera *Waldeckia*, *Maxilliphimedia* and *Parandania* were remarkably collapsible. The cuticle may be thinner and stabilizing elements less developed than in the smaller, more rigid stomachs, for example of *Cyphocaris*.

Relative lengths of the stomach and its filter units

The stomachs show interspecific differences in length. In Fig. 2B the stomach lengths are shown in relation to body length with species arranged from longest to shortest stomachs.

The four species with the largest stomachs are tissue ingestors, *Waldeckia* being necrophagous, *Parandania* and *Maxilliphimedia* feeding on coelenterates and *Epimeria* being carnivorous. *Paraceradocus*, *Byblis* and *Anchiphimedia* have a shorter but similar relative stomach length and all feed on detritus. The carnivorous *Eusirus*, *Cyphocaris* and *Aeginoides* have the smallest stomachs but they may have a different mode of digestion or feeding frequency compared with the first group which stores food in their distensible stomachs.

The relative lengths of the rough filters reflect,

with a few exceptions, the conditions of the relative stomach lengths (Fig. 2C). The reverse is true for the area of the fine filter. The stomachs with the longest rough filter have the smallest fine filter areas.

Armature of the lateralia

Comparing the lateralia of tissue ingestors (Fig. 3A) and detrital feeders (Fig. 3B) there seems to be no significant differences in the armature which might be related to food utilized. Almost all lateralia have the following common features: stout medial setae and laterally long and slender setae. Exceptions to this trend are three species which have a reduced setation on the lateralia. For example the lateralia of the caprellid *Aeginoides* carry only a few stout setae. This might be due to the extremely small lateralia surface. The iphimediids *Anchiphimedia* and *Maxilliphimedia* have very different food preferences, the former feeding on detritus and the latter on Cnidaria. Nevertheless, they have similar lateralia.

Discussion

The technique of processing fore-guts with hot potassium hydroxide solution is a noteworthy method for the study of fore-guts. On the other hand the integrity of the stomach is somewhat disturbed. By removing the muscles some structures slightly change their position, but not their lengths, thus the stomach volume may deviate from anatomical condition. For this reason only strongly chitinized stomach structures were compared. With histological series the stomach volume and with SEM studies the total area of pores and intersetal spaces of the filter elements will be determined and compared in a future project.

The remarkable diversity of lengths and proportions of the stomachs are difficult to interpret because exact data of the ecology for most species are missing. The adaptive pressures which

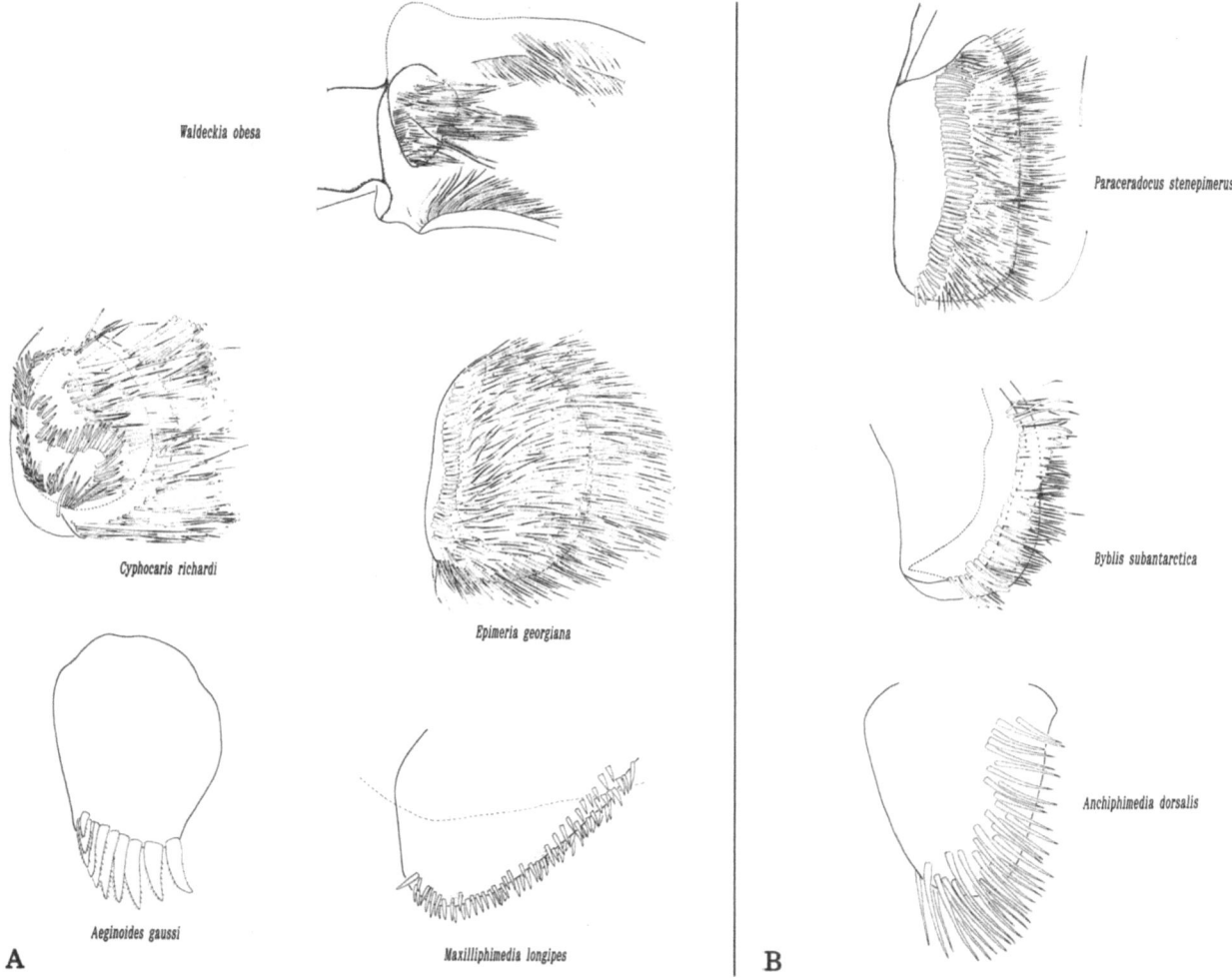

Fig. 3 A−B. Right lateralia of Antarctic amphipods. A: Lateralia of tissue ingesting species. B: Lateralia of detrital feeders.

influenced the evolution of the fore-guts may have been highly diverse. The following hypotheses might show relations between autecological data concerning feeding and stomach morphology.

The three major factors with respect to feeding appear to be as follows:

A) Intervals of feeding: extremeties range from almost continuous feeding to long intervals and storing of food.
B) Kind of preferred food: 1) The consistence of the ingested food may be solid, mucous or easily disintegratable. 2) The nutritious value per volume may differ considerably; the tissue

of medusae has certainly a lower nutritious value than muscle tissue. 3) The size of the morsels is bigger in species with cutting mandibles, medially excavated mandibular bodies and lacking molars compared to species with a normal mandibular morphology.
C) Speed of digestion: Dependent on the consistency and kind of food the digestion speed may differ.

All factors are interdependent.

A large stomach size is found in the necrophagous species *Waldeckia obesa*. Very likely, the intervals of feeding are long and the stomach acts as a food reservoir. Some specimens, maintained in aquaria, survived 18 months without feeding, which indicates that apart from the food storing capabilities these animals have additional physiological adaptation to endure long periods without feeding (Coleman, unpublished data). Dahl (1979) found differences in the fore-gut size of two *Orchomene* species with different potentials of storing food in the fore-gut. It is not clear if the observed extended rough filter length is an adaptation or simply a characteristic of stomach enlargement.

The area of the fine filter of *Waldeckia* is very small. It is large enough, however, to cope with the roughly filtered food, as the speed of digestion in such food storing stomachs is certainly low. On the other hand, for short feeding intervals, constant feeding or fast digestion, a small stomach may be positively selected. A large fine filter area may be developed in these stomachs to avoid obstruction of the pores, because they may be in contact with a large amount of fine particles per unit of time.

The kind of food certainly has an important influence on the stomach morphology. Species feeding on solid food have a larger stomach and a longer rough filter compared with species ingesting food consisting of small particles. The soft mucous food of the coelenterate feeders, *Parandania* and *Maxilliphimedia*, probably is disintegrated almost completely during digestion so that a relatively small fine filter is sufficient. The stomach cuticle of these species is very soft, perhaps a result of only feeding on soft food. The lumen of the esophagus is large in *Maxilliphimedia*, but reduced to a V-like slit in *Parandania*. The armature of the lateralia is reduced in *Maxilliphimedia*. It is not clear if this could be an adaptation to mucous food.

The nutritional value per volume could influence the stomach size. The mesogloea of medusae certainly has a low nutritional value in relation to its volume. A large stomach, as in *Parandania*, could be an adaptation to this. To get the same amount of energy, *Parandania* must eat more food than other species utilizing highly nutritive food.

The stomach morphology might be influenced by the size of the food morsels. With big food pieces being digested more slowly, a larger storage stomach with a relatively small fine filter could be advantageous. Big food particles could have some influence on the lumen of the esophagus. As observed in detrital feeders, many sandgrains and detritus in the food might be the reason that the stomach cuticle appears to be stronger. The stomachs of detrital feeders are possibly smaller than in the food specialists because food is taken up more or less continuously and probably leaves the stomach after only a quick filtration. These stomachs, therefore, have limited food storage function. A large area of the fine filter appears to be advantageous to deal with a large amount of fine particles in the food.

The armature of the lateralia seems to have no relation to the food source. More extensive comparative studies of lateralia will give a clearer insight into the adaptive significance of these structures for utilizing certain kinds of foods.

The lateralia structures might offer a good trait for phylogenetic reconstructions.

Acknowledgements

The author is grateful to the Alfred Wegener Institut für Polarforschung (Bremerhaven, FRG) for logistic help during the collection of the animals and to Dr. J. W. Wägele for critical discussions. I thank Mrs. E. Nelson, Dr. J. L. Barnard and Mr. W. Snyder for improving the manuscript. The study was supported by a grant from the Deutsche Forschungs-Gemeinschaft.

References

Andres, H. G., 1984. Neue Vertreter der antarktisch verbreiteten Gattung *Paraceradocus* Stebbing, 1899 (Crustacea: Amphipoda: Gammaridae). Mitt. hamb. zool. Mus. Inst. 81: 85–107.

Arnaud, P. M., 1970. Frequency and ecological significance of necrophagy among the benthic species of Antarctic coastal waters. In M. V. Holgate (ed), Antarctic ecology Vol. 1. Academic Press. London New York: 256–267.

Arnaud, P. M., 1977. Adaptations within the Antarctic marine ecosystem. In G. A. Llano (ed.), Adaptations within Antarctic marine ecosystem. In G. A. Llano (ed.), Adaptations within Antarctic ecosystems. Proc. 3rd SCAR Symp. Antarct. Biol. Houston: 135–158.

Barnard, K. H., 1930. Amphipoda: British Antarctic ('Terra Nova') Expedition, 1910. Nat. Hist. Rep., Zool., 8: 307–454.

Bone, D. G., 1972. Aspects of the biology of the Antarctic amphipod Bovallia gigantea Pfeffer at Signy Island, South Orkney Island. Br. Antarct. Surv. Bull. 27: 105–122.

Bregazzi, P. K., 1972. Life cycle and seasonal movements of Cheirimedon femoratus (Pfeffer) and Tryphosella kergueleni (Miers) (Crustacea: Amphipoda). Br. Antarct. Surv. Bull. 30: 1–34.

Bregazzi, P. K., 1973. Locomotor activity rhythms in Tryphosella kergueleni (Miers) and Cheirimedon femoratus (Pfeffer) (Crustacea: Amphipoda). Br. Antarct. Surv. Bull. 33 & 34: 17–32.

Chevreux, E., 1905. Description d'un amphipoda (Cyphocaris Richardi nov. sp.) provenant des peches dernière au filet a grande ouverture de la dernière campagne du yacht Princesse-Alice (1904). Bull. Mus. Ocean., Monaco 24: 1–5.

Chevreux, E., 1912. Diagnoses d'Amphipodes nouveaux. Deuxième Expedition dans l'Antarctique, dirigée par le Dr Charcot, 1908–1910. Bull. Mus. Hist. nat., Paris 18: 208–218.

Coleman, C. O., 1989a. On the nutrition of two Antarctic Acanthonotozomatidae (Crustacea: Amphipoda): Gut contents and functional morphology of mouthparts. Pol. Biol. 9: 287–294.

Coleman, C. O., 1989b. Gnathiphimedia mandibularis K. H. Barnard, 1930, an Antarctic amphipod (Acanthonotozomatidae, Crustacea) feeding on Bryozoa. Antarct. Sci. 1: 343–344.

Coleman, C. O., 1989c. Burrowing, grooming, and feeding behaviour of Paraceradocus, an Antarctic amphipod genus (Crustacea). Polar Biology 10: 43–48.

Coleman, C. O., 1990a. Bathypanoploea schellenbergi Holman & Watling, 1983, an Antarctic amphipod (Crustacea) feeding on Holothuroidea. Ophelia 31: 197–205.

Coleman, C. O., 1990b. Anatomy of the alimentary canal of Parandania boecki (Stebbing, 1888) (Stegocephalidae, Amphipoda, Crustacea) from the Antarctic Ocean. J. Nat. Hist. 24: 1573–1585.

Coleman, C. O., in press. Redescription of Anchiphimedia dorsalis K. H. Barnard, 1930 (Crustacea, Amphipoda, Iphimediidae) from the Antarctic and functional morphology of mouthparts. Zool. Scr.

Dahl, E., 1979. Deep-sea carrion feeding amphipods. Evolutionary patterns in niche adaptations. Oikos 33: 167–175.

Hopkins, T. L., 1985. The zooplankton community of Croker Passage, Antarctic Peninsula. Pol. Biol. 4: 161–170.

Kanneworff, E. & W. Nicolaisen, 1969. The stomach (foregut) of the amphipod Bathyporeia sarsi Watkin. Ophelia 6: 211–229.

Knox, G. A. & J. K. Lowry, 1974. A comparison between the benthos of the Southern Ocean and the North Polar Ocean with special reference to the Amphipoda and the Polychaeta. In Dunbar, M. J. (ed.), Polar Oceans. Proceedings of the Polar Oceans conference, Mc Gill University, Montreal, May 1974: 423–462.

Martin, A. L., 1964. The alimentary canal of Marinogammarus obtusatus (Crustacea, Amphipoda). Proc. zool. Soc. Lond. 143: 525–544.

Moore, P. G. & P. S. Rainbow, 1989. Feeding biology of the mesopelagic gammaridean amphipod Parandania boecki (Stebbing, 1888) (Crustacea: Amphipoda: Stegocephalidae) from the Atlantic Ocean. Ophelia 30: 1–19.

Rakusa-Suszczewski, S., 1972. The biology of Paramoera walkeri (Stebbing) (Amphipoda) and the Antarctic subfast ice community. Pol. Arch. Hydrobiol. 19: 11–36.

Richardson, M. G. & T. M. Whitaker, 1979. An Antarctic fast-ice food chain, observations on interactions of the amphipod Pontogeneia antarctica Chevreux with ice-associated microalgae. Br. Antarct. Surv. Bull. 47: 107–115.

Schellenberg, A., 1926. Die Caprelliden und Neoxenodice caprellinoides n. g. n. sp. der deutschen Südpolar-Expedition 1901–1903. Deutsche Südpolar-Exped., vol. 18: 465–476.

Schellenberg, A., 1931. Gammariden und Caprelliden des Magellangebietes, Südgeorgien und der Westantarktis. Further zoological Results of the Swedish Antarctic Expedition 1901–1903. 2(6): 1–290.

Scheloske, H. W., 1976. Vergleichend-morphologische und funktionelle Untersuchungen am Magen von Asellus aquaticus (L.) (Asellidae, Isopoda). Zool. Jb. Anat. 95: 519–573.

Slattery, P. N. & J. S. Oliver, 1986. Scavenging and other feeding habits of lysianassid amphipods (Orchomene spp.) from McMurdo Sound, Antarct. Pol. Biol. 6: 171–177.

Stebbing, T. R. R., 1888. Report on the Amphipoda collected by H. M. S. Challenger during the years 1873–76. Great Britain, Report on the scientific Results of the Voyage of H. M. S. Challenger during the years 1873–76. Zool., vol. 29: 1–1737 (in three volumes).

Stockton, W. L., 1982. Scavenging amphipods from unter the Ross Ice Shelf, Antarctica. Deep-Sea Res. 29 (7A): 819–835.

Thiem, E., 1942. Untersuchungen über den Darmkanal und die Nahrungsaufnahme von Synurella ambulans (Crustacea, Amphipoda). Z. Morph. Ökol. Tiere 38: 63–79.

Walker, A. O., 1906. Preliminary descriptions of new species of Amphipoda from the 'Discovery' Antarctic Expedition, 1902–1904. Annals and Magazine of Natural History Series 7 17: 452–458.

Hydrobiologia **223**: 11–25, 1991.
L. Watling (ed.), VIIth International Colloquium on Amphipoda.
© 1991 *Kluwer Academic Publishers.*

Methods for the study of amphipod swimming: behavior, morphology, and fluid dynamics

Michel A. Boudrias
Marine Biology Research Division, A-002, Scripps Institution of Oceanography, La Jolla, CA 92093 USA

Abstract

A thorough hydrodynamic approach to the study of swimming in amphipods demands a multipronged attack. A possible first step would be to gather swimming behavior data and determine the biomechanics and kinematics of pleopod beat. This requires careful observation of the swimming modes, swimming speeds, body positions and other aspects of behavior and limb motion that are crucial to swimming. Secondly, it is important to describe the morphology of the body and swimming appendages. Detailed drawings of body shape and design, skeletomusculature, condylic structure, and setal density and distribution on the pleopods and pereopods, are the tools required to ascribe hydrodynamic function to specific limb and body morphology. Finally, the information gathered from behavioral observations bolstered by functional morphology studies is applied to fluid dynamic calculations of drag, lift, and thrust. The theoretical calculations are then compared with empirical determinations of drag, wake generation, vortex shedding frequency, and flow patterns around an amphipod. The fluid dynamic facet of this research is the most challenging and requires an excellent grasp of the fundamental concepts of fluid flow and access to some highly technical equipment. The proposed tripartite approach for the study of amphipod swimming is by no means an exhaustive review of all the techniques that can be employed to quantify amphipod swimming. It will nevertheless permit a rigorous and systematic study of amphipod swimming.

Introduction

Amphipoda move about by walking or swimming. Virtually all species of Gammaridean amphipods can swim, some travelling long distances at a continuous rate, while others escape predators or change microhabitats with short bursts of rapid swimming. Detailed research on the locomotion of these organisms may provide insights into critical components of their ecology. By determining how far, how fast, and how effectively amphipods swim, their distribution patterns, dispersal potential, behavior as predators or prey in a food web, and aspects of their autecology will be more comprehensible. On an individual animal scale, a rigorous hydrodynamic study of swimming can

provide data on the biomechanics of pleopods which affect swimming activity, gill ventilation, and metabolic energy demands.

Yet with all the potential benefits of locomotory research on amphipods, few studies have dealt with the issue, especially in a hydrodynamic framework. Research on ecology involving swimming performance is more widespread, with papers on the bio-energetics of deep-sea scavengers (Yayanos, 1978, 1981; Smith & Baldwin, 1982), their vertical distribution (Smith & Baldwin, 1984), their feeding strategies (Ingram & Hessler, 1983), and their *in situ* swimming velocity (Laver *et al.*, 1985). Shallow water studies include work on endogenous rhythms and rheotropism (Fincham, 1972), sexual dimorphism and its effects on drag constraints and swimming (Adams & Greenwood, 1987; Naylor & Adams, 1987), vertical excursions and swimming in cold-water lysianassids (Sainte-Marie & Brunel, 1985; Sainte-Marie, 1986), and swimming activity in interstitial forms (Staude, 1986). The only methodical study of amphipod swimming with a hydrodynamic basis is F. Vogel's (1985) work on swimming in talitrids, yet it contains some grave miscalculations that affect its interpretation. In essence, the hydrodynamics and biomechanics of swimming in amphipods have not been detailed.

There seems to be a paucity of swimming studies on crustaceans as a whole in comparison to other aquatic organisms. In part, this may be the result of the wide range of Reynolds number affecting each crustacean, from 0.1 for the setae on the propulsive limbs to 10 000 for the whole body. Reynolds number is a dimensionless number that balances viscous forces, determined by the dynamic viscosity coefficient, with inertial forces, calculated by a combination of characteristic length (total length, body diameter, limb length for e.g.) and swimming speed or current speed (S. Vogel, 1981). The complex shapes of most crustaceans also present daunting obstacles for a realistic hydrodynamic analysis. Finally, the complexity of multiple propulsive elements used for swimming and the interactions between the swimming limbs, functioning in a viscous environment, and the body, moving in inertial flow, render mathematical analysis difficult. These hydrodynamical constraints lead to a simplification of the research on the mechanics and dynamics of swimming in Crustacea.

Thorough hydrodynamic research on swimming in amphipods is possible, but a series of interdisciplinary studies is required, including:

(1) Live observations and video or cinematographic data of the amphipods swimming in their natural environment. This behavioral component sets up the framework for more detailed studies by providing data on the kinematics of body movements and on swimming speeds. In addition, biomechanical data for both directional limbs (pereopods, uropods, and/or telsons) and propulsive limbs (pleopods) and determination of pleopod beat rates can be gleaned from behavioral analyses.

(2) Functional morphology of pleopods, encompassing information on muscle mass, skeletomusculature, and other specific features critical to swimming performance. Morphological studies on the anatomy of body design, on changes in body shape related to swimming behavior, and on the design of directional limbs complement the behavioral work and permit comparisons with amphipods from different environments or of different sizes. These morphological analyses can be accomplished using preserved specimens, thus enlarging the database to include species on which little behavioral data can be gathered.

(3) Fluid dynamics of flow around the body, the swimming appendages, and the interactive effects between the propulsive limbs and the body. This engineering facet of swimming in amphipods must take into account a suite of hydrodynamic variables such as the Reynolds number, the Strouhal number, and the overall characteristics of water flow. An understanding of the fluid dynamic forces acting on amphipods is needed to determine the thrust, drag, and lift and their effects on the efficiency of amphipod locomotion.

Some may argue, with good reason, that a physiological component should be added. The physiology of locomotion, which may involve the neuronal control of limb motion, the biochemistry of muscle function, and the energetics of swimming, is a completely different approach to the study of swimming in amphipods. Such a reductionist view of an animal will reveal many important aspects related to its swimming which can be incorporated in a thorough hydrodynamic analysis. For example, swimming efficiencies can only be calculated accurately by using data on muscle physiology. However, this conceptual view is not an exhaustive review of all the methods needed to elucidate the hydrodynamics of locomotion in amphipods. It simply presents one approach to a systematic study of swimming in amphipods. The following sections will explore in more detail the main components of this technique: behavior, morphology, and fluid dynamics.

Behavior and Biomechanics

Amphipods have invaded virtually all habitats and encompass as a group a whole suite of lifestyles. There are, for example, parasitic cyamids, planktonic hyperiids, troglobitic ingolfiellids and kelp-dwelling caprellids. Indeed, within the suborder Gammaridea, we find kelp-dwellers, burrowers, tubicolous, troglobitic, and epibenthic forms. It thus becomes essential to determine, for any given species, which mode of life is prevalent, its main habitat, and the important of swimming to its ecology.

Methodology

The methods needed to collect swimming behavior and biomechanical data range from simple, inexpensive observation to technically complex and expensive high-speed video or film. Visual assessment of swimming behaviors are dependent on the size of the animal but should include observations *in situ*, untethered activity in aquaria or flumes, as well as tethered experiments in appropriately scaled containers (Table 1). One must study locomotion under conditions hydrodynamically similar to the habitat conditions in which the amphipod lives. Habitat flow regimes determine if an aquarium with gentle flow is sufficient or if a more sophisticated system, such as a flume with controllable current velocity is necessary. As well, swimming behaviors *in situ* are necessary to groundtruth the experimental data. Similarly, untethered activity in aquaria or flumes provide data on the typical categories of swimming behaviors in more controlled environments. For species that are difficult to observe *in situ* without technical assistance, e.g. deep-sea scavengers, *in situ* video systems (Laver *et al.*, 1985) can provide valuable data on body position, turning, and in some cases, pleopod beat frequency, at least in the respiratory mode. A sizable amount of information can be gleaned from videos of amphipod swimming, especially in large species like *Eurythenes gryllus*, which commonly reach 10–12 cm in body length. Even still photographs of amphipods *in situ* can be helpful in deciphering the body position and degree of flexion an amphipod maintains while coming to bait or swimming away from a predator.

Experiments using tethered animals must be approached cautiously to avoid misinterpretations due to unnatural conditions. However, these experiments are especially useful for small species because the tethered amphipod can 'swim' in a small bowl while one observes the details of body position, body movements, and limb beat through a binocular dissecting microscope. The addition of a normal speed, high quality videorecorder attached to the microscope augments the quality and quantity of analyses by providing the opportunity for repeated measure-

Table 1. List of amphipod species observed, their depth of occurrence, and the method of observation.

Species	Depth	Method
Eogammarus confervicolus	intertidal	live; aquarium
Eohaustorius washingtonianus	intertidal/subtidal	aquarium
Paramaera mohri	intertidal/subtidal	live; aquarium; flume
Euonyx laqueus	1800 m	aquarium; flume
Eurythenes gryllus	2000–6000 m	video *in situ*
Eurythenes obesus	1000–2000 m	video in aquarium
Large vent species (?)	2470 m	live *in situ* (sub)
Perampithoe humeralis	on kelp blades subtidal	aquarium; video
Ampithoe sp.	on kelp blades subtidal	aquarium; video

14

ments of behaviors. This is the only way to garner details of the pleopod beat. If a high-speed video or film system is available, one can gain more precise data on the biomechanics and kinematics of pleopod motion, the beat frequency, and, with the use of non-toxic dyes, the flow patterns generated by the propulsive system of the amphipod.

With this in mind one can attain the objectives of behavioral studies which are (1) to delimit the type of motion an amphipod can accomplish, (2) to categorize the swimming modes used, and (3) to evaluate the swimming speeds, body positions, and body movements used in various swimming modes.

Swimming modes

To develop the appropriate structure for a hydrodynamic analysis of swimming in a chosen species, the potential swimming modes and range of motion must be broken down into useful categories. For example, an amphipod may swim in a continuous or sustained mode in comparison to short bursts of activity followed by gliding or stopping, defined as an acceleration mode. The classification of swimming modes, as done by Webb (1984) for fish, and the frequency of swimming will then dictate how important swimming is in the everyday ambit of the animal. As an example from my research, the deep-sea necrophagous scavenger *Eurythenes gryllus* is a sustained swimmer, similar to a tuna in Webb's (1984) classification, while the kelp curler *Perampithoe humeralis* is an accelerator swimming in short rapid bursts, much like a pike (Webb, loc. cit.), to escape predators or move from one kelp blade to another.

Swimming motion

What types of motion are exhibited? Do amphipods swim in a smooth fashion or does the pattern resemble the stop and go movement of copepods (Strickler, 1975) or kick and glide

characteristic of scombrid fishes (Webb, 1975, 1988; Webb & Weihs, 1983)? One can outline and quantify the major swimming behaviors, such as the amount of time swimming with the current or against it; how often an amphipod turns, darts side to side, or skips along the bottom of a stream; or any other swimming movement that reflects the natural activities of the animal (Boudrias, 1987). These data will be essential for fluid dynamic experiments of water flow around the amphipods and will determine the magnitude of the effect of acceleration and turning.

Swimming speeds

In conjunction with gathering data on swimming modes and range of motion, determination of swimming speeds is *de rigueur*. Knowledge of swimming speeds will permit better definition of the principal swimming mode, either sustained (same average speed over long time periods and distances) or burst swimming with periodic acceleration. But even more significant for any hydrodynamic analysis, swimming speeds are essential in the determination of Reynolds number, drag on the body and propulsive limbs, lift, and all other determinants of fluid forces acting on the amphipod. Because of its dimensionless nature, Reynolds number (Re) can be used to compare animals of different sizes or swimming at different speeds. Thus it becomes an important scaling tool. Swimming velocity, which incorporates the direction of swimming, should also be estimated, but for many species it is much more difficult to quantify. In any case, the evaluation of swimming speed must be undertaken to permit any possibility of a thorough hydrodynamic study.

Body and limb positions

The collection of data on body position and body movement during swimming is the next step in a behavioral analysis of locomotion. The degree of flexion of the abdomen, the position of the pereopods when the animal is swimming unidirectionally, and the position of all other limbs (antennae, gnathopods, uropods, and telson) that might influence water flow around the body must be established. Dorso-ventral flexion of the abdomen must be analyzed carefully to determine if the 'tail flick' is used only as an initial propulsive boost or if it is working in concert with pleopod beating throughout a swimming sequence.

In addition, examining the percentage of time spent with the body stretched out completely, or conversely curled up, will influence predictions of streamlines around the body, calculations of drag and lift, and the effects on vortex shedding. By paying particular attention to limb position, I was able to more accurately determine the exact body shape of a swimming amphipod (Boudrias, 1988). It is through observation of body position that one begins to see that amphipods do not have such unwieldy shapes. All the species I have studied so far (Table 1) control the arrangement of antennae, gnathopods, pereopods, uropods, and telson to minimize the area of body exposed to flow (Boudrias, 1988). They are capable of tucking in the first four pairs of pereopods, holding them within the thoracic cavity created by the coxal plates, and adducting the last 3 pairs close to the body to present an almost perfect airfoil shape to any oncoming flow (Fig. 1).

It is also through body position data that one can begin to understand how amphipods turn and which limbs control direction of motion. For small species, less that 1 cm in total body length, it is often difficult to see the position of each limb. However, even basic data on the amount of abdomenal flexion, when the body flexes, and how smoothly directional change is accomplished are sufficient for setting up the hydrodynamic framework. The extent of limb motion and which appendages are deployed for swimming and turning can then be deduced from functional morphology (see next section).

Biomechanics of pleopod beat

As I have already alluded to in the methodology section, magnified videotape data of swimming can produce results on the biomechanics of pleopod beat. With *in situ* images or direct visual observation, it is challenging to collect accurate data on the range of motion, the mechanics of beat, and the beat frequency of amphipod pleopods. But with a videorecorder attached to a dissecting microscope it is easier to discover details of motion of the pleopods, the main propulsive and current producing limbs.

The basic pattern of motion for pleopods has been described by Bousfield (1973) and F. Vogel (1985). The pleopods work in pairs, functioning as a wide paddle on the power stroke and collapsing the setal fan on the return stroke. The

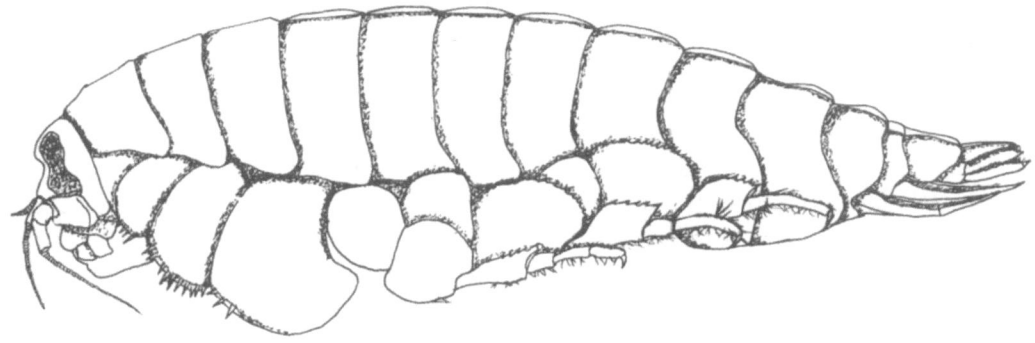

Fig. 1. Lateral view of *Eurythenes gryllus* in swimming position. All anterior pereopods are tucked in the thoracic cavity while the last 3 pereopods are adducted to the body, maintaining a streamlined shape. The pleopods are not visible because they are hidden away within the tail cone. Scale bar = 5 mm.

mechanics of this motion have been described for the lophogastrid mysid *Gnathophausia ingens* and apply reasonably well to amphipods (Hessler, 1985). In addition, the beating frequency of these current producing limbs falls within a narrow range of 1–3 beats/sec for several species (Waterman, 1961; Yayanos, 1978, 1981). Though the basics of pleopod motion may remain quite uniform among a wide variety of genera and families, the details become very important in a rigorous analysis of swimming.

For instance, pleopods, alone or sometimes in conjunction with abdomenal flexion (F. Vogel, 1985), produce not only the thrust for locomotion but the respiratory current that carries oxygen to the thoracic gills of amphipods. Much of the published work on pleopod beat rates is in the context of respiration (Waterman loc. cit.; Yayanos loc. cit.; George, 1979a, b) including Dahl's (1979) paper on the flow around a gammarid amphipod. So caution is required in any interpretation of the mechanics of pleopod beating, on their frequency, and on the current patterns they produce. The basic movement of the pleopods may be similar for the respiratory beat and the swimming beat but particulars of the motion vary. The determination of the mechanics of their motion and beat rate must be studied in the proper context, i.e. respiration with the animal at rest or locomotion with the animal moving freely or tethered but in its swimming mode. For either case, one needs to illustrate how the pleopods function and answer some important questions: how widely they open at full expansion during the power stroke, when and for what length of time are they extended, how they bend and close down on the recovery stroke, how fast they beat and whether this changes with variations in their immediate surroundings?

The accuracy of data collection for pleopod motion is imperative because they are the main, if not the only, propulsive elements. Because they produce a current, their motion and frequency of beating will affect the flow structure around the amphipod, interact with the oncoming current, and create additional vortices around the posterior end of the animal. For a hydrodynamically correct interpretation of swimming and water flow around amphipods, we absolutely need to understand the mechanics of pleopod activity.

Functional morphology

Live observations and videotape analysis of swimming behaviors provide a great deal of data on the swimming modes, type of motion, and body positions and movements of swimming crustaceans. Biomechanical investigation of pleopod motion helps to explain how the propulsive limbs of an amphipod move and generate the thrust needed for locomotion. However, these studies would be incomplete without cognition of the overall body design, and the skeletomusculature and architectonics of the swimming and directional appendages.

Methodology

Research in this field requires very little equipment but generates a great deal of valuable information. The basic tools needed are two good microscopes, a compound for small animals or minutiae, a dissecting for most other work, a camera lucida for accurate drawing, staining and clearing chemicals, and a precise measuring device (micrometer, electronic ruler). Of course, serial sectioning, scanning electron microscopy (SEM), and transmission electron microscopy (TEM) are excellent techniques for detailed work on muscles, setae, and fine scale sensory structures but one can still obtain a stringent morphological analysis without resorting to these finer techniques.

Accurate drawings of the overall body shape, and the shape and dimensions of directional and propulsive limbs at several magnifications and using several different views are required for comparative purposes and for fluid calculations of drag, lift, and thrust generation. With morphological analysis of limb structure, one can delimit the range of motion of the appendages, discover how they are built, and use these data in hydrodynamic testing of limb design. The particular methods for limb design evaluation will be described in the ultimate part of this section in functional morphology.

Body design

In the analysis of body design, several measurements are required. To begin with, a dorsal view of the amphipod in its typical swimming position, as determined from the behavioral observations, is essential to compute hydrodynamic ratios (Fig. 2). For many species the body will be completely stretched out so total length must be gauged directly or drawn and then measured. Though taxonomists and other amphipod biologists measure total length from the tip of the rostrum to the tip of the telson (Barnard, 1969; Bousfield, 1973), this will yield incorrect body lengths for hydrodynamic purposes. Water flow affects all protruding limbs so body and limb position data will establish the outer limits of the amphipod and the 'hydrodynamic length' of the species. Hydrodynamic length is defined as the total length of the body affected by water flow. It can encompass part of or the entire length of the first antennae and extent as far caudally as the tip of the third uropods. In some cases, as in *Perampithoe humeralis* (Fig. 3), the last pair of pereopods (Pr 7) reach past the uropod-telson complex and become the posterior-most body part defining hydrodynamic length (Boudrias, 1989). It is essential to know the body position of the amphipod to determine its actual limits of hydrodynamic length. I realize this creates a rather nebulous and potentially sloppy term but small changes in total length cause significant changes in hydrodynamic ratios affecting the interpretation of swimming efficiency. Secondly, one needs to calibrate the body diameter at the thickest point on the body. Thirdly, a lateral view is required to determine

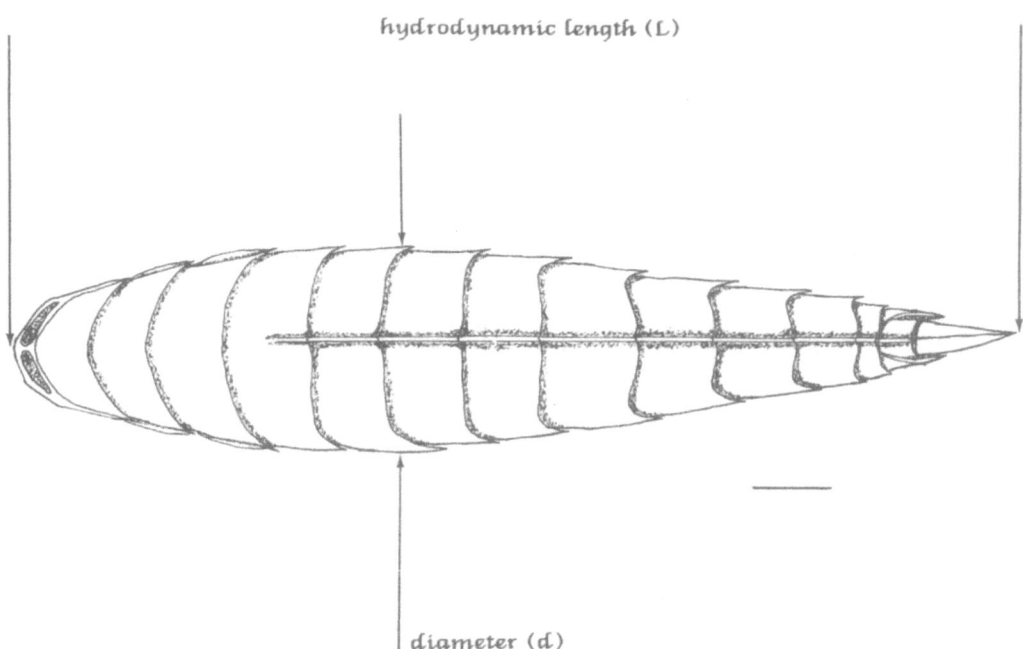

Fig. 2. Dorsal view of *Eurythenes gryllus* in swimming position. The antennae are not visible as they are bent ventrally, and the uropods are folded away within the concave telson resulting in a very streamlined shape. The double line along the middle of the animal is a keel that may enhance flow around the body. The d/L ratio represents the diameter (d) at the thickest portion of the body divided by the hydrodynamic length (L). A d/L ratio of 0.21 is close to the optimal drag reduction profile thickness for airfoils (Hertel 1966). Scale bar = 5 mm.

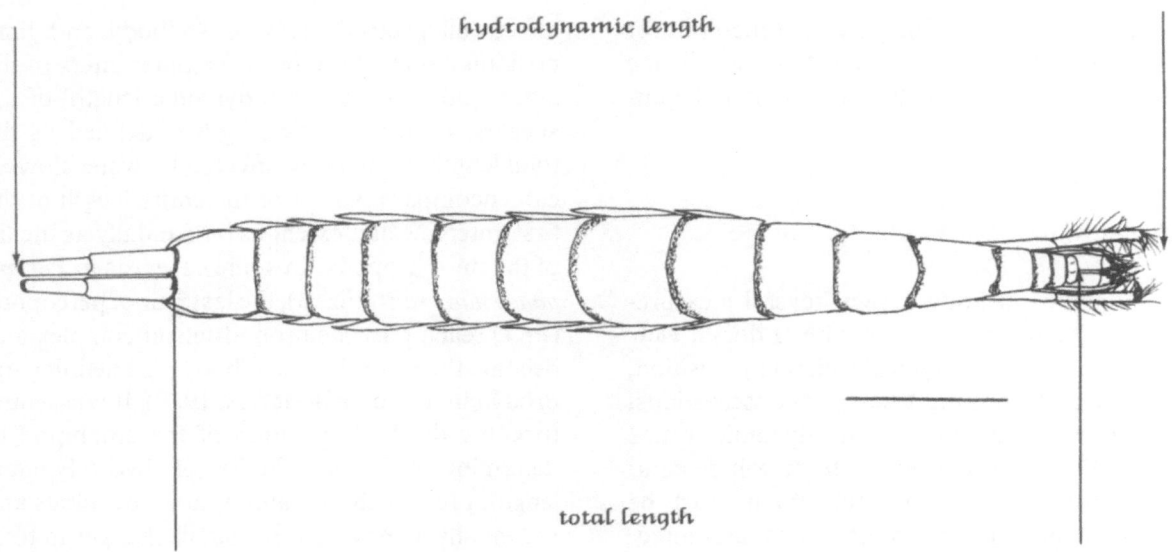

hydrodynamic length

total length

Fig. 3. Dorsal view of *Perampithoe humeralis*, a kelp-dwelling amphipod. This view of an amphipod in its swimming mode shows that part of the antennal peduncle, all of the uropods, and a portion of the last 2 segments of pereopod 7 make up the hydrodynamic length of this species. The added length of the stretched out appendages affects the calculation of the d/L ratio and the flow around the body. Scale bar = 5 mm.

body depth at the thickest point. Finally, one must measure body weight in conjunction with length and diameter to calculate frontal area, projected surface area, body volume, and body density. All views and drawings should be repeated for the different body positions employed in sustained swimming, burst or escape swimming, and turning.

With these drawings, one can begin to quantify differences in body design using hydrodynamic ratios, such as profile thickness from dorsal and lateral views, position of maximum thickness, and slimness ratios which are akin to the aerodynamic properties used for airfoils (Hertel, 1966). These aerodynamic methods can be transferred to swimming hydrodynamics because of the airfoil-like shape of most gammaridean amphipods. These ratios can then be used in computations of body drag and lift coefficients (see Blevins 1984 for hydrodynamic formulae). The use of ratios standardizes the data and permits intraspecific comparisons between life stages and sexes and interspecific comparisons between animals of vastly different body shapes and sizes.

Amphipods must be considered 3-dimensional airfoils in theoretical calculations of body drag or lift. This is why drawings and profile thickness measurements must include dorsal views, to incorporate flow characteristics in the X-Y plane, and lateral views, for flow in the X-Z plane. Additionally, in most species, a 'tail cone' is formed from a combination of limbs typically including the third pair of pleopods, the uropods and occasionally the telson. Thus ventral views can be used to determine the extent of the water jet produced when the pleopods complete their power stroke and close off the bottom of the 'tail cone'. Ventral views can also be useful to get an inkling of the water flow between the thoracic coxal plates. One must know the preferred body positions of the amphipod and use this information to determine what views are most important for a complete hydrodynamic analysis of flow around its body. This is the only way to discover the effects of body design.

Notwithstanding the need for good behavioral data, one of the major advantages of functional morphology studies is the availability of preserved specimens. For many unusual species, or for some common species that live in inaccessible habitats (e.g. the deep-sea), it is practically impossible to collect any *in situ* swimming behavior data. But with the appropriate drawings, some basic knowledge of their ecology, and judicious use of hydrodynamic ratios for comparisons, one can infer a great deal about the swimming performance and abilities of these amphipods from an analysis of body design. One must always be cautious when working with preserved material and infering function from specific morphological traits, but I think it is worthwhile to study the body design and compare it to closely related species or to relate the functional deductions made to the ecology of the animal.

Limb design

Once some information on body design and body positions while swimming has been amassed, data on the propulsive and directional limbs should be obtained. One must determine the limits of motion of the limbs, their design characteristics, and the skeletomusculature that controls limb motion. Here again the basic technique of accurate morphological drawings with a camera lucida provides ample data. While detailed drawings from many viewing angles and at different magnifications, one can: (1) elucidate setation patterns of pleopods with respect to the number, length of setae, and degree of plumosity; (2) describe the hook structure that holds the pleopod pair together on its power stroke; and (3) determine the shape, setation pattern, length, and width of all essential directional limbs (Pereopods 5, 6, and/or 7, uropods 2, 3, and the telson) and of the peduncle of all pleopods. The need for good behavioral data once again becomes apparent since it is important to known which directional limbs are used and how shape and design differences affect the turning abilities of an amphipod. The research on both directional and propulsive limbs should include drawings of limbs still attached to the body to gain perspective on limb layout with reference to body shape and body position while swimming and individual drawings of limbs at higher magnification for more precise information on shape and architectonics.

To more fully analyze pleopod biomechanics, a complete description of the skeletal structure of the condyles, apodemes, and arthrodial membrane at the peduncle-body joint and at the peduncle-ramal joints is necessary. To enhance the details of cuticular structure, musculature may be cleared with potassium hydroxyde (KOH) (Hessler, 1982). The cuticular structures can be viewed stained, with methylene blue, Erlich's triple stain, or chlorazol black, and/or unstained.

Morphological investigations of cuticular structure only provide part of the puzzle. To fully ascribe function to a jointed limb, one should dissect the limb and establish the origins and insertions of all major muscles. Muscle attachment points, condylic array, and the angle and extent of flexion must be determined, as well as the number of muscles controlling each pleopod segment, and the distribution of extrinsic and intrinsic muscles. Though pleopods are considered a conservative taxonomic character, close examination has revealed some structural complexities and highly specialized muscular arrangements (Boudrias, 1988). The number of muscles controlling the pleopods, their placement in the pleon, and the interaction between muscles and cuticular conformation may prove to be an important tool in deducing the relative swimming efficiency of various amphipod species.

Finer scale techniques may be useful for a complete analysis of functional morphology of swimming appendages. Through the use of scanning electron microscopy (SEM) one can more accurately calculate the spacing of setae on the rami and, more importantly for fine scale flow effects, the spacing and length of setules on the setae. In addition, detailed information on the condylic structure of the pleopod joints and the hooks and spines joining pleopod pairs on its power stroke can be ascertained from SEM research. Some of the more unusual cuticular structures in the pleonal wall (Boudrias, pers. obs.) could also be better elucidated with SEM.

Finally, some may criticize the lack of even finer scale studies on limb morphology or muscle function but I must reiterate that this conceptual view is only one way, albeit a quite rigorous way, of studying the biomechanics and fluid dynamics of amphipod swimming. For researchers with a more reductionist approach, transmission electron microscope (TEM) of fine serial sections of pleopods may well provide good data on muscle arrangement and condylic structure but it does require a more extensive technical set-up. Biochemical techniques to study muscle function could also be profitable. Simple and inexpensive techniques have a more general appeal and will still give solid results.

Fluid dynamics

The final piece of the puzzle revolves around the concepts of fluid dynamics that relate to crustacean locomotion. The mere mention of fluid dynamics, integral equations, calculus, and complex mathematics often scares biologists. The discipline of fluid dynamics is complex and requires many years of experience and a facility with numbers. But with the help of mathematically inclined colleagues and the patience to spend the time learning the major concepts of water flow in simple but realistic situations, biologists can gain access to a whole new world of interesting possibilities (see S. Vogel, 1981). For amphipod researchers who want to attempt a fluid dynamic study of their animal, they must concentrate on the flow conditions relevant to their animal and place limits on the fluid dynamic variables they need to assess. The main goals to achieve in a study of amphipod swimming would be to: (1) describe the flow patterns and vortex shedding frequency around the whole animal and the propulsive limbs under steady state conditions; (2) calculate and measure the drag on the body and limbs, the lift produced while swimming, and the thrust and propulsive power of the pleopods; and (3) add on some unsteady flow effects and evaluate the importance of acceleration bursts in relation to steady swimming.

Methodology

To investigate the flow influencing amphipod locomotion, two approaches to the fluid dynamics of swimming can be combined. Morphological and behavioral data on body shape and design show that amphipods have rather simple and symmetrical airfoil shapes. This permits theoretical calculations of the fluid dynamic forces affecting swimming in these crustaceans. But theoretical data are not enough and must be grounded with experimental data on the drag, lift, and thrust generated by the pleopods. In addition to purely empirical results on drag, flow patterns around the whole animal and around the propulsive appendages must be described and quantified. Each of these approaches to fluid dynamics and amphipod swimming will be explained in subsequent parts.

Theoretical calculations

Fluid dynamic force can be computed with the standard equation:

$$\text{Force} = \tfrac{1}{2} * \text{density of fluid } (\rho) * \text{constant} * \text{area} * (\text{velocity})^2$$

(Hoerner, 1965). The force can be either drag or lift, the constant becoming then the drag (C_d) or lift (C_l) coefficient, which is a function of body shape and Reynolds number. Values of C_d and C_l can be determined from formulae defined in Hertel (1966) or Blevins (1984). The area used in these calculations must always be specified clearly because it can denote cross-sectional, frontal, projected, or wetted surface area. The hydrodynamic ratios calculated in the functional morphology section are critical in determining the correct area and shape function to use. The changes in the shape of the amphipod resulting from the different body positions while swimming, turning, or stopping must be accounted for in theoretical calculations. The morphological work becomes especially important when trying to estimate the area represented by the pleopods. The drag coefficient for pleopods can only be determined by assuming the pleopods behave as a fan-shaped paddle on the power stroke and a collapsed fan on the recovery stroke.

It is important to distinguish the effects of drag on the body compared to those on the pleopods. The drag on the amphipod body is wholly a resistive force that impedes locomotion. The drag on the pleopods is the propulsive force that generates thrust and permits forward motion. Amphipods, as most other crustaceans, are drag-based locomotors moving their propulsive appendages in a direction parallel to the direction of motion, i.e. the pleopods wave back and forth on an anterior-posterior axis. Thrust is generated on the power stroke when the pleopods reach their maximum width producing high drag, and pushing water posteriorly. The pleopods collapse on the recovery stroke to reduce the resistive drag as they return to their position at the anterior end of the pleon. The net thrust is actually the difference in drag produced posteriorly on the power stroke and resistance to forward motion in the recovery stroke. Thrust and propulsive power are maximized when drag on the pleopod pair is maximal on the caudally-directed power stroke and minimal on the rostrally-directed recovery stroke. A thorough study of drag-based propulsion in labriform fishes has dealt with the effect of fin shape, the mechanics of propulsion, and the theoretical calculations of thrust and power (Blake, 1981, 1983). Many of the concepts put forth by Blake (loc. cit.) can be modified for amphipods and applied to the theoretical evaluations of drag, thrust, and power production by pleopods since they function very similarly to fins in their power stroke. I will not define the equations needed nor all the theory behind them. It is essential however to realize that pleopods do not rotate as fish fins, or even as insect legs (Blake, 1986; Nachtigall, 1977) and estimates of drag on the return stroke must account for these morphological differences.

The airfoil-like shape of amphipods creates lift when the animal is swimming. Aside from estimating this lift, which is strongly affected by the angle of attack of the amphipod relative to water flow, one can define the circulation around the amphipod body and the consequences of bound and shed vortices. These theoretical calculations, similar to those of circulation around a wing, are based on the angular velocity around the body and the radius of its idealized, airfoil shape (see Weihs-Fogh 1975 for details and equations). Theoretical knowledge of vortex patterns generated and the circulation around the amphipod will then be compared with empirical studies of wake visualization and vortex shedding.

The final theoretical calculation concerns the effect of added mass, a result of acceleration or burst swimming. The added mass of entrained fluid for a body in unsteady motion is a fixed amount which depends on size, volume, shape, mode of motion, and the density of the fluid (Batchelor, 1967; Blake, 1979, 1980; Daniel, 1984). Because all amphipods will have a component of added mass at some time in their swimming cycle, either at the start as they accelerate from rest to reach a steady velocity or throughout as they accelerate and decelerate in a kick and glide mode, evaluation of this effect is critical in understanding the realistic fluid dynamics of swimming in amphipods. Added mass calculations for pleopods, the main thrust generators which move in unsteady fashion during their beat cycle, are also required.

Flow visualization

The description of streamlines and flow patterns around a swimming amphipod is an essential prerequisite to determine the fluid realm around the body or limbs and to elucidate the importance of laminar, turbulent, smooth, or rough flow conditions. In addition, quantification of the frequency of vortex shedding and visualization of the wake produced at the posterior end of the animal should be attempted. In all cases, flow conditions must be replicated by either placing live animals in a large aquarium and filming the flow patterns generated by their swimming, or by building large scale models of amphipods and towing them in still water at appropriate speeds to generate streamlines. It is important to remember that water flow past a still animal and an animal moving in still water produce the same flow patterns if the relative velocities of water to animal remain unchanged.

Flow visualization techniques allow one to gather streamline, wake generation, and vortex shedding data simultaneously. The simplest way to collect this type of data is to move the amphipods through a suspension of neutrally buoyant particles (aluminum shavings, *Artemia* eggs, sea urchin eggs) and videotape the resultant motion of the particles, the wake left by the animal, and the number of vortices produced at a given speed. If the velocity of particles is less than 8–10 cm sec^{-1}, their trajectories can be plotted frame by frame to yield streamlines (Leonard *et al.*, 1988). At higher velocities the blur of particles is also useful as it gives both magnitude and direction of particle path. This frame by frame analysis using an ordinary videorecorder is a perfectly acceptable though tedious and time consuming way of gathering flow visualization data. A more accurate, but technically complex and expensive method, has been proposed by Gharib & Willert (1988). The number of vortices shed are used in calculations of Strouhal frequency, a good descriptor of flow regimes.

Flow patterns around pleopods is a much more complex issue. Even on large amphipods, pleopods are quite small and are composed of multiple cylinders arranged in an intricate array. The best solution for determining the flow around and through these setose appendages is the construction of larger scale models. Theoretical evaluation of flow through setose appendages has been done for infinitely long, simple arrays of cylinders (or setae) (Cheer & Koehl, 1987a, b) but both theoretical and experimental treatment of more realistic flow situations, i.e. 3-dimensional flow of finite, complex arrays, are just beginning.

All flow visualization experiments require some expertize in fluid dynamics and possesion of flow tanks, videosystems, and materials to build large scale models. The experimental approach to fluid dynamics of swimming in amphipods is the most difficult step for the uninitiated. It almost always requires collaboration with engineers and fluid mechanicians to ensure a proper understanding of the underlying concepts and techniques. Nevertheless, the benefits outweigh the difficulties and without empirical data the theory cannot be vindicated and the hydrodynamic framework is wobbly.

Empirical determination of drag

Theoretical calculations of drag on the body and pleopods will set the limits for the empirical measurements. Again this will require a rather extensive technical set-up, including a flow tank, a method to control flow or animal speed, special measuring devices such as strain gauges, and a computer system dedicated to gathering data and controlling the experimental conditions. The various experimental conditions should cover the range of swimming speeds observed, steady versus unsteady conditions, different levels of acceleration, and the whole range of behaviorally and morphologically defined body and limb positions. The experimental protocol, for either drag on the whole body or only on the pleopods, involves measuring the amount of force needed to deflect a force beam attached to a strain gauge. The amount of deflection recorded on the strain gauge can be calibrated with known weights and forces can then be resolved. More thorough technical details may be found in the literature (e.g. Cowles *et al.*, 1986; Sandeman, 1989; Williams, 1988).

A simpler but less accurate approach to empirical determination of drag on amphipod bodies is by a sedimentation method (Morgan & Hayes, 1977). The relationship between terminal sinking velocity, hence the force of gravity, and body drag, the opposing resistive or friction force, yields some empirical data on body drag. With the proper insertion of calibrated weights along different segments of the body, one can measure tangential drag, parallel to the long axis of the body, and normal drag, the force perpendicular to the long axis. Preserved specimens can be used but attention to changes in body density due to preservation must be corrected (Morgan & Hayes, loc. cit.).

I cannot emphasize strongly enough the need for empirical corroboration of theoretical results. Fluid dynamic theory has a long history and has been tested in a multitude of situations but conditions are never quite the same. It is essential to base drag calculations on reality and this can only be done by empirical calibration. Small changes in one of the terms of the drag (or lift) equation can greatly affect interspecific comparisons or calculations of swimming efficiency. For example, Cowles *et al.* (1986) found that the measured drag on *Gnathophausia ingens*, a lophogastrid mysid, was not a function of velocity squared but velocity to the power of 1.5–1.8. This type of empirical result has a profound effect on the theoretical evaluation of fluid dynamic forces affecting locomotion. Marine biologists must overcome their uneasiness with mathematics and fluid dynamics to couch their research in appropriate terms. Though the experimental aspect of fluid dynamics is daunting and expensive, these diligent studies are indispensable in solidifying the hydrodynamic framework I have devised for research on swimming in amphipods.

Closing comments

A research program combining behavioral, morphological, and fluid dynamic approaches is an ambitious undertaking. It requires some experience at observing amphipods, following their movements, and quantifying their behavior patterns. It means long hours at the microscope, studying the details of body design, limb design, and functional aspects of amphipod anatomy. Artistic and mechanical abilities are helpful in presenting the morphological data in a coherent manner. Finally, it demands the effort of learning fluid dynamic theory, of becoming familiar with the language and concepts of fluid flow, and of building specific measuring devices to answer the questions relevant to the animal. This rigorous hydrodynamic study of amphipod swimming is a very demanding and challenging task.

Notwithstanding the technical difficulties and the time required to achieve the primary goals, the benefits of the research extend beyond the somewhat limited field of amphipod locomotion. The detailed studies of pereopods, pleopods, uropods, and telson will further our knowledge of amphipod morphology and biomechanics. We can learn how distinct these appendages are and how they may be used to differentiate species functionally. Data on the biomechanics and kinematics of pleopod beat is crucial for any physiological study on respiration and energy budgets. Flow visualization experiments provide us with the patterns of water flow around an amphipod and may further our understanding of amphipod responses to flow, distribution of their sensory organs, and the effects of complex, 3-dimensional airfoil shapes on wake production and vortex shedding. Finally, the comparative information gleaned from the hydrodynamic ratios and the flow patterns around differently shaped amphipod species may yield some results on the dispersal abilities of certain species, the role of vertically migrating amphipods in energy transfer between pelagic and benthic communities, and the intraspecific differences between sexes with respect to mate location ad guarding, feeding, and swimming activity.

This tripartite conceptual view of amphipod swimming is by no means the ultimate method. Many other factors can be added to this hydrodynamic framework to improve its form. I only hope that this ambitious agenda can be used by other amphipod biologists to begin a comprehensive study of swimming in this successful group of crustaceans. On an even grander scale, I want to encourage other crustacean biologists to adapt these methods for their groups so the field of crustacean locomotion can move forward at a brisker place.

References

Adams, J. & P. J. Greenwood, 1987. Loading constraints, sexual selection, and assortative mating in peracarid Crustacea. J. Zool. Lond. 211: 35–46.

Barnard, J. L., 1969. The families and genera of marine Gammaridean Amphipoda. Bull. U.S. Nat. Mus. 271: 535 pp.

Batchelor, G. K., 1967. An introduction to fluid dynamics. Cambridge University Press, London.

24

Blake, R. W., 1979. The mechanics of labriform locomotion I. Labriform locomotion in the angelfish (*Pterophyllum eimekei*): an analysis of the power stroke. J. exp. Biol. 82: 255–271.

Blake, R. W., 1980. The mechanics of labriform locomotion II. An analysis of the recovery stroke and the overall fin-beat cycle propulsive efficiency in the angelfish. J. exp. Biol. 85: 337–342.

Blake, R. W., 1981. Influence of pectoral shape on thrust and drag in labriform locomotion. J. Zool., Lond. 194: 53–66.

Blake, R. W., 1983. Mechanics of drag-based mechanisms of propulsion in aquatic vertebrates. IN M. H. Day (ed.), Vertebrate Locomotion. Academic Press, London, Symp. Soc., Lond. 48: 29–52.

Blake, R. W., 1986. Hydrodynamics of swimming in the water boatman, *Cenocorixa bifida*. Can. J. Zool. 64: 1606–1613.

Blevins, R. D., 1984. Applied fluid dynamics handbook. Van Nostrand Reinhold Company, New York.

Boudrias, M. A., 1987. Effects of current velocity on the swimming behavior of a gravel beach amphipod *Paramaera mohri*. Am. Zool. 27: 81A (Abstract 416).

Boudrias, M. A., 1988. Locomotion in deep-sea amphipods: body design and functional morphology of swimming appendages. Am. Zool. 28: 3A (Abstract 7).

Boudrias, M. A., 1989. Swimming limb mechanics in *Ampithoe humeralis* (Crustacea: Amphipoda). Am. Zool. 29: 130A (Abstract 580).

Bousfield, E. L., 1973. Shallow-water Gammaridean Amphipoda of New England. Cornell University Press. London, 312 pp.

Cheer, A. Y. L. & M. A. R. Koehl, 1987a. Fluid flow through filtering appendages of insects. IMA J. Math. Applied Med. & Biol. 4: 185–199.

Cheer, A. Y. L. & M. A. R. Koehl, 1987b. Paddles and rakes: fluid flow through bristled appendages of small organisms. J. Theor. Biol. 129: 17–39.

Cowles, D. L., J. J. Childress & D. L. Gluck, 1986. New method reveals unexpected relationships between velocity and drag in the bathypelagic mysid *Gnathophausia ingens*. Deep-Sea Res. 33: 865–880.

Daniel, T. L., 1984. Unsteady aspects of aquatic locomotion. Am. Zool. 24: 121–134.

Fincham, A. A., 1972. Rhythmic swimming and rheotropism in the amphipod *Marinogammarus marinus* (Leach) J. exp. mar. Biol. Ecol. 8: 19–26.

George, R. Y., 1979a. Behavioral and metabolic adaptations of polar and deep-sea crustaceans: a hypothesis concerning physiological basis for evolution of cold adapted crustaceans. In A. B. Williams (Ed.), Symposium on the composition and evolution of crustaceans in the cold and temperate waters of the world ocean, Bull. Biol. Soc. Wash. 3: 283–296.

George, R. Y., 1979b. What adaptive strategies promote immigration and speciation in deep-sea environment. Sarsia 64: 61–65.

Gharib, M. & C. Willert, 1988. Particle tracing: Revisited.

AIAA-88-3776-CP. 1st National Fluid Dynamics Congress, Cincinnati, Ohio: 1935–1943.

Hertel, H., 1966. Structure, form, and movement. Reinhold Publishers, New York.

Hessler, R. R., 1982. The structural morphology of walking mechanisms in Eumalacostracan crustaceans. Phil. trans. r. Soc. London. 296B: 245–298.

Hessler, R. R., 1985. Swimming in Crustacea. Trans. r. Soc. Edinburgh 76: 115–122.

Hoerner, S. F., 1965. Fluid-dynamic drag. S. F. Hoerner, Midland Park, New Jersey.

Ingram, C. L. & R. R. Hessler, 1983. Distribution and behavior of scavenging amphipods from the central North Pacific. Deep-Sea Res. 30: 683–706.

Laver, M. B., M. S. Olson, J. L. Edelman & K. L. Smith Jr., 1985. Swimming rates of scavenging deep-sea amphipods recorded with a free-vehicle video camera. Deep-Sea Res. 32: 1135–1142.

Leonard, A. B., J. R. Strickler & N. D. Holland, 1988. Effects of current speed on filtration during suspension feeding in *Oligometra serripinna* (Echinodermata: Crinoidea). Mar. Biol. 97: 111–125.

Morgan, E. & S. V. Hayes, 1977. The swimming of *Nymphon gracile* (Pycnogonida): the vertical lift forces generated while swimming at constant depth. J. exp. Biol. 71: 187–203.

Nachtigall, W., 1977. Swimming mechanics and energetics of locomotion of variously sized water beetles: Dysticidae, body length 2 to 35 mm. IN T. J. Pedley (ed.), Scale effects in animal locomotion. Academic Press, London: 269–283.

Naylor, C. & J. Adams, 1987. Sexual dimorphism, drag constraints, and male performance in *Gammarus duebeni* (Amphipoda). Oikos 48: 23–27.

Sainte-Marie, B., 1986. Feeding and swimming of lysianassid amphipods in a shallow cold-water bay. Mar. Biol. 91: 219–229.

Sainte-Marie, B. & P. Brunel, 1985. Suprabenthic gradients of swimming activity by cold-water gammaridean amphipod Crustacea over a muddy shelf in the Gulf of Saint-Lawrence. Mar. Ecol. Prog. Ser. 23: 57–69.

Sandeman, D. C., 1989. Physical properties, sensory receptors, and tactile reflexes of the antenna of the australian freshwater crayfish *Cherax destructor*. J. exp. Biol. 141: 197–217.

Smith, K. L., Jr. & R. J. Baldwin, 1982. Scavenging deep-sea amphipods: effects of food odor on oxygen consumption and a proposed metabolic strategy. Mar. Biol. 68: 287–298.

Smith, K. L., Jr. & R. J. Baldwin, 1984. Vertical distribution of the necrophagous amphipod *Eurythenes gryllus* in the North Pacific: spatial and temporal variation. Deep-Sea Res. 31: 1179–1196.

Staude, C. P., Jr., 1986. Systematics and behavioral ecology of the amphipod genus *Paramaera* Miers (Gammaridea : Eusiroidea : Pontogeniidae) in the eastern North Pacific. Ph. D. Thesis. University of Washington, 324 pages.

Strickler, J. R., 1975. Swimming of planktonic *Cyclops* species (Copepoda, Crustacea): pattern, movements and their control. In T. Y.-T. Wu, C. J. Brokaw & C. Brennen (eds), Swimming and flying in nature. Vol. 2, Plenum Press, New York: 599–613.

Vogel, F., 1985. Das Schwimmen der Talitridae (Crustacea: Amphipoda): Funktionsmorphologie, Phanomenologie und Energetik. Helgolander Meeresunters. 39: 303–339.

Vogel, S., 1981. Life in moving fluids. Princeton University Press, Princeton, New Jersey, 352 pp.

Waterman, T. H. (ed.), 1961. The Physiology of Crustacea. Volumes 1 & 2. Academic Press, New York.

Webb, P. W., 1975a. Hydrodynamics and energetics of fish propulsion. Bull. fish. Res. Bd Can. 190: 1–159.

Webb, P. W., 1984. Form and function in fish swimming. Sci. Am. 251: 72–82.

Webb, P. W., 1988. Simple physical principles and vertebrate aquatic locomotion. Am. Zool. 28: 709–725.

Webb, P. W. & D. Weihs, 1983. Fish biomechanics. Praeger Publishers. New York. 398 pp.

Weis-Fogh, T., 1975. Flapping flight and power in birds and insects, conventional and novel mechanisms. IN T. Y.-T. Wu, C. J. Brokaw & C. Brennen (eds.), Swimming and flying in nature. Vol. 2, Plenum Press, New York: 729–762.

Williams, T. A., 1988. A physical model of aquatic locomotion mimics changes during *Artemia*'s larval development. Am. Zool. 28: 142A (Abstract 755).

Yayanos, A. A., 1978. Recovery and maintenance of live amphipods at a pressure of 580 bars from an ocean depth of 5700 meters. Science 200: 1056–1059.

Yayanos, A. A., 1981. Reversible inactivation of deep-sea amphipods (*Paralicella capresca*) by a decompression from 601 bars to atmospheric pressure. Compar. Biochem. Physiol. 69A: 563–565.

Hydrobiologia **223**: 27–34, 1991.

Is the oostegite structure of amphipods determined by their phylogeny or is it an adaptation to their environment?

D. H. Steele
Dept. of Biology, Memorial University of Newfoundland, St. John's, Nfld., Canada A1B 3X9

Key words: oostegite structure, ecology, phylogeny

Abstract

The oostegites of amphipods attached to peraeopods 2–5 are of two main types – broad with relatively short marginal setae and narrow with long marginal setae. Broad oostegites are found in other peracarids and are considered the primitive type. Amphipods with broad oostegites tend to have smaller eggs than those with narrow oostegites. It is concluded that following the evolution of the major amphipod groups, the oostegites were modified as egg sizes have changed as part of the reproductive strategies of the species within these groups.

Introduction

While making observations on the numbers and sizes of the eggs in various amphipod species it became apparent that the structure of the oostegites forming the brood pouch varied considerably. Such variations have been previously noted by Leite *et al.* (1986) and mentioned as a taxonomic character by Bousfield (1979). In general two main types can be distinguished:

1) Broad, with relatively short marginal setae which extend almost to the base (Fig. 1). The broad oostegites essentially form the brood pouch. Similar broad oostegites are found in other peracarids and this type probably represents the primitive type.

2) Narrow, with long marginal setae (Fig. 2). The oostegites mainly form a support for the setae. The brood pouch is therefore an open structure formed primarily by the long setae. This type of oostegite seems to be unique to amphipods. If the coxal plates are large, as in stegocephaloids or lysianassoids, the proximal setae may be short or

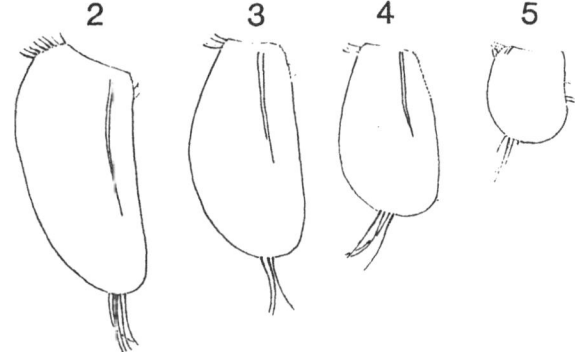

Fig. 1. Oostegites of *Gammaracanthus loricatus* (Sabine, 1821). Only the proximal and a few of the distal marginal setae have been figured on each oostegite.

missing and the coxal plates help to enclose the brood pouch laterally. If the coxal plates are small, as in some of the corophioids, the long setae extend to the proximal portion of the oostegite.

There is considerable variation in the two main types and some amphipods have a mixture of both. In all the intermediate types that I observed

28

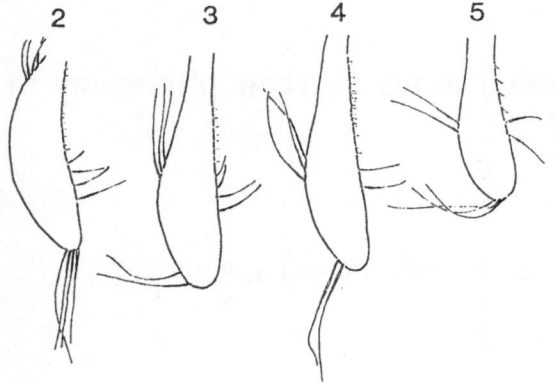

Fig. 2. Oostegites of *Pseudalibrotus litoralis* (Kroyer, 1845).

the anterior oostegites are broad and the posterior ones are narrow. Leite *et al.* (1986) also divided amphipod oostegites into two groups but their classification is based on the type and abundance of the marginal setae and there is little agreement between their classification and mine.

At first it appeared that the structure of the oostegites would serve as a character that might link together different groups of amphipods and make it possible to unravel their phylogeny. Subsequent observations, however, indicated that this approach was much too simplistic and that oostegite structure could be related as much to morphological and physiological constraints as phylogeny.

Oostegite type and egg size

Steele & Steele (1975) showed that the egg diameters of *Gammarus sensu lato* species of amphipods are positively correlated with the body length at which the female parent matures. Subsequent observations of other gammaroid, pontoporeoid, lysianassoid, eusiroid and stegocephaloid amphipods have also shown a similar positive correlation (egg diameters varying from 0.25 to 2.4 mm

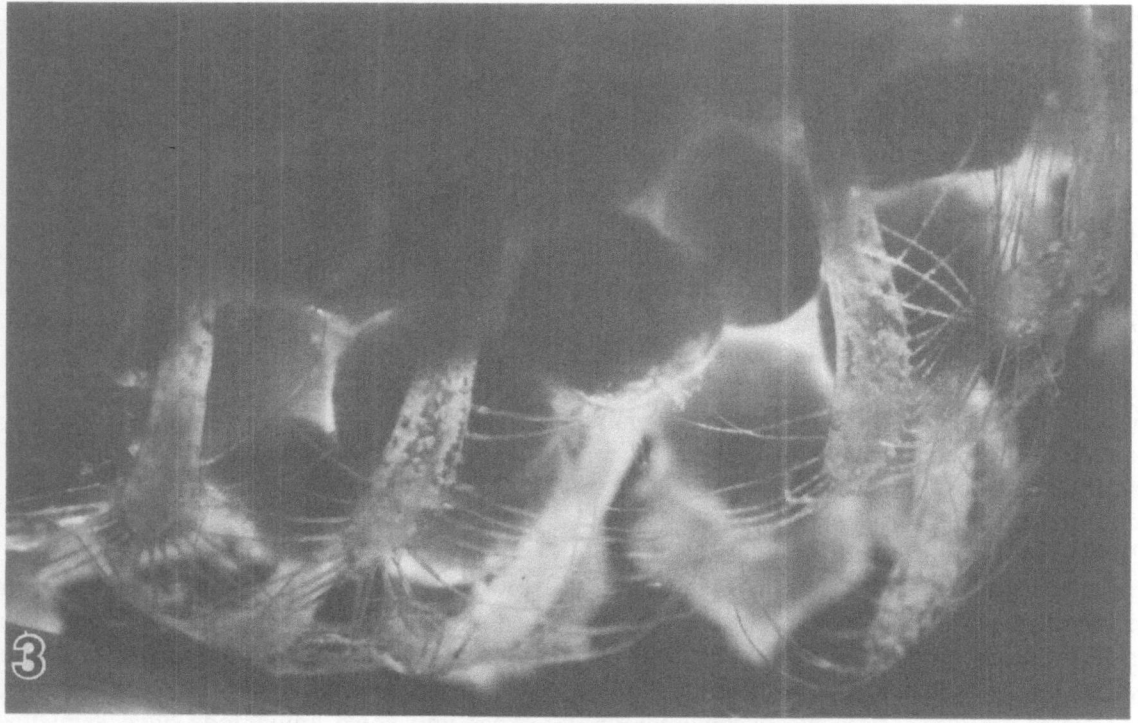

Fig. 3. Large eggs in the brood pouch (narrow oostegites) of *Socarnes bidenticulatus* (Bate, 1835) (coxal plates have been removed).

and female body sizes at maturity varying from 4 to 40 mm) (Steele & Steele, in press). However, these later observations also showed that within the general correlation there were significant variations. In particular, it was noted that oostegite type was also correlated with egg size. Thus lysianassoid, stegocephaloid, pontoporeoid and talitroid species with narrow oostegites and which mature at a large size, have much larger eggs than gammaroid and eusiroid species of the same large size at maturity, but which have broad oostegites. There is no significant difference in the egg diameters of species maturing at a small size.

The reasons for these correlations are two fold. If the brood pouch is constructed of narrow oostegites and long setae (Fig. 3) small eggs will tend to fall out more readily than will large ones. When the female parent is of small size and adjacent oostegites are close together there will be no problem for eggs of any size, but as the body size of the female increases and the distance between adjacent oostegites becomes larger small eggs can fall out more readily (Fig. 4). Therefore small eggs require broad oostegites to retain them.

On the other hand, respiration of the egg becomes more of a problem as its diameter increases. Oxygen consumption per unit weight declines although total oxygen consumption increases, with increasing egg size. The surface area/volume decreases with increasing egg size (Fig. 5). With increase in size, eggs will require a better circulation of water through the brood pouch to ensure an adequate suppy of oxygen for respiratory purposes. This is made possible by a brood pouch which has an open structure. Additional support for this argument comes from the mysids in which the brood plates are large and the female expands and contracts the brood pouch to pump water in and out (personal observations, Tattersall & Tattersall, 1951). This action insures an adequate circulation of water around the eggs.

Therefore a correlation between egg size and brood pouch structure in the amphipods would be expected. If egg size varies as a component of

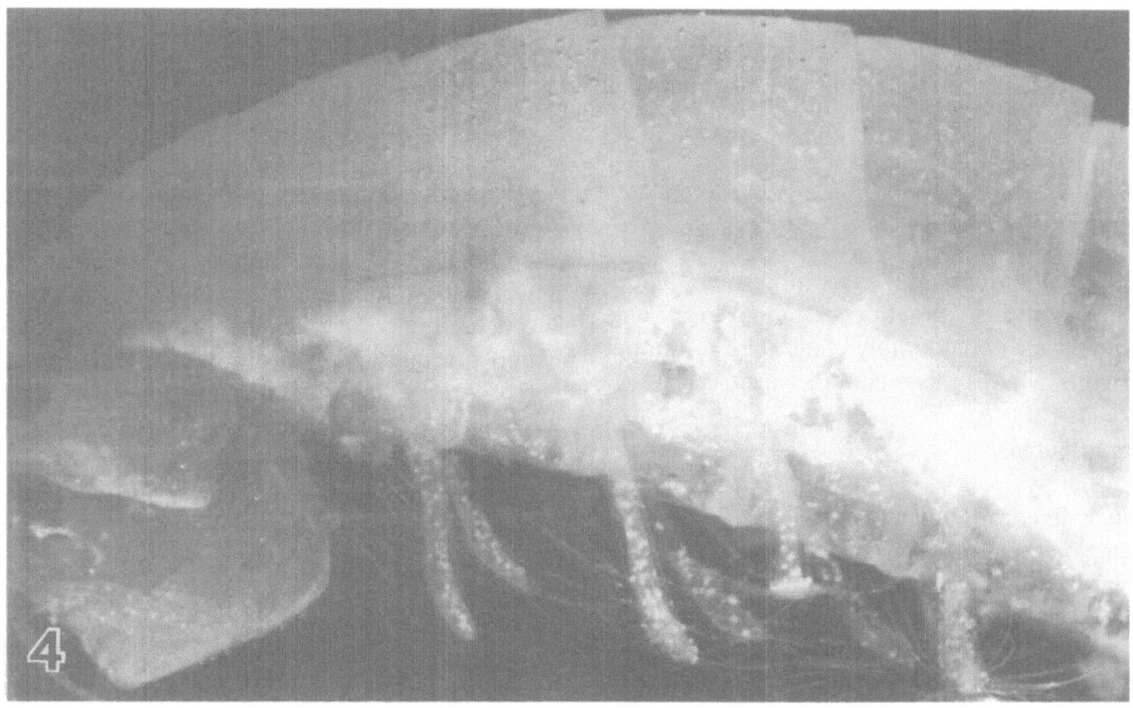

Fig. 4. Brood pouch of *Socarnes bidenticulatus* with eggs removed.

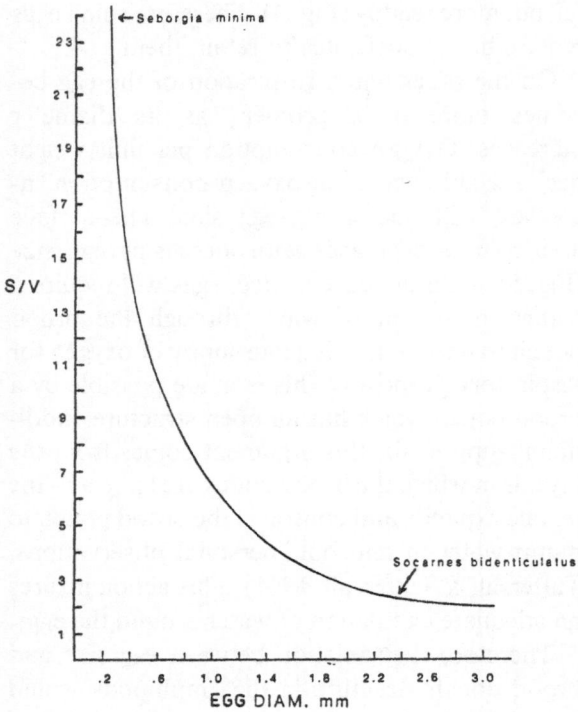

Fig. 5. Surface area/volume ratio for eggs of varying diameters.

Duration of development

Because of the correlation between egg size and metabolic rate, large eggs take longer to hatch than do small eggs at the same temperature (Steele & Steele, 1973). A long development period provides extended parental care and also the opportunity to time the release of the young. Where there is a short growing season such as at high latitudes, a long ovarian diapause and a large egg such as is found in *Gammarus setosus*, result in eggs produced in the autumn hatching and the young released the following spring (Steele & Steele, 1975).

Size of hatched young

The size of hatched young is correlated with the size of the egg. Thus if large young are better able to survive after being released from the brood pouch, for example if food is scarce, large eggs are adaptive (Steele & Steele, in press). On the other hand, if the probability of survival is low as it is at high environmental temperatures, the production of large numbers of small eggs will be adaptive.

Size at maturity

As discussed above, egg size and size at maturity are positively correlated in gammaridean amphipods. Thus, if large adult size occurs, large eggs will also be found. For example, at low environmental temperatures, the growth, reproductive and mortality rates are reduced, with the result that a large size at maturity is adaptive (Steele & Steele, 1975). However, since large egg size may also be adaptive (see Nos. 2 & 3), it is impossible to state which is the cause and which is the effect.

Fresh water

The eggs of fresh water crustaceans, such as the decapods, are generally larger than those of ma-

the reproductive strategy and an adaptation to environmental conditions, the brood pouch will also vary.

Variation in egg size

Although all amphipods, except for some hyperiids, have complete direct development in the egg, egg sizes vary considerably. There are a number of reasons why egg size should vary.

Size and number

If only a fixed amount of material is invested in the eggs by the female parent, the size of the eggs will vary with their number (Smith & Fretwell, 1974; Steele & Steele, 1975; Steele & Steele, in press; Wildish, 1980). Any factor promoting an increase in number will decrease size and vice versa.

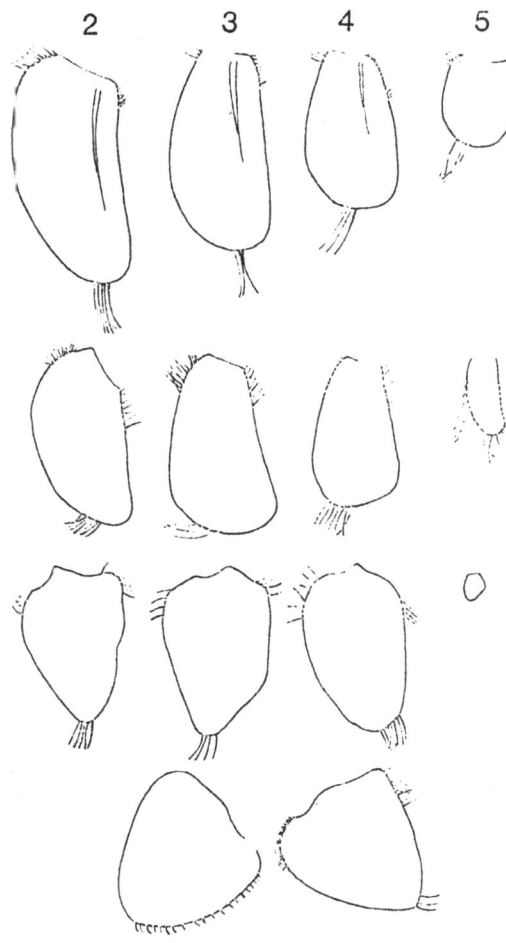

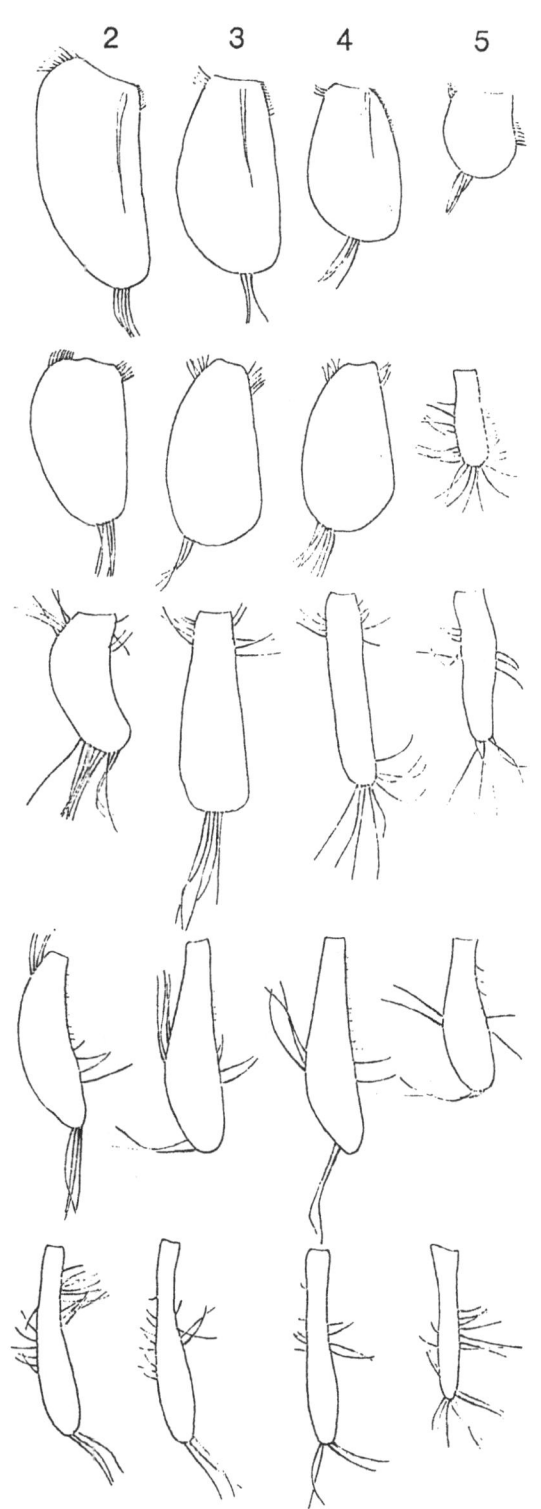

Fig. 6. Grades of development of broad oostegites – *Gammaracanthus*, *Ischyrocerus*, *Dryopedos*, and *Caprella* (from top to bottom).

rine species and this is often related to a reduction of larval stages (Rabalais & Gore, 1985). However, fresh water amphipods also tend to have larger eggs than the marine species (Steele & Steele, in press) even though this cannot be related to the mode of embryonic development, since all amphipods have direct development without any larval stages. A more general explanation for the larger eggs of fresh water crusta-

Fig. 7. Grades of development of narrow oostegites – *Gammaracanthus*, *Calliopius*, *Gammarus*, *Pseudalibrotus*, and *Stegocephalus* (from top to bottom).

Table 1. Distribution of oostegite types among species in various amphipod families (after Bousfield 1983)

Superfamily	Family	Broad	Intermediate	Narrow
Lysianassoidea	*Lysianassidae*	0	0	3
	Uristidae	0	0	18
Ampeliscoidea	*Ampeliscidae*	0	0	5 (2)*
Phoxocephaloidea	*Urothoidae*	0	0	1
	Phoxocephalidae	0	0	1 (9)
Stegocephaloidea	*Stegocephalidae*	0	0	7
Oedicerotoidea	*Oedicerotidae*	0	0	2 (9)
Hadzioidea	*Melitidae*	0	0	2 (9)
Liljeborgoidea	*Liljeborgidae*	0	0	2 (5)
Pontoporeoidea	*Pontoporeidae*	2	1	2
	Haustoriidae	0	0	3 (10)
Gammaroidea	*Acanthogammaridae*	1	0	0
	Gammaridae	0	15	(2)
Talitroidea	*Hyalidae*	2 (2)	0	0
	Hyalellidae	1	0	0
	Talitridae	0	0	2 (5)
Corophioidea	*Ampithoidae*	0	0	2
	Isaeidae	4 (8)	0	0
	Ischyroceridae	2 (6)	0	0
	Aoridae	2 (8)	0	0
	Corophiidae	1 (7)	0	2 (7)
	Podoceridae	2 (4)	0	0
Caprelloidea	*Aeginellidae*	(4)	0	0
	Caprellidae	3 (6)	0	0

* The datum in brackets was obtained from the literature.

ceans is that they will be subject to less osmotic stress than small eggs due to their reduced surface area/volume ratio (Fig. 5).

Thus there are a number of possible reasons why eggs in amphipods can be expected to vary in size. Such variations in egg size should be accompanied by a corresponding change in the oostegites.

Evolutionary trends in oostegite structure

There are two possible models for the evolution of oostegites in amphipods. Broad oostegites maintain or increase their width, but tend to be reduced in numbers (Fig. 6). The oostegite on peraeopod 5 is reduced in size so as to become nonfunctional. The oostegites on peraeopods 3 and 4 become the largest. This trend culminates in the caprellids with a brood pouch formed only of oostegites on peraeopods 3 and 4. This restriction in numbers and a reduction to those on peraeopods 3 and 4 can be related to the fact that the primary afferent water current produced by the pleopod beat of an amphipod at rest occurs between the coxal plates.

The narrow type oostegite generally persists on peraeopods 2 to 5 although there are some exceptions (Fig. 7). The oostegites seem to have become narrowed from the posterior to anterior. Amphipods such as *Calliopius*, gammaroids and *Priscillina* which have a mixture have broad anterior and narrow posterior oostegites.

It should be noted that these models are considered to be grades of development rather than direct lines of descent. It would seem doubtful if a brood pouch constructed of narrow oostegites could evolve into one with a reduced number of broad oostegites or vice versa, but the primitive broad type could evolve in either direction.

Distribution of oostegite types

Both oostegite types are widely dispersed throughout the amphipods (Table 1) without any obvious correlation with relationships deduced using other characters (Bousfield, 1979, 1983; Barnard & Karaman, 1980). Secondly, in well defined amphipod groups such as the lysianassoids, stegocephaloids, oedicerotoids and ampeliscoids only one type of oostegite is known. In the taxonomically more problematic groups such as the pontoporeoids, corophioids and gammaroids where constant realignment is being proposed by different authors, both types have been recorded.

In the pontoporeoids, for example, broad oostegites of the primitive type (*Pontoporeia*) as well as intermediates (*Priscillina*) and narrow oostegites (*Haustorius*) can all be found suggesting either that narrow oostegites evolved within this group

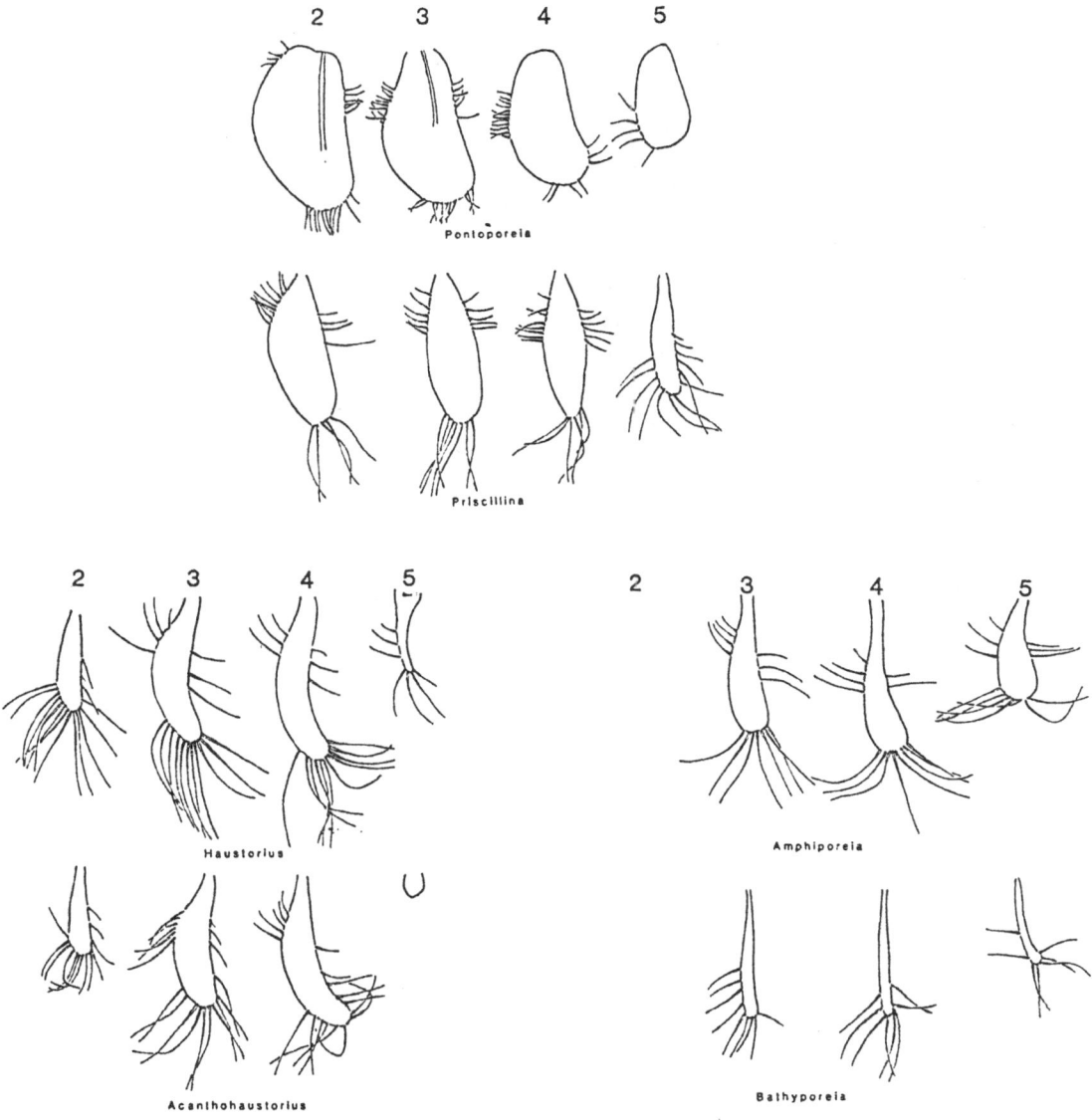

Fig. 8. Postulated evolutionary trends of oostegites in pontoporeoid amphipods.

of amphipods (Fig. 8) or that some species have been misclassified. Note that *Amphiporeia* has lost oostegite 2 whereas *Acanthohaustorius* has almost lost oostegite 5.

Gammaroids are also of interest since broad oostegites are found in some families, but within the Gammaridae itself (genera and subgenera of *Gammarus sensu latu*), mostly intermediate oostegites are found. In some species the oostegites have been reported in the literature all to be narrow (Cole, 1985). This is a similar pattern to that observed in the pontoporeoids and suggests that narrow oostegites have evolved independently within this group.

The present analysis indicates that the amphipods should not be classified into two major divisions on the basis of the structure of their oostegites. The evolution of oostegite structure appears to have occurred after the evolution of the major groups of amphipods, as an adaptation to environmental factors and as part of their reproductive strategies. Oostegite structure may therefore be useful in unravelling evolution within an amphipod group such as the pontoporeoids or gammaroids and can be used to support the phylogeny deduced from the analysis of more traditional characters.

References

Barnard, J. L. & G. S. Karaman, 1980. Classification of gammarid Crustacea. Crustaceana, Suppl. 6: 5–16.

Bousfield, E. L., 1979. A revised classification and phylogeny of amphipod crustaceans. Trans. r. Soc. Can. (1978) Ser. 4, 16: 343–390.

Bousfield, E. L., 1983. An updated phyletic classification and palaeohistory of the Amphipoda. In F. R. Schram (ed.), Crustacean Issues Vol. 1: Crustacean phylogeny, A. A. Balkema: 257–277.

Cole, G. A., 1985. Analysis of the *Gammarus-pecos* complex (Crustacea, Amphipoda) in Texas and New Mexico, USA. J. Ariz.-Nev. Acad. Sci. 20: 93–103.

Leite, F. P., Y. Wakabara & A. S. Tararam, 1986. On the morphological variations in oostegites of gammaridean species (Amphipoda). Crustaceana 51: 77–94.

Rabalais, N. N. & R. H. Gore, 1985. Abbreviated development in decapods. In A. M. Wenner (ed.), Crustacean Issues Vol. 2: Larval growth, A. A. Balkema: 67–126.

Smith, C. C. & S. D. Fretwell, 1974. The optimal balance between size and number of offspring. Am. Nat. 108:499–506.

Steele, D. H. & V. J. Steele, 1973. The biology of *Gammarus* (Crusteacea, Amphipoda) in the northwestern Atlantic. VII. The duration of embryonic development in five species at various temperatures. Can. J. Zool. 51: 995–999.

Steele, D. H. & V. J. Steele, 1975. The biology of *Gammarus* (Crustacea, Amphipoda) in the northwestern Atlantic. XI. Comparison and Discussion. Can. J. Zool. 53: 1116–1126.

Steele, D. H. & V. J. Steele, in press. Morphological and environmental restraints on egg production in amphipods. In A. M. Wenner (ed.), Crustacean Issues Vol. 5: Crustacean Egg Production, A. A. Balkema.

Tattersall, W. M. & O. S. Tattersall, 1951. British Mysidacea. Ray Society, London, 460 pp.

Wildish, D., 1980. Reproductive bionomics of two sublittoral amphipods in a Bay of Fundy estuary. Int. J. Invertebr. Reprod. 2: 311–320.

Hydrobiologia **223**: 35–42, 1991.
L. Watling (ed.), VIIth International Colloquium on Amphipoda.
© 1991 *Kluwer Academic Publishers.*

The distribution and frequency of the type II microtrichs in some gammaridean amphipods

V.J. Steele
Department of Biology, Memorial University of Newfoundland, St. John's, Nfld, A1B 3X9 (709) 737-7929, Canada

Abstract

The distribution and frequency of the type II microtrichs was determined with the aid of the scanning electron microscope in several gammarid species from different habitats in the marine environment. The type II microtrichs are recessed in shallow cuticular depressions on the tergum and occur in dorsal and/or lateral groups on the somites and the telson. The frequency of these microtrichs varies directly with the body length for any one species, and in the different gammaridean taxa studied. The commensal and supralittoral gammarids have none or very small numbers. The average numbers per somite are higher in the intertidal species of *Gammarus*, then increase in the subtidal *Calliopius*, and reach very large numbers in the benthic *Anonyx* and *Halirages*. On the average, the microtrichs on the pleon outnumber those on the peraeon somites. Large-sized *Halirages* and *Anonyx* have very large numbers on pleon somite 11. The possible significance of these differences is discussed.

Introduction

Oshel *et al.* (1988) classified integumental microtrichs in amphipods into two types on the basis of their socket morphology. This paper deals with the distribution and abundance of the type II microtrichs. The microstructure of this type of microtrich is a subject of a separate publication. In the present study the type II microtrichs were studied using the scanning electron microscope (SEM), and their numbers on each body segments were determined in species which occupy different habitats in the marine environment. Platvoet (1985) reported the type II microtrichs as 'side-line' organs in gammarids of the genera *Gammarus* and *Echinogammarus* with the aid of the SEM.

Materials and methods

All specimens were collected in Newfoundland. *Anonyx lilljeborgi* Boeck, 1870, *A. makarovi* Gurjanova, 1962, *A. ochoticus* Gurjanova, 1962, and *Halirages fulvocinctus* (M. Sars, 1858) were caught in baited traps by R. Hooper in Bonne Bay in July and October 1988. *Metopa pusilla* Sars, 1892 was found on a hydroid by R. Hooper in Bonne Bay in July 1988. The other species were collected by the author: *Talorchestia longicornis* (Say, 1818) at Cow Head in July 1986, *Gammarus duebeni* Lilljeborg, 1852 and *G. obtusatus* (Dahl, 1938) at St. Philips in December 1987, *Gammarus oceanicus* Segerstråle, 1947, *Orchestia gammarellus* (Pallas, 1766), *Gammarellus angulosus* (Rathke, 1843), and *Calliopius laeviusculus* (Krøyer, 1838) in Bay Bulls in August 1988.

The specimens were fixed in Karnovsky's glutaraldehyde-paraformaldehyde fixative with 0.2 M cacodylate buffer at pH 7.4 for 12 h or longer, then washed in buffer and placed in buffered 2% osmium tetroxide for 3–8 h at room temperature. A graded ethanol series was used for dehydration.

For scanning electron microscopy the specimens were infiltrated with liquid carbon dioxide after dehydration, critical-point dried from carbon dioxide, mounted on aluminum stubs, and sputter gold-coated with an Edwards S150A sputter coater. The specimens were viewed with a Hitachi S570 scanning electron microscope operated at a maximum accelerating voltage of 20 kV. Exposures were made on Polaroid Positive/Negative film type 665.

Results

The body of amphipods consists of 7 peraeon and 6 pleon somites. For convenience these have been numbered 1 to 13, equivalent to peraeon 1–7 and pleon 1–6.

Morphology

Type II microtrichs are distinguished from the other microtrich types by their circular sockets. They are recessed in shallow depressions whose cuticular decorations are species-specific (see Figs. 1c, 2a, 3a, 5, 8, 9). The microtrichs have the same length in small and large individuals of the same species, but their length varies amongst the families studied. The smallest are found in *Orchestia* (4 µm, see Fig. 1c), *Gammarus* and *Gammarellus*. In *Anonyx* they are $5\frac{1}{2}$–$8\frac{1}{2}$ µm in length (see Fig. 8), and in *Halirages* 8–12 µm (see Fig. 9). They increase in length posteriorly in the lateral and dorsal groups (see Fig. 8). In the groups that are aligned with the body axis the microtrichs are at right angles to it (see Figs. 2a, b, 3a, b, 5, 6).

Distribution

Type II microtrichs occur in dorsal and/or lateral groups (see Figs. 1, 2, 3, 5, 6) on all somites and on the telson. However, none could be found on the commensal *Metopa pusilla*, and the supralittoral *Talorchestia longicornis*.

Typically there are 4 groups on each body somite: 2 lateral and 2 dorsal. The location and spacing of the microtrichs show taxonomic as well as individual variations. On the peraeon the lateral groups are often V-shaped (see Fig. 3a), but on somite 1 and 2 they may have a different pattern (see Fig. 2a). The lateral groups on the

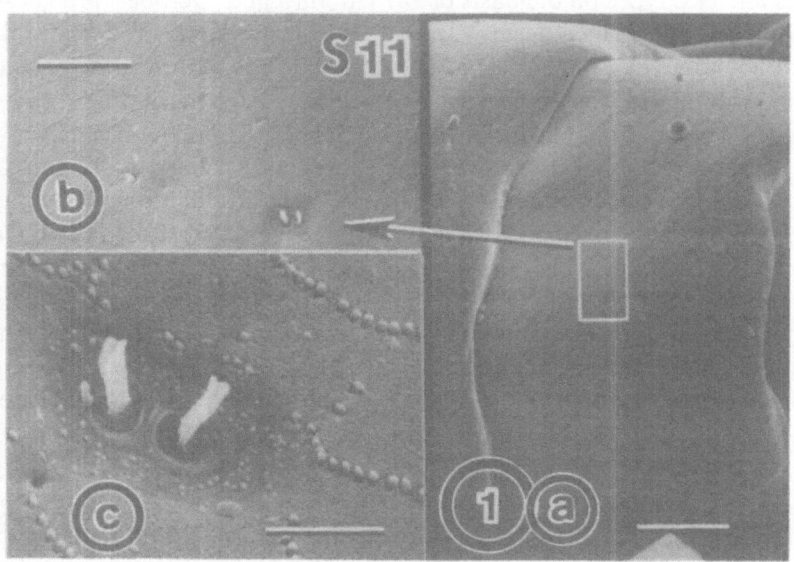

Fig. 1. Orchestia gammarellus. The left lateral group (a) on somite 11 (S11) consists of 2 type II microtrichs (b) in circular sockets which are recessed in a shallow cuticular depression (c). Scale bars: a. 0.2 mm, b. 20 µm, c. 5 µm.

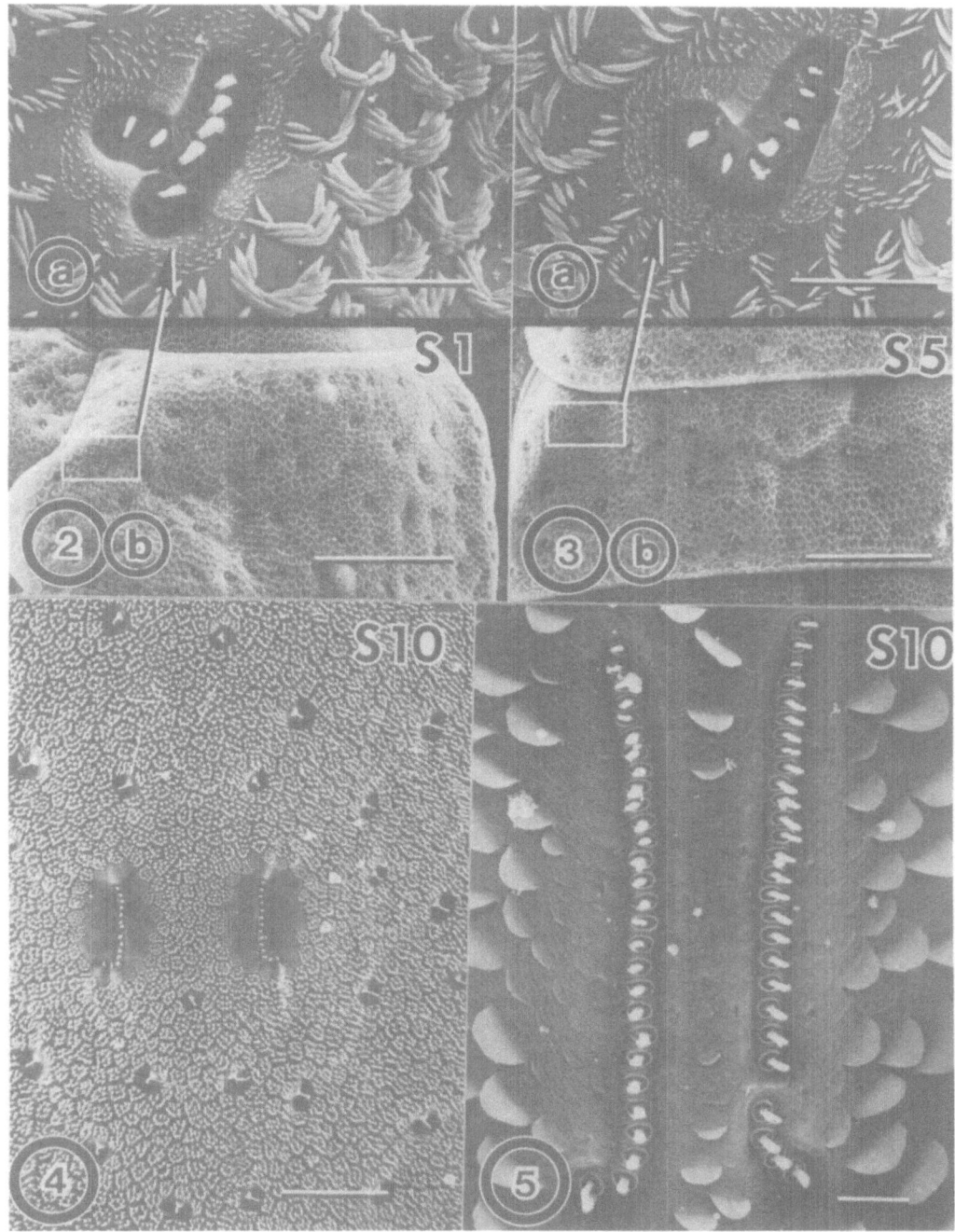

Fig. 2. Gammarellus angulosus. T-shaped lateral group (a) located on somite 1 (S1) (b). Scale bars: a. 20 μm, b. 200 μm.

Fig. 3. Gammarellus angulosus. V-shaped lateral group (a) located on somite 5 (S5) (b). Scale bars: a. 20 μm, b. 200 μm.

Fig. 4. Gammarus oceanicus. Dorsal groups on pleon somite 10 (S10). Scale bar: 100 μm.

Fig. 5. Calliopius laeviusculus. Dorsal groups on pleon somite 10 (S10). Scale bar: 10 μm.

38

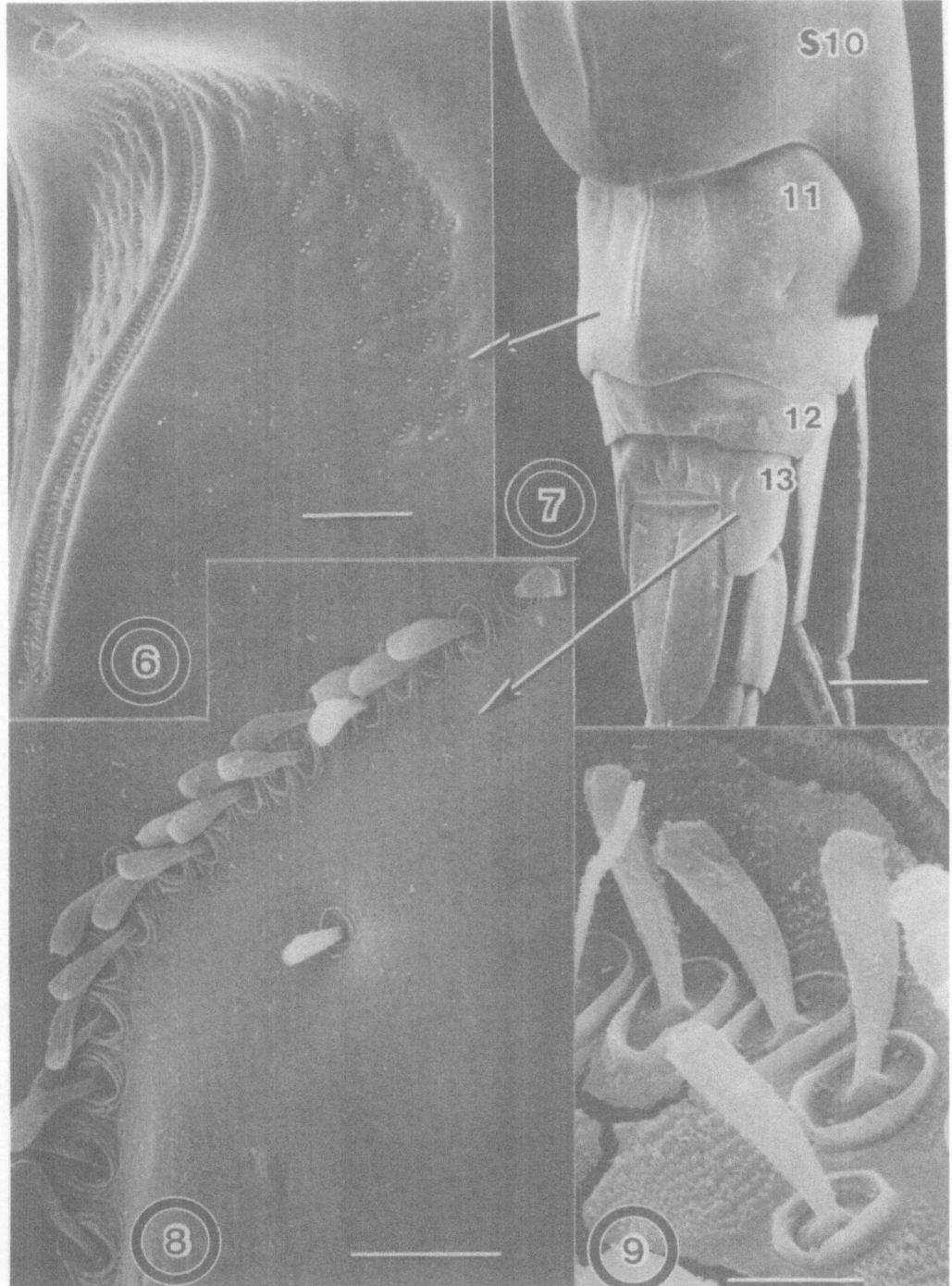

Fig. 6. Anonyx lilljeborgi. Central groups along the carina on pleon somite 11 are flanked by small groups of type II microtrichs. Scale bar: 10 μm.

Fig. 7. Anonyx lilljeborgi. Pleon somites 10 (S10), 11, 12, 13. The center of somite 11 is enlarged in Fig. 6, and the lateral groups on somite 13 in Fig. 8. Somite 12 is not carinate and lacks the central groups. Scale bar: 500 μm.

Fig. 8. Anonyx lilljeborgi. Lateral group on somite 13. The microtrichs gradually increase in length posteriorly. Scale bar: 10 μm.

Fig. 9. Halirages fulvocinctus. Type II microtrichs from a lateral group on somite 8 (12 m long). Cuticle is damaged by the electron beam. Scale bar: 5 μm.

pleon (somites 8, 9, 10) are postero-dorsal with respect to the lateral peraeon groups. The lateral groups on somites 11, 12, 13 are expanded laterally in some species: in *Anonyx* on somites 11 and 13, in *Calliopius* on 12, in *Halirages* on 11 and 12. In *Orchestia gammarellus*, there are no lateral groups on the pleon somites.

The dorsal groups are J-shaped (on the left side, mirror image of J on the right side) on all body somites (see Figs. 4, 5). The dorsal groups are absent from somite 12 in all the species examined, and from somite 1 in some species (*Calliopius laeviusculus*, *Gammarus oceanicus*, *Orchestia gammarellus*, *Halirages fulvocinctus*). They are also absent from somites 3–7 in individual *Halirages* (as shown in Table 2).

The distances between the dorsal groups (as shown in Table 1) are proportional to the size of the animal. The maximum distances are on the peraeon: in the *Gammarus* species on the first or second somite, in *Calliopius* on the third, in *Anonyx* and *Halirages* on the fourth. The distances between the left and the right groups decrease posteriorly to a few microns on one of the peraeon somites (*G. oceanicus*, *G. obtusatus*, *G. duebeni*) or on the pleon (ninth somite in *Halirages*, tenth in *G. lawrencianus* and *calliopius*, eleventh in *Anonyx*). In the carinate species the dorsal groups are parallel to the carinae (*Calliopius*, *Anonyx* see Figs. 5, 6, 7). In *Calliopius* the central group forms a grid on somite 13. In *Anonyx* (see Fig. 6), *Halirages*, and *Gammarus*

oceanicus, the central linear groups on somite 11 are flanked by an array of small groups. Somite 12 lacks the dorsal groups. *Gammarus duebeni* and *Orchestia gammarellus* have none on the telson, where they are very numerous in *Halirages*. In other *Gammarus* species the space between the left and the right dorsal groups is wide behind the head, narrows on one of the peraeon somites, and then widens toward somite 10, before narrowing again toward the telson. In *Anonyx* and *Calliopius* the distance widens gradually from somite 1 to 5, and then narrows to somite 11. In *Calliopius* it is broad anteriorly and narrows to a few microns on somite 7. *Halirages* has no dorsal groups on somites 5 and 6.

Abundance

The number of microtrichs varies directly with body length, as shown for *Gammarus obtusatus*, *G. oceanicus*, *G. duebeni*, *Calliopius laeviusculus*, *Anonyx lilljeborgi*, *A. makarovi*, and *Halirages fulvocinctus* (as shown in Table 2). A newly hatched *G. obtusatus* has half the number found in a specimen four times longer, and the number per somite is nearly the same in the tiny animal.

The frequency of the type II microtrichs varies in the different gammaridean taxa studied. Table 2 lists the total counts per somite and the average numbers per peraeon and pleon somite in some of the species included in this study.

Table 1. Distances between dorsal groups of microtrich type II in μm.

Species	Body length mm	Somite													
		1	2	3	4	5	6	7	8	9	10	11	12	13	T
Gammarus duebeni	14	250	240	210	200	160	100	110	120	160	220	200	0	20	0
G. lawrencianus	7	0	50	50	50	50	50	50	50	50	30	40	0	40	40
G. obtusatus	11	0	240	240	220	190	120	120	140	140	140	80	0	40	80
G. oceanicus	12	130	120	100	90	70	60	50	60	70	70	40	0	20	40
Calliopius laeviusculus	10	0	430	580	540	480	100	25	25	20	15	20	0	60	60
Anonyx lilljeborgi	8	60	70	120	160	150	150	100	60	40	20	5	0	30	30
A. makarovi	20	200	250	340	440	480	400	340	220	70	40	15	0	50	70
Halirages fulvocinctus	10	0	100	110	130	0	0	120	60	40	50	40	0	50	50

Table 2. Frequency of type II microtrichs per somite and telson.

Species	Body length mm	Peraeon somite							Average no. per Peraeon SOM	Pleon somite						Average no. per Pleon SOM	T
		1	2	3	4	5	6	7		8	9	10	11	12	13		
Commensal:																	
Metopa pusilla		0	0	0	0	0	0	0	0	0	0	0	0	0	0	0	0
Supralittoral:																	
Talorchestia longicornis		0	0	0	0	0	0	0	0	0	0	0	0	0	0	0	0
Orchestia gammarellus	8–10	2	4	4	4	6	8	8	4	4	4	4	2	0	0	2	0
Intertidal:																	
Gammarus obtusatus	2.5	6	8	8	6	8	8	8	7	8	8	8	6	6	8	7	4
	11.0	8	18	13	14	18	20	18	16	18	20	20	12	4	10	14	6
	14.0	10	22	20	17	20	21	20	19	20	20	18	14	6	14	15	6
Gammarus duebeni	8.5	24	25	20	20	23	23	28	23	22	24	22	32	10	34	24	0
	14.0	24	26	24	24	26	28	30	26	24	22	24	40	10	32	25	0
Gammarus oceanicus	12.0	18	26	26	24	30	26	28	25	27	30	34	49	10	35	31	12
	20.0	43	33	29	29	29	32	38	33	36	40	46	126	4	29	46	12
Gammarus lawrencianus	6.5	18	27	26	25	24	28	28	26	30	28	36	65	8	16	30	10
Subtidal:																	
Calliopius laeviusculus	4.5	12	30	24	24	26	26	36	26	38	40	74	26	12	28	34	28
	6.0	22	24	36	31	16	18	26	25	24	22	46	74	22	64	42	46
	7.0	20	48	38	28	36	46	52	38	59	53	79	50	74	84	67	73
	10.0	26	58	50	48	44	52	64	49	72	74	115	84	68	122	89	71
Anonyx ochoticus	10.0	34	36	36	36	36	36	34	35	32	42	60	108	20	37	48	13
Anonyx lilljeborgi	5.5	20	15	21	24	22	22	24	21	30	33	47	91	6	16	37	4
	6.5	30	23	26	25	27	29	44	29	52	56	61	128	8	32	57	4
	8.0	28	32	30	30	28	32	32	30	38	46	50	90	18	35	46	6
	9.0	44	39	40	40	41	41	41	41	44	40	68	536	18	50	126	8
	12.0	74	74	76	72	72	68	86	74	92	106	130	562	24	48	160	15
Anonyx makarovi	20.0	82	78	86	92	88	88	100	88	114	120	124	656	23	82	187	22
	26.0	120	108	112	114	115	106	110	112	136	133	132	1524	26	108	346	28
Halirages fulvocinctus	4.5	10	26	10	10	10	8	10	12	24	24	28	20	10	8	19	10
	5.0	12	26	12	12	10	18	26	17	30	30	42	34	10	10	26	14
	10.0	22	47	42	42	22	20	20	35	73	104	112	116	31	28	77	16
	11.0	30	66	22	24	22	52	78	42	98	90	144	236	72	63	118	49
	16.0	32	80	26	22	22	50	66	42	80	82	120	349	92	48	125	40

The commensal amphipod *Metopa pusilla* lacks type II microtrichs. Among the supralittoral gammarids, *Talorchestia longicornis* has none, and *Orchestia* has a very small number.

In the intertidal species, the numbers per somite are similar to those in *Gammarus oceanicus* and *G. duebeni* in animals which are 8.5–14 mm long. The numbers per somite are less in *Gammarus obtusatus* and higher in *G. lawrencianus* of similar size. The average number per somite decreases on the pleon in *G. obtusatus*, equals the number on peraeon somites in *G. duebeni*, and is larger in *G. lawrencianus* and *G. oceanicus*, particularly on somite 11.

In *Calliopius*, from the subtidal fringe, the average numbers per somite are higher than for the intertidal *Gammarus*. In *Calliopius* the numbers of the microtrichs increase on the pleon in proportion to body length (as shown in Table 2). The greatest increase occurs on somites 10 and 13.

In the benthic *Anonyx* and *Halirages*, the average number is higher on the pleon. The most marked increase occurs in large animals. In *Anonyx ochoticus* (body length 10 mm) the increase in frequency per pleon somite is the same as in *G. oceanicus* (body length 20 mm, as shown in Table 2). In *Anonyx* and *Halirages* pleon somite 11 makes the main contribution to the large average number per pleon somite. In these species the telson has only few type II microtrichs, as in *Gammarus*.

Discussion

Although some authors have speculated that the type II microtrichs may be mechanoreceptors on the basis of their arrangement along the lateral and dorsal surfaces in amphipods (Oshel *et al.*, 1988; Platvoet, 1985), study of their ultrastructure indicates that the 2 neurons that innervate each type II microtrich sensillum lack any structures that could be interpreted as mechanoreceptive (unpublished observations). Their circular sockets are not innervated. The structural details of their dendrites are consistent with chemoreceptive function, as has been demonstrated in insects

(Keil & Steinbrecht, 1984). The final resolution of their function rests in electrophysiological experimentation, which could be carried out successfully in those species which have large numbers of these microtrichs along the carinae and on pleon somite 11.

In view of the lack of mechanoreceptory morphology, a mechanoreceptive function is impossible. The placement of the type II microtrichs along the lateral and dorsal body surface would enable the animals to detect chemical stimuli in the water currents during swimming and at rest. During swimming water currents affecting the animal are parallel to the body axis (Dahl, 1977). All the type II microtrich groups are aligned with the body axis, but the majority of the microtrichs are perpendicular to it. Their shallow recessing on the body surface may allow them to sample the swiftly moving fluid flowing past the tergum.

When the animals are at rest in the curled-up position, their tergum is under the influence of pleopod currents (Schellenberg, 1942; Barnard, 1969; Dahl, 1977) from both the anterior and posterior direction. The strong ventral efferent current heads into the slow but steady posterodorsal afferent current, creating a region of turbulence along the pleon. A large number of chemoreceptors on this part of the body would constitute a tasting organ. The antennae are in fact quite short in the lysianassids, a group renowned for their chemosensory capability.

The extraordinary ability of *Anonyx sarsi* to detect odour trails from carrion is well known, and its behaviour in this respect has been studied. Sainte-Marie (1986a) reported that the second cohort *A. sarsi* (15–28 mm) respond faster to feeding opportunities than the first (5–9 mm). The reaction time (20–40 s) of the second cohort was in fact faster than expected (50 s) given the current speed and the distance from the bait to the animals. Sainte-Marie (1986b) also observed that *A. sarsi* travels up to 2 km in 12 h forays. It has been suggested that deep-sea scavengers might use current-borne clues to locate bait (Dayton & Hessler, 1972), and that detection of odour trails might occur over distances of 1–2 km (Ingram & Hessler, 1983). The rapid response of the larger lysianassid is consistent with the hypothesis that they are better equipped by having very large numbers of specialized chemosensory sensilla to detect odour plumes from carrion/bait. These sensilla – microtrichs type II – are concentrated on the posterodorsal body surface, which is in the mainstream of its swimming currents when the body is extended, and in the path of the pleopod-generated microcurrents when it is curled up.

References

Barnard, J. L., 1969. The families and genera of marine Gammaridean Amphipoda. Bull. U.S. natn. Mus. 271: 1–535.

Dahl, E., 1977. The amphipod functional model and its bearing upon systematics and phylogeny. Zool. Scr. 6: 221–228.

Dayton, P. K. & R. R. Hessler, 1972. Role of biological disturbance in maintaining diversity in the deep sea. Deep-Sea Res. 19: 199–208.

Ingram, C. L. & R. R. Hessler, 1983. Distribution and behavior of scavenging amphipods from the central North Pacific. Deep-Sea Res. 30: 683–706.

Keil, T. A. & R. A. Steinbrecht, 1984. Mechanosensitive and olfactory sensilla of insects. In R. C. King & H. Akai (eds), Insect ultrastructure. Vol. 2. Plenum Press, New York: 477–516.

Oshel, P. E., V. J. Steele & D. H. Steele, 1988. Comparative morphology of amphipod microtrich sensilla. Crustaceana, Suppl. 13: 100–106.

Platvoet, D., 1985. Side-line organ in gammarids (Crustacea, Amphipoda). Beaufortia 35: 129–133.

Sainte-Marie, B., 1986a. Effect of bait size and sampling time on the attraction of the lysianassid amphipods *Anonyx sarsi* Steele and Brunel and *Orchomenella pinguis* (Boeck). J. exp. mar. Biol. Ecol. 99: 63–77.

Sainte-Marie, B., 1986b. Feeding and swimming of lysianassid amphipods in shallow cold-water bay. Mar. Biol. 91: 219–229.

Schellenberg, A., 1942. Flohkrebse oder Amphipoda. Tiervelt Dtl. 40: 1–252.

Hydrobiologia **223**: 43–45, 1991.
L. Watling (ed.), VIIth International Colloquium on Amphipoda.
© 1991 *Kluwer Academic Publishers.*

What can vicariance biogeographic models tell us about the distributional history of subterranean amphipods?

John R. Holsinger
Dept. of Biological Sciences, Old Dominion University, Norfolk, Virginia 23529, USA

Abstract

Because of taxonomic diversity, geographic isolation, and other considerations, subterranean groundwater amphipods would appear to make excellent candidates for biogeographic studies. Limted dispersal ability in combination with local endemism makes it likely that vicariance models will generally offer better explanations for present distribution patterns of subterranean amphipods than scenarios based on centers of origin and dispersal. Vicariance biogeography demands a knowledge of both phylogeny and area relationships, which are typically shown on biological area cladograms. To date most biogeographic studies on subterranean amphipods have been limited to cladograms of single taxonomic groups. Although useful in showing possible relationships between areas and nested subsets of taxa, these single taxon studies do not consider covariant patterns among different groups. However, in order to be fully effective, future biogeographic research will have to focus on analyses of congruence between biological area cladograms of amphipod taxa and other subterranean crustacean groups, such as isopods. To date many covariant distributions among groups of subterranean crustaceans have been recognized but not yet analyzed for congruence.

To date, approximately 700 species (representing 130 genera and 34 families) of amphipod crustaceans have been described from subterranean groundwater habitats throughout the world. A majority of these species are restricted to life in hypogean groundwaters and are, by definition, stygobionts. By virtue of their restriction to secluded groundwater environments and for the reasons outlined below, subterranean amphipods would appear to be excellent candidates for biogeographic studies (Holsinger 1986). (1) They are becoming well known taxonomically, due in large part to the diligent collecting and taxonomic activities of recent years. (2) Dispersal abilities are generally limited, primarily because these organisms lack a free-swimming larval stage and most live in shallow water and are closely 'tied' to the substrate. (3) Ranges are typically narrowly circumscribed, resulting in numerous locally endemic species. (4) Many taxa represent old phyletic lineages that have probably lived in stable groundwater habitats (refugia) for long periods of time, and as a result there are apparently many relicts — both phylogenetic and distributional.

Given the combination of limited dispersal and local endemism of stygobiont amphipods, vicariance models should have a much greater probability of providing satisfactory explanations for present distribution patterns than scenarios based on centers of origin and dispersal. The general principle of vicariance biogeography holds that common distribution patterns (i.e., covarying patterns) are more likely caused by vicariance than by random, uncorrelated dispersal (see Wiley, 1988). Dispersal, however, is not eliminated as a cause but it is relegated to a second-order explanation.

In order to do vicariance biogeographic analysis on amphipods, or any other group for that matter, we need to have good data on both phylogeny and the areas occupied by the taxa being analyzed. Hypothesized phylogenies must be in the form of cladograms and data on areas should incorporate as much pertinent information on geomorphology and geological structure as possible. Although the availability of such data are rapidly increasing for subterranean amphipods, vicariance biogeographic studies on these organisms to date have been very limited and confined to single taxonomic groups.

To my knowledge, all published biogeographic studies on subterranean amphipods utilizing biological area cladograms have so far attempted to show only correlations of the 'stranding' of putative ancestors of freshwater stygobionts with times of marine regressions or transgressions or island emergence (see Stock & Rondé-Broekhuizen, 1986; Boutin & Coineau, 1988), and/or relationships between nested subsets of taxa and geographic isolation (see Holsinger, 1986; Notenboom, 1988). Outside subterranean waters, Meyers (1988) also has used vicariance biogeography to explain the geographic distribution of epigean marine amphipods of the subfamily Aorinae.

With respect to the first program listed above, the origin of freshwater stygobiont amphipods from marine sources in several different parts of the world is believed to have occurred through stranding of putative ancestors in areas formerly exposed to or covered by shallow marine embayments, or on islands having undergone emergence above sea level at some time in the past (Stock, 1980; Holsinger & Longley, 1980; Holsinger, 1986). Stranding is a vicariance hypothesis because it involves the isolation of faunas through geological events, which fragmented more continuous distributions that existed earlier. Many stygobionts of marine origin live at present in freshwater habitats, where they presumably adapted to newly opened niches following regression of sea water or tectonic uplifting in Late Cretaceous and Tertiary times (Stock, 1980; Holsinger, 1986). Correlation of stranding with marine regressions, although not always precise, has to date been one of the most important applications of vicariance biogeography theory to the study of stygobiont amphipod distribution. It has been especially useful in explaining the complex patterns of sty-gobiont amphipod distributions in parts of the peri-Mediterranean and greater Caribbean-Gulf of Mexico regions.

Future vicariance biogeographic research on subterranean amphipods has several potentially interesting possibilities. However, to be fully effective it will have to focus on analyses of congruence between the area cladograms of various amphipod taxa and those of other groups of subterranean crustaceans. As Wiley (1988) has pointed out, using vicariance biogeography to look for congruence between area cladograms of different monophyletic taxa is a 'strong sense' application. Methods for making these comparisons include reduced area cladograms, component analysis or parsimony analysis (see Humphries et al., 1988). Utilization of these methods of analysis in future research on the biogeography of subterranean amphipods may prove to be quite rewarding and potentially more powerful than anything attempted in past studies of these organisms.

The many interesting covariant distributional patterns that exist among remipedes, bathynellaceans, thermosbaenaceans, amphipods, isopods, mysids, and decapods in subterranean groundwater waters in different parts of the world have probably resulted from events that were common to the evolutionary history of members of these taxa in a given area. Furthermore, it is highly probable that many of the events we assume have affected distribution, such as stranding, subterranean stream capture, spring failure, etc., were vicariant.

Three broad patterns of distribution showing remarkable similarities between amphipods and other crustacean groups have been identified previously and are good candidates for future, detailed vicariance biogeographic analyses utilizing comparison of area cladograms. (1) The widespread overlap between the ranges of many species of crangonyctid amphipods and asellid isopods in subterranean groundwaters of the Holarctic region has never been carefully analyzed biogeographically. Both of these families are believed to be very old freshwater crustacean groups, and the majority of their species are local endemics that inhabit subterranean groundwaters. These families appear to have much in common, both biogeographically and ecologically.

(2) A broad overlap in the distributions of hadziid amphipods and cirolanid isopods, and to a lesser extend mysids and thermosbaenaceans, in the subterranean groundwaters of a region extending from south-central Texas to the Yucatan Peninsula of southern Mexico appears to be a generalized pattern involving areas formerly exposed to marine embayments. The colonization of hypogean freshwaters in this region has been attributed to the stranding of putative marine ancestors during marine regressions.

(3) The ranges of the hadziid amphipod *Bahadzia* and four other small, monotypic genera of stygobiont crustaceans (remipedes, thermosbaenaceans, isopods, shrimps) in the West Indian region are nearly congruent, suggesting a similar distributional history for these taxa. Although the results of a preliminary track analysis have been published (Holsinger, 1989), a follow-up study to compare area cladograms of these taxa currently awaits additional distribution data and phylogenetic analyses of the non-amphipod groups.

References

Boutin, C. & N. Coineau, 1988. *Pseudoniphargus maroccanus* n. sp. (subterranean amphipod), the first representative of the genus in Morocco. Phylogenetic relationships and paleobiogeography. Crustaceana, Suppl. 13: 1–19.

Holsinger, J. R., 1986. Zoogeographic patterns of North American subterranean amphipod crustaceans. In R. H. Gore & K. L. Heck (eds), Crustacean Biogeography (Crustacean Issues 4). A.A. Balkema, Rotterdam: 85–106.

Holsinger, J. R., 1989. Preliminary zoogeographic analysis of five groups of crustaceans from anchialine caves in the West Indian region. Proc. 10th Int. Congr. Speleol., Budapest 1: 25–26.

Holsinger, J. R. & G. Longley, 1980. The subterranean amphipod crustacean fauna of an artesian well in Texas. Smithson. Contr. Zool. 308, 1–62.

Humphries, C. J., P. Y. Ladiges, M. Roos & M. Zandee, 1988. Cladistic biogeography. In A. A. Myers & P. S. Giller (eds), Analytical Biogeography, Chapman & Hall, London: 371–404.

Myers, A. A., 1988. A cladistic and biogeographic analysis of the Aorinae subfamily nov. Crustaceana, Suppl. 13: 167–192.

Notenboom, J., 1988. Phylogenetic relationships and biogeography of the groundwater-dwelling amphipod genus *Pseudoniphargus* (Crustacea), with emphasis on the Iberian species. Bijdr. Dierk. 58 (2): 159–204.

Stock, J. H., 1980. Regression model evolution as exemplified by the genus *Pseudoniphargus* (Amphipoda). Bijdr. Dierk. 50: 104–144.

Stock, J. H. & B. L. M. Rondé-Broekhuizen, 1986. A new species of *Pygocrangonyx*, an amphipod genus with African affinities, from Fuerteventura. Bijdr. Dierk. 56 (2): 247–266.

Wiley, E. O , 1988. Vicariance biogeography. Ann. Rev. Ecol. Syst. 19: 513–542.

Hydrobiologia **223**: 47–68, 1991.
L. Watling (ed.), VIIth International Colloquium on Amphipoda.
© 1991 *Kluwer Academic Publishers.*

Actual state of gammaridean amphipoda taxonomy and catalogue of species from Chile

Exequiel Gonzalez
Departamento de Biologia Marina, Facultad de Ciencias del Mar, Universidad Católica del Norte, P.O. Box 117 Coquimbo-Chile

Key words: Chile, amphipoda, gammaridea, taxonomy, catalogue

Abstract

The taxonomy of gammaridean amphipods is poorly known in Chile. The lack of literature has discouraged investigation of this group, which has been overlooked for years. This has led to the exclusion of amphipods from biological and ecological studies done in our country. I present here a preliminary report on the actual state of the group with reference mainly to its taxonomy, ecological studies and a species catalogue. I have only included species from continental Chile north of 56 °S, Juan Fernandez Archipelago and Isla de Pascua. Species from Chilean Subantarctic and Antarctic waters are excluded, because they are well covered by other authors. There are 168 known for Chile, 36 of them undescribed or with their status not clear. This report is based on a revision of the species recorded for Chile, analysis of samples from the Chilean coast and the work of ecologists who are using the group as material for study.

Introduction

Gammaridean amphipods are a group poorly studied in Chile. Since 1960, less than a dozen species have been described, of those only two were by local zoologists (Gallardo, 1962; Varela, 1983). Before that date, no species were described by Chilean specialists.

Species described before 1960 were collected by expeditions to the Southern Ocean during the 19th century. The new species or records are from old samples, not always well labeled. Types and paratypes are not available in the places where they were originally deposited. Localities are uncertain and accordingly records are questionable. Many of the geographical distributions say only 'Chile' (e.g., Nicolet, 1849). Several species have not been collected in Chile since the time of original publication. Others are included in cosmopo-

litan species groups, where their status is not yet clear.

The present work reviews the status of ecological studies done after 1950 and the taxonomy of the species recorded for continental Chile North of 56 °S, Juan Fernandez Archipelago and Isla de Pascua.

I have not included species from Chilean Subantarctic or Antarctic waters, published in detail by Lowry & Bullock (1976).

The problem of taxonomy of marine gammaridean amphipods become important around 1960, with the growing interest in marine sciences. There were important collecting efforts conducted by different universities. Intertidal and subtidal samples from the Central and Northern region of the country were gathered. Identification was difficult because of the insufficient knowledge of our amphipod fauna and the lack of pertinent literature. Descriptions were in old expedition reports or journals which were not always available, or if they were, the species were poorly described and not well depicted. Further, there has been very little study of our fauna and east Pacific amphipods in general.

In response to those needs some universities began to put together the literature and to interest students in the taxonomy of the group. At the end of the seventies I began to gather information to produce a list of the species recorded for Chile. The present report is based on that information.

Ecological studies which include amphipods identified to species level are not abundant. Dahl (1952) was the first, giving some data in his study of the ecology of sandy beaches. Other ecological work was done in relation to *Ampelisca araucana* and *Orchestoidea tuberculata*. Recently some experimental ecology has begun on *Hyale* (Buschmann & Santelices, 1987), after I reviewed the taxonomy of the genus (González, 1990).

In the catalogue, many species are listed to genus level, those are probably new species for science. There are also new records or range extensions, like some species of *Hyale*. Each synonymy has been chosen according to the author that gives the latest information and a synonymic revision of the species. I also include some references to authors that give illustrations of the species.

Family arrangements follows a combination of Barnard's (1969b) and Bousfield's (1983) work. Some discrepancies arise especially in Talitroidea, and others. Some new work on Lysianassids by Lowry has also brought new arrangements within this family; these have not been included. The families are ordered alphabetically. Asterisks, on the right side of a locality in the geographic distribution indicate new records for Chile, the specimens are in my own collection.

History of taxonomic knowledge of Amphipoda

The description of new species for Chile began at the end of the first half of the 19th century. Productivity was very low and only a couple of species described during that time used Chilean material. Most of the species from that period which are known today are records from later years (Table 1). The principal works were by Milne Edwards (1840), Kroyer (1845) and Nicolet (1849), the last part of an enormous effort in knowing the Chilean fauna and flora by the naturalist Claudio Gay. It was published in Spanish with fairly good drawings for that time. It recorded 10 gammaridean species, from which only two are now valid as they were described, five having been already described and three have not been found since the publication date (*Ampithoe chilensis* Nicolet, 1849; *Gammarus chilensis* Nicolet, 1849; *Orchestia gayi* Nicolet, 1849). A significant increase in species number occurred in the second half of the century. This was mainly due to the work of Dana (1852, 1853 & 1855) and Stebbing (1888). During the first half the 20th Century, descriptions and new records were very numerous due to the work of Schellenberg (1931, 1935), who is responsible for 82 new species or records for Chile. Almost all the species described were based on material originally collected in Chile. The second half of this

Table 1. Authors involved in the descriptions of new species and new records for Chile (mainly chilean material).

Authors	Year	No of spp.	New spp.
Milne Edwards	1840	1	2
Kroyer	1845	3	3
Nicolet	1849	10	3
Dana	1852, 1853 & 55	25	9
Bate	1862	17	2
Stebbing	1906	34	19
Walker	1913	1	1
Schellenberg	1931, 1935	127	82
Ruffo	1949	28	7
Barnard	1960, 1979, 1980	11	0
Gallardo	1962	1	1
Andres	1975, 1979	47	1
Barnard & Barnard	1980	2	1
Bousfield	1982	3	0
Barnard & Clark	1982, 1984	3	2
Varela	1983	5	1
Lowry	1984	1	1
Gonzalez	1986	1	0
Clark & Barnard	1986, 1987, 1988	3	3
Lowry & Stoddart	1987	1	1
Conlan	1990	2	1

century has seen only 11 new species described in 49 years. In this period, only 24 new records also were registered. The most relevant work is by Andres (1975), which includes 18 new records for Chile, and one new species. A few new descriptions have been produced in the last ten years, due mainly to the exchange of material with world specialists.

Geographic distribution of species

It is important to mention the geographic distribution of Gammaridea Amphipoda along the Chilean coast. There is an increase of species toward the south (Table 2). This is not due to biological, biogeographic or oceanographic reasons, but rather to a higher pressure of sampling in the southern part of the country. Most of the species found in the north over the last 20 years are new records for the country.

Taxa with problems that need resolution

This section presents a review of the major groups with known taxonomic problems. In other groups, the lack of work does not allow definitive statements of problems needing resolution.

Within the talitroideans, *Hyale* species are the most abundant along the Central and Northern Chilean coast. Five species are known, living together in almost all sites where they have been collected. Two of them are new records, *H. maroubrae* Stebbing, 1899 and *H. media* (Dana, 1953&55). For a third species, *H. rubra* (Thomson, 1879), its range is extended to the south as far as Coquimbo (29° 56′ S 71° 21′ W)

Table 2. Distribution of records according to geographic area.

Area	Latitude	No of spp.
Arica – Coquimbo	18°–29° S	25
Coquimbo – Chiloe	30°–41° S	57
Chiloe – Cabo de Hornos	42°–56° S	117

and to the north to Iquique (20° 12′ S 70° 10′ W). *H. rubra* was cited only for Antofagasta (23° 38′ S 70° 25′ W) by Barnard (1979). *H. hirtipalma* (Dana, 1852) and *H. grandicornis* (Kroyer, 1845) are the most common species, although after 1950 are only mentioned for Chile by Andres (1975). Besides *Hyale*, we have found a new species of *Parhyalella* (in the process of description) and an unknown species in the Talitridae for which we have only one specimen. Others species in the group also need to be reviewed, particularly the orchestids. *Orchestia scutigerula* Dana, 1853&55 probably should be placed in a new genus (Bousfield, personal communication). *O. selkirki* Stebbing, 1888 and *O. gayi* Nicolet, 1849 have not been recognized since their original descriptions.

Hyalella species represent another complex and appear to be a very interesting group of species exhibiting slight morphological variation, at least in the forms known to Chile, which I have had the opportunity to examine from small streams in Antofagasta and Valdivia, 2500 kilometers apart.

Transorchestia chiliensis (Milne Edwards, 1840) had until now a disjunct distribution. It has been collected in Antofagasta (23° 38′ S 70° 25′ W), Juan Fernandez Archipelago (34° 00′ S 78° 00′ W) and south of 39 ° S. The morphological variation of the group in relation to the specimens from New Zealand suggest an interesting subject of study.

Paracorophium hartmannorun Andres, 1975 was redescribed by Gonzalez (1986), because the specimens of the original description were shown to be juveniles.

Ampelisca. The genus has four known species for Chile: *Ampelisca macrocephala* f. *gracilicauda* Schellenberg, 1931; *A. composita* Schellenberg, 1931; *A. macrocephala* f. *dentifera* Schellenberg, 1931; and *A. araucana* Gallardo, 1962. From my analysis of samples of *A. macrocephala* f. *gracilicauda*, it is apparent that *A. gracilicauda* is a separate species. *Ampelisca araucana* urgently needs to be redescribed. I have also found a new species, which is in the process of description with the redescription of *A. gracilicauda* and *A. araucana*.

Elasmopus. This genus has two species known for Chile: *E. chilensis* Walker, 1913, and *E. rapax* Costa, 1853. The morphological variation of this group is quite large. Specimens are being studied to determine valid species and variation within the genus.

Aora. Two species are known for Chile, *A. anomala* Schellenberg, 1926 and *A. typica* Kroyer, 1845. The distribution and the correct identification of the different forms collected need some detailed analysis.

Gibberosus. The genus is for the first time reported from Chile and the species is probably new.

Ampithoe. Several specimens have been collected in intertidal and subtidal samples of continental Chile and Isla de Pascua. They may be assigned to *A. chilensis* Nicolet, 1849, not recorded for Chile since the original description.

Other problems. Several other species have been collected in which some clarification of identity is needed such as *Eudevenopus, Heterophoxus* spp., *Proharpinia, Liljeborgia,* Dexaminidae, Ischyroceridae, Lysianassidae, etc. The main problem consists in determining the status of forms similar to those collected far from type localities.

Ecological or biological work on Amphipoda

The first ecological work citing Amphipoda at the species level was by Dahl (1952). He recorded seven species that were collected in the course of the Lund University Expedition to Chile in 1948, *Parharpinia fuegiensis* (Schellenberg, 1931) (= *Fuegiphoxus fuegiensis*); *Metharpinia cornuta* (Schellenberg, 1931) (= *Microphoxus cornutus*); *Bathyporeiapus magellanicus* Schellenberg, 1931; *Monoculopsis vallentini* Stebbing, 1914; *Monoculodes* sp., *Metoediceros fuegiensis* Schellenberg, 1931; and *Orchestoidea tuberculata* Nicolet, 1849. All occurr in sandy beaches along the Chilean coast. These species are the only records, besides the one of Lowry and Stoddart (1987), mentioned as having been collected by the Lund Expedition.

Orchestoidea tuberculata has been the species most frequently mentioned in the literature with respect to ecological studies. Duarte (1974) showed for this species the consumption of 24.8 mg d^{-1}ind^{-1} of *Durvillea antarctica.* Jaramillo *et al.* (1980) studied locomotion on the beach. Several other authors have published studies on the descriptive ecology of sandy beaches including some reference to other species of gammaridean amphipods (Varela, 1983).

During the last five years research on biology and experimental biology of *Hyale* has begun. Buschmann and Santelices (1987) studied three species of *Hyale* using different intertidal levels. They were seen to eat algae, preferring mature cystocarp tissue. A positive effect of this feeding is the liberation of higher numbers of spores. Germination and growth rate indicate that grazer-mediated release does not affect further spore development. A fraction of the ingested spores survives passage through the amphipod digestive tract.

There is a group of students at Universidad Católica del Norte, Chile studying the distribution and abundance of *Hyale* in Coquimbo (29° 56′ S 71° 21′ W) and such things as optimal factors for laboratory experiments, feeding habits, reproductive biology of *H. maroubrae,* including duration of precopula, numbers of eggs, incubation time and embryogenesis.

Some work is also occurring at Universidad Austral in Valdivia, Chile (38° 50′ S 73° 15′ W) on *Paracorophium hartmannorum,* mainly on sex determination and population biology.

Catalogue

AMPELISCIDAE

Ampelisca araucana Gallardo, 1962
    A. araucana Gallardo, 1962.
    Distribution in Chile: Iquique, Antofagasta*, Bahia la Herradura*, Concepcion.

Ampelisca composita Schellenberg, 1931
    A. composita Schellenberg, 1931; J.L. Barnard, 1958b.
    Distribution in Chile: Estrecho de Magallanes.

Ampelisca gracilicauda Schellenberg, 1931
  A. macrocephala f. *gracilicauda* Schellenberg, 1931; Schellenberg, 1935; Andres, 1975.
  Distribution in Chile: Antofagasta*, Coquimbo*, Bahia la Herradura*, Valparaiso, Corral, Canal Desertores, Estrecho de Magallanes.

Ampelisca macrocephala f. *dentifera* Schellenberg, 1931
  A. macrocephala f. *dentifera* Schellenberg, 1931.
  Distribution in Chile: Estrecho de Magallanes.

AMPHILOCHIDAE

Amphilochus marionis Stebbing, 1888
  A. marionis Stebbing, 1888;
  Gitanopsis marionis Schellenberg, 1926; Bellan-Santini & Ledoyer, 1974; J.L. Barnard, 1958b;
  ? *Gitanopsis marionis* Schellenberg, 1931 (questioned by J.L. Barnard, 1972c);
  Amphilochus marionis J.L. Barnard, 1972b; Lowry & Bullock, 1976.
  Distribution in Chile: Canal Beagle, Estrecho de Magallanes.

Gitanopsis squamosa (Thomson, 1880)
  Amphilochus squamosus Thomson, 1880;
  Gitanopsis squamosa Schellenberg, 1926; Schellenberg, 1931; J.L. Barnard, 1958b; Andres, 1975; Lowry & Bullock, 1976; J.L. Barnard, 1972b.
  Distribution in Chile: Niebla, Puerto Pantalon, Estrecho de Magallanes.

Gitanopsis sp.
  Gitanopsis sp. Disalvo *et al.*, 1988.
  Distribution in Chile: Isla de Pascua.

AMPITHOIDAE

Ampithoe chilensis Nicolet, 1849
  Amphithoe chilensis Nicolet, 1849;
  Ampithoe chilensis Bate, 1862; Stebbing, 1906
  ? *A. chilensis* J.L. Barnard, 1958b.
  Distribution in Chile: Chile.

Ampithoe sp.
  Ampithoe sp. Disalvo *et al.*, 1988
  Distribution in Chile: Isla de Pascua.

Ampithoe ramondi Audouin, 1826
  A. ramondi Audouin, 1826; J.L. Barnard, 1955; J.L. Barnard, 1970; Disalvo *et al.*, 1988.
  Distribution in Chile: Isla de Pascua.

Peramphithoe femorata (Kroyer, 1845)
  Amphithoe femorata Kroyer, 1845; Schellenberg, 1931; Chilton, 1921; Schellenberg, 1935;
  Amphithoe femorata Stebbing, 1906; J.L. Barnard, 1958b; Andres, 1975; Kreibhom de Paternoster & Escofet, 1976; Lowry & Bullock, 1976;
  Amphithoe brevipes Dana, 1852; Dana 1853&55;
  Amphithoe brevipes Stebbing, 1906;
  ? *Amphithoe peregrina* Dana, 1853&55;
  Peramphithoe femorata Conlan & Bousfield, 1982.
  Distribution in Chile: Iquique, Valparaiso, Isla Juan Fernandez, Tumbes, Niebla, Puerto Bueno, Puerto Churuca, Punta Arenas, Porvenir, Puerto Pantalon, Isla Picton, Navarino, Cabo de Hornos, Estrecho de Magallanes, Chile.

ANAMIXIDAE

Anamixis sp.
  Anamixis sp. Disalvo *et al.*, 1988.
  Distribution in Chile: Isla de Pascua.

CEINIDAE

Ceina platei Schellenberg, 1935
  C. platei Schellenberg, 1935; J.L. Barnard, 1958b; J.L. Barnard, 1972b.
  Distribution in Chile: Isla Juan Fernandez.

COLOMASTIGIDAE

Colomastix fissilingua Schellenberg, 1926
  C. fissilingua Schellenberg, 1926; Schellenberg, 1931; J.L. Barnard, 1958b; Lowry & Bullock, 1976; Holman & Watling, 1983.

Distribution in Chile: Ultima Esperanza, Puerto Condor.

COROPHIIDAE

Aora anomala Schellenberg, 1926
    A. typica f. *anomala* Schellenberg, 1926; Schellenberg, 1931;
    A. anomala J.L. Barnard, 1972b; Lowry & Bullock, 1976.
Distribution in Chile: Ultima Esperanza, Punta Arenas, Puerto Esperanza.

Aora typica Kroyer, 1845
    A. typica Kroyer, 1845; Bate, 1862; Stebbing, 1906 (in part); Chilton, 1921; J.L. Barnard, 1958b; Schellenberg, 1935; Andres, 1975.
    Lalaria longitarsis Nicolet, 1849;
Distribution in Chile: Coquimbo*, Bahia la Herradura*, Isla Juan Fernandez, Niebla.

Corophium bonelli (Milne Edwards, 1830)
    Corophia bonelli Milne Edwards, 1830;
    Corophium bonelli Stebbing, 1906; Shoemaker, 1947; J.L. Barnard, 1958b;
    C. pseudacherusicum Schellenberg, 1931; Lowry & Bullock, 1976.
Distribution in Chile: Punta Arenas, Isla Picton.

Corophium insidiosum Crawford, 1937
    C. insidiosum Crawford, 1937; J.L. Barnard, 1958b; J.L. Barnard, 1970; J.L. Barnard, 1971; Andres, 1975.
Distribution in Chile: Concepcion, Niebla, Puerto Montt.

Corophium uenoi Stephensen, 1932
    C. uenoi Stephensen, 1932; J.L. Barnard, 1952; J.L. Barnard, 1958b; Andres, 1975.
Distribution in Chile: Rio Andalien, Concepcion, Bahia San Vicente, Niebla.

Ericthonius brasiliensis (Dana, 1853&55)
    Pyctilus brasiliensis Dana 1853&55;
    Ericthonius brasiliensis Stebbing, 1906; Schellenberg, 1931; Schellenberg, 1935; J.L. Barnard, 1958b; J.L. Barnard, 1969a; Andres, 1975
Distribution in Chile: Iquique, Taltal, Caleta Chascos*, Tongoy*, Valparaiso, Chile (Region Norte).

Gammaropsis longicornis Walker, 1906
    G. longicornis Walker, 1906b; J.L. Barnard, 1973; Lowry & Bullock, 1976
    Eurystheus longicornis Schellenberg, 1931; J.L. Barnard, 1958b.
Distribution in Chile: Estrecho de Magallanes.

Gammaropsis (Gammaropsis) dentifer (Haswell, 1880)
    Moera dentifera Haswell, 1880b;
    Gammaropsis dentifera Stebbing, 1899a;
    Eurystheus dentifer Stebbing, 1906; Schellenberg, 1931; Schellenberg, 1935; J.L. Barnard, 1958b;
    Gammaropsis (Gammaropsis) dentifer J.L. Barnard, 1969b (by implication);
    Gammaropsis dentifer Bellan-Santini & Ledoyer, 1974; Andres, 1975;
    Gammaropsis (Gammaropsis) dentifer Lowry & Bullock, 1976.
Distribution in Chile: Caleta Chascos*, Bahia la Herradura*, Valparaiso, Niebla, Punta Arenas, Puerto Pantalon, Isla Picton, Canal Beagle, Isla Nueva.

Gammaropsis (Gammaropsis) longitarsus (Schellenberg, 1931)
    Eurystheus longitarsus Schellenberg, 1931; J.L. Barnard, 1958b;
    Gammaropsis (Gammaropsis) longitarsus J.L. Barnard, 1969b (by implication); Lowry & Bullock, 1976.
Distribution in Chile: Isla Picton, Canal Beagle.

Gammaropsis (Gammaropsis) monodi (Schellenberg, 1931)
    Eurystheus monodi Schellenberg, 1931; J.L. Barnard, 1958b;
    Gammaropsis (Gammaropsis) monodi J.L. Barnard, 1969b (by implication); Lowry & Bullock, 1976.
Distribution in Chile: Canal Beagle, Estrecho de Magallanes.

Gammaropsis (Gammaropsis) triodon (Schellenberg, 1926)

> *Eurystheus triodon* Schellenberg, 1926; Schellenberg, 1931; J.L. Barnard, 1958b; *Gammaropsis (Gammaropsis) triodon* J.L. Barnard, 1969b (by implication); Lowry & Bullock, 1976.

Distribution in Chile: Estrecho de Magallanes.

Haplocheria balssi Schellenberg, 1931

> *H. balssi* Schellenberg, 1931; J.L. Barnard, 1958b; Lowry & Bullock, 1976.

Distribution in Chile: Canal Desertores, Punta Arenas, Bahia Inutil.

Haplocheria barbimana (Thomson, 1879)

> *Gammarus barbimanus* Thomson, 1879; *Haplocheria barbimana* Stebbing, 1906; J.L. Barnard, 1972b; Lowry & Bullock, 1976; Schellenberg, 1931; J.L. Barnard, 1958b; Bellan-Santini, 1972.

Distribution in Chile: Puerto Bueno, Ultima Esperanza, Punta Arenas, Bahia Inutil.

Lembos fuegiensis (Dana, 1853&55)

> *Gammarus fuegiensis* Dana, 1853&55; *Lembos fuegiensis* Stebbing, 1906; Stebbing, 1914; Schellenberg, 1931; K.H. Barnard, 1932; J.L. Barnard, 1958b; Lowry & Bullock, 1976

Distribution in Chile: Punta Arenas, Bahia Inutil, Puerto Esperanza, Puerto Pantalon.

Paracorophium hartmannorum Andres, 1975

> *P. hartmannorum* Andres, 1975; Andres, 1979; Karaman, 1979; Gonzalez, 1986; *P. chilensis* Varela, 1983.

Distribution in Chile: Salinas de Puyaye, Constitucion, Rio Andalien, Concepcion, Bahia San Vicente, Valdivia, Mehuin, Niebla.

Photis reinhardi Kroyer, 1842

> *P. reinhardi* Kroyer, 1842; Stebbing, 1906; J.L. Barnard, 1958b; Andres, 1975.

Distribution in Chile: Taltal, Isla Mocha, Canal Desertores, Chile (Region Norte).

DEXAMINIDAE

Atylus homochir dentatus (Schellenberg, 1931)

> *Nototropis homochir* f. *dentatus* Schellenberg, 1931; *Atylus homochir dentatus* Lowry & Bullock, 1976

Distribution in Chile: Estrecho de Magallanes, Isla Lennox.

Atylus villosus Bate, 1862

> *A. villosus* Bate, 1862; J.L. Barnard, 1958b; Andres, 1975; Bellan-Santini & Ledoyer, 1974; Lowry & Bullock, 1976. ? *Nototropis villosus* Schellenberg, 1931;

Distribution in Chile: Valparaiso, Estrecho de Magallanes, Isla Lennox.

Paradexamine nana Stebbing, 1914

> *P. nanus* Stebbing, 1914; *P. nana* Schellenberg, 1931; J.L. Barnard, 1958b; Lowry & Bullock, 1976.

Distribution in Chile: Puerto Condor, Isla Lennox.

Paradexamine pacifica (Thomson, 1879)

> *Dexamine pacifica* Thomson, 1879; *Paradexamine pacifica* Stebbing, 1899b; Stebbing, 1906; Schellenberg, 1931; J.L. Barnard, 1958b; J.L. Barnard, 1972a; Lowry & Bullock, 1976.

Distribution in Chile: Corral.

Paradexamine sp.

> *Paradexamine* sp. Andres, 1975

Distribution in Chile: Niebla.

Polycheria acanthocephala Schellenberg, 1931

> *P. acanthocephala* Schellenberg, 1931; J.L. Barnard, 1958b.

Distribution in Chile: Estrecho de Magallanes.

Polycheria antarctica (Stebbing, 1875)

> *Dexamine antarctica* Stebbing, 1875; *Polycheria antarctica* Schellenberg, 1931; J.L. Barnard, 1958b; Andres, 1975; Lowry & Bullock, 1976.

Distribution in Chile: Coquimbo, Puerto Montt, Ultima Esperanza

54

DULICHIIDAE

Podocerus cristatus rotundatus Schellenberg, 1931
    P. cristatus rotundatus Schellenberg, 1931;
    J.L. Barnard, 1962a; Lowry & Bullock,
    1976.
    Distribution in Chile: Estrecho de Magallanes.

EUSIRIDAE

Atyloella magellanica (Stebbing, 1888)
    Atylopsis magellanicus Stebbing, 1888;
    Pontogeneia magellanica Stebbing, 1906;
    Atyloella magellanica Schellenberg, 1931;
    K.H. Barnard, 1932; Schellenberg, 1935;
    J.L. Barnard, 1958b; Andres, 1975; Bellan-
    Santini & Ledoyer, 1974; Lowry & Bullock,
    1976.
    Distribution in Chile: Concepcion, Lota, Punta
Arenas, Porvenir, Isla Picton, Canal Beagle,
Isla Nueva, Estrecho de Magallanes.

Cleonardo longipes Stebbing, 1888
    C. longipes Stebbing, 1888; Stebbing, 1906;
    J.L. Barnard, 1958b; J.L. Barnard, 1969b.
    Distribution in Chile: Chile, abyssal.

Chosroes incisus Stebbing, 1888
    C. incisus Stebbing, 1888; Stebbing, 1906;
    J.L. Barnard, 1958b; Schellenberg, 1931;
    Lowry & Bullock, 1976.
    Distribution in Chile: Estrecho de Magallanes.

Eusiroides monoculoides (Haswell, 1880)
    Atylus monoculoides Haswell, 1880a;
    Eusiroides monoculoides Stebbing, 1906; J.L.
    Barnard, 1958b; J.L. Barnard, 1971; J.L.
    Barnard, 1972b; Lowry & Bullock, 1976
    ? *E. monoculoides* Schellenberg, 1931
    (questioned by J.L. Barnard, 1972b);
    Distribution in Chile: Bahia Inutil.

Eusirus antarcticus Thomson, 1880
    E. cuspidatus var. *antarcticus* Thomson,
    1880;
    E. antarcticus Stebbing, 1906; Schellenberg,
    1931; J.L. Barnard, 1958b; Lowry &
    Bullock, 1976
    E. antarcticus f. *typica* Schellenberg, 1931;
    E. antarcticus f. *walkeri* Schellenberg, 1931;

Distribution in Chile: Bahia Inutil, Bahia Borja,
Puerto Esperanza, Canal Beagle, Estrecho de
Magallanes.

Gondogeneia antarctica (Chevreux, 1906)
    Pontogeneia antarctica Chevreux, 1906;
    Schellenberg, 1931; K.H. Barnard, 1932;
    J.L. Barnard, 1958b;
    Gondogeneia antarctica J.L. Barnard, 1972a;
    Lowry & Bullock, 1976.
    Distribution in Chile: Isla Nueva, Isla Lennox,
Estrecho de Magallanes.

Gondogeneia gracilicauda (Schellenberg, 1931)
    Pontogeneia gracilicauda Schellenberg, 1931;
    J.L. Barnard, 1958b;
    ? *Gondogeneia gracilicauda* J.L. Barnard,
    1972a (generic status not verified);
    G. gracilicauda Lowry & Bullock, 1976.
    Distribution in Chile: Puerto Pantalon.

Gondogeneia macrodon (Schellenberg, 1931)
    Pontogeneia macrodon Schellenberg, 1931;
    J.L. Barnard, 1958b;
    ? *Gondogeneia macrodon* J.L. Barnard, 1972a
    (generic status not verified);
    G. macrodon Lowry & Bullock, 1986
    Distribution in Chile: Puerto Pantalon, Isla
Lennox, Estrecho de Magallanes.

Gondogeneia simplex (Dana, 1852)
    Iphimedia simplex Dana, 1852; Dana,
    1853&55;
    Paramoera simplex Stebbing, 1906;
    Pontogeneia simplex Schellenberg, 1931;
    Schellenberg, 1935; J.L. Barnard, 1958b;
    Andres, 1975;
    Gondogeneia simplex J.L. Barnard, 1972a;
    Lowry & Bullock, 1976.
    Distribution in Chile: Tumbes, Concepcion,
Punta Arenas, Puerto Pantalon, Estrecho de
Magallanes.

Gondogeneia ushuaiae (Schellenberg, 1931)
    Pontogeneia ushuaiae Schellenberg, 1931;
    J.L. Barnard, 1958b;
    ? *Gondogeneia ushuaiae* J.L. Barnard, 1972a
    (generic status not verified);
    G. ushuaiae Lowry & Bullock, 1976
    Distribution in Chile: Estrecho de Magallanes.

Austroregia batei (Cunningham, 1871)
  ? *Atylus batei* Cunningham, 1871
  ? *Halirages batei* J.L. Barnard, 1958b;
  H. batei Stebbing, 1906; Lowry & Bullock,
  1976.
  Distribution in Chile: Estrecho de Magallanes.

Austroregia huxleyana (Bate, 1862)
  Atylus huxleyanus Bate, 1862;
  Halirages huxleyanus Stebbing, 1914; Schellenberg, 1931; J.L. Barnard, 1958b; Lowry & Bullock, 1976; not *H. huxleyanus* Stebbing, 1906
  Austroregia huxleyana J.L. Barnard, 1989.
  Distribution in Chile: Porvenir, Isla Nueva, Isla Lennox.

Austroregia regis (Stebbing, 1914)
  Bovallia regis Stebbing, 1914;
  Halirages huxleyanus Stebbing, 1888; Stebbing, 1906;
  H. regis K.H. Barnard, 1932; J.L. Barnard, 1958b; Lowry & Bullock, 1976
  H. stebbingi Schellenberg, 1931
  Austroregia regis J.L. Barnard, 1989.
  Distribution in Chile: Estrecho de Magallanes, Porvenir, Puerto Condor, Puerto Eugenia, Isla Nueva, Isla Lennox.

Paramoera brachyurus Schellenberg, 1931
  P. brachyurus Schellenberg, 1931; J.L. Barnard, 1958b; Lowry & Bullock, 1976.
  Distribution in Chile: Punta Arenas, Puerto Toro.

Paramoera fissicauda (Dana, 1852)
  Amphitoe fissicauda Dana, 1852;
  Iphimedia fissicauda Dana, 1853&55;
  Atylus fissicauda Bate, 1862;
  Pontogeneia fissicauda Stebbing, 1906;
  Paramoera fissicauda Schellenberg, 1931; Schellenberg, 1935; J.L. Barnard, 1958b; Bellan-Santini & Ledoyer, 1974; Lowry & Bullock, 1976.
  Distribution in Chile: Tongoy*, Valparaiso, Concepcion, Niebla, Ultima Esperanza, Puerto Toro, Paramo, Punta Arenas, Puerto Angosto, Punta Carreras, Estrecho de Magallanes.

Paramoera pfefferi Schellenberg, 1931
  P. pfefferi Schellenberg, 1931; Schellenberg, 1935; J.L. Barnard, 1958b; Andres, 1975; Lowry & Bullock, 1976.
  Distribution in Chile: Concepcion, Tumbes, Punta Arenas, Puerto Angosto, Puerto Laguna, Isla Nueva, Isla Lennox, Estrecho de Magallanes.

Rhachotropis sp.,
  Rhachotropis sp. Andres, 1975
  Distribution in Chile: Punta Galera.

Tylosapis dentatus (Stebbing, 1888)
  Atylopsis dentatus Stebbing, 1888;
  A. dentata Stebbing, 1906; Schellenberg, 1931; J.L. Barnard, 1958b;
  Tylosapis dentatus Thurston, 1974; Lowry & Bullock, 1976.
  Distribution in Chile: Ultima Esperanza, Isla Picton, Canal Beagle.

EXOEDICEROTIDAE

Bathyporeiapus magellanicus Schellenberg, 1931
  B. magellanicus Schellenberg, 1931; J.L. Barnard, 1958b; Escofet, 1971; Lowry & Bullock, 1976; Varela, 1983; J.L. Barnard & Thomas, 1988.
  Distribution in Chile: Mehuin, Punta Arenas, Isla Lennox.

Exoediceropsis chiltoni Schellenberg, 1931
  E. chiltoni Schellenberg, 1931; J.L. Barnard, 1958b; Lowry & Bullock, 1976.
  Distribution in Chile: Estrecho de Magallanes.

Metoediceros fuegiensis Schellenberg, 1931
  M. fuegiensis Schellenberg, 1931; J.L. Barnard, 1958b; Lowry & Bullock, 1976.
  Distribution in Chile: Punta Arenas.

GAMMARIDAE

Elasmopus chilensis Walker, 1913
  E. chilensis Walker, 1913.
  Distribution in Chile: Arica, Tocopilla, Coquimbo*.

Elasmopus rapax Costa, 1853
  E. rapax Costa, 1853; Stebbing, 1906; Schellenberg, 1931; Schellenberg, 1935; J.L. Barnard, 1958b; Andres, 1975.
Distribution in Chile: Valparaiso, Region Norte.

Elasmopus sp.
  Elasmopus sp. Disalvo *et al.*, 1988
Distribution in Chile: Isla de Pascua.

Gammarus chilensis Nicolet, 1849
  G. chilensis Nicolet, 1849;
  ? *G. chilesis* J.L. Barnard, 1958b (Cannot be in genus *Gammarus*, = obscure taxon).
Distribution in Chile: Chile.

Maera eugeniae Schellenberg, 1931
  M. eugeniae Schellenberg, 1931; J.L. Barnard, 1958b; Lowry & Bullock, 1976
Distribution in Chile: Estrecho de Magallanes.

Maera inaequipes (Costa, 1847)
  Amphithoe truncatipes Costa, 1847 (in White, 1847);
  Maera inaequipes Stebbing, 1906; J.L. Barnard, 1958b; Andres, 1975 (Probably misidentified).
Distribution in Chile: Chile: (Region Norte).

Maera quadrimana (Dana, 1853&55)
  Gammarus quadrimanus Dana, 1853&55;
  Maera quadrimana J.L. Barnard, 1955; J.L. Barnard, 1970; Disalvo *et al.*, 1988
Distribution in Chile: Isla de Pascua.

Melita gayi (Nicolet, 1849)
  Amphitoe gayi Nicolet, 1849;
  Melita gayi Stebbing, 1906; Schellenberg, 1931; Schellenberg, 1935; J.L. Barnard, 1958b.
Distribution in Chile: Coquimbo*, Bahia la Herradura*, Chile.

HYALELLIDAE

Hyalella azteca (Saussure, 1858)
  Ampithoe dentata Saussure, 1858;
  Hyalella knickerbockeri Bate, 1862; Schellenberg, 1931; J.L. Barnard, 1958b; J.L. & C.M. Barnard, 1983;

H. dentata Smith, 1874;
  H. knickerbockeri f. *inermis* Schellenberg, 1935.
Distribution in Chile: Quilpue (freshwater), Valdivia (freshwater), Puerto Montt (freshwater), Punta Arenas (freshwater).

Hyalella gracilicornis (Faxon, 1876)
  Allorchestes gracilicornis Faxon, 1876;
  Amphithoe andina Philippi, 1860;
  Hyalella inermis Stebbing, 1906; J.L. Barnard, 1958b;
  H. gracilicornis J.L. & C.M. Barnard, 1983.
Distribution in Chile: Chile (freshwater), Antofagasta* (freshwater).

Hyalella patagonica (Cunningham, 1871)
  Allorchestes patagonicus Cunningham, 1871; Stebbing, 1906;
  Hyalella patagonica Schellenberg, 1931; J.L. Barnard, 1958b; J.L. & C.M. Barnard, 1983.
Distribution in Chile: Gente Grande (freshwater), Punta Arenas (freshwater), Isla Picton (freshwater), Canal Beagle (freshwater).

Parhyalella sp.
  Parhyalella sp. Andres, 1975.
Distribution in Chile: Coquimbo*, Bahia la Herradura*, Rio Andalien, Concepcion, Bahia San Vicente.

HYALIDAE

Hyale grandicornis (Kroyer, 1845)
  Orchestia grandicornis Kroyer, 1845;
  Nicea lucasii Nicolet, 1849; Bate, 1862;
  Allorchestes verticillata Dana, 1852; Dana, 1853&55; Bate, 1862;
  A. peruviana Dana, 1852;
  ? *A. novizealandiae* Dana, 1852; Stebbing, 1906;
  Hyale goetschi Schellenberg, 1935; Andres, 1975;
  H. grandicornis Stebbing, 1906; Schellenberg, 1935; J.L. Barnard, 1958b; Andres, 1975; Hurley, 1957; Lowry & Bullock, 1976; J.L. Barnard, 1979; Gonzalez, 1990.
Distribution in Chile: Antofagasta*, Taltal, Coquimbo*, Bahia la Herradura*, Monte-

Coquimbo*, Bahia la Herradura*, Montemar*, Valparaiso, Isla Negra*, Punta el Lacho*, Pelancura*, Cobquecura*, Concepcion, Chile (Region Norte).

Hyale hirtipalma (Dana, 1852)
Allorchestes hirtipalma Dana, 1852; Dana, 1853&55; Bate, 1862;
Hyale hirtipalma Stebbing, 1906; Schellenberg, 1931; Schellenberg, 1935; J.L. Barnard, 1958b; Andres, 1975; Lowry & Bullock, 1976; Gonzalez, 1990.
Distribution in Chile: Coquimbo*, Los Molles*, Montemar*, Valparaiso, Isla Negra*, Punta el Lacho*, El Tabo*, Pelancura*, Cobquecura*, Concepcion, Rio Andalien, Puerto Montt, Puerto Toro, Punta Arenas, Estrecho de Magallanes, Isla Lennox.

Hyale maroubrae Stebbing, 1899
H. maroubrae Stebbing, 1899c; Stebbing, 1906; J.L. Barnard, 1958b; Hurley, 1957; Gonzalez, 1990.
Distribution in Chile: Coquimbo*, Bahia la Herradura*, Montemar*, Punta El Lacho*, Isla Juan Fernandez*.

Hyale media (Dana, 1853&55)
Allorchestes media Dana, 1853&55;
Hyale media Stebbing, 1906; J.L. Barnard, 1958b; Gonzalez, 1990.
Distribution in Chile: Coquimbo*, Bahia la Herradura*, Punta El Lacho*, Pelancura*, Tierra del Fuego(?).

Hyale rubra (Thomson, 1879)
Nicea rubra Thomson, 1879;
Hyale rubra Stebbing, 1906; Hurley, 1957; J.L. Barnard, 1958b; J.L. Barnard, 1979; Gonzalez, 1990.
Distribution in Chile: Iquique*, Antofagasta, Coquimbo*, Isla Juan Fernandez*.

IPHIMEDIIDAE

Iphimedia multidentata (Schellenberg, 1931)
Panoplea multidentata Schellenberg, 1931
P. multidentata; J.L. Barnard, 1958b; Lowry & Bullock, 1976.
Distribution in Chile: Golfo Corcovado, Isla Picton, Estrecho de Magallanes.

Pariphimedia normani (Cunningham, 1871)
Iphimedia normani Cunningham, 1871; Stebbing, 1906;
Pariphimedia normani Stebbing, 1914; Schellenberg, 1931; J.L. Barnard, 1958b; Lowry & Bullock, 1976.
Distribution in Chile: Puerto Esperanza, Estrecho de Magallanes.

Pseudiphimediella glabra (Schellenberg, 1931)
Pariphimediella glabra Schellenberg, 1931; J.L. Barnard, 1958b; Lowry & Bullock, 1976.
Pseudiphimediella glabra Coleman & Barnard, 1991.
Distribution in Chile: Ultima Esperanza, Punta Arenas, Estrecho de Magallanes.

Pseudiphimediella nodosa (Dana, 1852)
Iphimedia (Acanthosoma) nodosa Dana, 1852;
I. nodosa Dana, 1853&55; Bate, 1862; Stebbing, 1906;
P. nodosa Schellenberg, 1931; J.L. Barnard, 1958b; Lowry & Bullock, 1976; Coleman & Barnard, 1991.
Distribution in Chile: Punta Arenas, Porvenir, Bahia Inutil, Tierra del Fuego.

ISCHYROCERIDAE

Ischyrocerus sp.
Ischyrocerus sp. Andres, 1975.
Distribution in Chile: Niebla.

Jassa falcata (Montagu, 1808)
Cancer gammarus falcatus Montagu, 1808;
Jassa falcata Schellenberg, 1931; Schellenberg, 1935; J.L. Barnard, 1958b; Lowry & Bullock, 1976.
Distribution in Chile: Valparaiso, Puerto Pantalon, Estrecho de Magallanes.

Jassa marmorata Holmes, 1903
J. marmorata Holmes, 1903; Stebbing, 1906; Conlan, 1990.
Distribution in Chile: Bahia la Herradura, Tongoy*.

Jassa slatteryi Conlan, 1990
J. slatteryi Conlan, 1990.
Distribution in Chile: Bahia la Herradura, Tongoy*.

Pseudischyrocerus denticauda Schellenberg, 1931
    P. denticauda Schellenberg, 1931; J.L. Barnard, 1958b; Lowry & Bullock, 1976.
Distribution in Chile: Isla Nueva.

Ventojassa frequens (Chilton, 1883)
    Podocerus frequens Chilton, 1883;
    Jassa frequens Stebbing, 1906; Schellenberg, 1931; Schellenberg, 1935; J.L. Barnard, 1958b; Andres, 1975;
    Ventojassa frequens J.L. Barnard, 1972b.
Distribution in Chile: Iquique, Valparaiso, Chile (Region Norte).

Ventojassa georgiana (Schellenberg, 1931)
    Parajassa georgiana Schellenberg, 1931; K.H. Barnard, 1932; J.L. Barnard, 1958b;
    Ventojassa georgiana J.L. Barnard, 1973; Lowry & Bullock, 1976.
Distribution in Chile: Estrecho de Magallanes.

LEUCOTHOIDAE

Leucothoe spinicarpa (Abildgaard, 1789)
    Gammarus spinicarpus Abildgaard, 1789;
    Leucothoe spinicarpa Haswell, 1885; Schellenberg, 1931; J.L. Barnard, 1958b; Holman & Watling, 1983; Lowry & Bullock, 1976.
Distribution in Chile: Isla Lennox, Estrecho de Magallanes.

Leucothoe sp.
    Leucothoe sp. Disalvo *et al.*, 1988
Distribution in Chile: Isla de Pascua.

LILJEBORGIIDAE

Liljeborgia longicornis (Schellenberg, 1931)
    Lilljeborgiella longicornis Schellenberg, 1931;
    Liljeborgia longicornis K.H. Barnard, 1932; J.L. Barnard, 1958b; Lowry & Bullock, 1976.
Distribution in Chile: Estrecho de Magallanes.

Liljeborgia macrodon Schellenberg, 1931
    Lilljeborgia macrodon Schellenberg, 1931;
    Liljeborgia macrodon J.L. Barnard, 1958b; Holman & Watling, 1983; Lowry & Bullock, 1976.
Distribution in Chile: Bahia Inutil, Puerto Esperanza.

Liljeborgia octodentata Schellenberg, 1931
    Lilljeborgia octodentata Schellenberg, 1931;
    Liljeborgia octodentata J.L. Barnard, 1958b; Holman & Watling, 1983; Lowry & Bullock, 1976.
Distribution in Chile: Punta Arenas; Puerto Condor, Bahia Inutil, Estrecho de Magallanes.

Liljeborgia sp.
    Liljeborgia sp.
Distribution in Chile: Bahia la Herradura*, Valparaiso*.

LYSIANASSIDAE

Acontiostoma marionis Stebbing, 1888
    A. marionis Stebbing, 1888; Stebbing, 1906; J.L. Barnard, 1958b; Lowry & Bullock, 1976; Lowry & Stoddart, 1983;
    A. magellanicum Stebbing, 1888; Stebbing, 1906;
Distribution in Chile: Bahia Posesion, Estrecho de Magallanes.

Amaryllis macrophthalma Haswell, 1880
    A. macrophthalmus Haswell, 1880a;
    A. macrophthalma Stebbing, 1906; Schellenberg, 1931; J.L. Barnard, 1958b; J.L. Barnard, 1972a; Andres, 1975; Lowry & Bullock, 1976.
Distribution in Chile: Concepcion, Punta Lavapie, Puerto Bueno, Punta Arenas, Puerto Condor, Bahia Inutil, Puerto Eugenia, Isla Nueva, Navarino, Estrecho de Magallanes.

Aristias antarcticus Walker, 1906
    A. antarcticus Walker, 1906a; Schellenberg, 1931; J.L. Barnard, 1958b; Lowry & Bullock, 1976.
Distribution in Chile: Canal Beagle.

Cyphocaris anonyx Boeck, 1871
    C. anonyx Boeck, 1871; Stebbing, 1906; J.L. Barnard, 1958b; Hurley, 1963.
Distribution in Chile: Pacifico Sur.

Erikus dahli Lowry & Stoddart, 1987
    E. dahli Lowry & Stoddart, 1987
Distribution in Chile: Seno de Reloncavi, Canal de Chacao, Punta Abtao, Punta Tenaum.

Eurythenes gryllus (Lichtenstein, 1822)
　　Gammarus gryllus Lichtenstein, 1822;
　　Lysianassa magellanica Bate, 1862;
　　Eurythenes gryllus Stebbing, 1906; J.L.
　　Barnard, 1958b; Lowry & Bullock, 1976.
Distribution in Chile: Cabo de Hornos,
Estrecho de Magallanes.

Lysianopsis subantarctica (Schellenberg, 1931)
　　Aruga subantarctica Schellenberg, 1931; J.L.
　　Barnard, 1958b;
　　Lysianassa subantarctica J.L. Barnard,
　　1969b; Lowry & Bullock, 1976;
　　Lysianopsis subantarctica Lowry & Stoddart,
　　1984.
Distribution in Chile: Puerto Condor.

Orchomenella (Orchomenopsis) chilensis (Heller,
1865)
　　Anonyx chilensis Heller, 1865;
　　Orchomenopsis abyssorum Stebbing, 1906 (in
　　part);
　　O. chilensis f. *chilensis* Schellenberg, 1931;
　　Orchomenella chilensis J.L. Barnard, 1958b;
　　Hurley, 1965;
　　Orchomene chilensis Andres, 1975; Lowry &
　　Bullock, 1976;
　　Orchomenella (Orchomenopsis) chilensis De
　　Broyer, 1984.
　　Distribution in Chile: Valparaiso, Puerto
　　Bueno, Punta Arenas, Puerto Esperanza,
　　Puerto Pantalon, Estrecho de Magallanes.

Pachychelium schellenbergi Lowry, 1984
　　P. schellenbergi Lowry, 1984
　　P. antarcticum Schellenberg, 1931;
　　? *P. davinis* K.H. Barnard, 1932;
Distribution in Chile: Ultima Esperanza, Punta
Arenas, Isla Picton.

Paralysianopsis odhneri Schellenberg, 1931
　　P. odhneri Schellenberg, 1931; K.H. Bar-
　　nard, 1932; J.L. Barnard, 1958b; Andres,
　　1975; Lowry & Stoddart, 1984.
　　Distribution in Chile: Golfo Corcovado.

Parawaldeckia kidderi (Smith, 1876)
　　Lysianassa kidderi Smith, 1876;
　　Nannonyx kidderi Stebbing, 1906;

Parawaldeckia kidderi Schellenberg, 1931;
J.L. Barnard, 1958b; J.L. Barnard, 1969b;
Andres, 1975; J.L. Barnard & Hurley, 1975;
Lowry & Bullock, 1976; Lowry & Stoddart,
1983.
Distribution in Chile: Puerto Montt, Calbuco,
Punta Arenas, Estrecho de Magallanes.

Socarnes unidentatus Schellenberg, 1931
　　S. unidentatus Schellenberg, 1931; J.L.
　　Barnard, 1958b.
Distribution in Chile: Estrecho de Magallanes.

Stenia magellanica Dana, 1852
　　S. magellanica Dana, 1852; Stebbing, 1906
　　(genera dubia); Lowry & Bullock, 1976
　　Anonyx fuegiensis Bate, 1862;
　　? *Stenia magellanica* J.L. Barnard, 1958b.
Distribution in Chile: Tierra del Fuego.

Stephensenia haematopus Schellenberg, 1928
　　S. haematopus Schellenberg, 1928; Schellen-
　　berg, 1931; J.L. Barnard, 1958b; Lowry &
　　Bullock, 1976.
Distribution in Chile: Paramo, Estrecho de
Magallanes.

Stomacontion kergueleni (Stebbing, 1888)
　　Acontiostoma kergueleni Stebbing, 1888;
　　Stomacontion pepinii Stebbing, 1906 (in part);
　　S. kergueleni Schellenberg, 1931; J.L. Bar-
　　nard, 1958b; Lowry & Bullock, 1976.
Distribution in Chile: Bahia Borja, Estrecho de
Magallanes.

Stomacontion pepinii (Stebbing, 1888)
　　Acontiostoma pepinii Stebbing, 1888;
　　Stomacontion pepinii Schellenberg, 1931; J.L.
　　Barnard, 1958b; Lowry & Bullock, 1976;
　　Lowry & Stoddart, 1983.
Distribution in Chile: Puerto Condor, Bahia
Borja.

Tryphosella paramoi (Schellenberg, 1931)
　　Tmetonyx paramoi Schellenberg, 1931; J.L.
　　Barnard, 1958b;
　　Tryphosella paramoi J.L. Barnard, 1969b;
　　Lowry & Bullock, 1976.
Distribution in Chile: Paramo, Estrecho de
Magallanes.

Tryphosella schellenbergi (Schellenberg, 1931)
   Tmetonyx serratus Schellenberg, 1931; Schellenberg, 1935; J.L. Barnard, 1958b; Andres, 1975;
   Tryphosella serratus J.L. Barnard, 1969b;
   T. schellenbergi Lowry & Bullock, 1976.
Distribution in Chile: Valparaiso, Gente Grande, Bahia Inutil, Estrecho de Magallanes.

Tryphosella serrata (Schellenberg, 1931)
   Tryphosa serrata Schellenberg, 1931; J.L. Barnard, 1958b;
   Tryphosella serrata J.L. Barnard, 1969b (senior homonym, by implication); Lowry & Bullock, 1976.
Distribution in Chile: Bahia Inutil, Puerto Eugenia, Isla Nueva, Navarino, Estrecho de Magallanes.

Tryphosites chevreuxi Stebbing, 1914
   Tryphosites chevreuxi Stebbing, 1914; Schellenberg, 1931; K.H. Barnard, 1932; J.L. Barnard, 1958b; Schellenberg, 1935; Andres, 1975; Lowry & Bullock, 1976.
Distribution in Chile: Valparaiso, Ultima Esperanza, Gente Grande, Punta Arenas, Puerto Condor, Bahia Inutil, Bahia Borja, Puerto Eeugenia, Isla Nueva, Navarino, Isla Lennox, Estrecho de Magallanes.

Tryphosites sp.
   Tryphosites sp. Andres, 1975
Distribution in Chile: Golfo de Ancud, Canal Desertores.

Tryphosoides falcata Schellenberg, 1931
   T. falcata Schellenberg, 1931; J.L. Barnard, 1958b; Lowry & Bullock, 1976.
Distribution in Chile: Paramo, Estrecho de Magallanes.

Uristes serratus Schellenberg, 1931
   U. serratus Schellenberg, 1931; J.L. Barnard, 1958b; Lowry & Bullock, 1976.
Distribution in Chile: Puerto Esperanza, Estrecho de Magallanes.

Uristes subchelatus (Schellenberg, 1931)
   Uristoides subchelatus Schellenberg, 1931; J.L. Barnard, 1958b;

Uristes subchelatus J.L. Barnard, 1962b; Lowry & Bullock, 1976.
Distribution in Chile: Porvenir, Bahia Inutil.

MEGALUROPIDAE

Gibberosus sp.
   Gibberosus sp.
Distribution in Chile: Bahia la Herradura*.

OEDICEROTIDAE

Bathymedon palpalis K.H. Barnard, 1916
   B. palpalis K.H. Barnard, 1916; J.L. Barnard, 1958b; Andres, 1975.
Distribution in Chile: Punta Galera.

Bathymedon sp.
   Bathymedon sp. Andres, 1975
Distribution in Chile: Punta Galera, Quedal.

Bathymedon sp. A
   Bathymedon sp. Andres, 1975
Distribution in Chile: Punta Galera.

Monoculopsis vallentini Stebbing, 1914
   M. vallentini Stebbing, 1914; Schellenberg, 1931; J.L. Barnard, 1958b; Lowry & Bullock, 1976.
Distribution in Chile: Bahia la Herradura*, Valparaiso*, Punta Arenas, Isla Picton.

Oediceroides lahillei Chevreux, 1911
   O. lahillei Chevreux, 1911; Schellenberg, 1931; J.L. Barnard, 1958b; Lowry & Bullock, 1976.
Distribution in Chile: Canal Beagle, Estrecho de Magallanes.

PARAMPHITHOIDAE

Metepimeria acanthurus Schellenberg, 1931
   M. acanthurus Schellenberg, 1931;
   Epimeria acanthurus K.H. Barnard, 1932;
   Metepimeria acanthurus J.L. Barnard, 1958b; Lowry & Bullock, 1976.
Distribution in Chile: Puerto Condor.

PARDALISCIDAE

Pardaliscoides tenellus Stebbing, 1888
   P. tenellus Stebbing, 1888; Stebbing, 1906;
   J.L. Barnard, 1958b; J.L. Barnard, 1969b.
   Distribution in Chile: Chile, Abyssal.

PHOXOCEPHALIDAE

Fuegiphoxus abjectus J.L. & C.M. Barnard, 1980
   Fuegiphoxus abjectus J.L. & C.M. Barnard,
   1980.
   Distribution in Chile: Bahia Inutil.

Fuegiphoxus fuegiensis (Schellenberg, 1931)
   Parharpinia fuegiensis Schellenberg, 1931;
   Paraphoxus fuegiensis J.L. Barnard, 1958b;
   J.L. Barnard, 1960; Lowry & Bullock, 1976.
   Fuegiphoxus fuegiensis J.L. & C.M. Barnard,
   1980.
   Distribution in Chile: Valparaiso, Ultima
   Esperanza, Punta Arenas, Puerto Condor,
   Bahia Inutil, Puerto Esperanza, Navarino,
   Estrecho de Magallanes.

Harpiniopsis fulgens J.L. Barnard, 1960
   H. fulgens J.L. Barnard, 1960; Andres, 1975;
   J.L. Barnard & Drummond, 1978.
   Distribution in Chile: Punta Galera, Quedal.

Heterophoxus oculatus (Holmes, 1908)
   Harpinia oculatus Holmes, 1908;
   Heterophoxus oculatus J.L. Barnard, 1958b;
   J.L. Barnard, 1960; Andres, 1975; J.L.
   Barnard & Drummond, 1978.
   Distribution in Chile: Bahia la Herradura*,
   Punta Galera, Quedal.

Heterophoxus videns Barnard, 1930
   H. videns K.H. Barnard, 1930; Schellenberg,
   1931; J.L. Barnard, 1958b; Andres, 1975;
   Bellan-Santini, 1972; Lowry & Bullock,
   1976; J.L. Barnard & Drummond, 1978.
   Distribution in Chile: Valparaiso, Valdivia,
   Corral, Puerto Toro, Punta Arenas, Puerto
   Eugenia, Estrecho de Magallanes.

Metharpinia longirostris Schellenberg, 1931
   M. longirostris Schellenberg, 1931; Schellenberg, 1935; J.L. Barnard, 1980

Paraphoxus longirostris J.L. Barnard, 1958b;
   J.L. Barnard, 1960; Andres, 1975.
   Distribution in Chile: Valparaiso, Punta
   Lavapie, Isla Mocha, Corral, Niebla, Estrecho
   de Magallanes.

Microphoxus cornutus (Schellenberg, 1931)
   Metharpinia cornuta Schellenberg, 1931;
   Paraphoxus cornutus J.L. Barnard, 1960;
   Lowry & Bullock, 1976;
   Microphoxus cornutus J.L. Barnard, 1980.
   Distribution in Chile: Puerto Montt, Punta
   Arenas.

Phoxorgia sinuata (K.H. Barnard, 1932)
   Parharpinia villosa Schellenberg, 1931; (not
   Haswell, 1880a); Schellenberg, 1935;
   P. sinuata K.H. Barnard, 1932;
   Paraphoxus sinuatus J.L. Barnard, 1958b;
   J.L. Barnard, 1960; Lowry & Bullock, 1976;
   P. villosus Andres, 1975;
   Phoxorgia sinuata J.L. & C.M. Barnard,
   1980.
   Distribution in Chile: Bahia la Herradura*,
   Valparaiso, Punta Arenas, Bahia Inutil, Isla
   Nueva, Estrecho de Magallanes.

Proharpinia antipoda Schellenberg, 1931
   P. antipoda Schellenberg, 1931; J.L. Bar-
   nard, 1958b; Lowry & Bullock, 1976; J.L.
   Barnard & Drummond, 1978.
   Distribution in Chile: Ultima Esperanza,
   Puerto Toro, Puerto Angosto, Puerto Condor,
   Isla Picton.

Proharpinia sp.
   Proharpinia sp.
   Distribution in Chile: Bahia la Herradura*.

Proharpinia stephenseni (Schellenberg, 1931)
   Heterophoxus stephenseni Schellenberg,
   1931;
   Proharpinia stephenseni J.L. Barnard, 1958a;
   J.L. Barnard, 1958b; J.L. Barnard, 1960;
   Lowry & Bullock, 1976; J.L. Barnard &
   Drummond, 1978.
   Distribution in Chile: Estrecho de Magallanes.

Pseudharpinia sp.
   Pseudharpinia sp. Andres, 1975
   Distribution in Chile: Quedal, Golfo de Ancud.

62

Pseudharpinia dentata Schellenberg, 1931
  P. dentata Schellenberg, 1931; J.L. Barnard, 1958b; Lowry & Bullock, 1976; J.L. Barnard & Drummond, 1978.
  Distribution in Chile: Estrecho de Magallanes.

PHOXOCEPHALOPSIDAE

Phoxocephalopsis gallardoi J.L. Barnard & Clark, 1984
  P. gallardoi J.L. Barnard & Clark, 1984.
  Distribution in Chile: Estrecho de Magallanes.

Phoxocephalopsis mehuinensis Varela, 1983
  P. mehuinensis Varela, 1983.
  Distribution in Chile: Mehuin, Playa Muicolpue.

Phoxocephalopsis zimmeri Schellenberg, 1931
  P. zimmeri Schellenberg, 1931; Lowry & Bullock, 1976; J.L. Barnard & Clark, 1984.
  Distribution in Chile: Punta Arenas.

PLATYISCHNOPIDAE

Eudevenopus gracilipes (Schellenberg, 1931)
  Platyischnopus gracilipes Schellenberg, 1931 (in part); Schellenberg, 1935; J.L. Barnard, 1958b; Andres, 1975;
  Eudevenopus gracilipes Thomas & J.L. Barnard, 1983.
  Distribution in Chile: Bahia la Herradura*, Valparaiso, Estrecho de Magallanes.

PLEUSTIDAE

Stenopleustes gracilis (Holmes, 1903)
  Apherusa gracilis Holmes, 1903; Stebbing, 1906;
  Stenopleustes gracilis J.L. Barnard, 1958b; Andres, 1975 (not Sars, 1895).
  Distribution in Chile: Concepcion, Punta Lavapie, Isla Mocha.

SEBIDAE

Seba saundersii Stebbing, 1875
  S. saundersii Stebbing, 1875; Stebbing, 1888; Stebbing, 1906; J.L. Barnard, 1958b; Holman & Watling, 1983;

  S. saundersii f. *saundersii* Schellenberg, 1931; Lowry & Bullock, 1976.
  Distribution in Chile: Canal Beagle, Estrecho de Magallanes.

Seba subantarctica Schellenberg, 1931
  S. subantarctica Schellenberg, 1931; J.L. Barnard, 1958b; Andres, 1975; Lowry & Bullock, 1976; Holman & Watling, 1983.
  Distribution in Chile: Puerto Montt, Calbuco, Canal Beagle, Estrecho de Magallanes.

STEGOCEPHALIDAE

Andaniotes corpulentus (Thomson, 1882)
  Anonyx corpulentus Thomson, 1882;
  Andaniotes corpulentus Stebbing, 1906; Schellenberg, 1931; J.L. Barnard, 1958b; Lowry & Bullock, 1976.
  Distribution in Chile: Ultima Esperanza, Puerto Condor, Canal Beagle.

STENOTHOIDAE

Metopoides compactus (Stebbing, 1888)
  Metopa compacta Stebbing, 1888;
  Metopoides compactus Stebbing, 1906; Lowry & Bullock, 1976;
  M. compacta Schellenberg, 1931;
  Proboloides compacta J.L. Barnard, 1958b.
  Distribution in Chile: Estrecho de Magallanes.

Metopoides crenatipalmatus (Stebbing, 1888)
  Metopa crenatipalmatus Stebbing, 1888;
  Proboloides crenatipalmatus Stebbing, 1906; J.L. Barnard, 1958b;
  Metopoides crenatipalmatus J.L. Barnard, 1964; Lowry & Bullock, 1976.
  Distribution in Chile: Estrecho de Magallanes.

Metopoides longicornis Schellenberg, 1931;
  M. longicornis Schellenberg, 1931;
  Proboloides longicornis J.L. Barnard, 1958b;
  Metopoides longicornis J.L. Barnard, 1969b (by implication); Andres, 1975; Lowry & Bullock, 1976.
  Distribution in Chile: Punta Galera, Golfo Corcovado.

Metopoides magellanicus (Stebbing, 1888)
  Metopa magellanica Stebbing, 1888
  Metopoides magellanicus Stebbing, 1906;
  Lowry & Bullock, 1976;
  M. magellanica Schellenberg, 1931;
  Proboloides magellanica J.L. Barnard, 1958b.
  Distribution in Chile: Estrecho de Magallanes.

Metopoides parallelocheir (Stebbing, 1888)
  Metopa parallelocheir Stebbing, 1888
  Metopoides parallelocheir Stebbing, 1906;
  Schellenberg, 1931; Lowry & Bullock, 1976;
  Proboloides parallelocheir J.L. Barnard, 1958b.
  Distribution in Chile: Estrecho de Magallanes.

Probolisca elliptica (Schellenberg, 1931)
  Metopella elliptica Schellenberg, 1931;
  Probolisca elliptica J.L. Barnard, 1958b;
  Lowry & Bullock, 1976.
  Distribution in Chile: Canal Beagle.

Probolisca nasutigenes (Stebbing, 1888)
  Metopa nasutigenes Stebbing, 1888;
  Metopella nasutigenes Stebbing, 1906; J.L. Barnard, 1958b;
  Probolisca nasutigenes J.L. Barnard, 1962b;
  Andres, 1975; Lowry & Bullock, 1976.
  Distribution in Chile: Isla Mocha, Punta Galera.

Probolisca ovata (Stebbing, 1888)
  Metopa ovata Stebbing, 1888
  Metopella ovata Stebbing, 1906; Schellenberg, 1931;
  Probolisca ovata Gurjanova, 1938; J.L. Barnard, 1958b; Lowry & Bullock, 1976.
  Distribution in Chile: Estrecho de Magallanes.

Stenothoe sp.
  Stenothoe sp. Disalvo *et al.*, 1988
  Distribution in Chile: Isla de Pascua.

SYNOPIIDAE

Bruzelia sp.,
  Bruzelia sp. Andres, 1975.
  Distribution in Chile: Punta Galera.

TALITRIDAE

Orchestia gayi Nicolet, 1849
  O. gayi Nicolet, 1849
  ? *O. gayi* J.L. Barnard, 1958b.
  Distribution in Chile: Chile.

Orchestia scutigerula Dana, 1853&55
  O. scutigerula Dana, 1853&55; Bate, 1862;
  Schellenberg, 1931; J.L. Barnard, 1958b.
  Talorchestia scutigerula Stebbing, 1906
  Distribution in Chile: Gente Grande, Puerto Toro, Punta Arenas, Tierra del Fuego, Isla Picton, Isla Lennox, Estrecho de Magallanes.

Orchestia selkirki Stebbing, 1888
  O. selkirki Stebbing, 1888; Stebbing, 1906;
  Schellenberg, 1935; J.L. Barnard, 1958b.
  Distribution in Chile: Isla Juan Fernandez.

Orchestoidea tuberculata Nicolet, 1849
  O. tuberculata Nicolet, 1849; Bate, 1862;
  Stebbing, 1906; Andres, 1975; J.L. Barnard, 1958b; Bousfield, 1982; Varela, 1983.
  Talitrus chilensis Nicolet, 1849;
  T. ornatus Dana, 1852;
  Talitronus insculptus Dana, 1852;
  Orchestia (Talitrus) insculpta Dana, 1853&55;
  Distribution in Chile: Antofagasta*, Huasco*, Coquimbo*, Los Molles*, Valparaiso, El Tabo*, Constitucion, Mehuin, Corral, Playa Muicolpue, Cucao, Quellon Viejo.

Protorchestia nitida (Dana, 1852)
  Orchestia nitida Dana, 1852; Dana, 1853&55; Bate, 1862; Stebbing, 1906;
  Schellenberg, 1931; J.L. Barnard, 1958b;
  Orchestia fuegensis Bate, 1862;
  Protorchestia nitida Bousfield, 1982
  Distribution in Chile: Ultima Esperanza, Punta arenas, Puerto Angosto, Tierra del Fuego, Navarino, Cabo de Hornos, Estrecho de Magallanes.

Talorchestia quoyana (Milne Edwards, 1840)
  Talitrus brevicornis Milne Edwards, 1840;
  Orchestia quoyana Milne Edwards, 1840;
  O. brevicornis Nicolet, 1849;
  Talorchestia quoyana Stebbing, 1906; J.L. Barnard, 1958b.
  Distribution in Chile: Chile.

64

Transorchestia chiliensis (Milne Edwards, 1840)
   Orchestia chiliensis Milne Edwards, 1840;
   O. chilensis Nicolet, 1849;
   O. chilensis Dana, 1853&55; Bate, 1862;
   Stebbing, 1906; Chilton, 1921; Schellenberg,
   1931; Schellenberg, 1935; Andres, 1975;
   Varela, 1983; J.L. Barnard, 1958b.
   Transorchestia chiliensis Bousfield, 1982
Distribution in Chile: Antofagasta*, Isla Juan
Fernandez, Mehuin, Puerto Montt, Ultima
Esperanza, Tierra del Fuego, Navarino, Cabo
de Hornos, Estrecho de Magallanes, Chile,
(Region Norte), Chile.

UROHAUSTORIIDAE

Huarpe escofeti J.L. Barnard & Clark, 1982
   H. escofeti J.L. Barnard & Clark, 1982.
Distribution in Chile: Estrecho de Magallanes.

UROTHOIDAE

Urothoe falcata Schellenberg, 1931
   U. falcata Schellenberg, 1931; J.L. Barnard,
   1958b; Andres, 1975; Lowry & Bullock,
   1976.
Distribution in Chile: Punta Galera, Estrecho
de Magallanes.

ZOBRACHOIDAE

Chono angustiarum Clark & J.L. Barnard, 1987
   Chono angustiarum Clark & J.L. Barnard,
   1987.
Distribution in Chile: Estrecho de Magallanes.

Tonocote introflexidus Clark & J.L. Barnard, 1988
   T. introflexidus Clark & J.L. Barnard, 1988
Distribution in Chile: Estrecho de Magallanes.

Tonocote magellani Clark & J.L. Barnard, 1986
   T. magellani Clark & J.L. Barnard, 1986
Distribution in Chile: Estrecho de Magallanes.

Acknowledgements

I wish to thank Dr. Patricio Sanchez and Dr. Juan Carlos Castilla of the Grupo de Ecologia Marina, Pontificia Universidad Catolica de Chile. They encouraged me to work on amphipods and made possible my visit to Smithsonian Institution in Washington D.C., USA. I am also indebted to Pablo Schmiede for his constant help in my beginnings in this field. I would also like to thank Dr. Thomas E. Bowman and Dr. J. Laurens Barnard for their facilities while I was at the Crustacean Division of the National Museum of Natural History in Smithsonian. To Smithsonian Institution for the funds that made possible my work there. I am also grateful to all who made possible this catalogue, especially to Gaston Gutierrez who wrote the programs in DBASE3 making the process of listing species much easier. I am also indebted to Dr. Louis DiSalvo who provided the samples from Isla de Pascua, to Nelson Reyes, Sergio Gonzalez and Domingo Lancelloti who collected a lot of the new records included in this list.

References

Abildgaard, P. C., 1789. Zoologiae Danicae seu animalium Daniae et Norvegiae Rarior, AC Minus Notorum, descriptiones et historia, auctore O.F. Muller 3: 66–67.

Andres, H. G., 1975. Zur Verbreitung eulitoraler Gammaridea (Amphipoda, Crustacea) an dem von Kaltwasserströmen beeinflussten Küsten Südamerikas sowie Angaben über sublitorale Gammaridea vor der chilenischen Küste. Dissertation zur Erlangung des Doktorgrades des Fachbereichs Biologie der Universität Hamburg. 139 pp.

Andres, H. G., 1979. [*Paracorophium hartmannorum*, new species from the eulittoral of the chilean Pacific coast (Crustacea: Amphipoda)]. Mitt. hamb. zool. Mus. Inst. 76: 381–386. (In German).

Barnard, J. L., 1952. Some Amphipoda from Central California. Wasmann J. Biol. 10: 9–36.

Barnard, J. L., 1955. Gammaridean Amphipoda (Crustacea) in the collections of Bishop Museum. Bernice P. Bishop Mus. Bull. 215: 1–46.

Barnard, J. L., 1958a. Revisory notes on the Phoxocephalidae (Amphipoda), with a key to the genera. Pac. Sci. 12: 146–151.

Barnard, J. L., 1958b. Index to the families, genera, and species of the Gammaridean Amphipoda (Crustacea). Allan Hancock Found. Publ. Occas. Pap. 19: 1–145.

Barnard, J. L., 1960. The Amphipod family Phoxocephalid in the Eastern Pacific Ocean, with analysis of other species and notes for a revision of the family. Allan Hancock Found. Pac. Exp. 18: 175–368.

Barnard, J. L., 1962a. Benthic marine amphipoda of

Southern California: Families Aoridae, Photidae, Ischyroceridae, Corophiidae, Podoceridae. Pac. Nat. 3: 1–72.

Barnard, J. L., 1962b. Benthic marine amphipoda of Southern California: Families Amphilochidae, Stenothoidae, Argissidae, Hyalidae. Pac. Nat. 3: 116–163.

Barnard, J. L., 1964. Marine Amphipoda of Bahia de San Quintin, Baja California. Pac. Nat. 4: 55–139.

Barnard, J. L., 1969a. Gammaridean Amphipoda of the rocky intertidal of California: Monterrey Bay to La Jolla. U.S. natn. Mus. Bull. 258: 1–230.

Barnard, J. L., 1969b. The families and genera of marine Gammaridean Amphipoda. U.S. natn. Mus. Bull. 271: 1–535.

Barnard, J. L., 1970. Sublittoral Gammaridea (Amphipoda) of the Hawaiian Islands. Smithson. Contr. Zool. 34: 1–286.

Barnard, J. L., 1971. Keys to the Hawaiian Marine Gammaridea, 0–30 meters. Smithson. Contr. Zool. 58: 1–135.

Barnard, J. L., 1972a. Gammaridean Amphipoda of Australia, Part I. Smithson. Contr. Zool. 124: 1–94.

Barnard, J. L., 1972a. Gammaridean Amphipoda of Australia, Part I. Smithson. Contr. Zool. 124: 1–94.

Barnard, J. L., 1972b. The Marine Fauna of the New Zealand; Algae living littoral Gammaridea (Crustacea Amphipoda). N. Z. Dep. Sci. Ind. Res. Bull. 210. N. Z. oceanogr. Inst. Mem. 62: 1–216.

Barnard, J. L., 1973. Revision of Corophiidae and related families (Amphipoda). Smithson. Contr. Zool. 151: 1–27.

Barnard, J. L., 1979. Littoral Gammaridean Amphipoda from the Gulf of California and the Galapagos Islands. Smithson. Contr. Zool. 271: 1–149.

Barnard, J. L., 1980. Revision of *Metharpinia* and *Microphoxus* (Marine Phoxocephalid Amphipoda from the Americas). Proc. biol. Soc. Wash. 93: 104–235.

Barnard, J. L., 1989. Rectification of *Halirages regis* and *H. huxleyanus* (Crustacea: Amphipoda), from marine Antarctica, with description of new genus, *Austroregia*. Proc. biol. Soc. Wash. 102: 701–715.

Barnard, J. L. & D. E. Hurley, 1975. Redescription of *Parawaldeckia kidderi* (Smith) (Amphipoda, Lysianassidae). Crustaceana 29: 68–73.

Barnard, J. L. & M. M. Drummond, 1978. Gammaridean Amphipoda of Australia, Part III: Phoxocephalidae. Smithson. Contr. Zool. 245: 1–551.

Barnard, J. L. & C. M. Barnard, 1980. Two new Phoxocephalid genera, *Fuegiphoxus* and *Phoxorgia*, from Magellanic South America (Amphipoda: Crustacea). Proc. biol. Soc. Wash. 93: 849–874.

Barnard, J. L. & J. Clark, 1982. *Huarpe escofeti*, new genus, new species, a burrowing marine Amphipod from Argentina (Crustacea Amphipoda, Urohaustoriidae). J. Crustacean Biol. 2: 281–295.

Barnard, J. L. & C. M. Barnard, 1983. Freshwater Amphipoda of the world. Part I and Part II. Hayfield Associates. Mt. Vernon, Virginia. 830 pp.

Barnard, J. L. & J. Clark, 1984. Redescription of *Phoxocephalopsis zimmeri* with a new species, and establishment of the family Phoxocephalopsidae (Crustacea, Amphipoda) from Magellanic South America. J. Crustacean Biol. 4: 85–105.

Barnard, K H., 1916. Contribution to the Crustacean fauna of South Africa. 5. The Amphipoda. Ann. s. afr. Mus. 15: 105–302.

Barnard, K. H., 1930. Crustacea, Part XI. Amphipoda. British Antarctic ('Terra Nova') Expedition, 1910. Zoology 8: 307–454.

Barnard, K. H., 1932. Amphipoda. Discovery Report 5: 1–326.

Bate, C. S., 1862. Catalogue of the specimens of Amphipodous Crustacea in the collection of the British Museum. Br. Mus. (Natural History) London 399 pp.

Bellan-Santini, D., 1972. Invertebres marines des XII et XV Expeditions Antarctiques Francaises en Terra Adelie 10. Amphipodes Gammariens. Tethys Supp. 4: 157–238.

Bellan-Santini, D. & M. Ledoyer, 1974. Gammariens (Crustacea – Amphipoda) des Iles Kerguelen et Crozet. Tethys 5: 635–708.

Boeck, A., 1871. Crustacea Amphipoda borealia et arctica. [separata from:] Forhandlinger i Videnskabs. Selskabet i Christiania, 1870; 83–280 pp.

Bousfield, E. L., 1982. The amphipod superfamily Talitroidea in the North Eastern Pacific region. I. Family Talitridae: Systematics and distributional ecology. Natn. Mus. nat. Sci. (Ottawa) Publ. biol. Oceanogr. 1: 1–73.

Buschmann, A. & B. Santelices, 1987. Micrograzers and spore release in *Iridaea laminarioides* Bory (Rhodophyta: Gigartinales). J. exp. mar. Biol. Ecol. 108: 171–179.

Chevreux, D., 1906. Diagnoses D'amphipodes nouveaux provenant de l'expedition Antartique du Francais. 2. Metopidae – Iphimedidae. Bull. Soc. zool. Fr. 31: 37–40.

Chevreux, D., 1911. Sur quelques amphipodes des isles Sandwich du Sud. An. Mus. nac. Bs. As. 21: 403–407.

Chilton, C., 1883. Further additions to our knowledges of the New Zealand Crustaces. Trans. Proc. N. Z. Inst. 15: 69–86.

Chilton, C., 1921. A small collection of Amphipoda from Juan Fernandez. In, C. Skotsberg (ed.). The natural history of Juan Fernandez and Eastern Island. Almquist & Wiksells, Uppsala, Sweeden. 3: 81–92.

Clark, J. & J. L. Barnard, 1986. *Tonocote*, a new genus and species of Zobrachoidae from Argentina (Crustacea: Marine Amphipoda). Proc. biol. Soc. Wash. 99: 225–236.

Clark, J. & J. L. Barnard, 1987. *Chono angustiarum*, a new species of Zobrachoidae (Crustacea: Amphipoda) from Magellan Strait, with revision of Urohaustoriidae. Proc. biol. Soc. Wash. 100: 75–88.

Clark, J. & J. L. Barnard, 1988. *Tonocote introflexidus*, a new species of marine amphipod from Argentina (Crustacea: Gammaridea: Amphipoda). Proc. biol. Soc. Wash. 101: 354–365.

Coleman, C. O. & J. L. Barnard, 1991. Redescription of two species of *Pseudiphimediella* from the Southern Ocean (Amphipoda: Iphimediidae). Proc. biol. Soc. Wash. 104: 76–90.

66

Conlan, K. E., 1990. Revision of the crustacean amphipod genus *Jassa* Leach (Corophioidea: Ischyroceridae). Can. J. Zool. 68: 2031–2975.

Conlan, K. E. & E. L. Bousfield, 1982. Studies on Amphipod Crustaceans of the Northeastern Pacific Region. I. 2. The Amphipod superfamily Corophoidea in the Northeastern Pacific Region. Family Ampithoidae. Systematics and distributional ecology. Natn. Mus. nat. Sci. (Ottawa) Publ. biol. Oceanogr. 10: 41–75.

Costa, A., 1853. Relazione sulla memoria del Dottor Achille Costa, di reserche su crostacei anfipodi del regno di Napoli. Rend. Soc. Reale Borbanica, Accad. Sci., Anno 1853, 166–178 pp.

Crawford, G. I., 1937. A review of the Amphipod genus *Corophium*, with notes on the British species. J. mar. biol. Ass. U. K. 21: 589–630.

Cunnigham, R. O., 1871. Notes on the Reptiles, Amphibia, Fishes, Mollusca and Crustacea obtained during the voyage of H.M.S. 'Nassau' in the years 1866–69. Trans. linn. Soc. Lond. 27: 465–502.

Dahl, E., 1952. Some aspects of the ecology and zonation of the fauna on sandy beaches. Oikos 4: 1–27.

Dana, J. D., 1852. Conspectus Crustaceorum quae in Orbis Terrarum Circumnavigatione, Carolo Wilkes e Classe Republicae Faederate Duce, Lexit et descripsit Jacobus D.Dana. Pars III (Amphipoda. No 1). Proc. am. Acad. Arts Sci. 2: 201–220.

Dana, J. D., 1853 & 1855. Crustacea, Part II. United States Exploring Expedition during the years 1838–42 Under the command of Charles Wilkes USN 14: 689–1618. Atlas of 96 pls.

De Broyer, C., 1984. Evolution du complexe *Orchomene* Boeck (Amphipoda, Lysianassidae). Ann. Soc. r. zool. Belg. 114(Suppl.1): 197–198.

DiSalvo, L. H., J. E. Randall & A. Cea, 1988. Ecological reconnaissance of the Easter Island sublittoral marine environment. Natn. geogr. Res. 4(4): 451–473.

Duarte, W. E., 1974. *Orchestoidea tuberculata* Nicolet, 1849 como organismo desintegrador de algas (Crustacea, Amphipoda, Talitridae). Not. mens. Mus. Hist. nat. Chile 19(220–221): 3–9.

Escofet, A. M., 1971. Amphipoda marinos de la Provincia de Buenos Aires II. Observaciones sobre el genero *Bathyporeiapus* Schellenberg (Gammaridea: Oedicerotidae), con la descripcion de *Bathyporeiapus ruffoi* sp.nov.. Neotropica 17: 107–115.

Faxon, W., 1876. Exploration of Lake Titicaca, by Alexander Agassiz and S. W. Garman. IV. Crustacea. Bull. Mus. Comp. Zool. 3: 361–375.

Gallardo, A., 1962. Descripción de una nueva especie de Ampelisca (Amphipoda). *Ampelisca araucana*. Gayana, Zool. 7: 1–11.

Gonzalez, E., 1986. A new record of *Paracorophium hartmannorum* Andres, 1975, from the chilean coast, with a description of the adult (Amphipoda: Corophiidae). Proc. biol. Soc. Wash. 99: 21–28.

Gonzalez, E., 1990. The genus *Hyale* in Chile (Crustacea: Amphipoda). Spixiana 13(3) (in press).

Gurjanova, E., 1938. Amphipoda, Gammaroidea Zalilov Siaukhu I Sudzukhe (Yaponskoe More). [Amphipoda Gammaroidea of Siaukhu and Sudzukoe Bays (Sea of Japan)]. Report Japan Sea Hydrobiological Expedition. Zool. Inst. Acad. Sci. USSR 1: 241–404.

Haswell, W. A., 1880a. On Australian Amphipoda. Proc. linn. Soc. N. s. W. 4: 245–279.

Haswell, W. A., 1880b. On some additional new genera and species of Amphipodous Crustaceans. Proc. linn. Soc. N. s. W. 4: 319–350.

Haswell, W. A., 1885. Notes on the Australian Amphipoda. Proc. linn. Soc. N. s. W. 10: 95–114.

Heller, C., 1865. Crustacean. Reise der Osterreichischen Fregatte Novara um die Erde in den Jahren 1857–1859. 2: 1–280.

Holmes, S. J., 1903. Synopses of North American Invertebrates 28. The Amphipoda. Am. Nat. 37: 267–269.

Holmes, S. J., 1908. The Amphipoda collected by the U.S. Bureau of Fisheries Steamer 'Albatros' off the West Coast of North America, in 1903 and 1904, with descriptions of a new genera and species. Proc. U. S. natn. Mus. 35: 489–543.

Holman, H. & L. Walting, 1983. Amphipoda from the Southern Ocean: Families Colomastigidae, Dexaminidae, Leucothoidae, Liljeborgiidae and Sebidae. Biology of the Antarctic Seas XIII. Antarct. Res. Ser. 38: 215–262.

Hurley, D. E., 1957. Studies on the New Zealand Amphipodan fauna No 14. The genera *Hyale* and *Allorchestes* (Family Talitridae). Trans. r. Soc. N. Z. 84: 903–933.

Hurley, D. E., 1963. Amphipoda of the Family Lysianassidae from the West coast of North and Central America. Allan Hancock Found. Publ. Occas. Pap. 25: 1–160.

Hurley, D. E., 1965. A re-description of *Orchomenella chilensis* (Heller) (Crustacea Amphipoda: Family Lysianassidae) from the original material collected by the 'Novara' in chilean waters. Trans. r. Soc. N. Z. 6: 183–188.

Jaramillo, E., W. Stotz, C. Bertran, J. Navarro, C. Roman, & C. Varela, 1980. Actividad locomotriz de *Orchestoidea tuberculata* (Amphipoda, Talitridae) sobre la superficie de una playa arenosa del Sur de Chile (Mehuin, Prov. de Valdivia). Stud. neotrop. Fauna Environ. 15: 9–33.

Karaman, G. S., 1979. Revision of the genus *Paracorophium* Stebb. with description of *P. chelatum*, n. sp. and genus Chaetocorophium, n. gen. (Fam. Corophiidae) (Contribution to the knowledge of Amphipoda 100). Glas. Republ. Zavoda Zast. Prirode-Prirodnjackoc Muzeja Titograd 12: 87–100.

Kreibhom de Paternoster, I. & A. M. Escofet, 1976. La fauna de anfipodos asociados a los bosques de *Macrocystis pyrifera* en el Chubut: *Ampithoe femorata* (Kroyer) (Ampithodae) y *Bircenna fulva* Chilton (Eophliantidae). Physis (A) 35: 77–91.

Kroyer, H. N., 1845. Karcinologiske Bidrag. Naturh. Tiddskr. 1: 283–345, 403.

Kroyer, H. N., 1842. Une Nordiske Slaegter og Arter of Amfipodernes Order, henhorende til Familien Gammarina (Forelobigt uddrag af et Storre Arbejde). Naturh. Tiddskr. 4: 141–166.

Lichtenstein, H., 1822. In: Mandt MW, Observationes in historiam naturalem et anatomian comparatam in itinere Groenlandica factae. Dissertatio inauguralis quam consensu et auctoritae gratiosi medicorum ordinis in Universitate Literaria berolinensi ut summi in medicina et chirurgia honores rite sibi concedantur die xxii. M. Julii A. MDCCCXXII. H.L.Q.S., Publice defendet auctor Martinus Gullelmus Mandt Beyenburgensis.

Lowry, J. K., 1984. Systematics of the Pachynid group of Lysianassoid Amphipoda (Crustacea). Rec. aust. Mus. 36: 51–105.

Lowry, J. K. & S. Bullock, 1976. Catalogue of the Gammariean Amphipoda of the Southern Ocean. Bull. r. Soc. N. Z. 16: 1–187.

Lowry, J. K. & H. E. Stoddart, 1983. The shallow-water gammaridean Amphipoda of the subantarctic islands of New Zealand and Australia: Lysianassoidea. J. r. Soc. N. Z. 13: 279–294.

Lowry, J. K. & H. E. Stoddart, 1984. Redescription of Schellenberg's types of *Lysianopsis subantarctica* and *Paralysianopsis odhneri* (Amphipoda, Lysianassidae). Crustaceana 47: 98–108.

Lowry, J. K. & H. E. Stoddart, 1987. A new South American genus and species in the amaryllid group of lysianassoid Amphipoda. J. Nat. Hist. 21: 1303–1309.

Milne Edwards, H., 1830. Extrait de Recherches pour servir a l'histoire naturelle des Crustaces Amphipodes. Ann. Sci. nat. 20: 353–399.

Milne Edwards, H., 1840. Histoire naturelle des crustacés, comprenant L'anatomie, La Physiologie et la classification de ces animaux. Roret, Paris 3: 1–638.

Montagu, G., 1808. Descriptions of several marine animals found on the South Coast of Devonshire. Trans. linn. Soc. Lond. 9: 81–114.

Nicolet, H., 1849. Anfipodos. In, Gay, C. Historia Fisica y Politica de Chile. Zoologia 3: 226–256. Paris.

Philippi, R. A., 1860. Reise durch die Wueste Atacama auf Befehl der chilenischen regierung im sommer 1853–54 unternommen und beschreiben. Halle: Eduard Anton, 192 pp.

Sars, G. O., 1895. Amphipoda. In an account of the Crustacea of Norway. Alb. Cammermeyers, Chistiania and Copenhagen. 1: 1–711 pp.

Saussere, H., 1858. Memoire sur divers crustaces nouveaux des Antilles et du Mexique. Mem. Soc. Phys. Hist. nat. 14P. 2: 417–496.

Schellenberg, A., 1926. Die gammariden der Deutschen Südpolar Expedition 1901–1903. Deutsch. Südpolar Exped. 18: 235–414.

Schellenberg, A., 1928. *Stephensenia haematopus* n.g. n.sp., Eine grabende Lysianassidae. Zool. Anz. 79: 285–289.

Schellenberg, A., 1931. Gammariden and Caprelliden des Magellangebietes, Südgeorgiens und der Westantarktis. Further Zool. Res. Swedish Antarct. Exped. 1901–1903 2: 1–290.

Schellenberg, A., 1935. Fauna chilensis, Amphipoden von Chile und Juan Fernandez. Zool. Jb. Ab. Syst. Okol. Geogr. Tiere 67: 225–234.

Smith, S. L., 1874. The Crustacea of the freshwaters of the Unites States, A: Synopsis of the higher freshwater Crustacea of the Northern United States, Appendix F. Natural History. Report of the Commissioner for 1872 and 1973. U. S. Comm. Fish Fisheries 2: 637–661.

Smith, S. I., 1876. Crustaceans. In: Kidder JH (ed), Contributions to the Natural History of Kerguelen Island. U. S. natn. Mus. Bull. 3: 57–62.

Stebbing, T. R. R., 1875. On some new exotic sessile-eyed Crustacea. Ann. Mag. nat. Hist. Series 4, 15: 184–188.

Stebbing, T. R. R., 1888. Report on the Amphipoda collected by H. M. S. Challenger during the years 1873–76. In Great Britain Report on the Scientific results of the voyage of H. M. S. Challenger during the years 1873–76. Zoology 29: 1–1737.

Stebbing, T. R. R., 1899a. Revision of the Amphipoda. Ann. Mag. nat. Hist., Series 7, 3: 350.

Stebbing, T. R. R., 1899b. Revision of the Amphipoda (continued). Ann. Mag. nat. Hist. Series 7, 4: 205–211.

Stebbing, T. R. R., 1899c. Amphipoda from the Copenhagen Mus. and other sources. Part 2. Trans. linn. Soc. Lond., Series 2 (Zoology) 7: 395–432.

Stebbing, T. R. R., 1906. Amphipoda I. Gammaridae. Das Tierreich. 21: 1–806.

Stebbing, T. R. R., 1914. Crustacea from the Falkland Island collected by Mr. Rupert Vallentin, F.L.S. Part 2. Proc. zool. Soc. Lond. 1: 341–378.

Stephensen, K., 1932. Some new amphipods from Japan. Annot. zool. Jpn 13: 487–501.

Thomson, G. M., 1879. New Zealand Crustacea, with descriptions of new species. Trans. Proc. N. Z. Inst. 11: 230–248.

Thomson, G. M., 1880. New species of Crustacea from New Zealand. Ann. Mag. nat. Hist., Series 5, 6: 1–6.

Thomson, G. M., 1882. Additions to the Crustacean fauna of New Zealand. Trans. Proc. N. Z. Inst. 14: 1–230.

Thomas, J. D. & J. L. Barnard, 1983. The Platyischnopidae of America (Crustacea: Amphipoda). Smithson. Contr. Zool. 375: 1–33.

Thurston, M. H., 1974. Crustacea Amphipoda from Graham Land and the Scotia Arc, collected by operation Tabarin and the Falkland Islands dependencies survey, 1944–59. Br. Antarct. Surv. Sci. Rep. 85: 1–893

Varela, C., 1983. Anfipodos de las playas de arena del sur de Chile (Bahía de Manquillahue, Valdivia). Stud. neotrop. Fauna Eviron. 18: 25–52.

Walker, A. O., 1906a. Preliminary descriptions of new species of Amphipoda from the 'Discovery' Antarctic Expedition, 1902–1904. Ann. Mag. nat. Hist., Series 7, 17: 452–458.

Walker, A. O., 1906b. Preliminary descriptions of new species of Amphipoda from the 'Discovery' Antarctic Expedition, 1902–1904. Ann. Mag. nat. Hist., Series 7, 18: 150–154.

Walker, A. O., 1907. Crustacea 3. Amphipoda. Natn. Antarct. Exped. 1901–1904, London (Zoology) 3: 1–39.

Walker, A. O., 1913. A new Amphipod from the Pacific coast of South America. Rev. chil. Hist. nat. 17: 242.

White, A., 1847. List of the Specimens of Crustacea in the collection of the British Museum. E. Newman. London. 143 pp.

Appendix

Location of major geographic features in Chile.
(Latitude and longitude of localities whose record indicates a wide region, were assigned to a middle point of that region).

Abyssal, Chile: (37° 00′ S 83° 00′ W)
Antofagasta (23° 38′ S 70° 25′ W)
Arica (28° 55′ S 70° 24′ W)
Bahia Borja (53° 32′ S 72° 29′ W)
Bahia Inutil (53° 30′ S 70° 10′ W)
Bahia la Herradura (29° 58′ S 71° 22′ W)
Bahia Posesion (52° 18′ S 68° 57′ W)
Bahia San Vicente (36° 44′ S 73° 09′ W)
Cabo de Hornos (55° 58′ S 67° 15′ W)
Caleta Chascos (27° 40′ S 71° 00′ W)
Calbuco (41° 46′ S 73° 08′ W)
Canal Beagle (55° 10′ S 66° 15′ W)
Canal Desertores (42° 24′ S 72° 57′ W)
Canal de Chacao (41° 48′ S 73° 31′ W)
Cobquecura (36° 06′ S 72° 49′ W)
Concepcion (36° 41′ S 73° 02′ W)
Constitucion (35° 20′ S 72° 25′ W)
Coquimbo (29° 56′ S 71° 21′ W)
Corral (39° 52′ S 73° 25′ W)
Cucao (42° 43′ S 74° 14′ W)
El Tabo (33° 31′ S 71° 43′ W)
Estrecho de Magallanes (53° 00′ S 70° 00′ W)
Gente Grande (53° 03′ S 70° 16′ W) (freshwater)
Golfo Corcovado (42° 57′ S 72° 57′ W)
Golfo de Ancud (42° 00′ S 73° 00′ W)
Huasco (28° 29′ S 71° 16′ W)
Iquique (20° 12′ S 70° 10′ W)
Isla Juan Fernandez (34° 00′ S 78° 00′ W)

Isla Lennox (55° 20′ S 67° 00′ W)
Isla Mocha (38° 16′ S 73° 39′ W)
Isla Negra (33° 25′ S 71° 43′ W)
Isla Nueva (55° 10′ S 66° 30′ W)
Isla de Pascua (27° 00′ S 109° 00′ W)
Isla Picton (55° 05′ S 66° 50′ W)
Los Molles (32° 14′ S 71° 33′ W)
Lota (37° 06′ S 73° 10′ W)
Mehuin (39° 27′ S 73° 16′ W)
Montemar (32° 58′ S 71° 30′ W)
Navarino (55° 10′ S 67° 40′ W)
Niebla (39° 51′ S 73° 24′ W)
Pacifico Sur (55° 18′ S 109° 20′ W)
Paramo (53° 10′ S 70° 54′ W)
Pelancura (33° 32′ S 71° 43′ W)
Playa Muicolpue (40° 35′ S 73° 46′ W)
Porvenir (53° 18′ S 70° 24′ W)
Puerto angosto (53° 13′ S 73° 22′ W)
Puerto Bueno (50° 29′ S 74° 13′ W)
Puerto Churuca (53° 02′ S 73° 55′ W)
Puerto Condor (53° 21′ S 72° 38′ W)
Puerto Esperanza (54° 07′ S 70° 59′ W)
Puerto Eugenia (54° 55′ S 67° 16′ W)
Puerto Laguna (54° 55′ S 70° 34′ W)
Puerto Montt (41° 30′ S 73° 10′ W)
Puerto Pantalon (54° 40′ S 67° 30′ W)
Puerto Toro (53° 07′ S 71° 59′ W)
Punta Abtao (41° 49′ S 73° 22′ W)
Punta Arenas (53° 10′ S 70° 54′ W)
Punta Carreras (53° 36′ S 70° 55′ W)
Punta Galera (39° 59′ S 73° 54′ W)
Punta El Lacho (33° 31′ S 71° 43′ W)
Punta Lavapie (37° 08′ S 73° 38′ W)
Punta Tenaum (42° 20′ S 73° 22′ W)
Quedal (40° 54′ S 74° 00′ W)
Quellon viejo (43° 07′ S 73° 37′ W)
Quilpue (33° 04′ S 71° 28′ W) (freshwater)
Rio Andalien (36° 41′ S 73° 02′ W)
Salinas de Puyaye (32° 24′ S 71° 25′ W)
Seno de Reloncavi (41° 33′ S 73° 02′ W)
Taltal (25° 24′ S 70° 29′ W)
Tierra del Fuego (54° 00′ S 70° 00′ W)
Tocopilla (22° 05′ S 70° 13′ W)
Tongoy (30° 16′ S 71° 35′ W)
Tumbes (36° 41′ S 73° 02′ W)
Ultima Esperanza (50° 30′ S 71° 50′ W)
Valdivia (38° 50′ S 73° 15′ W)
Valparaiso (33° 09′ S 71° 37′ W)

Hydrobiologia **223**: 69–77, 1991.
L. Watling (ed.), VIIth International Colloquium on Amphipoda.
© 1991 Kluwer Academic Publishers.

Gammaridean and caprellidean fauna from Brazil

Yoko Wakabara[1], Airton S. Tararam[1], Maria Teresa Valério-Berardo[1], Wania Duleba[1] &
Fosca P. Pereira Leite[2]
[1] *Inst. Oceanográfico da U.S.P., Praça do Oceanográfico, 191, Butantã, CEP 05508, São Paulo, Brazil*;
[2] *Departamento de Zoologia, Instituto de Biologia, UNICAMP, Cx. Postal 6109, Campinas, CEP 13081,
São Paulo, Brazil*

Key words: Gammaridea, Caprellidea, Brazil, review, distribution

Abstract

This review of Brazilian Amphipod works is arranged in a chronological list and points out the significant works for each one of the four arbitrary periods. Besides that, this paper deals with the distribution of 83 species of Brazilian Gammaridea and Caprellidea in relation to bathymetry, substrate, latitude and thermal region.

Introduction

This paper present a review of Brazilian Amphipoda and analyse the distribution of the species identified in our previous works, in relation to bathymetry, substrate, latitude and thermal region. Despite our extensive surveys Amphipoda from wide areas of the Brazilian coast are still rather unknown, mainly those from the Northern States. Moreover, Amphipoda from soft bottom and deeper waters have been less investigated than those of hard bottom and shallow waters.

Material and methods

This review of Brazilian Amphipoda works was based on literature available to us covering the period 1840 to date.

The material examined was collected along the Brazilian coast from 01° 20′ S to 35° 20′ S, part of it provided by cooperative research projects, part by the Instituto Oceanográfico's own projects and a smaller part from material sent to us for identification, that originating from States of Northeast, Southeast and South.

For the studies on bathymetrical and substrate distributions, species were grouped according to the following habitats: intertidal, from intertidal down to 65 m, from intertidal down to 180 m, from shallow water to 250 m, from shallow water down to 790 m, and muddy sand, Algae, Phanerogamae, animal substrate (Anthozoa, Hydrozoa, Bryozoa, Polychaeta, Mytilidae and Ascidiacea).

In relation to latitudinal distribution, species were grouped in 4 ranges: a) 20°–25° S, b) 01°–25° S, c) 18°–35° S, d) 01°–35° S. For the thermal regional distribution data were collected from the available literature of the species. One might expect that these distributions may be modified as explorations proceed and the material already collected has been examined. In this work data on material obtained no later than 1988 is presented.

Table 1. Review of Brazilian Amphipoda works (1840–1989).

Year	Author(s)	Subject	South Atlantic latitude	Publication
1840	Milne Edwards	Taxonomy–Gammaridea	Brazilian coast	Roret, Paris, 3: 1–638
1842	Krøyer	Taxonomy–Caprellidea	22° 57' S	Syst. Naturhist. Tidsskr., 4: 490–518, 585–616
1853	Dana	Taxonomy–Gammaridea and Caprellidea	Brazilian coast	U.S. Expl. Exped., 14: 689–1618
1890	Mayer	Taxonomy–Caprellidea	22° 57' S	Fauna Flora Golfe Neapel, 17: 1–157
1915	Müller	Taxonomy–Gammaridea	27° 30' S	Fritz Müller Werke, Briefe und Leben, Jena, Gustav Fisher, 1(1): 200–263
1916	Walker	Taxonomy–Gammaridea	22° 57' S	Ann. Mag. nat. Hist. ser. 8, 17: 343–346
1932	Shoemaker	Taxonomy–Gammaridea	22° 57' S	J. Wash. Acad. Sci., 22(7): 184–187
1933	Shoemaker	Taxonomy–Gammaridea	23° 49' S	Am. Mus. Novit., 598: 1–24
1938	Schellenberg	Taxonomy–Gammaridea	07° 48' S–09° 40' S	Zool. Jb., 71: 203–218
1940	Oliveira	Taxonomy–Gammaridea and Caprellidea	22° 57' S	Mems Inst. Oswaldo Cruz, 35: 137–151, 375–377
1942	Schubart	Taxonomy–Gammaridea	07° 48' S	Bolm Mus. nac., Rio de Janeiro, 14(17): 22–61
1951	Oliveira	Taxonomy–Gammaridea	22° 57' S	Bolm Inst. paul. Oceanogr., 2(2): 1–17
1953	Oliveira	Taxonomy–Gammaridea	22° 57' S	Mems Inst. Oswaldo Cruz, 51: 289–376
1954	Oliveira	Taxonomy–Gammaridea	22° 57' S	Mems Inst. Oswaldo Cruz, 52(3/4): 603–615
1955	Oliveira	Taxonomy–Gammaridea	22° 57' S	Mems Inst. Oswaldo Cruz, 53: 313–317
1955–56	Ruffo	Taxonomy–Gammaridea	24° 00' S–25° 02' S	Memorie Mus. civ. Stor. nat. Verona, 5: 115–124
1968	McCain	Taxonomy–Caprellidea	Brazilian coast	Bull. U.S. natn. Mus., 278: 1–147
1969	Wakabara	Taxonomy–Gammaridea	23° 30' S	M. Sc. Dissert. Inst. oceanogr. USP, 52 p.
1971	Escofet	Taxonomy–Gammaridea	32° 12' S	Neotropica, 17: 107–115
1971	Quitete	Taxonomy–Caprellidea	01° 21' S	Atas Soc. Biol. Rio de Janeiro, 14(5/6): 161–164
1971	Quitete	Taxonomy–Caprellidea	07° 48' S–20° 36' S	Atas Soc. Biol. Rio de Janeiro, 14(5/6): 189–192
1972	Quitete	Taxonomy–Caprellidea	08° 21' S–23° 00' S	Atas Soc. Biol. Rio de Janeiro, 15(3): 165–168
1972	Wakabara	Taxonomy–Gammaridea	03° 23' S–38° 05' S	PhD. thesis Inst. Biociencias USP, 87 p.
1974	Soares	Taxonomy–Gammaridea	07° 48' S	Cienc. Cult., supl., 26(7): 356
1976	Leite	Biology–Gammaridea	23° 30' S	M. Sc. Dissert. Inst. oceanogr. USP, 74 p.
1977	Carvalho	Taxonomy–Gammaridea	22° 57' S–23° 00' S	M. Sc. Dissert. Univ. Fed. Rio de Janeiro, 121 p.
1977	Tararam	Ecology–Gammaridea	23° 30' S	M. Sc. Dissert. Inst. Oceanogr. USP, 73 p.
1977	Soares	Ecology–Gammaridea	07° 48' S	Cienc. Cult., supl., 29(7): 808
1977	Soares	Ecology–Gammaridea	07° 48' S	Cienc. Cult., supl., 29(7): 808
1977	Wakabara & Leite	Taxonomy–Ecology–Gammaridea	13° 00' S–20° 41' S	Crustaceana, 33(1): 90–96
1978	Soares	Ecology–Gammaridea	07° 48' S	Cienc. Cult., supl., 30(7): 591
1978	Soares	Ecology–Gammaridea	07° 48' S	Cienc. Cult., supl., 30(7): 591
1978	Tararam et al.	Biology–Gammaridea	24° 12' S	Bull. mar. Sci., 28(4): 782–786
1979	Soares	Taxonomy–Gammaridea	03° 51' S	Cienc. Cult., supl., 31(7): 689
1979	Soares	Taxonomy–Gammaridea	07° 48' S	Trabhs oceanogr. Univ. Fed. PE, 14: 93–104
1980	Soares	Ecology–Gammaridea	07° 48' S	Trabhs oceanogr. Univ. Fed. PE, 15: 263–276
1980	Soares	Gammaridea of the fish stomach content	07° 48' S	Cienc. Cult., supl., 32(7): 844

Table 1. (continued).

Year	Author(s)	Subject	South Atlantic latitude	Publication
1980	Tararam	Autoecology–Gammaridea	24° 12′ S	PhD. thesis Inst. oceanogr. USP, 99 p.
1980	Leite et al.	Ecology–Gammaridea	23° 30′ S	Bolm Inst. oceanogr., S Paulo, 29(2): 297–299
1981	Takeda	Biology–Caprellidea	23° 30′ S	M. Sc. Dissert. Inst. oceanogr. USP, 94 p.
1981	Tararam & Wakabara	Ecology–Gammaridea	23° 30′ S	Mar. Ecol. Prog. Ser., 5: 157–163
1981	Tararam et al.	Ecology–Gammaridea	24° 12′ S	Sem. reg. Ecologia, S. Carlos: 305–321
1981	Leite	Biology–Gammaridea	23° 30′ S	PhD. thesis Inst. oceanogr. USP, 177 p.
1981	Soares et al.	Gammaridea of the fish stomach content	07° 48′ S	Anais III Encontro de Zool. Nordeste: 155–159
1982	Wakabara et al.	Gammaridea and Caprellidea of the fish stomach content	31° 00′ S–36° 00′ S	Mar. Biol., 68: 67–70
1982	Tararam & Wakabara	Gammaridea of the fish stomach content	24° 12′ S	Bolm Inst. oceanogr., S Paulo, 31(2): 1–3
1982	Ribeiro	Gammaridea of the fish stomach content	29° 21′ S–33° 41′ S	M. Sc. Dissert. Inst. oceanogr. USP, 97 p.
1983	Wakabara et al.	Ecology–Gammaridea and Caprellidea	24° 12′ S	J. crustacean Biol., 3(4): 602–607
1985	Tararam et al.	Biology–Gammaridea	24° 12′ S	Bolm Inst. oceanogr., S Paulo, 33(2): 193–199
1985	Petti et al.	Ecology–Gammaridea	24° 00′ S	Congr. Bras. Zool., 12, Campinas: 59
1985	Jacobi	Ecology–Gammaridea	24° 00′ S	Congr. Bras. Zool., 12, Campinas: 59
1985	Valério-Berardo & Wakabara	Taxonomy–Gammaridea	19° 38′ S–23° 19′ S	Congr. Bras. Zool., 12, Campinas: 60
1985	Dutra	Biology–Caprellidea	25° 32′ S	Congr. Bras. Zool., 12, Campinas: 60
1986	Leite et al.	Biology–Gammaridea	23° 30′ S–24° 12′ S	Crustaceana, 5(1): 77–94
1986	Soares	Taxonomy–Gammaridea	00° 46′ S–36° 29′ S	M. Sc. Dissert. Univ. Fed. PE, 197 p.
1986	Tararam et al.	Ecology–Caprellidea	24° 12′ S	Crustaceana, 5(2): 183–187
1986	Valério-Berardo	Taxonomy–Gammaridea	07° 45′ S–38° 05′ S	M. Sc. Dissert. Inst. oceanogr. USP, 82 p.
1987	Wakabara et al.	Taxonomy–Gammaridea and Caprellidea	01° 20′ S–38° 05′ S	Mini-Simp. Biol. Mar, 6, São Sebastião, 22
1987	Jacobi	Biology–Gammaridea and Caprellidea	24° 00′ S	Mar. Ecol. Prog. Ser., 35: 51–58
1987	Jacobi	Taxonomy–Gammaridea and Caprellidea	23° 49′ S	Mini-Simp. Biol. Mar., 6, São Sebastião, 14
1987	Sá Rego	Taxonomy–Gammaridea	22° 57′ S–24° 00′ S	Iheringia. Ser. Zool., Porto Alegre, (66): 141–147
1988	Wakabara et al.	Taxonomy–Gammaridea	18° 29′ S–36° 29′ S	Relat. int. Inst. oceanogr. Univ. S Paulo, 23: 1–10
1988	Valério-Berardo et al.	Ecology–Gammaridea	07° 45′ S–38° 05′ S	Congr. Bras. Zool., 15, Curitiba: 127
1988	Wakabara et al.	Ecology–Gammaridea and Caprellidea	01° 20′ S–38° 05′ S	Mini-Simp. Biol. Mar., 7, São Sebastião, 40
1988	Masunari & Dubiaski-Silva	Biology–Caprellidea	25° 30′ S	Congr. Bras. Zool., 15, Curitiba: 104
1988	Barnard & Thomas	Taxonomy–Gammaridea	22° 57′ S	Proc. biol. Soc. Wash., 101(2): 366–374
1988	Barnard & Thomas	Taxonomy–Gammaridea	22° 57′ S	Proc. biol. Soc. Wash., 101(3): 614–621
1989	Leite	Biology–Gammaridea	23° 30′ S	Congr. Bras. Zool., 16, João Pessoa: 224
1989	Leite	Biology–Gammaridea	23° 30′ S	Congr. Bras. Zool., 16, João Pessoa: 225
1989	Leite	Biology–Gammaridea	23° 30′ S	Congr. Bras. Zool., 16, João Pessoa: 234
1989	Leite & Wakabara	Biology–Gammaridea	23° 30′ S	Bull. mar. Sci., 45(1): 85–97
1989	Masunari & Dubiaski-Silva	Biology–Caprellidea	25° 50′ S	Congr. Bras. Zool., 16, João Pessoa: 215

Results and general considerations

Studies on Brazilian amphipods (Table 1) may be divided chronologically in four periods of time and for each one we point out the remarkable works.

The first period, covering 1840 to 1950, was characterized by a paucity of works, those few written by classical authors such as Dana (1853), Walker (1916), Shoemaker (1933) and Schellenberg (1938). In the following period, 1951–1970, Brazilian amphipodologists started to work in this field, specializing in taxonomy, namely Oliveira (1951, 1953, 1954, 1955) and Wakabara (1969). In the decade 1971–1980, studies on Caprellidean taxonomy started by Quitete (1971a, b, 1972) and Gammaridean taxonomy continued by Wakabara (1972) and Wakabara & Leite (1977). Biology and autoecology of Gammaridea also started at this time (Leite, 1976; Tararam, 1980). In the last decade, 1981–1989, important papers were published on Gammaridean taxonomy (Valério-Berardo, 1986; J. L. Barnard & Thomas, 1988a, b; Wakabara et al., 1988), biology of Gammaridea (Leite & Wakabara, 1989), Amphipoda as fish prey (Wakabara et al., 1982; Tararam & Wakabara, 1982; Ribeiro, 1982), biology of Caprellidea (Takeda, 1981; Masunari & Dubiaski-Silva, 1988, 1989) and ecology of Gammaridea (Tararam & Wakabara, 1981; Tararam et al., 1986; Jacobi, 1987)

Presently, we are examining only the data of amphipod specimens that we have already identified, and so far we can verify this amphipod fauna comprises 83 species belonging to 59 genera.

Regarding the bathymetrical distribution (Table 2), 39 species were found occurring exclusively in the intertidal region, a fact which characterizes it as the richest in species. Twenty three more species can live there, since they were found inhabiting the intertidal down to a 180 meter depth. Deeper regions seem to shelter fewer species, but it should be noted that those regions are still not well explored and many samples are yet to be examined.

Nine groups of species were considered in re-lation to the substrates from where they were collected (Table 3). It is worth mentioning that except for the groups occurring exclusively in muddy sand, Algae and Phanerogamae, the remaining species were found occupying diverse substrates. If we add the number of those species living on diverse types to those of the single substrate, we might have different figures, so they area: 58 species living in muddy sand, 40 living on Algae, 6 species living on Phanerogamae and 17 species occurring on animal substrate.

Latitudinal data of Gammaridean and Caprellidean species recorded from along the Brazilian coast are listed in Table 4 and the results revealed that 24 species occur in a narrow band, SE (20°–25° S), 13 species were collected in the N–SE region (01°–25° S), 19 species are distributed in the SE–S (18°–35° S) and 8 species were found in a wide band of latitude, N–S region (01°–35° S).

Although the present data might be influenced by exploratory and even taxonomic deficiences, the thermal distribution of the Brazilian Amphipoda fauna can be outlined. As one might expect, obviously, there is the presence of abundant (over 81%) of Tropical, Warm and Cold Temperate elements (Table 4). We consider here as 'widespread' species those distributed in such widely dispersed localities as Tropical, Warm and Cold Temperate regions as far as cold waters. Such species are in smaller percentage (12%) and species of genus *Caprella* are a good example of such distribution.

Elements from cooler waters although least in percentage (3.6%), occur along the Brazilian coast. Palacio (1982), in his revision of Brazilian zoogeography considers the area between Espirito Santo and Rio Grande do Sul States a hydrographically, faunistically, and zoogeographically transitional zone where many Patagonian species are found. In this area, which he named the Paulinean Province (Lat. 20° 30′ S–29° 30′ S), under the influence of Brazilian and Falkland Currents, the coexistence of warm and cold water species is possible. Furthermore, according to Palacio (*op. cit.*) species inhabiting colder waters are more adapted to a wider range of annual

Table 2. Bathymetrical distribution of Brazilian Amphipoda species.

A – Intertidal region

1. *Caprella danileviskii*
2. *Caprella dilatata*
3. *Caprella equilibra*
4. *Caprella scaura*
5. *Paracaprella pusilla*
6. *Phtisica marina*
7. *Ampithoe ramondi*
8. *Cymadusa filosa*
9. *Sunamphitoe pelagica*
10. *Amphilochus neapolitanus*
11. *Aora spinicornis*
12. *Lembos hypacanthus*
13. *Lembos unicornis*
14. *Atylus minikoi*
15. *Batea catharinensis*
16. *Cerapus tubularis*
17. *Corophium acherusicum*
18. *Grandidierella bonnieroides*
19. *Parhyalella whelpleyi*
20. *Hyale media*
21. *Parhyale hawaiensis*
22. *Ericthonius brasiliensis*
23. *Jassa falcata*
24. *Microjassa macrocoxa*
25. *Leucothoe denticulata*
26. *Ceradocus paucidentatus*
27. *Elasmopus pectenicrus*
28. *Elasmopus spinidactylus*
29. *Melita mangrovi*
30. *Quadrivisio lutzi*
31. *Bathyporeiapus bisetosus*
32. *Bathyporeiapus ruffoi*
33. *Gammaropsis atlantica*
34. *Gammaropsis chelifera*
35. *Podocerus brasiliensis*
36. *Podocerus fulanus*
37. *Tethygeneia longleyi*
38. *Stenothoe valida*
39. *Orchestia darwini*

B – From intertidal down to 65 m

1. *Liljeborgia dubia* (6–65)
2. *Listriella titinga* (1–27)
3. *Elasmopus brasiliensis* (0–50)
4. *Elasmopus rapax* (0–42)
5. *Mallacoota subcarinata* (0–24)
6. *Melita orgasmos* (0–32)
7. *Gibberosus myersi* (1–56)
8. *Resupinus coloni* (1–15)
9. *Heterophlias seclusus* (0–58)
10. *Gammaropsis togoensis* (0–50)
11. *Photis brevipes* (5–53)
12. *Photis longicaudata* (0–65)
13. *Tiburonella viscana* (0–4)

C – From intertidal down to 180 m

1. *Ampelisca brevisimulata* (1–176)
2. *Lembos smithi* (0–140)
3. *Phoxocephalopsis zimmeri* (0–120)
4. *Dulichiella appendiculata* (0–166)
5. *Maera grossimana* (0–90)
6. *Maera hirondellei* (0–98)
7. *Maera quadrimana* (0–176)
8. *Monoculodes nyei* (1–166)
9. *Chevalia aviculae* (0–136)
10. *Tiron tropakis* (0–136)

D – From shallow water to 250 m

1. *Ampelisca pugetica* (18–194)
2. *Pseudomegamphopus barnardi* (56–150)
3. *Maera inaequipes* (50–140)
4. *Synchelidium americanum* (20–54)
5. *Westwoodilla rectirostris* (24–166)
6. *Ampelisciphotis podophthalma* (16–64)
7. *Cheiriphotis megacheles* (18–30)
8. *Gammaropsis sophiae* (40–150)
9. *Gammaropsis thompsoni* (12–166)
10. *Heterophoxus videns* (17–25)
11. *Microphoxus cornutus* (18–25)
12. *Pseudotiron longicaudatus* (136–250)
13. *Synopia ultramarina* (41–60)
14. *Syrrhoe crenulata* (97–224)

E – From shallow water down to 790 m

Liljeborgia quinquedentata

seasonal conditions than tropical ones. Brazilian colder water species (*Gammaropsis thompsoni, Heterophoxus videns* and *Oediceroides cinderella*) were collected at latitudes more southerly than 23° S, thus they are distributed in the Paulinean Province. Moreira (1976) recorded two Isopoda species, *Serolis polaris* Richardson, 1911 and *S. elliptica* Sheppard, 1933 from Cold Temperate

Table 3. Distribution of species according to substrates.

A – Muddy sand

1. *Ampelisca brevisimulata*
2. *Ampelisca cristata*
3. *Ampelisca panamensis*
4. *Ampelisca pugetica*
5. *Lembos smithi*
6. *Pseudomegamphopus barnardi*
7. *Cerapus tubularis*
8. *Grandidierella bonnieroides*
9. *Phoxocephalopsis zimmeri*
10. *Lepechinella auca*
11. *Liljeborgia dubia*
12. *Liljeborgia quinquedentata*
13. *Listriella titinga*
14. *Maera hirondellei*
15. *Maera inaequipes*
16. *Resupinus coloni*
17. *Bathyporeiapus bisetosus*
18. *Bathyporeiapus ruffoi*
19. *Monoculodes nyei*
20. *Oediceroides cinderella*
21. *Synchelidium americanum*
22. *Westwoodilla longimana*
23. *Westwoodilla rectirostris*
24. *Ampelisciphotis podophthalma*
25. *Cheiriphotis megacheles*
26. *Gammaropsis sophiae*
27. *Gammaropsis thompsoni*
28. *Photis brevipes*
29. *Heterophoxus videns*
30. *Microphoxus cornutus*
31. *Pseudharpinia dentata*
32. *Tiburonella viscana*
33. *Pseudotiron longicaudatus*
34. *Synopia ultramarina*
35. *Syrrhoe crenulata*

B – Algae

1. *Caprella danileviskii*
2. *Caprella dilatata*
3. *Sunamphitoe pelagica*
4. *Lembos hypacanthus*
5. *Lembos unicornis*
6. *Batea catharinensis*
7. *Hyale media*
8. *Ceradocus paucidentatus*
9. *Elasmopus spinidactylus*
10. *Maera quadrimana*
11. *Quadrivisio lutzi*
12. *Gammaropsis atlantica*
13. *Gammaropsis chelifera*

C – Muddy sand and Algae

1. *Amphilochus neapolitanus*
2. *Aora spinicornis*
3. *Atylus minikoi*
4. *Corophium acherusicum*
5. *Maera grossimana*
6. *Mallacoota subcarinata*
7. *Melita mangrovi*
8. *Melita orgasmos*
9. *Gibberosus myersi*
10. *Heterophlias seclusus*
11. *Chevalia aviculae*
12. *Gammaropsis togoensis*
13. *Photis longicaudata*
14. *Tethygeneia longleyi*
15. *Stenothoe valida*

D – Phanerogamae

Paracaprella pusilla

E – Muddy Sand and Phanerogamae

1. *Parhyalella whelpleyi*
2. *Orchestia darwini*

F – Animal substrate

1. *Caprella equilibra*
2. *Phtisica marina*
3. *Microjassa macrocoxa*
4. *Podocerus brasiliensis*
5. *Podocerus fulanus*

G – Muddy sand, Algae, Animal substrate

1. *Ampithoe ramondi*
2. *Jassa falcata*
3. *Dulichiella appendiculata*
4. *Elasmopus brasiliensis*
5. *Elasmopus rapax*
6. *Tiron tropakis*

H – Algae, Phanerogamae, Animal substrate

1. *Caprella scaura*
2. *Cymadusa filosa*
3. *Parhyale hawaiensis*

I – Algae and Animal substrate

1. *Ericthonius brasiliensis*
2. *Leucothoe denticulata*
3. *Elasmopus pectenicrus*

Table 4. Latitudinal distribution of Amphipoda on Brazilian coast.

SE region (20°–25° S)

1. *Caprella equilibra* (23° 48′–24° 00′)
2. *Caprella scaura* (23° 30′–23° 48′)
3. *Ampithoe ramondi* (22° 57′–25° 02′)
4. *Cymadusa filosa* (23° 25′–24° 12′)
5. *Sunamphitoe pelagica* (23° 25′–24° 12′)
6. *Amphilocus neapolitanus* (23° 30′–25° 02′)
7. *Aora spinicornis* (23° 30′–23° 48′)
8. *Atylus minikoi* (23° 30′–25° 02′)
9. *Batea catharinensis* (23° 30′–23° 48′)
10. *Phoxocephalopsis zimmeri* (23° 30′–25° 02′)
11. *Hyale media* (22° 57′–24° 12′)
12. *Parhyale hawaiiensis* (23° 25′–25° 02′)
13. *Ericthonius brasiliensis* (22° 57′–24° 12′)
14. *Jassa falcata* (22° 57′–25° 02′)
15. *Leucothoe denticulata* (22° 57′–23° 48′)
16. *Listriella titinga* (23° 10′–25° 02′)
17. *Elasmopus spinidactylus* (23° 30′–24° 07′)
18. *Maera inaequipes* (22° 48′–23° 50′)
19. *Gibberosus myersi* (21° 42′–23° 30′)
20. *Gammaropsis atlantica* (23° 30′–24° 12′)
21. *Tiburonella viscana* (23° 30′–25° 02′)
22. *Podocerus brasiliensis* (23° 25′–23° 48′)
23. *Tethygeneia longleyi* (23° 30′–25° 02′)
24. *Stenothoe valida* (22° 57′–25° 02′)

N–SE region (01°–25° S)

1. *Lembos smithi* (13° 00′–23° 48′)
2. *Lembos unicornis* (07° 30′–23° 22′)
3. *Grandidierella bonnieroides* (01° 20′–25° 02′)
4. *Ceradocus paucidentatus* (09° 42′–20° 41′)
5. *Elasmopus brasiliensis* (13° 00′–24° 07′)
6. *Elasmopus rapax* (07° 45′–23° 48′)
7. *Malacoota subcarinata* (08° 16′–23° 58′)
8. *Quadrivisio lutzi* (01° 20′–25° 02′)
9. *Synchelidium americanum* (18° 29′–19° 50′)
10. *Heterophlias seclusus* (13° 00′–20° 41′)
11. *Chevalia aviculae* (07° 45′–23° 19′)
12. *Synopia ultramarina* (18° 33′–19° 50′)
13. *Orchestia darwini* (01° 20′–25° 02′)

SE–S region (18°–35° S)

1. *Pseudomegamphopus barnardi* (21° 42′–34° 34′)
2. *Liljeborgia dubia* (19° 05′–35° 05′)
3. *Liljeborgia quinquedentata* (19° 02′–33° 47′)
4. *Maera hirondellei* (22° 22′–35° 05′)
5. *Melita orgasmos* (23° 10′–35° 05′)
6. *Resupinus coloni* (21° 46′–25° 02′)
7. *Bathyporeiapus ruffoi* (24° 30′–32° 30′)
8. *Monoculodes nyei* (18° 34′–34° 25′)
9. *Westwoodilla rectirostris* (18° 33′–34° 25′)
10. *Ampelisciphotis podophthalma* (18° 44′–34° 34′)

11. *Cheiriphotis megacheles* (23° 30′–33° 58′)
12. *Gammaropsis chelifera* (20° 19′–27° 05′)
13. *Gammaropsis sophiae* (22° 10′–35° 05′)
14. *Gammaropsis togoensis* (18° 34′–33° 47′)
15. *Gammaropsis thompsoni* (22° 48′–34° 25′)
16. *Photis brevipes* (18° 44′–33° 58′)
17. *Pseudotiron longicaudatus* (23° 19′–30° 43′)
18. *Syrroe crenulata* (23° 03′–30° 43′)
19. *Tiron tropakis* (18° 29′–25° 13′)

N–S region (01°–35° S)

1. *Ampelisca brevisimulata* (09° 42′–34° 21′)
2. *Ampelisca pugetica* (09° 42′–34° 34′)
3. *Corophium acherusicum* (07° 45′–27° 35′)
4. *Dulichiella appendiculata* (07° 36′–34° 25′)
5. *Elasmopus pectenicrus* (03° 23′–26° 15′)
6. *Maera grossimana* (16° 02′–33° 15′)
7. *Maera quadrimana* (03° 23′–30° 46′)
8. *Photis longicaudata* (07° 45′–34° 32′)

and Cold Waters (South Sandwich Is., Falkland Is. and South Patagonia) along the Brazilian Southern coast. J. L. Barnard (1970), describing the mechanism of Amphipoda dispersion, considers that tubicolous and phycophylous species have a greater chance of dispersion than other species. However, Brazilian colder water species so far known have none of those two habits, so we believe that currents might have carried the species from long distances to the Brazilian coast.

Finally, we can say that, although there are a lot

Table 5. Thermal distribution (%) of Brazilian Amphipoda species (T: Tropical, WT: Warm Temperate, CT: Cold Temperate, C: Cold, W: Widespread).

Thermal region (data from the literature)	No. of species	%
T	8	
T + WT	37	
WT	12	81.9
WT + CT	9	
T + WT + CT	4	
W (T,WT,CT,C)	10 }	12.0
WT + C	1	
WT + CT + C	1	3.6
CT + C	1	

of studies on Brazilian Amphipoda, only 40 of them are effectively published papers. Furthermore, an immense area of the coast is still unsurveyed, in terms of the amphipodan fauna and many of our samples of Ampeliscidae, Haustoriidae, Ischyroceridae, Lysianassidae, Oedicerotidae, Phoxocephalidae and Synopiidae are not yet entirely investigated. Most of them will be the subjects of our future works, thus contributing to further development of Amphipoda studies in Brazil.

References

Barnard, J. L., 1970. Sublittoral Gammaridea (Amphipoda) of the Hawaiian Islands. Smithson. Contr. Zool. 34: 1–286.

Barnard, J. L. & J. D. Thomas, 1988a. *Vadosiapus copacabanus*, a new genus and species of Exoedicerotidae from Brazil (Crustacea, Amphipoda). Proc. biol. Soc. Wash. 101: 366–374.

Barnard, J. L. & J. D. Thomas, 1988b. Ipanemidae, new family, *Ipanema talpa*, new genus and species, from the surf zone of Brazil (Crustacea: Amphipoda: Haustorioidea). Proc. biol. Soc. Wash. 101: 614–621.

Dana, J. D., 1853. Crustacea. United States Exploring Expedition, during the years 1838, 1839, 1840, 1841, 1842, under the command of Charles Wilkes, U.S.N., Philadelphia, C. Sherman, 14(2): 805–1021.

Jacobi, C. M., 1987. Spatial and temporal distribution of Amphipoda associated with mussel beds from the Bay of Santos (Brazil). Mar. Ecol.-Prog. Ser. 35: 51–58.

Leite, F. P. P., 1976. Estádios de crescimento e aspectos de reprodução de *Hyale media* (Crustacea, Amphipoda, Hyalidae) da fauna vágil de *Sargassum cymosum*. Dissertação de mestrado. Universidade de São Paulo, Instituto Oceanográfico. 74 p.

Leite, F. P. P. & Y. Wakabara, 1989. Aspects of marsupial and post-marsupial development of *Hyale media* (Dana) 1853 (Hyalidae, Amphipoda). Bull. mar. Sci. 45: 85–97.

Masunari, S. & J. Dubiaski-Silva, 1988. Distribuição dos Amphipoda Caprellidae (Crustacea) nos fitais de Caiobá, Paraná. In: Congresso Brasileiro de Zoologia, 15, Curitiba, 1988. Resumos. Curitiba, Sociedade Brasileira de Zoologia, p. 104.

Masunari, S. & J. Dubiaski-Silva, 1989. Flutuação da densidade das populações de *Caprella penantis* Leach, 1814 (Crustacea: Amphipoda: Caprellidae) nos fitais de Caiobá, PR. In: Congresso Brasileiro de Zoologia, 16, João Pessoa, 1989. Resumos. João Pessoa, Sociedade Brasileira de Zoologia, p. 215.

Moreira, P. S., 1976. Crustacea Isopoda collected during the Oc/S 'Almirante Saldanha' cruise in southern South America. I. Species of *Serolis* (Flabellifera, Serolidae). Bolm Inst. oceanogr., S Paulo 25: 113–130.

Oliveira, L. P. H., 1951. The genus *Elasmopus* on the coast of Brazil with descriptions of *Elasmopus besnardi* n. sp. and *E. fusimanus* n. sp. (Crustacea, Amphipoda). Bolm Inst. paul. Oceanogr. 2: 1–17.

Oliveira, L. P. H., 1953. Crustacea Amphipoda do Rio de Janeiro. Mems Inst. Oswaldo Cruz 51: 289–376.

Oliveira, L. P. H., 1954. Nova espécie de crustáceo Amphipoda da Baía de Guanabara: *Ampelisca soleata*. Mems Inst. Oswaldo Cruz 52: 603–615.

Oliveira, L. P. H., 1955. *Phoxocephalus capuciatus*, nova espécie de Crustacea Amphipoda, Phoxocephalidae. Mems Inst. Oswaldo Cruz 53: 313–317.

Palacio, F. J., 1982. Revision zoogeografica marina del sur del Brasil. Bolm Inst. oceanogr., S Paulo 31: 69–92.

Quitete, J. M. P. A., 1971a. *Paracaprella digitanus*, nova espécie de Caprellidae da costa brasileira (Crustacea: Amphipoda). Atas Soc. Biol., Rio de Janeiro 14: 161–164.

Quitete, J. M. P. A., 1971b. *Fallotritella montoucheti*, nova espécie de Caprellidae da costa brasileira (Crustacea: Amphipoda). Atas Soc. Biol., Rio de Janeiro 14: 189–192.

Quitete, J. M. P. A., 1972. *Hemiaegina costai*, nova espécie de Caprellidae da costa brasileira (Crustacea: Amphipoda). Atas Soc. Biol., Rio de Janeiro 15: 165–168.

Ribeiro, M. A. G., 1982. Crustacea (em especial Amphipoda) do conteúdo estomacal de Sciaenidae da plataforma continental do Brasil (Lat. 29° 21' S e 33° 41' S). Dissertação de mestrado. Universidade de São Paulo, Instituto Oceanográfico, 97 p.

Schellenberg, A., 1938. Brasilianische amphipode mit biologischen Bemerkungen. Zool. Jb. 71: 203–218.

Shoemaker, C. R., 1933. Amphipoda from Florida and the West Indies. Am. Mus. Novit. 598: 1–24.

Takeda, A. M., 1981. Aspectos do crescimento e da alimentação de *Caprella scaura typica* Mayer, 1890. Dissertação de mestrado. Universidade de São Paulo, Instituto Oceanográfico, 94 p.

Tararam, A. S., 1980. Alimentação e distribuição de *Hyale media* (Crustacea-Amphipoda) do fital da Praia do Poço, Itanhaém (SP), com observações sobre a predação da espécie por alguns peixes da região. Tese de doutorado. Universidade de São Paulo, Instituto Oceanográfico, 99 p.

Tararam, A. S. & Y. Wakabara, 1981. The mobile fauna – especially Gammaridea – of *Sargassum cymosum*. Mar. Ecol.-Prog. Ser. 5: 157–163.

Tararam, A. S. & Y. Wakabara, 1982. Notes on the feeding of *Blennius cristatus* Linnaeus from a rocky pool of Itanhaém, São Paulo State, Bolm Inst. oceanogr., S Paulo 31: 1–3.

Tararam, A. S., Y. Wakabara & F. P. P. Leite, 1986. Vertical distribution of amphipods living on algae of a Brazilian intertidal rocky shore. Crustaceana 51: 183–187.

Valério-Berardo, M. T., 1986. Sistemática e distribuição das espécies da familia Corophiidae (Crustacea, Amphipoda) da costa brasileira (7° 45' S a 38° 05' S). Dissertação de mestrado. Universidade de São Paulo, Instituto Oceanográfico, 82 p.

Wakabara, Y., 1969. Sôbre alguns Gammaridea (Crustacea–Amphipoda) da região de Ubatuba. Dissertação de mestrado. Universidade de São Paulo, Instituto Oceanográfico, 52 p.

Wakabara, Y., 1972. Espécies da família Gammaridae (Crustacea–Amphipoda), entre as latitudes 03°23′ S–38°05′ S do Atlântico Ocidental. Tese de dcutorado. Universidade de São Paulo, Instituto de Biociências, 87 p.

Wakabara, Y. & F. P. P. Leite, 1977. *Heterophlias seclusus* Shoemaker, 1933 (Amphipoda, Phliantidae) from the Brazilian coast. Crustaceana 33: 90–96.

Wakabara, Y., E. Kawakami de Rezende & A. S.Tararam, 1982. Amphipods as one of the main food components of three Pleuronectiformes from the continental shelf of South Brazil and North Uruguay. Mar. Biol. 68: 67–70.

Wakabara, Y., A. S. Tararam, M. T. Valério-Berardo & F. P. P. Leite, 1988. Liljeborgiidae (Amphipoda-Gammaridea) from the southern coast of Brazil. Relat. int. Inst. oceanogr. Univ. S Paulo 23: 1–10.

Walker, A. O., 1916. Edriophtalma from South America. Ann. Mag. nat. Hist. ser. 8, 17: 343–346.

Hydrobiologia **223**: 79–80, 1991.
L. Watling (ed.), VIIth International Colloquium on Amphipoda.
© 1991 *Kluwer Academic Publishers.*

Amphipods from hydrotechnical structures in the north-western part of the Black Sea

R.P. Alexeev
Odessa Branch, Institute of Biology of the Southern Seas, Academy of Sciences of the U.S.S.R., Odessa 270011, USSR

Abstract

Marine hydrotechnical structures, such as wave breakers and artificial reefs made of various materials and built on sandy bottom of shallow waters, serve as an additional gradient zone promoting the development of life. Over a period of five years after the structures were built, amphipod fauna in these regions was enriched from 5 to 23 specimens.

As a result of the building of hydrotechnical structures, such as wave breakers and artificial reefs, in the north-western part of the Black Sea, the normal bottom biocoenoses of exposed shores and bays were covered by sand washing in from the construction sites. The marginal nearshore biocoenosis, to a greater degree, performed the role of a 'nursery' for different species of amphipods due to good aeration and warming of the water masses (Kaminskaya *et al.*, 1977, 1978).

The construction of wave-breakers and artificial reefs made of limestone, granite, concrete, and used automobile tires deprived the crustaceans of their usual substrates, but on the other hand produced a near marginal effect at the boundary of the hard and liquid bodies, where intensive surface phenomena take place. As a result favourable conditions for the life of hydrobionts, in particular, for amphipods occur.

In hydrobiology the marginal effect means that the species composition and abundance of organisms increase on the boundaries of biocoenoses in the zones where there is a sharp gradient, or ecotones. In this respect, artificial hydrotechnical structures play the role of an additional active surface, a new high gradient zone in the sea promoting the development of life. In conditions of sandy and muddy-sandy deposits, they have a powerful influence on the biota including the am-

Table 1. List of species occurring on soft bottoms prior to building hydrotechnical structures.

Ampelisca diadema A. Costa
Stenothoe monculoides (Montagu)
Gammarus (Marinogammarus) olivii (Milne-Edwards)
Pontogammarus maeoticus (Sowinskyi)
Corophium bonellii (Milne-Edwards)
Corophium volutator (Pallas)

phipod fauna. More often the biocoenoses of sandy deposits become richer because of the new gradient zones.

In areas prior to the building of hydrotechnical structures, 5–6 species of amphipods (Table 1) were observed. Within 5 years after their construction up to 23 species appeared (Table 2). Most of these species produce three or four generations a year. Some species reproduce all year round due to mild winters when the coast is not covered with ice. These amphipods include *Marinogammarus olivii* (M-Edwards) which dominate in abundance and biomass in all seasons of the year.

Specimens of *Gammarus aequicauda* (Martynov) are the largest, fertile males reaching more than 20 mm in length. The smallest in size are female *Microprotopus minutus* Sowinskyi, 2.5–3 mm in length.

Table 2. List of species found after the construction of hydrotechnical structures.

Ampelisca diadema A. Costa
Bathyporeia guilliamsonia (Bate)
Stenothoe monoculoides (Montagu)
Colomastix pusilla Grube
Perioculoides longimanus (Bate & Westwood)
Gammarus aequicauda (Martynov)
Gammarus insensibilis Stock
Gammarus (Marinogammarus) olivii (Milne-Edwards)
Melita palmata (Montagu)
Chaetogammarus ischnus major Carausu
Dikerogammarus villosus (Sowinsky)
Dikerogammarus haemobaphes (Eichwald)
Pontogammarus maeoticus (Sowinsky)
Dexamine spinosa (Montagu)
Hyale pontica Rathke
Hyale perieri (Lucas)
Microdeutopus gryllotalpa A. Costa
Microprotopus minutus Sowinsky
Amphithoe vaillanti Sowinsky
Jassa ocia (Bate)
Erichthonius difformis (Milne-Edwards)
Corophium bonnellii (Milne-Edwards)
Corophium volutator (Pallas)

From December to February when the temperature of the sea near the shore falls to 3–5 °C, many *Bathyporea guilliamsoniana* (Bate) can be found here frequently inhabiting the base of the construction. However, fertility and abundance of species is low. Females, 4–5 mm in length produce 8–10 eggs, abundance is 24 specimens per square metre. The ratio of females to males is 2 : 1.

It should be noted that 3–4 species of amphipods can be found (*Ampelisca diadema* A. Costa dominating) at the base of the construction in the sand. All other species find food and shelter in the fouling biocoenosis, on the surface of underwater hydroconstructions, the main inhabitant of which is the mollusk, *Mytilus galloprovincialis* Lamarck.

In recent years changes have occurred in the amphipod populations of fouling communities of hydroconstructions as a result of which species of a small size, such as, *Stenothoe monoculoides* (Montagu), *Hyale pontica* Rathke, *Jassa ocia*

(Bate), and *Erichthonius difformis* M-Edwards now dominate.

This process probably is tied to the quality of marine water near the coast which today is becoming worse because of industrial waste water discharge and considerable recreational loading on the coastline.

In contrast to those parts of the sea with hydrotechnical structures inhabited by 23 species of amphipods, only two species of amphipods (*A. diadema* and *P. maeoticus*), the abundance of which reaches 4–60 and even 100 thousand specimens per square metre due to the lack of competition and predators, inhabit the near shore sandy shallow water regions of the sea influenced by the Danube and Dneiper rivers.

Analysis of a large volume of information shows that building materials of hydrotechnical structures (concrete, limestone, granite and automobile tires) are fouled by sedentary organisms, and attract motile invertebrates, such as amphipods (Zaitsev & Alexeev, 1989).

Acknowledgements

The author expresses thanks to Dr. Vera Lisovskaya for assistance in preparing and translating the paper, and to Dr. Les Watling for editorial help.

References

Kaminskaya, L. D., R. P. Alexeev, E. V. Ivanova & I. A. Sinegub, 1977. Bottom fauna of coastline of Odessa Bay in conditions of hydroconstruction. Biol. Morya 43: 54–64. (In Russian).
Kaminskaya, L. D., R. P. Alexeev, E. V. Ivanova & I. A. Sinegub, 1978. Fouling of hydroconstructions in the Odessa Bay. Materials of workshop and seminar, Kiev. Naukova Dumka 1978: 235–237. (In Russian).
Zaitsev, Yu. P. & R. P. Alexeev, 1989. The influence of hydrotechnical constructions on the marine environment. Man and the biosphere: Ukrainian scientists in realization of UNESCO programmes, Kiev.Naukova Dumka 1989: 73–79. (In Russian).

Hydrobiologia **223**: 81–104, 1991.
L. Watling (ed.), VIIth International Colloquium on Amphipoda.
© 1991 *Kluwer Academic Publishers.*

Structure of a suprabenthic shelf sub-community of gammaridean Amphipoda in the Bay of Fundy compared with similar sub-communities in the Gulf of St. Lawrence

Andrée Chevrier[1], Pierre Brunel[1] & David J. Wildish[2]
[1] *Département de sciences biologiques, Université de Montréal, Montréal (Québec), Canada, H3C 3J7;*
[2] *Department of Fisheries and Oceans, Biological Station, St. Andrews, New Brunswick, Canada, EOG 2XO*

Key words: pelagic-benthic coupling, suprabenthos, community structure, gammaridean Amphipoda

Abstract

A conceptual model of the adaptive structure of offshore bottom communities through seasonal patterns of pelagic-benthic coupling is tested by comparing the success of suprabenthic, shelf sub-communities of gammarideans selected in three areas differing in trophic status: the eutrophic Baie des Chaleurs and two oligotrophic ecosystems, the Bay of Fundy and the Lower St. Lawrence Estuary.

The relative amount of primary production appears to affect both species richness and total density. It is higher in the very productive Baie des Chaleurs, implying tightly coupled pelagic-benthic links. In the Bay of Fundy, gammaridean density, although lower than in the Baie des Chaleurs, is higher than in the Lower St. Lawrence Estuary, but it appears to be characterized by smaller size at maturity or selection of smaller species of the same taxa. The Baie des Chaleurs community may be more biologically accommodated than those of the Lower St. Lawrence Estuary and the Bay of Fundy, which are physically less stable. In the latter, all trophic guilds appear less abundant, but scavengers-predators retain a similar relative success in all three ecosystems, presumably from being more remote from the seasonal pelagic food supply in the trophic chain. In both oligotrophic systems, slope species compete efficiently with shelf species in their preferred depth zone, presumably because of their natural adaptation to food scarcity.

The amplitude of vertical migrations by suprabenthic gammarideans appears to be higher in the Baie des Chaleurs, possibly to favor horizontal redistribution to new patches of 'first-quality food' on the very stable and homogeneous mud basin of the Baie. Trends of seasonal abundance for ten gammaridean species in the Bay of Fundy confirm the reality of swimming guilds previously observed in the Baie des Chaleurs.

Introduction

Interest in pelagic-benthic coupling has increased in recent years. Much of it centers on the proportion of pelagic primary production which is conveyed to the benthos as fecal pellets or sinking phytoplankton cells, and on energetic inputs to bottom production, at any depths in the oceans (Emerson *et al.*, 1986; Davies & Payne, 1984; Pfannkucke *et al.*, 1983; Graf *et al.*, 1983, 1984; Honjo *et al.*, 1982).

Another part of this interest focuses on the transmission of surface seasonal pulses of primary production to depths where constant, non-seasonal conditions prevail, and on the effects which such signals have on breeding periodicity. At abyssal depths, where most studies have taken place, both continuous and seasonal breeding has been found (Tyler, 1988; Rokop, 1977a, 1977b; George & Menzies, 1967, 1968). Almost nothing is known regarding pelagic-benthic coupling effects on the trophic structure and composition of deep-water bottom communities.

A conceptual model of the adaptive structure of offshore bottom communities through the seasonal patterns of pelagic-benthic coupling has been developed gradually by the second author (Brunel, 1983 and unpubl.) and his students (references below). The model may be summarized as follows. Secondary or tertiary benthic production and species 'success' (see below) should be less in oligotrophic than in eutrophic ecosystems (Mann, 1976). Seasonal peaks of primary production in temperate and subarctic surface waters are followed by similar food pulses which are rapidly transmitted to the bottom, mainly as fecal pellets (Schink, 1979; Billett et al., 1983; Wefer et al., 1982; Falkowski et al., 1983; Hargrave, 1985) or as copepods migrating downward during daylight (Brunel, 1979; Forward, 1988). Benthic species directly dependent on such food sources, i.e. seston-filtering, non-selective deposit and plankton feeders should benefit from adjusting their breeding season to such pulses, as do many pelagic species following the match-mismatch model (Heinrich, 1962; Cushing, 1972). The adaptive purpose is either to build up female reserves to put into yolk, or to release neonates and juveniles at a time when high-quality food is available. Species higher in the food web, such as opportunistic omnivores, bottom-feeding carnivores, scavengers and host-dependent commensals, should be less dependent on such seasonal food pulses than the others, and should either breed more continuously or during other seasons. Following the surface phytoplankton bloom, detritus particles are expected to become available to benthic invertebrates in a seasonal succession of decreasing size, quantity and quality; this succession can be conveniently separated into three stages: (1) The most nutritive

and large pellets, 'first-quality food', is deposited first and made available immediately on the sediment surface as a flocculent film (Billett et al., 1983), about which direct information is still very scanty. (2) After some lag time, poorly known but presumably longer in deep than in shallow water, a less nutritive fraction of this organic matter, 'second-quality food', becomes adsorbed on mineral grains and colonized by bacteria at or just below the sediment surface. (3) What is not used there is ultimately mixed downward by bioturbation into mud, a 'third-quality food' which is commonly very low in organic content and nutritive value. Planktonic copepods as live prey should become most available to bottom predators, through daily downward migrations (Brunel, 1979), when they proliferate shortly after the surface phytoplankton bloom; although very few data are available on their maximum seasonal availability near the bottom, it is expected to take place at about the same time as that of first-quality detritus, since copepods, as producers of fecal pellets for detritus feeders and of live prey for predators, simultaneously account for first-quality food to two different trophic guilds near the bottom.

The model has been tested mostly on populations of suprabenthic peracarid Crustacea, which offer five attractive features for study: (1) Because of their direct development and brooding habits, their young have no planktonic phase and therefore depend on sestonic food falls; the latter may also serve as breeding signals to adults; (2) because of their size-related bottom-swimming ability, the larger species are likely to be the first to feed on the larger particles in the flocculent deposits or on the downward-migrating copepods, and their population response is thus expected to be rapid; (3) they can be caught in much larger mud-free quantitative samples by plankton nets mounted on suprabenthic (Brunel et al., 1978) or epibenthic sleds than by any endobenthic device; (4) their breeding stages can be determined fairly easily without dissection using secondary sexual characters (oostegites, eggs and marsupial embryos in females, antennal length and ornamentation in males) and nuptial swarmings of mature males; (5) except on coral reefs (Gerber & Marshall, 1974; Hobson & Chess, 1979; Ohlhorst, 1982; Porter & Porter, 1977; Porter *et al.*, 1977) and in Scandinavian fjords (Wassmann, 1984, 1985; Christensen & Kanneworff, 1985; Hopkins & Gulliksen, 1978), most studies of trophic structure and energy transfer in bottom communities have ignored these elusive populations (Holme & MacIntyre, 1984; Lopez & Levinton, 1987), despite their considerable importance as essential energetic links to demersal fish (Brunel, 1965, 1979).

Three taxa with pelagic larvae, caridean shrimp, Euphausiacea and Chaetognatha, have also been considered since they are significant components of the suprabenthic community, where they can compete with peracarid Crustacea either for live swimming prey or for organic deposits, because of their size and swimming power. Direct information on the elusive and seasonal flocculent film is still very scanty, presumably because of its swift removal by swimming Crustacea. On the other hand, we assume that non-selective detritus-feeders do not occur among Crustacea, whose deep burrowing often serves hiding rather than feeding functions; third-quality organic matter deeply buried as mud is therefore ignored in our model-generated hypotheses.

The model has generated predictive hypotheses which have been tested by comparing benthic 'success' (see below) of suprabenthic populations in two ecosystems differing in their primary production regimes: the Baie des Chaleurs, in the western Gulf of St. Lawrence, has high annual production, a temperate-type phytoplankton bloom in April–May (Legendre, 1971; Steven, 1974) and a significant groundfish fishery; the Lower St. Lawrence Estuary has lower ($\frac{1}{4}$) production, a subarctic-type bloom in June–July (Steven, 1974; Sinclair *et al.*, 1976; Therriault & Levasseur, 1985) and very poor groundfish catches (Brunel, unpubl.). Intensive sampling (day and night, twice a month from May to October) has been achieved for several years at comparable benthic monitoring stations on flat muddy grounds in both ecosystems, at a depth of 119 m remote from most seasonal temperature

variations and in the same biogeographic province, to ensure maximum hydroclimatic, edaphic and faunistic homogeneity and similarity at benthic levels. An attempt has thus been made to distinguish, among the factors controlling the bottom community, between classic biogeographic (temperature-sensitive) and edaphic effects, on the one hand, and food pulses from the surface, on the other hand.

Species success has been measured as density, % density and rank in the community, combined with size at maturity; sampling intensity ensures that the relative importance of each species was correctly assessed. These indicators are admittedly imperfect substitutes for biomass, measurement of which was deemed too destructive of material valuable for species-level information, and less valuable than thorough data on species-level temporal variability as a key indicator of community structure. 'Success' remains a loose and relative concept, however, whether biomass or density is used as an estimate. It was felt nevertheless that mean half-year density and rank, based as they are on 41–117 samples taken during the more productive half of the year, adequately reflect the real contribution of each species to community structure, irrespective of daily and seasonal variations.

Hypotheses have been tested to date mostly in projects (a) on breeding and comparative success of several species belonging to a single trophic guild, such as the detritus-feeding Cumacea (Messier, 1974; Messier & Brunel, 1977 and in prep.) and (b) on life history strategies and success of single species representing different trophic guilds, such as detritus-feeders (Granger *et al.*, 1979; Sainte-Marie & Brunel, 1983), opportunistic omnivores (Lamarche & Brunel, 1987), scavengers (Gagnon, 1983; Gagnon & Brunel, in prep.) and zooplankton feeders (Desroches, 1985; Desroches & Brunel, 1986 and in prep.); most of these species have been selected because of their occurrence in both ecosystems, which share 45% of their species.

Major findings, from 28 species examined to date, can be summarized as follows: (1) Both communities contain an equivalent number of species (about 130, including 65–70 Gammaridea). (2) Most taxa suffer a significant decrease of mean half-year density in the Lower St. Lawrence Estuary, but this reduction varies greatly among taxa and trophic guilds. (3) Detritus and plankton feeders tend to delay their breeding season by about two months in the Lower St. Lawrence Estuary, thus following a similar delay in the phytoplankton bloom. (4) Shelf-adapted late breeders in the Baie des Chaleurs show reduced offshore success or move inshore in the Lower St. Lawrence Estuary, presumably because of poor adaptation to low or late food supply at depth. (5) The larger Cumacea, presumed feeders on large particles, breed earlier than the smaller species, presumed to be grain lickers and manipulators like *Cumella vulgaris* (Wieser, 1956). (6) The large copepod-feeding amphipod *Acanthostepheia malmgreni* in the Baie des Chaleurs is replaced in the Lower St. Lawrence Estuary by the smaller *Rhachotropis oculata*, which can only cope with oligotrophy through a 15-km breeding migration to copepod-rich inshore depths using strong transverse currents. (7) Detritus-feeding slope species, presumably more adapted to a lower food supply, manage to compete better with sublittoral species at shelf depths in the oligotrophic Lower St. Lawrence Estuary than in the Baie des Chaleurs. (8) Scavengers or more opportunistic species tend to breed continuously and (9) to enjoy equivalent success in both ecosystems. (10) Size at maturity, as examined in Cumacea, one gammaridean omnivore and plankton-feeding predators, tends to be smaller in the oligotrophic Lower St. Lawrence Estuary. (11) Daily vertical migrations of suprabenthos are less extensive in the Lower St. Lawrence Estuary than in the Baie des Chaleurs, which may be a consequence of stronger tidal flow or of less foraging competition for patchy and scarcer detritus resources (Sainte-Marie & Brunel, 1985) among the less dense populations of the Lower St. Lawrence Estuary.

Our objectives in the present paper were twofold: (a) We wanted to extend the comparisons to a third ecosystem, the outer Bay of Fundy, known for its oligotrophy (Emerson *et al.*, 1986) and lo-cated in the same biogeographic province, using a sampling design similar to that applied to the two Gulf ecosystems. (b) We planned to try a third way of falsifying the model by testing hypotheses on the population structure of a whole taxon, gammaridean amphipods, which dominate the suprabenthic community and contain several trophic guilds (Enequist, 1949; Brunel, 1968; Besner, 1976; Sainte-Marie & Brunel, 1985).

The outer Bay of Fundy ecosystem is characterized by average to low primary production, a late phytoplankton bloom and very low endobenthic and groundfish production, much of the primary production being converted, through strong tidal mixing, into krill, pelagic fish, cetaceans and inshore production rather than into deep offshore benthos (Emerson *et al.*, 1986). In this, it is much closer to the Lower St. Lawrence Estuary than to the Baie des Chaleurs, although in its annual primary production it seems to be intermediate.

From these features, from the findings summarized above for the two Gulf systems, and from 1986 exploratory sampling in the outer Bay of Fundy, the following predictions were made concerning the structure and composition of the suprabenthic community on comparable sediments at similar depths (119 m); they are tested in the present paper. (1) Species richness should be of the same order of magnitude (65–70 species) as in the two Gulf ecosystems. (2) Given comparable body sizes at maturity, density of the whole community and of its major trophic guilds should be equal to, or even lower than that of the Lower St. Lawrence Estuary, but certainly lower than that of the Baie des Chaleurs; any higher densities should be compensated for by smaller sizes at maturity. (3) The success of detritus and plankton feeders (called the sensitive guilds in this paper) should be smaller in the Bay of Fundy than in the two Gulf systems, if vertical mixing is strong enough to prevent most fecal pellets and copepods from reaching the benthic boundary layer. (4) If there is such scarcity of larger-size particles in the form of a flocculent deposited film and downward-migrating copepods at outer shelf depths, then small-size suprabenthic forms should be more important in the Bay of Fundy than in the two Gulf systems. (5) Scavengers and other opportunists (called here the insensitive guilds) should exhibit equivalent success in all three ecosystems. (6) The relative importance of slope species should be greater in the two most oligotrophic ecosystems, i.e. in the Bay of Fundy and Lower St. Lawrence, than in the eutrophic Baie des Chaleurs. (7) The amplitude of daily vertical migrations, as estimated by relative catches in the upper net and by relative night catches, should be smaller in the Bay of Fundy and Lower St. Lawrence than in the Baie des Chaleurs.

Environments

As shown in Fig. 1, of the three stations compared in this paper, two are located in the Gulf of St. Lawrence, i.e. Baie des Chaleurs (48° 18′ 07″ N; 64° 21′ 22″ W) and Lower St. Lawrence Estuary (48° 45′ 24″ N; 64° 49′ 36″ W), and one in the Bay of Fundy (44° 48′ 00″ N; 66° 35′ 30″ W). Each station

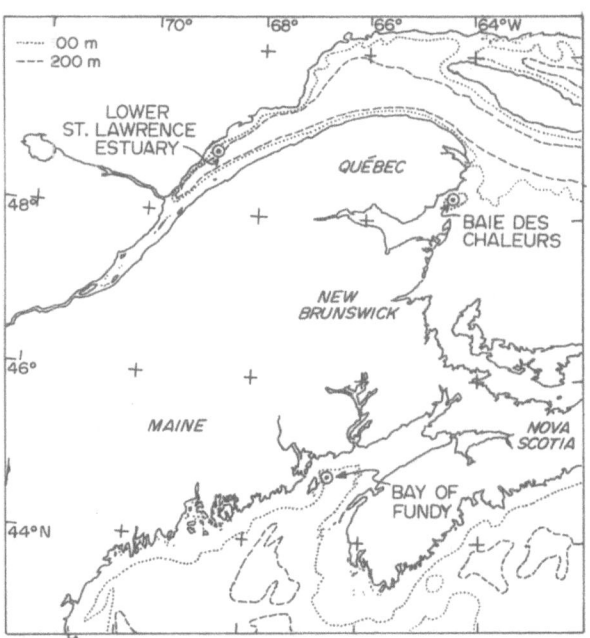

Fig. 1. Positions of the three monitoring stations in the Gulf of St. Lawrence and Bay of Fundy.

was situated on a flat, muddy shelf or basin at a depth of 119 m. The stations were at different distances from the continental slope: the Lower St. Lawrence Estuary station is just 6 km, the Chaleurs station, 90 km, and the Fundy station, 120 km from the 200 m isobath.

At both Gulf stations the bottom is soft mud: 74% silt, 14% sand and 12% clay in the Lower St. Lawrence Estuary; 1% sand, 49% silt and 50% clay in the Baie des Chaleurs. Water temperature is low and fairly constant: $\bar{x} = 3$ °C, range of 1–4 °C in the Lower St. Lawrence Estuary; $\bar{x} = 0.8$ °C, range of 0.4–1.6 °C in the Baie des Chaleurs, and the salinity range is 32–34‰ (Lauzier & Trites, 1958; Besner, 1976; Sainte-Marie & Brunel, 1985). These two stations differ in their primary production regimes. In the Lower St. Lawrence Estuary, the water column is less stable, primary production is lower and blooms occur in July–August, two months later than in the Baie (Steven, 1974; Legendre, 1971; Sinclair *et al.*, 1976). In the section of the Lower St. Lawrence Estuary where our station was located, the daily maximum production reaches 0.8 g C m^{-2} and the annual production is 94 g C m^{-2} (Therriault & Levasseur, 1985). In the Baie des Chaleurs, Legendre (1971) reports a daily maximum production of 3 g C m^{-2} and Steven (1974), Spence & Steven (1974) estimate the annual production at 380 g C m^{-2}.

The Fundy station is located in a net depositional area covered by so-called LaHave clay (Fader *et al.*, 1977). Granulometric results indicate 100% silt and clay (Peer *et al.*, 1980) and bottom photographs show the occasional presence of shell fragments (Wildish & Lobsiger, 1987, Wildish, unpubl.). Endobenthic gammaridean distribution in the Bay of Fundy has been described by Wildish & Dadswell (1985).

Our measurements from May to October 1987 indicate a salinity range of 31–34‰ and a bottom temperature (at approximately 110 m) increasing from 5.2 to 9.3 °C throughout the May–October period. Emerson *et al.* (1986) show similar results at a nearby station. We do not have direct information for our station in the winter but our results are also comparable with those obtained by

Hachey *et al.*, 1954 for the May–October period at the entrance of the Bay of Fundy; we believe, therefore, that their winter observations are similar to those which could be made at our station: bottom temperature at the entrance of the Bay remains at approximately 9–10 °C in November, drops slightly to 7–8 °C in December and below 4 °C only in February, remaining between 1 and 4 °C until May.

The primary production regime in the Lower Bay of Fundy was described by Emerson *et al.* (1986) as follows: daily primary production ranges from a low of 0.042 g C m^{-2} d^{-1} in early June to a high of 0.366 g C m^{-2} d^{-1} in late August with peaks in mid-July and late August; annual production is about 139 g C m^{-2} yr^{-1}.

Current speeds may reach 56 cm sec^{-1} in the southwestern Bay of Fundy (Wildish, 1984). The Bay of Fundy is well known for its large-amplitude tides (up to 15 m in the upper reaches) which are associated with energetic tidal currents. Emerson *et al.* (1986) 'conclude that the largest proportion of secondary production at the southwestern mouth of the Bay of Fundy is realized within the water column and is thus available to a predominantly pelagic fishery'. The authors refer to the possible loss of part of the euphotic zone primary production measured over the LaHave clay through lateral transport by tidal energy to the adjacent Scotian Shelf drift area, where it is utilized by dense horse mussel beds, as one potential explanation of the low endobenthic invertebrate production on LaHave clay.

In summary, as compared to the stations in the Baie des Chaleurs and the Lower St. Lawrence Estuary, the Bay of Fundy station features higher and more seasonally variable temperatures and stronger tidal currents. Coarser sediments are found in the Lower St. Lawrence Estuary although all three stations occur on essentially soft-bottom environments. The Bay of Fundy and the Lower St. Lawrence Estuary primary production patterns are similar: both are far less productive than that of the Baie des Chaleurs, and the major bloom occurs in the summer, 2–3 months later than in the Baie.

Material and methods

The Bay of Fundy suprabenthic samples were taken using a fourth model of the Macer-GIROQ sled described fully by Wildish *et al.* (in press). The new version supports two 0.5 mm-mesh standard plankton nets at 33–73 cm and 109–149 cm from the bottom, provided with flow meters and an opening-closing mechanism. The new sled therefore collects a pair of quantitative suprabenthic samples comparable to those taken with the earlier models (Brunel *et al.*, 1978) in the Baie des Chaleurs and the Lower St. Lawrence Estuary. The frame is however sturdier and allows using the apparatus on coarser and more heterogeneous substrates.

Samples from one station in the Bay of Fundy (Fig. 1) were taken once a month between May and October 1987 on the following dates: May 11–12, June 8–9, July 7–8, August 17–18, September 8–9, October 5–6. To study diurnal variations, four pairs of day samples were generally collected between 09:00 and 17:00 and four pairs of night samples between 21:00 and 01:00 (for a total of 50 upper-net and 50 lower-net samples, comprising 52 day and 48 night samples). Similar May–October, day-and-night time series of paired samples (Table 1) were available for the Baie des Chaleurs in 1969 (Brunel, unpubl., Sainte-Marie & Brunel, 1983, 1985) and the Lower St. Lawrence Estuary in 1970–71 (Besner, 1976; Sainte-Marie & Brunel, 1983).

Seawater samples were collected at a depth of 110 m with Nansen bottles for salinity determination by semi-automatic conductimetry. Temperature was measured at the same depth using reversing thermometers attached to the bottles.

Suprabenthic samples were sorted and the gammaridean Amphipoda identified to species. Depth preferences and feeding habits were determined from a review of the literature contained in Besner (1976), from Wildish & Dadswell (1985), and updated from several published sources.

Size at maturity of selected species was determined using a Wild/Censor electronic micro-length measuring device adapted to a Wild stereoscopic microscope. The animals were straightened dorsally and measured from the tip of the rostrum to the end of the telson. Mature females were identified by the presence of setigerous oostegites and maturity of males was determined by the presence of copulatory papillae and elongated second antennae. With the exception of *Anonyx lilljeborgi*, the species measured were selected because of their adequate frequency in all three ecosystems, their feeding habits (omnivores and detritus feeders), their easily observed secondary sexual characters and the availability of measurements from the two Gulf ecosystems. It would have been interesting to include a scavenger-predator species but none was present in sufficient numbers in the three regions.

Results

Table 1 summarizes available information on total half-year density and total number of species of gammarideans in both nets at the three stations. From the two years of sampling in the Lower St. Lawrence Estuary, it can be seen that these characteristics may vary at the same station from year to year. Since only one year of data is currently available for the Baie des Chaleurs and the Bay of Fundy, and since further sampling in the Lower St. Lawrence Estuary, in 1978–81, demonstrated drastic changes of unknown causes in the composition of the amphipod sub-community (Dauvin & Brunel, in prep.), the present study must not be seen as a final attempt at characterizing the selected stations but rather as a further step in testing the model. An ongoing research program by Brunel is centered on annual variations over 2–8 years, depending on available collections from the three stations, and is designed to examine the robustness of these ecosystem differences.

Total density in the Bay of Fundy

As shown in Table 1, the average density of gammarideans in the Bay of Fundy is of 27.1 individuals per 100 m^3 in the upper net and 430.9 in

Table 1. Total half-year density, weighted for volume of water filtered, average unweighted half-year density and species richness of Gammaridea Amphipoda at the three stations, for the indicated year

	Bay of Fundy 1987	St. Lawrence Estuary 1970[c]	St. Lawrence Estuary 1971[c]	Baie des Chaleurs 1969[d]
Total density[a] ($n \cdot 100 \text{ m}^{-3}$)[b]				
Upper net	29.55	18.87	15.1	171.8
Lower net	419.97	288.53	107.6	1127.2
Both nets	449.52	307.4	122.7	1299
Average half-year density ($n \cdot 100 \text{ m}^{-3}$) (standard deviation)				
Upper net	27.1 (63.1)	12.9 (19.8)	17.5 (18.8)	174.9 (223.9)
Lower net	430.9 (310.2)	203.9 (160.5)	108.6 (79.7)	1173.7 (690.5)
Total number of species				
Upper net	33	38	26	38
Lower net	48	54	61	72
Both nets	50	56	66	76
Volume of water filtered (m^3)				
Upper net	4795	3550	10750	2536
Lower net	5048	3550	10500	3304
Both nets	9843	7100	21250	5840
Number of samples				
Upper net	50	29	55	19
Lower net	50	29	54	22

[a] 100 × Sum of number of individuals divided by Sum of water volumes filtered, for all day and night samples from May to October
[b] n = number of individuals
[c] From Besner 1976
[d] From Brunel unpubl.

the lower net. The standard deviations of 63.1 in the upper net and 310.2 in the lower net are due to large night-and-day and seasonal differences in density.

Table 2 indicates higher densities at night in both nets. In addition, the night/day ratios and the proportions in upper net further suggest that although the density remains low in the upper net, there is a definite increase of one order of magnitude at night. These trends are maintained when looking at the monthly results (Fig. 2 and 3), i.e.

Table 2. Total gammaridean half-year density in the lower and upper nets, weighted for volume of water filtered, Bay of Fundy, 1987

	Density		Proportion in upper net[a]
	Lower net $(n \cdot 100 \text{ m}^{-3})^{b}$	Upper net $(n \cdot 100 \text{ m}^{-3})$	
Night + day	420.0	29.6	0.07
Night	517.2	54.5	0.10
Day	319.5	4.1	0.01
Night/day ratio	1.6	13.2	

[a] $\dfrac{\text{Density in the upper net}}{\text{Density in both nets}}$

[b] n = number of individuals.

with the exception of June in the lower net and August in the upper net, the envelope of day values does not extend above the night envelope.

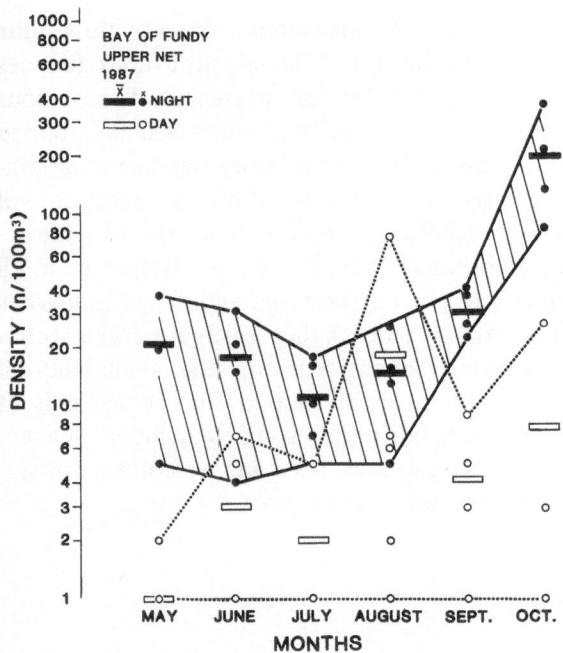

Fig. 3. Temporal variations of gammaridean total density in the upper net, Bay of Fundy, 1987 (n = number of individuals).

Density of major species and guilds in the Bay of Fundy

The contribution of the ten most abundant species to the structure of the gammaridean sub-community can be assessed by comparing their seasonal density variations in the lower net (Fig. 4–5) with those of total gammaridean density at the same level (Fig. 2). The species have been grouped according to similarities in the shape of their density curves. In Fig. 4, the overall seasonal trend is either downward (*Stenopleustes inermis*) or little apparent, the minimum occurs in September and the maximum generally in July. In Fig. 5, the overall seasonal trend is upward (except for *Monoculodes packardi*), the September low shifts to August, and the maximum occurs in October; all species show a density increase from September to October, and all but *Haploops fundiensis* and *Monoculodes packardi* show some increase from May to July.

The ten species are segregated according to their swimming activity, as studied by Sainte-

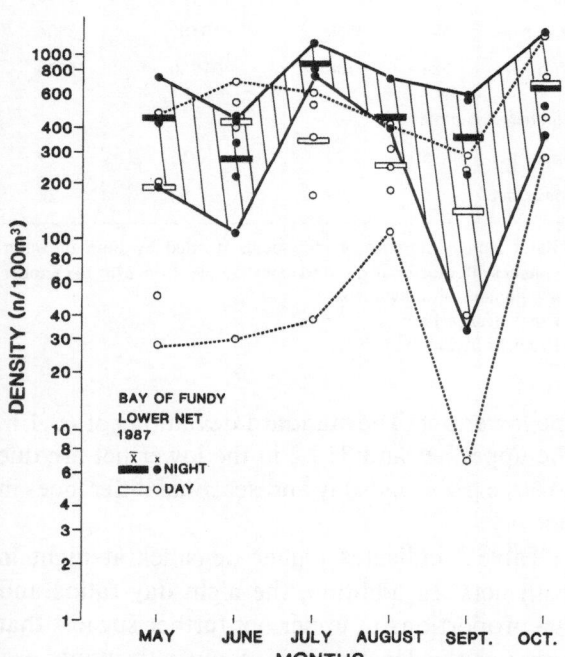

Fig. 2. Temporal variations of gammaridean total density in the lower net, Bay of Fundy, 1987 (n = number of individuals).

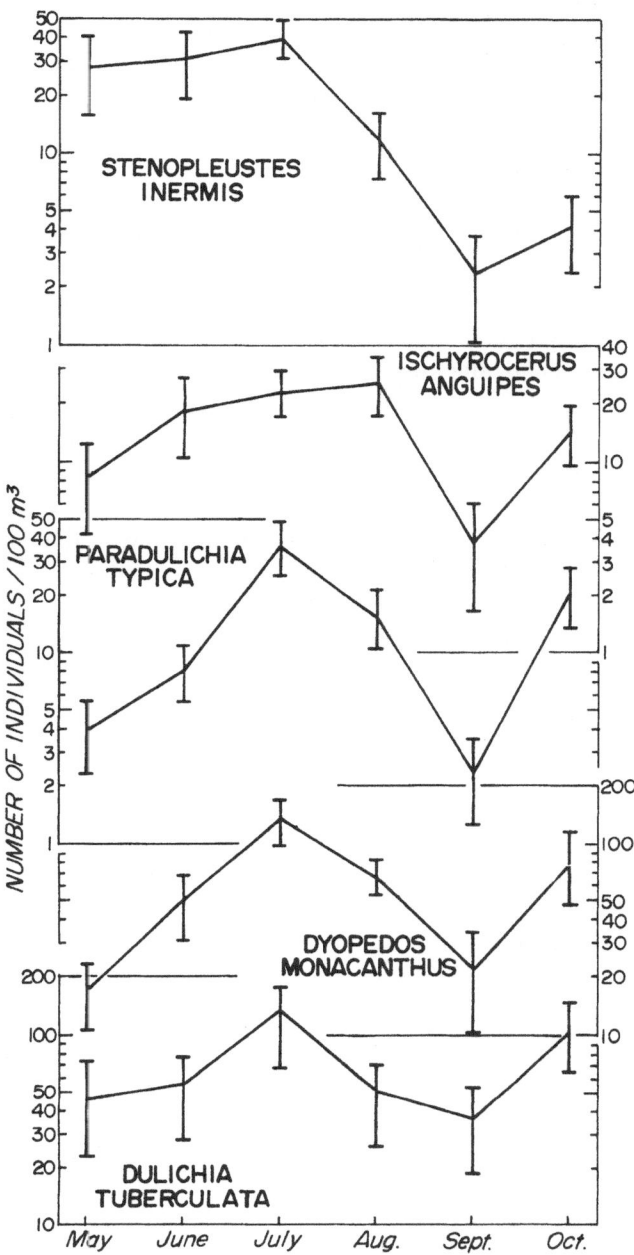

Fig. 4. Monthly average density in the lower net of five dominant gammaridean species belonging to the more epibenthic swimming guild, Bay of Fundy, 1987.

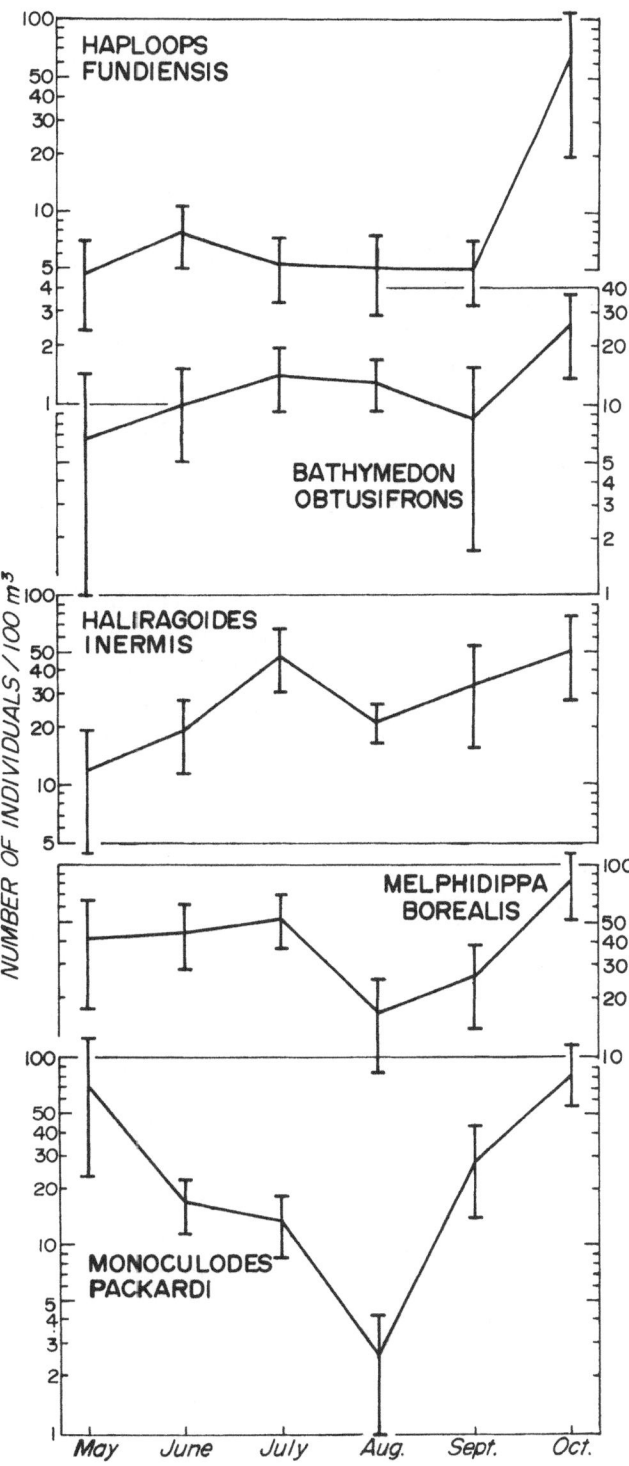

Fig. 5. Monthly average density in the lower net of five dominant gammaridean species belonging to the more suprabenthic swimming guild, Bay of Fundy, 1987.

Marie and Brunel (1985) on 51 species in the Baie des Chaleurs: the more epibenthic species, including the three Podoceridae, are grouped in Fig. 4, whereas Fig. 5 includes the more suprabenthic forms. This suggests the possibility of swimming guilds with some common seasonal pattern of swimming activity or some common swimming response to the oncoming sled. In fact, the significant maxima of *Monoculodes packardi* in May and of *Haploops fundiensis* in October can be traced in part to male nocturnal nuptial swarming, a behavior well known in the Ampeliscidae (Mills, 1967) and Cumacea (Messier, 1974). In October at night, *H. fundiensis* accounted for 20% (mature males for 17%) of gammaridean density in the lower net, and as much as 86% (67% for mature males) in the upper net. In May, nocturnal density of *M. packardi* was also very high (25%) in the lower net (18% for mature males) and twice as high (58%) in the upper net (44% for mature males).

To a large extent, swimming guilds account for the differences between the nets (Fig. 2–3) in seasonal trends of total density. It can be noted that the seasonal changes in density of epibenthic species in the lower net (Fig. 4) are quite similar to those of total gammaridean density in the same net (Fig. 2), whereas those of the more suprabenthic species in that net (Fig. 5) are quite similar to those of total gammaridean density in the upper net (Fig. 3), i.e. at a different level. In addition, the epibenthic guild represents 44.0% of the total density in the lower net and only 5.8% in the upper net, whereas the suprabenthic guild accounts for 33.3% in the lower net and as much as 87.3% in the upper net. This confirms for the Bay of Fundy the reality of swimming guilds observed in more detail in the Baie des Chaleurs (Sainte-Marie & Brunel, 1985).

Since densities are much higher in the lower net than in the upper net at all times, we have used the former to study the trophic structure of the sub-community in the three ecosystems. In the Bay of Fundy (Table 3 & appendix), the dominant trophic guild, the suspension-feeders, is comprised of four families, including the Podoceridae which is also the dominant family of the whole Fundy gammaridean sub-community. The next trophic guild, the detritus-feeders, is dominated by five small-size species of Oedicerotidae whereas the plankton-feeding guild (which is represented in the other two ecosystems) is completely absent. Other guilds account for only 2.3 to 4.3% of the total density.

Total density in the three ecosystems

From Table 1, one can see that the total half-year density in both nets is highest in the Baie des Chaleurs, followed by the Bay of Fundy and then by the Lower St. Lawrence Estuary. An analysis of variance performed on log-transformed data

Table 3. Gammaridean trophic guilds in the lower net, Bay of Fundy, 1987

Trophic guilds	Density ($n \cdot 100 \text{ m}^{-3}$)[a]	% density	Number of species	% species
Suspension feeders	217.84	51.9%	8	16.7%
Detritus feeders	141.37	33.7%	20	41.7%
Mixed suspension & detritus feeders	12.74	3.0%	2	4.2%
Plankton feeders	0	0.0%	0	0.0%
Omnivores	17.85	4.3%	5	10.5%
Commensals	9.65	2.3%	4	8.3%
Scavengers-predators	14.03	3.3%	6	12.5%
Unknown	5.37	1.3%	3	6.3%

[a] n = number of individuals.

shows a significant difference between the Baie des Chaleurs mean and those obtained in the other two regions. In addition, in the lower net, the Bay of Fundy mean is significantly ($\alpha = 0.5$) higher than both means obtained in the Lower St. Lawrence Estuary. There is no significant difference between the mean of the Bay of Fundy and those of the Lower St. Lawrence Estuary in the upper net.

Size at maturity

Table 4 indicates that size at maturity is smaller at the Fundy than at the other two stations for *Hippomedon propinquus* and the females of *Mono-*

culodes packardi. In addition, for the same taxa, species of smaller size are found in the Bay of Fundy. For example, *Anonyx debruyni* (generally 14–21 mm at maturity, Steele & Brunel, 1968) appears to replace, in the scavenger-predator niche, the larger *Anonyx makarovi* (average size at maturity 33.4 mm, Gagnon, 1983) which occurs at the two Gulf stations. In addition, there are no large-size Oedicerotidae (i.e. the larger species of the Lower St. Lawrence Estuary or the Baie des Chaleurs, such as *Arrhis phyllonyx*, *Paroediceros lynceus*, *Monoculodes longirostris*) and no gammaridean plankton-feeders in the Bay of Fundy, and all cumaceans therein are of small size and may be grain lickers or small-particle deposit-feeders.

Table 4. ANOVA comparison of the average size at maturity of selected gammarideans at the three stations ($\alpha = 0.05$).

Species		Males	Females
Hippomedon propinquus	Regions	E = C > F	E = C > F
	Average size at maturity (mm)	9.4 = 9.3 > 8.7	9.7 = 9.9 > 8.7
	Number of individuals	E = 87, C = 87, F = 190	E = 72, C = 72, F = 72
Monoculodes packardi	Regions	C > E = F	E > F
	Average size at maturity (mm)	6.3 > 5.7 = 5.5	5.7 > 5.2
	Number of individuals	C = 10, E = 13, F = 59	E = 10, F = 77
Bathymedon obtusifrons	Regions	C = F	C = F
	Average size at maturity (mm)	4.3 = 4.2	4.4 = 4.3
	Number of individuals	C = 39, F = 117	C = 24, F = 98
Anonyx lilljeborgi	Regions	E = F	E = F
	Average size at maturity (mm)	14.8 = 14.0	13.9 = 13.3
	Number of individuals	E = 29, F = 54	E = 20, F = 8

E = Lower St. Lawrence Estuary.
F = Bay of Fundy.
C = Baie des Chaleurs.

Species richness

The largest number of species was found at the least intensively sampled station, in the Baie des Chaleurs, whereas only 50 species occurred at the Fundy station where twice as much water was filtered by the sampling nets (Table 1). Species richness in the Lower St. Lawrence Estuary is intermediate for both years of sampling. To obtain 66 species in the Lower St. Lawrence Estuary in 1971, Besner (1976) had to triple the 1970 sampling effort, adding only ten additional species. It is well known that sampling effort affects estimates of diversity and species richness (Sanders, 1968; Smith & Grassle, 1977). However, from the data of Table 1, the Chaleurs station harbors the richest community and that of Fundy, the poorest. This is further confirmed by rarefaction curves (Fig. 6) which tend toward a maximum of approximately 45 species for the Bay of Fundy, 65 for the Lower St. Lawrence Estuary in 1971 and 70 for the Baie des Chaleurs.

Table 5. Usual depth preferences of the Fundy species sampled in either net, 1987

Category of depth preference	Number of species	% species
Shallow littoral (0–30 m)	3	6%
Shallow littoral & deep shelf (0–150 m)	1	2%
Deep shelf (30–150 m)	16	32%
Southern deep shelf (30–150 m)	4	8%
Deep shelf & slope (30–400 m)	8	16%
Slope (150–400 m & +)	9	18%
Shallow littoral & slope (0–30 & 150–400 m)	3	6%
Eurybathic (0–400 m & +)	2	4%
Unknown (incomplete identification)	4	8%

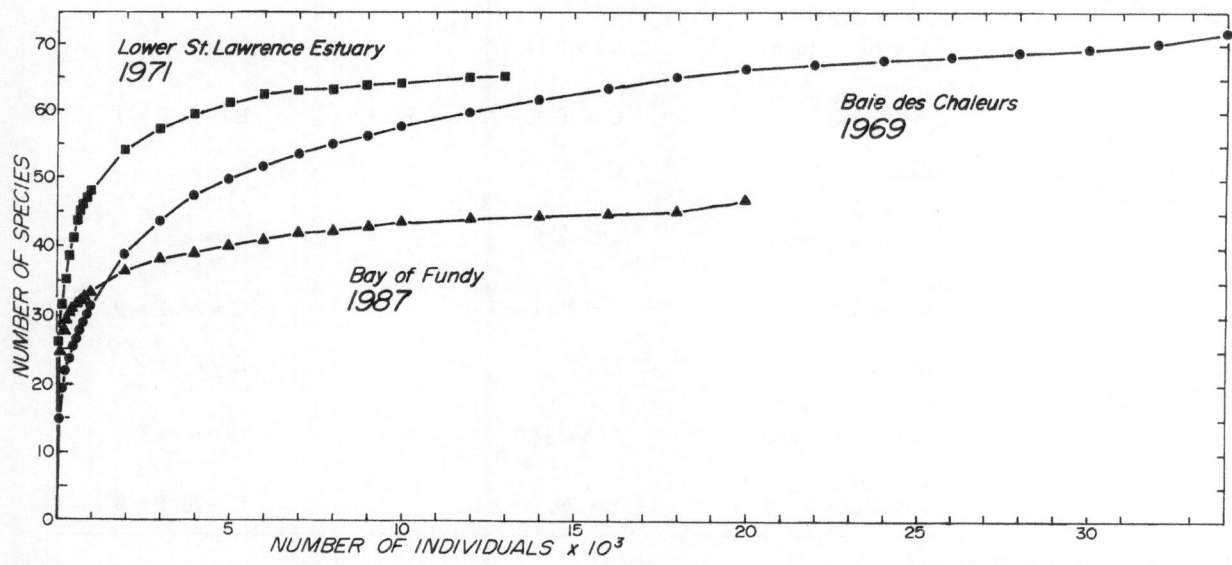

Fig. 6. Rarefaction curves predicting species richness of Amphipoda Gammaridea at the three monitoring stations from May to October.

Table 6. Total half-year density of peracarid and eucarid Crustacea in both nets, weighted for volume of water filtered, Bay of Fundy, 1987.

Taxon	Lower net		Upper net	
	Density $(n \cdot 100 \text{ m}^{-3})^a$	%	Density $(n \cdot 100 \text{ m}^{-3})$	%
Amphipoda gammaridea	419.97	44.3%	29.55	23.1%
Cumacea	278.61	29.4%	14.47	11.3%
Amphipoda Caprelloidea	102.29	10.8%	1.36	1.1%
Mysidacea	74.15	7.8%	14.54	11.4%
Euphausiacea	49.91	4.7%	66.26	51.8%
Isopoda	14.78	1.6%	0.56	0.4%
Tanaidacea	11.15	1.2%	0.23	0.2%
Decapoda Caridea	1.23	0.1%	0.83	0.6%

[a] n = number of individuals.

Slope species

The model also predicts invasion of the shelf by species typically found on the slope, and therefore better adapted to oligotrophic conditions. In order to verify that hypothesis we have classified the Fundy species in accordance with their usual depth preferences in typical North Atlantic environments (Table 5). True slope species represent 18% of the total number of species in the Bay of Fundy.

Density of all malacostracan Crustacea

As presented in Table 6, the Bay of Fundy malacostracan Crustacea are not represented in similar densities or proportions in the two nets. With the exception of euphausiids, they occur in greater densities in the lower net. Euphausiids occupy the first rank in the upper net whereas the lower net is dominated by gammarideans. Cumaceans follow gammarideans in both nets whereas caprellids, third in the lower net, virtually disappear from the upper net. Consequently, and in spite of a much lower total density than in the lower net, Mysidacea show a higher rank in the upper net. Other taxa occur in small densities in both nets.

Discussion

Species richness and success of gammarideans in the three ecosystems

The number of gammaridean species increases gradually with sampling effort in the Baie des Chaleurs (Fig. 6), which reflects a fine division of the resources among a large number of species, mainly Oedicerotidae (8 species) (Sainte-Marie & Brunel, 1985), apparently well adapted to the stable environment of the locality. This situation contrasts with the massive dominance of one species, *Rhachotropis oculata* (Besner, 1976), in the Lower St. Lawrence Estuary and of a different family in the Bay of Fundy, the Podoceridae, which is represented by only four species (appendix). It is interesting to note that the Baie des Chaleurs and the Lower St. Lawrence Estuary support approximately 130 species each, considering all suprabenthic taxa, occupying niches similar to those partitioned among gammarideans. It may well be that one could find a comparable total species richness in the Bay of Fundy if the specific composition of all non-gammaridean taxa had been studied in our samples.

The Baie des Chaleurs, showing homogeneous and stable physico-chemical characteristics, as well as high primary production, appears to bear the attributes of a biologically accommodated community (Sanders, 1968), i.e. large number of species co-existing in a physically stable environment. On the other hand, the Fundy and the Lower St. Lawrence Estuary communities are exposed to greater variations of temperature and stronger tidal currents, and are in this way similar to shallow-water marine communities which are also characterized by pronounced changes in physico-chemical conditions and an impoverished fauna. As predicted by the stability-time hypothesis (Sanders, 1968), we have found a lower number of species in our two oligotrophic soft-bottom communities, particularly in the Bay of Fundy which does support the smallest number of species, while presenting greater seasonal temperature variations and stronger tidal mixing than the Lower St. Lawrence Estuary. Both communities appear to be physically-controlled. However, when compared to shallow estuarine gammaridean sub-communities in the Bay of Fundy, the LaHave clay infaunal amphipod sub-community may appear more biologically accommodated (Wildish & Dadswell, 1985).

All three communities are probably equally young, from a historical biogeographic viewpoint. Their fauna must have immigrated from cold North Atlantic shelves or upper slopes after the Laurentide Ice Sheet had retreated sufficiently to expose their sea areas to normal oceanographic processes, some 13 to 14 000 years ago. The depths and tidal currents prevailing over the three communities, however, were probably quite different for several thousand more years, because of changes in global sea level and differences in continental isostasy. Because of coastal submergence of the Maritime provinces, it was not until about 6000 years ago that the sea level had risen sufficiently to establish the present strong tidal regime in the Bay of Fundy (Fader *et al.*, 1977) where, some 8400 years B.P., depths were still 60 m shallower than now (Dyke & Prest, 1987). On the other hand, because of continental rebound to the north, the Lower St. Lawrence Estuary in this period was at least 110 m deeper than now (Dyke & Prest, 1987) and perhaps as much as 135 m deeper (Dionne, 1990), so that our shelf community was at slope depths at that time.

The Baie des Chaleurs in the same period was some 20 m shallower, and may therefore have enjoyed a longer period of stability than the other two communities. Whereas all factors can account for the species richness and productivity of the Baie des Chaleurs community, it is more difficult to assess the intertwined effects of present (temperature, tidal current, and primary production patterns) and historical factors (age and evolution of the habitat) on the lower species richness and composition of the Bay of Fundy gammaridean sub-community, as compared with that of the Lower St. Lawrence Estuary.

Our data appear to support the productivity theory (Connell & Orias, 1964) which predicts: the more food produced, the greater the diversity. Unlike Sanders (1968), who reports on communities of high diversity in low-productivity but physically stable environments (e.g. the deep-sea), we found that our richest shelf community, in the Baie des Chaleurs, is at the same time eutrophic and stable, while the oligotrophic systems are more physically unstable. It is therefore impossible to distinguish the trophic from the physical factors in our three shelf ecosystems. Ideally, a further study should focus on a shelf ecosystem characterized by low productivity and high stability, as in the deep sea.

Our shelf communities, however, do not differ from many deep-sea communities which receive seasonal pulses of pelagic food. In the deep sea, such fall-outs create small-scale (mm to m) elusive patches of food, more widely spaced than those of shallow-water environments, and which give rise to high diversity (Grassle, 1989). Shelf bottom communities presumably offer a heterogeneous distribution of patches of food on a scale intermediate between those described by Grassle (1989) for shallow and deep-sea environments. In shallow environments, small-scale patches of food are likely obliterated by large-scale disturbances, thereby compromising the chances of the fauna reaching high diversity (Grassle, 1989). Our oligotrophic environments, the Bay of Fundy and the Lower St. Lawrence Estuary, are also more affected by tidal currents than the Baie des Chaleurs. The LaHave clay infaunal amphipods may suffer depopulation events (Wildish & Dadswell, 1985) which would also suggest instability. Even taking into account the spatial component of the deep sea model, we would expect low diversity in the oligotrophic shelf ecosystems. Small-scale sampling in the immediate vicinity of our stations would help to elucidate the contribution of spatial variability to diversity.

Total density of gammarideans is clearly highest in the Baie des Chaleurs (Table 1), where primary production is the highest and bottom water the coldest, two factors which may lead to the largest observed size at maturity (Table 4). Interpretation of differences between the two less productive ecosystems, however, is less straightforward. The higher density at the Bay of Fundy station is made up of individuals smaller than those of the Estuary station (Table 4), so that standing stock might be comparable in the two communities. Yet small size here does occur in water where temperature reaches highest values (up to 10 °C in November), which are likely to accelerate growth and limit size at maturity. On the other hand, despite somewhat higher primary production in surface waters of the Bay of Fundy, little of it may actually reach the LaHave clay bottom (Emerson *et al.*, 1986), so that trophic and thermal factors may again combine in accounting for small size. There is thus a clear need for more extensive comparisons on sizes and growth, if not on biomass, and on the taxa other than Gammaridea, in order to derive production data for the three communities, and estimate the relative 'success' of gammarideans.

Our results may be compared with those of Dayton & Oliver (1977) who found higher infaunal density in the eutrophic East McMurdo Sound (Antarctic) than in the oligotrophic adjacent West Sound. Food supply to the bottom community, as controlled by the regime of primary production, appeared to be the major factor affecting density. Similarly, in our study, high suprabenthic density and species richness in the Baie des Chaleurs soft-bottom community appear to depend largely on a rich pelagic food supply, much of which reaches the bottom.

The low suprabenthic density found on the silt-clay sediments of the Bay of Fundy (Table 1) corroborates results of Emerson *et al.* (1986) on the low macrobenthic infaunal productivity in the same area and supports their conclusion that secondary production here is realized in part within the water column and made available to the pelagic fishery; another part of primary production in the lower Bay of Fundy is diverted toward secondary production by horse mussels on shallower grounds covered by Scotian Shelf drift sediments. It is interesting to note that the Lower St. Lawrence Estuary also supports a mainly pelagic fishery and cetacean populations, whereas the very productive Baie des Chaleurs supported a significant groundfish fishery before overfishing brought about its decline.

Representation of gammaridean trophic guilds and possible competition from other malacostracan Crustacea

The trophic guild representation of the Bay of Fundy differs from that in both the Baie des Chaleurs and the Lower St. Lawrence Estuary. In the Baie des Chaleurs, the Oedicerotidae massively dominate, which propels the detritus-feeding guild far above any other, but still leaves some room for suspension and plankton feeders (Sainte-Marie & Brunel, 1985: Brunel, unpub.). All three trophic guilds are represented in the Lower St. Lawrence Estuary (Besner, 1976) but the order of dominance is different: plankton feeders, detritus feeders and finally, suspension feeders. Although each region has a unique arrangement of the three sensitive guilds, their total relative success is close and varies from 83% to 94% at our three stations.

The suspension feeding guild certainly dominates the Fundy community. The massive presence of the Podoceridae is possibly due to that of shell debris capable of supporting bryozoans and hydroids. Sediment 'whips', reported attached to hydroids by Moore & Earll (1985), serve as vantage points to the Podoceridae for suspension feeding. The Podoceridae are also known to live on algae, bryozoans, polychaete tubes, and sea urchin spines (Laubitz, 1977). Although active construction has not been observed, there is strong evidence that the Podoceridae build the whips themselves (Moore & Earll, 1985). We do not know whether the Baie des Chaleurs or the Lower St. Lawrence Estuary sediments support epifauna of a similar nature and abundance but the Podoceridae are definitely less successful, showing medium rank in the Lower St. Lawrence Estuary (Besner, 1976) and negligible frequency of occurrence in the Baie des Chaleurs (Sainte-Marie & Brunel, 1985). These conclusions on Podoceridae are supported by our results on caprellids (Table 6). They are known to occupy a niche similar to that of the Podoceridae (Moore & Earll, 1985) and are also very successful in the Bay of Fundy, while they are considered rare at the Lower St. Lawrence Estuary (Besner, 1976) and Chaleurs (Brunel, 1968) stations. It can be said that the suspension-feeding trophic guild as a whole, including the Podoceridae but also stronger swimming forms, such as the Melphidippidae, is more successful in the Bay of Fundy than in the Lower St. Lawrence Estuary or the Baie des Chaleurs. The guild appears to be favored by the strong tidal currents of the Bay of Fundy which most probably frequently resuspend the organic material that is available.

In the Baie des Chaleurs and the Lower St. Lawrence Estuary, detritus feeders compete successfully with their suspension-feeding counterparts for their shared resource of seston and resuspended material. The Oedicerotidae, of delicate morphology, forage easily in the uniform and smooth sediments of the Baie des Chaleurs where they reach a high density (Sainte-Marie & Brunel, 1985; Brunel, unpubl.). In the Lower St. Lawrence Estuary, coarser sediments appear to limit the success of that family, although it still dominates the detritus-feeding guild (Besner, 1976). The Fundy station harbors only five species of Oedicerotidae, three of which occur in very low densities, in spite of the very smooth mud (100% silt-clay). This might be due to competition for limited resources from suspension-feeders or to the presence of shell debris mixed with the sediments.

A look at detritus and suspension feeding gammarideans provides only a partial picture of the resource allocation since malacostracan crustaceans occupy similar niches. The available data do not yet allow us to delve deeply into this issue, but it is interesting to point out that the Cumacea enjoy in the Bay of Fundy (Table 5) a relative success comparable to that which they obtain in the Baie des Chaleurs and an absolute success definitely greater than that which they have in the Lower St. Lawrence Estuary. Fundy species are of very small size and hence are probably grain lickers. The Baie des Chaleurs and the Lower St. Lawrence Estuary support both grain lickers and large-size detritus foragers, the largest specimens being found in the Baie (Brunel & Messier, unpubl.). This might be related to the size and quality of seston reaching the bottom. The Baie would presumably provide a larger quality of first-quality food to the Oedicerotidae and large Cumacea, which are both very successful and effective foragers. On the other hand, from our preliminary observations, the oligotrophic Bay of Fundy appears to support smaller-size copepods and presumably fecal pellets of smaller size. These pellets would rapidly be incorporated in the sediments and be made available to the small grain-licking Cumacea. This might also explain why large-size Oedicerotidae are excluded from the Bay of Fundy. The Lower St. Lawrence Estuary does not appear to be a favorable environment to Cumacea or Oedicerotidae since both taxa remain at lower density there (Besner, 1976) than in the Bay of Fundy. It is possible however that larger size Cumacea can survive in the Lower St. Lawrence Estuary by occupying some of the niches left vacant by the Oedicerotidae.

The Lower St. Lawrence Estuary also stands out by the tremendous success of *Rhachotropis oculata*, a plankton-feeder. According to Desroches (1985), the species can dominate the community in the Lower St. Lawrence Estuary by using favorable currents to migrate seasonally to richer inshore waters for brooding and initial development of the juveniles. We can add that *R. oculata* is a much smaller species than the dominant convergent plankton feeder of Chaleurs, *Acanthostepheia malmgreni*. Since plankton feeders are absent from the Bay of Fundy (Table 3), could not the Bay be unable to support the food requirements of any plankton-feeding Gammaridea on account of both its oligotrophy and lack of a suitable inshore transport mechanism? The suprabenthic plankton feeding guild is left vacant by amphipods and mysidaceans and is apparently occupied by euphausiids, including *Meganyctiphanes norvegica*, a large and strong planktonic swimmer and a partial plankton feeder (Berkes, 1973). Euphausiids occur in high density in both nets (Table 6) and in proportions comparable to those observed in the Baie des Chaleurs (Brunel, unpubl.). With strong swimming abilities and swarming habits, euphausiids are apparently successfully taking advantage of the scarce resources at the suprabenthic level in the Bay of Fundy and possibly contributing to the exclusion of gammaridean plankton feeders.

Contrary to the model, we have found that opportunistic omnivores and commensal species are sensitive to the trophic conditions at all stations. The most successful omnivorous guild is found in the Lower St. Lawrence Estuary where *Hippomedon propinquus*, *Orchomenella pinguis* and *Anonyx lilljeborgi* enjoy high rank in the community (Besner, 1976). The success of the omnivorous guild, particularly *Hippomedon propinquus* (Lamarche & Brunel, 1987), in the Lower St. Lawrence Estuary is difficult to understand. Robert (1974) had also found that opportunistic molluscs predominated in the Lower St. Lawrence Estuary. This would suggest that the latter might even be a less predictable environment than the Bay of Fundy in terms of food sources to benthos. The commensal species are virtually absent from the Baie des Chaleurs (Brunel, unpubl.) but they account for 2.3% of the fauna in the Bay of Fundy (Table 3) and 3.5% in the Lower St. Lawrence Estuary (Besner, 1976). In addition, the dominant Amphilochidae of the Bay of Fundy are replaced in the Baie des Chaleurs by a similar family, the Stenothoidae, whereas the Lower St. Lawrence Estuary seems to accommodate both families. Although we know that both families live in association with other inverte-

brates, including hydroids and bryozoans, presumably for feeding (Besner, 1976), the available information on their ecology does not explain the observed difference in distribution. This question calls for more data on the ecology of each species.

Scavengers-predators enjoy a similar relative success in all three regions, as predicted. They reach 3.3% in the Bay of Fundy, a value which compares with the 2.7–2.9% obtained in the Baie des Chaleurs (Brunel, unpubl.) and the Lower St. Lawrence Estuary (Besner, 1976). However, the species involved differ. It appears that *Anonyx makarovi* of the Baie and Lower St. Lawrence Estuary stations (Gagnon, 1983) is replaced by the smaller-size slope species *Anonyx debruyni* at the Bay of Fundy station. In addition, *Tmetonyx cicada*, which shows similar success in the Bay of Fundy and the Lower St. Lawrence Estuary, but is virtually absent from the Baie des Chaleurs, is a slope species.

In short, our results indicate that overall relative success (total % density) of the sensitive guilds is the same in all three regions while absolute density of each guild is highest in the Baie des Chaleurs. However, the density and relative importance of each sensitive guild varies between the regions. This appears to be due to the slight variations in local physico-chemical conditions which occur concurrently with the differing primary production regimes. On the other hand, the production regimes do not appear to affect the relative success of the scavengers-predators guild but does affect total density.

Finally, in contradiction to the model, the omnivores and commensal guilds seem to be sensitive to the primary production patterns in terms of both relative and absolute success. The overall bottom community response to oligotrophy, discussed in the previous section in terms of total density, appears to be mainly related to the effect of food availability on detritus, suspension and plankton feeding guilds. This may simply reflect the fact that most gammarideans are either detritus- or suspension-feeders. Since most endobenthic forms also belong to the same trophic groups, we are tempted to suggest that the infauna at the Fundy and Chaleurs stations should follow trends similar to those found in the suprabenthic layer. It should also be true for the Lower St. Lawrence Estuary, if *Rhachotropis oculata* is excluded from the comparison. Again, to verify this hypothesis, taxa other than gammarideans should be considered.

Invasion of the shelf by slope species in oligotrophic environments

Our data indicate that slope species of various trophic guilds invade the Bay of Fundy shelf station (Table 5), in a proportion similar (18%) to that found in the Lower St. Lawrence Estuary (16%); both values are much greater than the 2% found in the Baie des Chaleurs (Besner, 1976). The slope is closer to the Lower St. Lawrence Estuary station (6 km) than to the Chaleurs (90 km) and Fundy (120 km) stations. In the Lower St. Lawrence Estuary this factor probably further enhances the ability of slope species to invade the shelf but it cannot contribute in the Bay of Fundy. Brunel & Messier (in prep.) have found a similar invasion of the shelf by slope cumaceans in the Lower St. Lawrence Estuary. It is also of interest to note that the dominant Mysidacea of the Fundy station appears to be *Pseudomma roseum*, another slope invader. Our results therefore appear to support the model prediction that slope species can compete efficiently with shelf species in their preferred depth zone in oligotrophic environments.

Amplitude of migrations

The species of a suprabenthic community perform migrations of various amplitudes which may translate into an increase or a decrease of density in the nets, i.e. the number of individuals captured at the sampled level depends on the extent of the epi- and endobenthic contributions to the suprabenthos at night (Besner, 1976) and on whether migrations bring individuals above the sampled level for a more or less prolonged period of time (Kaartvedt, 1989). With a measurable increase of

gammaridean densities at night in both nets (Table 2), our Bay of Fundy data suggest that migrations are occurring, at least to some extent, within the sampled level (i.e., 1.5 m from the bottom). This might be indicative of low-amplitude migrations but it could only be confirmed by concurrent sampling at the endobenthic, epibenthic and planktonic levels. For the Baie des Chaleurs, plankton data are available and confirm that the amplitude of migrations extends far above the two suprabenthic nets while available data at the endo-, epi- and suprabenthic levels suggest a lower amplitude in the Lower St. Lawrence Estuary (Besner, 1976; Sainte-Marie & Brunel, 1985; Brunel, unpubl.).

Various explanations have been considered to explain the existence of vertical migrations at night: (1) reduced nocturnal predation by visual predators (Hobson & Chess, 1976; Porter & Porter, 1977; Porter *et al.*, 1977); (2) use of vertical movement to eventually reach new epibenthic feeding sites by lateral transport (Sainte-Marie & Brunel, 1985); (3) nuptial swarming seasonally restricted to the breeding season of the species and most likely superimposed on the regular pattern of daily migrations (Sainte-Marie & Brunel, 1985). In the Bay of Fundy, nuptial swarming of *Haploops fundiensis* and *Monoculodes packardi* affect total density in the nets but this influence is confined to October, in the first case and May in the latter. Reduced predation at night and searching for better feeding conditions remain as possible explanations of migrations but do not explain the difference of migration amplitudes inferred from the data, between the Baie des Chaleurs and the other two ecosystems.

Sainte-Marie & Brunel (1985) suggested a density-dependent response to crowding and competition within the suprabenthic community of the Baie des Chaleurs, which would induce a greater amplitude of migrations for ultimate efficient horizontal dispersal. Kaartvedt (1989) points out that current velocities decrease exponentially towards the sediment, which means that a low amplitude of migrations would thereby limit vulnerability to long distance advection. If migration patterns in the suprabenthos are an ecological trade-off between short distance dispersal (within the habitat) and long-distance dispersal (out of the habitat), as suggested by Kaartvedt (1989), then migrations of high amplitude in the Baie des Chaleurs present little risk: dispersion will probably occur within the habitat since the mud basin on which the community lives is large and homogeneous and since currents are not as strong as in the Lower St. Lawrence Estuary and the Bay of Fundy. In addition, some lateral transport would facilitate full exploitation of the habitat in the crowded Baie des Chaleurs.

Conclusions

(1) Species richness varies in the three ecosystems, the richest community being found in the very productive Baie des Chaleurs. The primary production regime appears to affect species richness through pelagic-benthic coupling.

(2) Average half-year density was significantly smaller in the oligotrophic environments. In addition, the higher density in the Bay of Fundy, as compared to the Lower St. Lawrence Estuary, appears to be compensated by smaller size at maturity or the selection of smaller species of the same taxa.

(3) Total density of the detritus, suspension and plankton feeding guilds differs in the three regions while their total relative success is the same. However, relative representation of the guilds is unique to each region.

(4) The Bay of Fundy community appears to adapt to strong vertical mixing of water masses by supporting a very successful suspension feeding trophic guild.

(5) The scavenger-predator guild exhibits equivalent relative success in all three ecosystems.

(6) Contrary to the model prediction, opportunists (commensal and omnivorous species) are clearly more successful in the Lower St. Lawrence Estuary than in the other two ecosystems.

(7) In the two oligotrophic environments, slope species compete more efficiently with shelf species in their preferred depth zone than in the eutrophic Baie des Chaleurs.

(8) The data are not inconsistent with the prediction that the amplitude of migrations should be lower in the Bay of Fundy, as was found for the Lower St. Lawrence Estuary, than in the Baie des Chaleurs, but further sampling will be required to adequately test this hypothesis.

Acknowledgements

We wish to express our thanks to the Canadian Department of Fisheries and Oceans for providing ship time. Scholarships to A. Chevrier from the Fonds de Bourses de Sciences Biologiques, Université de Montréal and the Natural Sciences and Engineering Research Council of Canada, and a grant from the latter to P. Brunel, supported this work. We also wish to thank S. Cordeau, G. Chevrier, B. Frost, D. Gaulin, Simon Lachance, Sylvain Lachance, C. Legault, J. Tremblay and A. Wilson for collections at sea, sorting, identification or data entry. Part of this work was funded through a Summer Canada project. Contribution to the program of GIROQ (Groupe interuniversitaire de recherches océanographiques du Québec).

Appendix: Relative importance of species in the lower and upper nets, Bay of Fundy, 1987

Species	Lower net			Upper net			Family code[a]	Feeding habit[b]	Depth preference[c]
	Density ($n \cdot 100$ m^{-3})	% density	Rank	Density ($n \cdot 100$ m^{-3})	% density	Rank			
Dulichia tuberculata Boeck, 1871	72.01	17.145%	1	0.69	2.329%	5	PO	SF	DS
Dyopedos monacanthus (Metzger, 1875)	62.12	14.792%	2	0.56	1.905%	6	PO	SF	DS & S
Melphidippa borealis Boeck, 1871	42.69	10.165%	3	0.71	2.399%	4	MP	SF	DS & S
Monoculodes packardi (Boeck, 1871)	35.87	8.542%	4	2.44	8.257%	3	OD	DF	DS
Haliragoides inermis (G. O. Sars, 1882)	31.93	7.603%	5	3.48	11.785%	2	CA	DF	DS & S
Stenopleustes inermis Shoemaker, 1949	20.23	4.816%	6	0.15	0.494%	12	PL	DF	SDS
Haploops fundiensis Wildish & Dickinson, 1982	16.40	3.905%	7	18.89	63.938%	1	AM	SF	SDS
Paradulichia typica Boeck, 1871	15.19	3.618%	8	0.19	0.635%	10	PO	SF	E
Ischyrocerus anguipes Kroyer, 1838	15.19	3.618%	8	0.13	0.423%	13	IC	DF	SL
Bathymedon obtusifrons (Hansen, 1887)	12.78	3.042%	9	0.27	0.917%	8	OD	DF	DS
Ampelisca declivitatis Mills, 1967	12.44	2.962%	10	0.29	0.988%	7	AM	SF & DF	S
Hippomedon propinquus G. O. Sars, 1890	12.14	2.891%	11	0.13	0.423%	13	LY	OMN	DS
Anonyx debruyni Hoek, 1882	10.36	2.467%	12	0.10	0.353%	14	LY	SCA	S
Ericthonius fasciatus (Stimpson, 1853)	9.31	2.217%	13	0.04	0.141%	17	IC	SF	SDS
Gitanopsis arctica G. O. Sars, 1892	5.41	1.288%	14	0.02	0.071%	18	AP	COM	DS
Astyra abyssi Boeck, 1871	5.33	1.269%	15	0.17	0.565%	11	AS	?	S
Ischyrocerus megacheir Boeck, 1871	4.38	1.042%	16	0.21	0.706%	9	IC	DF	S
Monoculodes tesselata Schneider, 1884	4.36	1.038%	17	0.19	0.635%	10	OD	DF	DS
Harpinia sp.	4.24	1.009%	18	0.06	0.212%	16	PX	DF	?
Anonyx lilljeborgi Boeck, 1871	4.12	0.981%	19	0.10	0.353%	14	LY	OMN	E
Gitanopsis bispinosa (Boeck, 1871)	3.88	0.924%	20	0.08	0.282%	15	AP	COM	DS
Tmetonyx cicada (O. Fabricius, 1780)	3.45	0.821%	21	0.10	0.353%	14	LY	SCA	DS & S
Argissa hamatipes (Norman, 1869)	3.39	0.807%	22	0.02	0.071%	18	AR	DF	DS
Ischyrocerus spA	3.31	0.788%	23	0.02	0.071%	18	IC	DF	?
Halirages fulvocinctus (M. Sars, 1858)	1.94	0.462%	24	0.27	0.917%	8	CA	DF	DS & S
Stegocephalus inflatus Kroyer, 1842	1.51	0.358%	25	–	–	–	SG	OMN	DS & S
Metopella angusta Shoemaker, 1949	1.17	0.278%	26	–	–	–	SN	DF	DS
Phoxocephalus holbolli (Kroyer, 1842)	1.15	0.274%	27	0.02	0.071%	18	PX	DF	SL & S
Orchomenella minuta (Kroyer, 1846)	1.09	0.259%	28	–	–	–	LY	SCA	DS
Synchelidium sp.	0.75	0.179%	29	–	–	–	OD	DF	?
Epimeria loricata G. O. Sars, 1879	0.34	0.080%	30	0.08	0.282%	15	PA	COM	S
Photis reinhardi Kroyer, 1842	0.30	0.071%	31	–	–	–	IA	SF & DF	DS
Halice abyssi Boeck, 1871	0.28	0.066%	32	–	–	–	PD	DF	S
Centromedon pumilus (Lillj., 1865)	0.20	0.047%	33	–	–	–	LY	SCA	SL & S
Unciola inermis Shoemaker, 1945	0.14	0.033%	34	0.02	0.071%	18	AD	DF	SDS
Casco bigelowi (Blake, 1929)	0.14	0.033%	34	–	–	–	ML	DF	SL & S

							a	b	c
Melphidippa goesi Stebbing, 1899	0.08	35	0.019%	–	–	–	MP	SF	DS & S
Orchomenella pinguis (Boeck, 1861)	0.06	36	0.014%	–	–	–	LY	SCA	DS
Pleustes panopla (Kroyer, 1838)	0.06	36	0.014%	–	–	–	PL	DF	DS
Dyopedos arcticus (Murdoch, 1885)	0.04	37	0.009%	–	–	–	PO	SF	DS
Andaniopsis nordlandica (Boeck, 1871)	0.04	37	0.009%	–	–	–	SG	OMN	DS & S
Bathymedon saussurei (Boeck, 1871)	0.04	37	0.009%	0.02	0.071%	18	OD	DF	S
Calliopius laeviusculus (Kroyer, 1838)	0.04	37	0.009%	0.02	0.071%	18	CA	OMN	SL
Tryphosella nanoides (G. D. Sars, 1895)	0.02	38	0.005%	–	–	–	LY	SCA	S
Leucothoe spinicarpa (Abildgaard, 1789)	0.02	38	0.005%	–	–	–	LU	?	S
Metopa spA	0.02	38	0.005%	–	–	–	SN	COM	?
Acanthonotozoma serratum (O. Fabricius, 1780)	0.02	38	0.005%	–	–	–	AC	COM	DS
Syrrhoe crenulata Goes, 1866	0.02	38	0.005%	0.02	0.071%	18	SY	DF	DS
Ampelisca vadorum Mills, 1963	–	–	–	0.02	0.071%	18	AM	SF & DF	SL
Leptocheirus pinguis (Stimpson, 1853)	–	–	–	0.02	0.071%	18	AO	?	SL & DS

a: (family code) AC = Acanthonotozomatidae, AM = Ampeliscidae, AP = Amphilochidae, AO = Aoridae, AR = Argissidae, AS = Astyridae, CA = Calliopiidae, IA = Isaeidae, IC = Ischyroceridae, LY = Lysianassidae, LU = Leucothoidae, ML = Melitidae, MP = Melphidippidae, OD = Oedicerotidae, PA = Paramphithoidae, PD = Pardaliscidae, PL = Pleustidae, PO = Podoceridae, PX = Phoxocephalidae, SG = Stegocephalidae, SN = Stenothoides, SY = Synopiidae

b: (feeding habit) OMN = omnivore, SF = suspension feeder, DF = detritus feeder, SF & DF = mixed suspension and detritus feeder, SCA = scavenger-predator, COM = commensal, ? = unknown

c: (depth preference): see table 5

References

Berkes, F., 1973. Production and comparative ecology of euphausiids in the Gulf of St. Lawrence. Ph.D. thesis, Marine Sciences Centre, McGill University, Montréal, 188 pp.

Besner, M., 1976. Structure écologique annuelle des associations d'Amphipodes gammaridiens dans l'hyperbenthos et l'endobenthos d'un fond vaseux circalittoral de l'estuaire maritime du Saint-Laurent en 1970 et 1971. Mémoire M. Sc., Université de Montréal, Montréal, 103 pp.

Billett, D. S. M., R. S. Lampitt, A. L. Rice & R. F. C. Mantoura, 1983. Seasonal sedimentation of phytoplankton to the deep-sea benthos. Nature 302: 520–522.

Brunel, P., 1965. Food as a factor or indicator of vertical migrations of cod in the western Gulf of St. Lawrence. Int. Comm. Northw. Atl. Fish., Spec. Publ. 6: 439–448.

Brunel, P., 1968. The vertical migrations of cod in the southwestern Gulf of St. Lawrence, with special reference to feeding habits and prey distribution. Ph.D. thesis, McGill University, Montréal, 510 pp.

Brunel, P., 1979. Seasonal changes of daily vertical migrations in a suprabenthic cold-layer shelf community over mud in the Gulf of St. Lawrence. In E. Naylor & R. G. Hartnoll (eds.), Cyclical phenomena in marine plants and animals, Proc. 13th Europ. Mar. Biol. Symp. Pergamon Press, Oxford: 383–390.

Brunel, P., 1983. Breeding seasons and success of slope over shelf suprabenthic species on the shelf in a food-stressed subarctic ecosystem. Florida Inst. Technol., Benthic Ecology Meetings, Abstracts: 10.

Brunel, P., M. Besner, D. Messier, L. Poirier, D. Granger & M. Weinstein, 1978. Le traîneau suprabenthique Macer-GIROQ: appareil amélioré pour l'échantillonnage quantitatif étagé de la petite faune nageuse au voisinage du fond. Int. Revue ges. Hydrobiol. 63: 815–829.

Christensen, H. & E. Kanneworff, 1985. Sedimenting phytoplankton as a major food source for suspension and deposit feeders in the Oresund. Ophelia 24: 223–244.

Connell, J. H. & E. Orias, 1964. The ecological regulation of species diversity. Am. Nat. 98: 399–414.

Cushing, D. H., 1972. The production cycle and the numbers of marine fish. In R. W. Edwards & D. J. Garrod (eds.), Conservation and productivity of natural waters, Symp. Zool. Soc. London 29. Academic Press, London & N.Y.: 213–232.

Davies, J. M. & R. Payne, 1984. Supply of organic matter to the sediment in the northern North Sea during a spring phytoplankton bloom. Mar. Biol. 78: 315–324.

Dayton, P. K. & J. S. Oliver, 1977. Antarctic soft-bottom benthos in oligotrophic and eutrophic environments. Science 197: 55–58.

Desroches, M., 1985. Cycle de développement, écologie et succès de l'Amphipode Gammaridien planctonophage Rhachotropis oculata dans deux écosystèmes du golfe du Saint-Laurent. Mémoire M. Sc., Université de Montréal, Montréal, 47 pp.

Desroches, M. & P. Brunel, 1986. Rhachotropis oculata, Crustacé Amphipode prédateur et migrateur suprabenthique saisonnier dans l'estuaire du St-Laurent. In Assoc. can.-franç. Avanc. Sci., Annales 54: 466 (abstract).

Dionne, J.-C., 1990. Observations sur le niveau marin relatif à l'Holocène, à Rivière-du-Loup, estuaire du Saint-Laurent, Québec. Géogr. phys. Quatern. 44: 43–53.

Dyke, A. S. & V. K. Prest, 1987. Late Wisconsinan and Holocene history of the Laurentide Ice Sheet. Géogr. phys. Quatern. 41: 237–267.

Enequist, P., 1949. Studies on the soft-bottom amphipods of the Skagerak. Zool. Bidr. Uppsala 28: 297–492.

Emerson, C. W., J. C. Roff & D. J. Wildish, 1986. Pelagic-benthic energy coupling at the mouth of the Bay of Fundy. Ophelia 26: 165–180.

Fader, G. B., L. H. King & B. MacLean, 1977. Surficial geology of the eastern Gulf of Maine and Bay of Fundy., Mar. Sci. Pap. 19, Geol. Surv. Can. Pap. 76–17: 23 pp.

Falkowski, P. G., J. Vidal, T. S. Hopkins, G. T. Rowe, T. E. Whitledge & W. G. Harrison, 1983. Summer nutrient dynamics in the Middle Atlantic Bight: primary production and utilization of phytoplankton carbon. J. Plankton Res. 5: 515–537.

Forward, Jr., R. B., 1988. Diel vertical migration: zooplankton photobiology and behaviour. Oceanogr. Mar. Biol., Ann. Rev. 26: 361–393.

Gagnon, J.-M., 1983. Cycle de développement, écologie et succès d'Anonyx makarovi, Amphipode gammaridien nécrophage saisonnier dans le circalittoral de deux écosystèmes du golfe du Saint-Laurent. Mémoire M. Sc., Université de Montréal, Montréal, 49 pp.

Gerber, R. P. & N. Marshall, 1974. Reef pseudoplankton in lagoon trophic systems. In A. M. Cameron, B. M. Campbell, A. B. Cribb, R. Endean, J. S. Jell, O. A. Jones & F. H. Talbot (eds.), Proc. Second Int. coral Reef Symp., Part 1: 105–107.

George, R. & R. J. Menzies, 1967. Indication of cyclic reproductive activity in abyssal organisms. Nature 215: 878.

George, R. & R. J. Menzies, 1968. Further evidence for seasonal breeding cycles in deep sea. Nature 220: 80–81.

Graf, G., R. Schultz, R. Peinert & L.-A. Myer-Reil, 1983. Benthic response to sedimentation events during autumn to spring at a shallow-water station in the Western Kiel Bight I. Analysis of processes on a community level. Mar. Biol. 77: 235–246.

Graf, G., W. Bengtsson, A. Faubel, L.-A. Meyer-Reil, R. Schulz, H. Theede & H. Thiel, 1984. The importance of the spring phytoplankton bloom for the benthic system of Kiel Bight. Cons. int. Explor. Mer, Rapp. et Procès-verb. 183: 136–143.

Granger, D., P. Brunel & D. Messier, 1979. Cycle de développement de Leucon nasica (Crustacea, Cumacea) dans la nappe glaciale circalittorale de la baie des Chaleurs, golfe du Saint-Laurent, en 1968 et 1969. Can. J. Zool. 57: 95–106.

Grassle, J. F., 1989. Species diversity in deep-sea communities. Trends Ecol. Evol. 4: 12–15.

Hachey, H. B., F. Hermann & W. B. Bailey, 1954. The waters of the ICNAF convention area. Ann. Proc. Int. Comm. Northw. Atl. Fish. 4: 67–102.

Hargrave, B. T., 1985. Particle sedimentation in the ocean. Ecol. Model. 30: 229–246.

Heinrich, A. K., 1962. The life histories of plankton animals and seasonal cycles of plankton communities in the ocean. J. Cons. perm. int. Explor. Mer 27: 15–24.

Hobson, E. S. & J. R. Chess, 1979. Zooplankters that emerge from the lagoon floor at night at Kure and Midway Atolls, Hawaii. Fish. Bull. 77: 275–280.

Holme, N. A. & A. D. MacIntyre (eds), 1984. Methods for the study of marine benthos, IPB Handbook, 16. Blackwell Sci. Publ., Oxford, 387 pp.

Honjo, S., S. J. Manganini & J. J. Cole, 1982. Sedimentation of biogenic matter in the deep ocean. Deep-sea Res. 29: 609–625.

Hopkins, C. C. E. & B. Gulliksen, 1978. Diurnal vertical migration and zooplankton-epibenthos relationships in a north Norwegian fjord. Proc. 12th Europ. Symp. Mar. Biol. Pergamon Press, Oxford: 271–280.

Kaartvedt, S., 1989. Nocturnal swimming of gammaridean amphipod and cumacean Crustacea in Masfjorden, Norway. Sarsia 74: 187–193.

Lamarche, G. & P. Brunel, 1987. Cycle de développement, écologie et succès d'*Hippomedon propinquus* (Amphipoda Gammaridea) dans deux écosystèmes du golfe du Saint-Laurent. Can. J. Zool. 65: 3116–3132.

Laubitz, D. R., 1977. A revision of the genera *Dulichia* Kroyer and *Paradulichia* Boeck (Amphipoda, Podoceridae). Can. J. Zool. 55: 942–982.

Lauzier, L. M. & R. W. Trites, 1958. The deep waters in the Laurentian Channel. J. Fish. Res. Bd Can. 15: 1247–1257.

Legendre, L., 1971. Production primaire dans la Baie-des-Chaleurs (Golfe Saint-Laurent). Nat. can. 98: 743–773.

Lopez, G. R. & J. S. Levinton, 1987. Ecology of deposit-feeding animals in marine sediments. Quart. Rev. Biol. 62: 235–260.

Mann, K. H., 1976. Production on the bottom of the sea. In D. H. Cushing & J. J. Walsh (eds.), The ecology of the seas. W. B. Saunders, Philadelphia: 225–250.

Messier, D., 1974. Rythmes journaliers et succession démographique en 1971 et 1972 des Cumacés d'un fond circalittoral dans l'estuaire maritime du Saint-Laurent. Mémoire M. Sc., Université de Montréal, Montréal, 78 pp.

Messier, D. & P. Brunel, 1977. Etude d'une communauté circalittorale de Cumacés de l'estuaire maritime du Saint-Laurent. Assoc. can.-franç. Avanc. Sci., Annales 44: 158 (abstract).

Mills, E. L., 1967. The biology of an ampeliscid amphipod crustacean sibling species pair. J. Fish. Res. Bd Can. 24: 305–355.

Moore, P. G. & R. Earll, 1985. Sediment 'whips': amphipod artefacts from the rocky sublittoral in Britain. J. exp. Mar. Biol. Ecol. 90: 165–170.

Ohlhorst, S. L., 1982. Diel migration patterns of demersal reef plankton. J. exp. mar. Biol. Ecol. 60: 1–15.

Peer, D., D. J. Wildish, A. J. Wilson, J. Hines & M. J. Dadswell, 1980. Sublittoral macro-infauna of the lower Bay of Fundy. Can. Tech. Rep. Fish. aquat. Sci. 981: 74 pp.

Pfannkucke, O., R. Theeg & H. Thiel, 1983. Benthos activity, abundance and biomass under an area of low upwelling off Morocco, Northwest Africa. 'Meteor' Forschung-Ergebn., D: 85–96.

Porter, J. W. & K. G. Porter, 1977. Quantitative sampling of demersal plankton migrating from different coral reef substrates. Limnol. Oceanogr. 22: 553–556.

Porter, J. W., K. G. Porter & Z. Batac-Catalan, 1977. Quantitative sampling of Indo-Pacific demersal reef plankton. In D. L. Taylor (ed.), Proc. Third Int. Coral Reef Symp., Part I: 105–112.

Robert, G., 1974. The sublittoral Mollusca of the St. Lawrence Estuary, east coast of Canada. Ph.D. thesis, Institute of Oceanography, Dalhousie University, 178 pp.

Rokop, F. J., 1977a. Patterns of reproduction in the deep-sea benthic crustaceans: a re-evaluation. Deep-Sea Res. 24: 683–691.

Rokop, F. J., 1977b. Seasonal reproduction of the brachiopod *Frieleia halli* and the scaphopod *Cadulus californicus* at bathyal depths in the deep sea. Mar. Biol. 43: 237–246.

Sainte-Marie, B. & P. Brunel, 1983. Differences in life history and success between suprabenthic shelf populations of *Arrhis phyllonyx* (Amphipoda Gammaridea) in two ecosystems of the Gulf of St. Lawrence. J. Crust. Biol. 3: 45–69.

Sainte-Marie, B. & P. Brunel, 1985. Suprabenthic gradients of swimming activity by cold-water gammaridean amphipod Crustacea over a muddy shelf in the Gulf of Saint Lawrence. Mar. Ecol. Progr. Ser. 23: 57–69.

Sanders, H. L., 1968. Marine benthic diversity: a comparative study. Am. Nat. 102: 243–282.

Schink, D. R., 1979. Review of marine geochemistry. Rev. Geophys. Space Phys. 17: 1447–1473.

Sinclair, M., M. I. El-Sabh & J.-R. Brindle, 1976. Seaward nutrient transport in the Lower St. Lawrence Estuary. J. Fish. Res. Bd Can. 33: 1271–1277.

Smith, W. & J. F. Grassle, 1977. Sampling properties of a family of diversity measures. Biometrics 33: 283–292.

Spence, C. & D. M. Steven, 1974. Seasonal variation of the chlorophyll a: pheopigment ratio in the Gulf of St. Lawrence. J. Fish. Res. Bd Can. 31: 1263–1268.

Steele, D. H. & P. Brunel, 1968. Amphipoda of the Atlantic and Arctic coasts of North America: *Anonyx* (Lysianassidae). J. Fish. Res. Bd Can. 25: 943–1060.

Steven, D. M., 1974. Primary and secondary production in the Gulf of St. Lawrence. McGill Univ., Mar. Sci. Centre, Manuscr. Rep. 26: 116 pp.

Therriault, J.-C. & M. Levasseur, 1985. Control of phytoplankton production in the Lower St. Lawrence Estuary: light and freshwater runoff. Nat. can. 112: 77–96.

Tyler, P. A., 1988. Seasonality in the deep sea. Oceanogr. Mar. Biol., Ann. Rev. 26: 227–258.

Wassmann, P., 1984. Sedimentation and benthic mineralization of organic detritus in a Norwegian fjord. Mar. Biol. 83: 83–94.

Wassmann, P., 1985. Sedimentation of particulate material in two shallow, land-locked fjords in western Norway. Sarsia 70: 317–331.

Wefer, G., E. Suess, W. Balzer, G. Liebezeit, P. J. Muller, C. A. Ungerer & W. Zenk, 1982. Fluxes of biogenic components from sediment trap deployment in circumpolar waters of the Drake Passage. Nature 299: 145–147.

Wieser, W., 1956. Factors influencing the choice of substratum in *Cumella vulgaris* Hart (Crustacea, Cumacea). Limnol. Oceanogr. 1: 274–285.

Wildish, D. J., 1984. Secondary production of four sublittoral, soft-sediment amphipods in the Bay of Fundy. Can. J. Zool. 62: 1027–1033.

Wildish, D. J. & M. J. Dadswell, 1985. Sublittoral gammaridean amphipods of soft sediments in the Bay of Fundy. Proc. N.S. Inst. Sci. 35: 1–15.

Wildish, D. J. & U. Lobsiger, 1987. Three-dimensional photography of soft-sediment benthos, S.W. Bay of Fundy. Biol. Oceanogr. 4: 227–241.

Wildish, D. J., A. J. Wilson & B. Frost, in press. Benthic boundary layer macrofauna of Browns Bank, N.W. Atlantic, as potential prey of juvenile benthic fish. Can. J. Fish. Atl. Sci.

Hydrobiologia **223**: 105–117, 1991.
L. Watling (ed.), VIIth International Colloquium on Amphipoda.
© 1991 *Kluwer Academic Publishers.*

Amphipod crustaceans as an important component of zoobenthos of the shallow Antarctic sublittoral

K. Jażdżewski, W. Teodorczyk, J. Siciński & B. Kontek
*Department of Invertebrate Zoology and Hydrobiology, University of Łódź, 12/16 Banacha St.,
90-237 Łódź, Poland*

Key words: Amphipoda, benthos South Shetland Islands, Antarctica

Abstract

Benthic quantitative samples were taken in 1988 in the soft bottom sublittoral of Admiralty Bay (King George Island, South Shetlands) using a Tvärminne-type bottom sampler and SCUBA-diving technique at 7 successive stations situated at depths from 4 to 30 m.

Dominant animal groups in terms of abundance were Amphipoda, Polychaeta and Bivalvia, whereas in terms of biomass Echinoidea were also dominant. Amphipod crustaceans clearly dominated the zoobenthos at depths from 10 to 25 m (the numerical share surpassing 60%) with maximal abundance of abt. 17 000 ind m^{-2}; in terms of biomass at specific depths amphipods occupied the 1st, 2nd or 3rd place with maximal biomass of abt. 100 g m^{-2} where the maximal total biomass of zoobenthos reached 260 g m^{-2} (10 m).

Amphipoda were the most diversified group with some 35 taxa belonging to 14 families. Most species belonged to Eusiridae s.l. and Lysianassidae s.l. Dominant forms were *Pontogeneiella brevicornis*, *Prostebbingia gracilis*, *Schraderia gracilis*, *Hippomedon kergueleni*, *Orchomenella* cf. *ultima*, *Cardenio paurodactylus* and *Paraphoxus rotundifrons*.

Introduction

The qualitative and quantitative wealth of the Antarctic sublittoral benthos is well known and has been discussed by many authors (i.a. Hedgpeth, 1969; Dell, 1972; Arnaud, 1974; Knox & Lowry, 1977; White, 1984; Picken, 1985; Jażdżewski *et al.*, 1986; Wägele & Schminke, 1986; Gallardo, 1987). Maximal abundances in Antarctica can surpass 100 000 ind m^{-2} (Dayton & Oliver, 1977) and the biomass exceed several kg m (Beliaev & Ušakov, 1957; Propp, 1970; Hardy, 1972; White & Robins, 1972; Nakajima *et al.*, 1982; Jażdżewski *et al.*, 1986). Dominant groups, in terms of abundance, were usually indicated as being Polychaeta, Bivalvia and such crustaceans

as Amphipoda, Cumacea or Tanaidacea, whereas in terms of biomass Echinodermata, Polychaeta, Porifera, Bivalvia, Ascidiacea and Bryozoa dominated. However, despite the increasing amount of new data (Dhargalkar *et al.*, 1988; Mühlenhardt-Siegel, 1988, 1989; Tucker & Burton, 1988; Voss, 1988), taking into account the variety of Antarctic zoogeographic regions (Hedgpeth, 1969; Knox & Lowry, 1977; Siciński, 1986), the vast span of sublittoral depth range (down to 500–600 m) and the diversity of sublittoral habitats (Gruzov & Pushkin, 1970; Hardy, 1972; White & Robins, 1972; White, 1984; Picken, 1985), the information on the Antarctic benthos must be considered to be still inadequate and not detailed enough (White, 1984). In order to help remedy this insufficient knowledge of the Antarctic bottom fauna we have undertaken the qualitative study of the soft shallow sublittoral in the near vicinity of the Polish Antarctic Station (King George I, S. Shetlands), using a bottom sampler operated by SCUBA-divers. Our study is a continuation of earlier investigations in the same area (Jażdżewski *et al.*, 1986).

According to our specialization, in this paper we have focused our interests on Amphipoda.

In comparison to many other animal groups and geographic regions the Antarctic Amphipoda can be regarded as quite well known. Information on these crustaceans found to the south of 50° S before 1975 is given in the most useful catalogue of Lowry & Bullock (1976) where 524 taxa are presented. From that time a rich literature on Antarctic and Subantarctic amphipods has appeared. Many of these papers contained the descriptions of new taxa. Taking into account only some major papers of the most active Antarctic amphipodologists, in the last ten years some 60 new species from the region south of 50° S were described (Watling & Holman, 1980, 1981; Andres, 1981, 1982, 1983, 1985, 1986, 1988; Andres & Coleman, 1988; De Broyer, 1981, 1983, 1985; Holman & Watling, 1981, 1983a, b; Lowry & Stoddaert, 1983; Bellan-Santini & Ledoyer, 1986). Therefore it is reasonable to consider that at present the amphipod fauna of Antarctica *sensu lato* (south of 50° S) exceeds some 600 species.

Amphipoda in Antarctica exhibit a high degree of endemism with nearly 40% of endemic genera and as much as 90% of endemic species (Knox & Lowry, 1977). The families of greatest importance in Antarctica, both in number of species and abundance of specimens, are Lysianassidae (s.l.), Eusiridae (s.l.) and Acanthonotozomatidae. Phoxocephalidae and Stenothoidae are also well represented in Antarctic waters (Arnaud, 1974; Lowry & Bullock, 1976; Arnaud *et al.*, 1986).

Investigated area, methods and material

The sampling area was Admiralty Bay of King George Island (South Shetlands, West Antarcti-

ca) in the immediate vicinity of the Polish Antarctic Station 'H. Arctowski' (Fig. 1). Admiralty Bay is a large, fjord-like embayment with a maximum depth of 600 m and surface area of about 120 km². The hydrology of the Bay is rather well known owing to the papers by Presler (1980), Pruszak (1980), Rakusa-Suszczewski (1980), Samp (1980), Szafrański & Lipski (1982) and Lipski (1987). In general, temperature and salinity variations are small with ranges at the bottom from − 1.8 to 1.2 °C and from 33 to 34.5‰, respectively. Local freshwater inflows can lower the salinity only from the surface to a depth of at most 1–2 m. The levels of nutrients and oxygen content are high and pH is very stable, ranging from 8.1 to 8.4 (Lipski, 1987). Tidal range in Admirably Bay is about 3 m.

In our study area the stony and rocky bottom of the littoral occurs down to a depth of 3–4 m; to the depth of some 10 m patches of rocks and boulders occur on the generally soft bottom which consists of coarse sand underlain by gravel. Gradually, below 10 m depth, the bottom becomes sandy and muddy, but mixed with muddy sediments at all greater depths there is coarse gravel and stones deposited on the sea bed by ice.

The most common algae of the littoral belt are *Monostroma harrioti* Gain and *Adenocystis utricularis* (Bory) Skottsberg; in the sublittoral down to the depth of some 15 m the dominants are *Ascoseira mirabilis* Skottsberg, *Desmarestia menziesi* J. Agardh and *Hildebrandtia lecanellieri* Harriot, whereas in deeper parts of the sublittoral *Cystosphaera jacquinoti* (Montagne) Skottsberg and *Himantothallus grandifolius* (A. et E. S. Gepp) Skottsberg predominate (Zieliński, 1981; Furmańczyk & Zieliński, 1982). The phytal occupies sometimes vast stretches of the bottom, decaying algal remnants make larger and smaller patches of detritus on the sea floor, especially in bottom recesses.

Our samples were collected between 7 and 25 January 1988 by divers using a Tvärminne-type bottom sampler (Kangas, 1972) consisting of an open-bottom cylindrical metal vessel with a sampling surface of 565 cm². This gear was lowered by rope from the boat and quickly put by

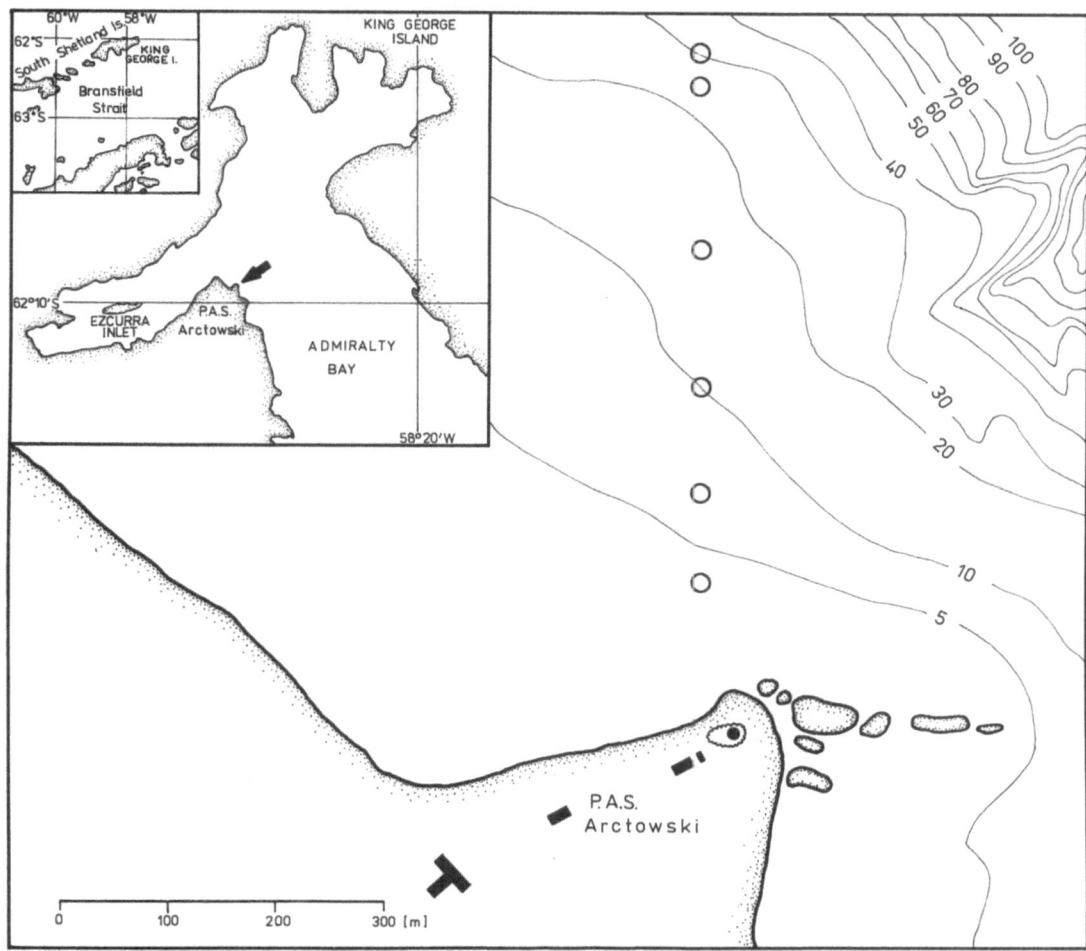

Fig. 1. Investigated area with isobaths and sampling stations (circles).

the diver into the sediment. Using a hand-operated cutting round metal plate with a handle (movable bottom of the vessel) a 5 cm thick portion of the bottom sediment was cut through a narrow slit after digging a pit along the outer wall of the sampler to enable the cutting edge to fit into the slit.

Along a transect aligned from Shag Point in the direction of Hennequin Point (Fig. 1, more or less Section 1 of the paper by Jażdżewski *et al.*, 1986), 7 sampling stations were selected at the following depths: 4, 6.5, 10, 15, 20, 25 and 30 m. At all these stations, except the last, three replicate sub-samples were taken (3×565 cm^2), whereas at

30 m – only two (2×565 cm^2). This method enabled sampling only the bottom soft enough to push the sampler into the sea floor. Intentionally we have avoided also the dense phytal zone; only very scarcely overgrown places or places with some decaying algae were sampled.

The sampler contents were sieved through a 0.5 mm mesh size screen and preserved in 4% formalin. In the laboratory animals were sorted into major taxonomic groups, specimens were counted and wet formalin weight determined after blotting on filter paper until dry. Colonial animals (Bryozoa, Hydrozoa) were only weighed. Data were extrapolated to 1 m^2.

Results and discussion

Figure 2 shows the percentage share of major groups of the bottom fauna. The area of the circles corresponds to the total abundance or biomass of the bottom fauna at each site. Maximal densities of benthic animals surpassed 25 000 ind m^{-2} and the mean for all samples was 14 400 ind m^{-2}. Biomass values ranged from about 80 to over 260 g m^{-2}; with general mean of 165 g m^{-2}. In all nearly 17 000 specimens of non-colonial animals were collected. Among them there were 1 192 (8%) specimens of Polychaeta, 4 305 (26%) of Bivalvia and 8 807 specimens of Amphipoda; therefore in terms of general abundance this last group constituted the major part of the material (over 50%).

The share of these three leading groups of the total collected biomass was: Polychaeta – abt. 14%, Bivalvia – abt. 16% and Amphipoda – abt. 29%. Other groups occurring in considerable numbers were Oligochaeta (abt. 6% of the total number of animals) and those important in general biomass were Echinoidea (abt. 17%), Isopoda (Serolidae) (abt. 8%) and Bryozoa (abt. 7%).

It is evident that three groups dominated in general: Amphipoda, Polychaeta and Bivalvia. Their significance varied among the stations, therefore at some of the stations, in abundance or in biomass, some other groups prevailed – for instance Isopoda at the depth of 4 m (in abundance), Echinoidea at the depth of 10 m (in biomass) and Bryozoa and Ascidiacea at the depth of 30 m (in biomass).

The numerical share of Amphipoda at particular depths varied between 9 and 69% (30 and 10 m, respectively), but at 4 successive depths (10, 15, 20 and 25 m) this share was higher than 60%. Due to the comparatively small size of amphipods their proportion of the biomass was less conspicuous but occasionally also remarkable – from the

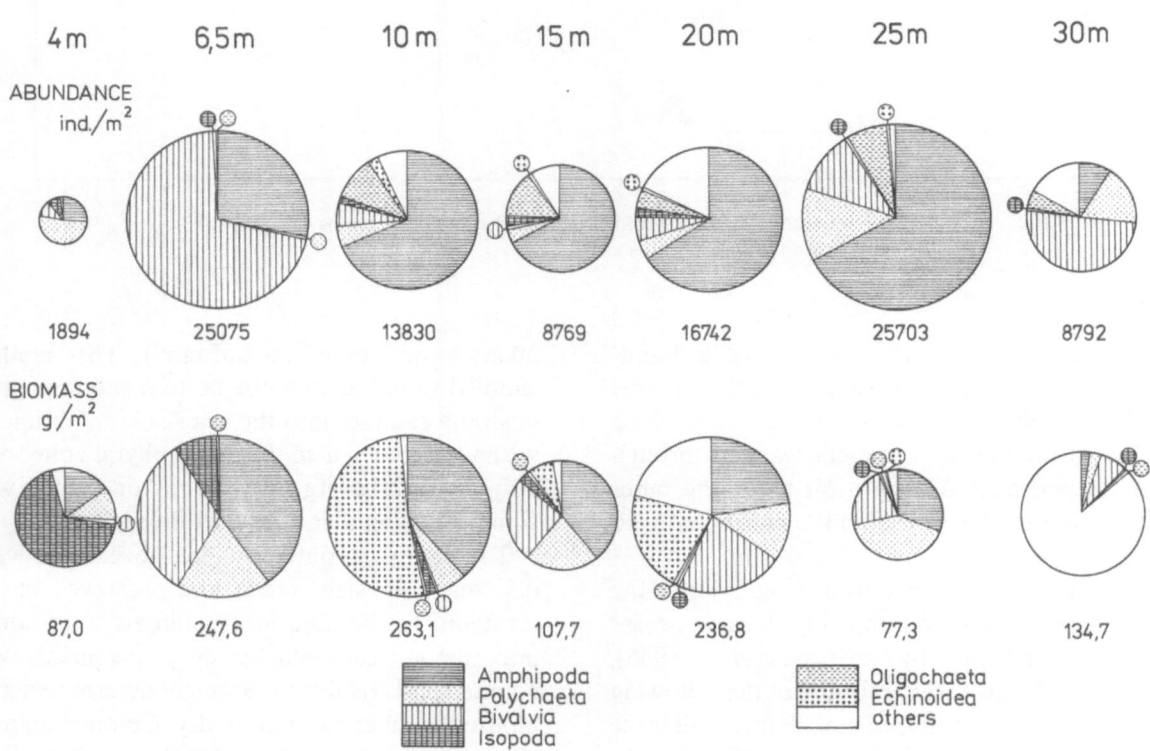

Fig. 2. The composition of zoobenthos of the soft bottom of the Admiralty Bay.

lowest of abt. 2% at the depth of 30 m, to nearly 40% at the depth of 6.5 m, with a mean for all depths of nearly 30%.

Absolute extreme values of amphipod densities varied from nearly 500 ind m^{-2} (4 m) up to nearly 17 000 ind m^{-2} (25 m), and the extremes for biomass were 3 g m^{-2} at 30 m and over 100 g m^{-2} at 10 m.

Polychaetes were very diversified: hitherto 30 species were determined (Sicinski, pers. comm.). Among them the most common and abundant were *Capitella capitata* (Fabr.) (s.l.), *Scoloplos marginatus* (Ehlers), *Microspio moorei* (Gravier) and *Travisia kerguelensis* McIntosh. This last species, due to its large size, clearly dominated in biomass. Among the bivalves the absolute dominants were *Mysella charcoti* (Lamy) and *Yoldia eightsi* (Couthouy).

The species composition of the amphipod material is presented in Table 1 and Fig. 3. One can clearly observe that the dominant families are Eusiridae s.l. and Lysianassidae s.l. The dominant species were *Prostebbingia gracilis*, *Pontogeneiella brevicornis* and *Hippomedon kergueleni*. Next in the importance were *Schraderia gracilis*, *Paraphoxus rotundifrons*, *Orchomenella* cf. *ultima* and *Cardenio paurodactylus*.

These species occurred in most of the samples and formed the bulk of the amphipod material both in terms of abundance and of biomass. Two other species were also very common – namely *Cheirimedon femoratus* and *Monoculodes scabriculosus*. However, the first contributed more significantly to the total number of Amphipoda only at the depth of 25 m and to the total biomass at the depths of 4 and 25 m. *M. scabriculosus* was present at all depths except 30 m and was found in large numbers only at a depth of 6 m.

The list of 35 amphipod taxa presented here may be regarded as a rich one. One should consider that the samples were taken from a comparatively monotonous bottom and in the rather narrow depth range of 4 to 30 m in 7 stations. This list complements the hitherto presented set of 43 amphipod species for Admiralty Bay (Chevreux, 1913; K. H. Barnard, 1932; Arnaud *et al.*, 1986) by a further 13 species, namely *Gitanopsis*

squamosa, *Wandelia crassipes*, *Atyloella magellanica*, *Oradarea edentata*, *Schraderia gracilis*, *Urothoe sp.*, *Jassa ingens*, *Jassa wandeli*, *Orchomenella* cf. *ultima*, *Methalimedon nordenskjoeldi*, *Oediceroides lahillei* and *Cardenio paurodactylus*. Most are rather common circumantarctic or West-antarctic species, but some of them deserve short comments.

Two burrowing lysianassids, the dominant *Hippomedon kergueleni* and a very frequent *Cheirimedon femoratus*, both of circumantarctic distribution and very common throughout Antarctica, are well known as most important necrophagous animals in the depths not exceeding 30 m (Bregazzi, 1972a, Presler, 1986). In greater depth of the Admiralty Bay they yield to other lysianassid necrophages – *Abyssorchomene plebs* (Hurley) and *Waldeckia obesa* (Chevreux) (Presler, 1986) that were totally absent in our samples.

Pontogeneiella brevicornis, on the other hand is a West-Antarctic and Subantarctic species, occurring abundantly in shallow sandy bottom, either bare or overgrown with algae. Thurston (1974a) observed *P. brevicornis* in large numbers in only a few of numerous samples taken at South Orkney Islands and suggested that this may be due to the possible swarming behavior of the species, but our quantitative samplings indicate rather that it is simply a very common and abundant species.

Prostebbingia gracilis is a widespread circumantarctic amphipod connected with the phytal zone and mainly with *Desmarestia* meadows where it occurs in incredible numbers (Jażdżewski, unpubl. data). High abundance of this phytophilous species in our materials taken mostly outside the phytal zone can be attributed to the fresh and old algal remnants occurring sometimes in considerable amounts in the sampled area, especially in some bottom recesses situated in the near vicinity of algal meadows.

The morphology and taxonomy of *Schraderia gracilis* was amply discussed by Thurston (1974a). He refrained from recognizing *S. calceolata* as a separate species since he found intermediate specimens among a vast material. In our

Table 1. Abundance and biomass of Amphipoda on the soft bottom of the shallow sublittoral of the Admiralty Bay (King George Island,

Depth	4 m				6.5 m				10 m			
Species	N/m²	%	B/m²	%	N/m²	%	B/m²	%	N/m²	%	B/m²	%
Acanthonotozomatidae												
Pariphimedia integricauda Chevreux												
Amphilochidae												
Gitanopsis squamosa (Thomson)									12	0.1	Z	X
Dexaminidae												
Paradexamine fissicauda Chevreux												
Eophliantidae												
Wandelia crassipes Chevreux												
Eusiridae s.l.												
Atyloella magellanica (Stebbing)												
Bovallia gigantea Pfeffer												
Djerboa furcipes Chevreux												
Eurymera monticulosa Pfeffer												
Gondogeneia sp. juv. (antarctica?)	12	2.5	Z	0.2								
Liouvillea oculata Chevreux												
Oradarea edentata K. H. Barnard									12	0.1	Z	X
Paramoera edouardi Schellenberg												
Paramoera hurleyi Thurston	6	1.2	Z	X					12	0.1	Z	X
Pontogeneiella brevicornis (Chevreux)	6	1.2	Z	0.2	991	14.1	46.0	46.7	2519	26.4	63.8	63.4
Prostebbingia gracilis (Chevreux)	6	1.2	Z	X					696	7.3	0.7	0.7
Schraderia gracilis Pfeffer									6	0.1	Z	X
Eusiridae indet.												
Ischyroceridae												
Jassa ingens (Pfeffer)												
Jassa wandeli Chevreux												
Jassa sp. juv.												
Lysianassidae s.l.												
Cheirimedon femoratus (Pfeffer)	24	4.9	1.0	6.2	41	0.6	1.3	1.3	12	0.1	Z	X
Hippomedon kergueleni (Miers)	372	77.8	14.0	91.0	4189	59.5	39.3	40.0	4077	42.8	33.1	32.9
Orchomenella rotundifrons K. H. Barnard												
Orchomenella cf. *ultima* (Bellan-Santini)									201	2.1	0.1	0.1
Orchomenella sp. juv.												
Oedicerotidae												
Methalimedon nordenskjoeldi Schellenberg												
Monoculodes scabriculosus K. H. Barnard	12	2.5	Z	0.1	1038	14.8	5.0	5.1	12	0.1	Z	X
Oediceroides lahillei Chevreux												
Phoxocephalidae												
Heterophoxus trichosus K. H. Barnard					12	0.2	1.1	1.1				
Paraphoxus rotundifrons (K. H. Barnard)	12	2.5	0.2	1.1	71	1.0	0.5	0.6	1882	19.7	2.3	2.3
Phoxocephalopsidae												
Phoxocephalopsis deceptionis Stephensen					271	3.9	3.3	3.4				
Stenothoidae												
Probolisca ovata (Stebbing)									18	0.2	Z	X
Proboloides sp.												
Synopiidae												
Cardenio paurodactylus Stebbing	29	6.2	0.1	0.8	407	1.9	1.9	1.9	71	0.7	0.5	0.5
Thaumatelsonidae												
Prothaumatelson nasutum Chevreux												
Urothoidae												
Urothoe sp.												
Other					18	Z	Z	X	6	0.1	Z	X
Total	479	100	15.4	100	7038	100	98.5	100	9536	100	100.7	100

South Shetlands). Explanations: X – less than 0,05%, Y – animal fragments, Z – less than 0,05 g.

15 m				20 m				25 m				30 m			
N/m²	%	B/m²	%	N/m²	%	B/m²	%	N/m²	%	B/m²	%	N/m²	%	B/m²	%
65	1.1	1.6	4.0	12	0.1	0.2	0.3	24	0.1	Z	0.1	9	1.1	Z	0.6
				53	0.5	Z	0.1	6	X	Z	X	27	3.4	Z	0.6
6	0.1	Z	X	6	0.1	0.1	0.2					9	1.1	0.2	8.5
				6	0.1	Z	X								
												27	3.4	0.5	15.9
				24	0.2	0.2	0.3	12	0.1	Z	0.2	9	1.1	Z	0.9
				6	0.1	0.1	0.2	18	0.1	0.5	2.0				
29	0.5	0.1	0.3	77	0.7	0.2	0.4								
				18	0.2	Z	X					Y	X	Z	X
6	0.1	0.6	1.6									9	1.1	Z	0.6
236	4.0	0.6	1.3	301	2.7	0.5	0.9	260	1.5	0.9	3.9	35	4.5	0.1	9.7
118	2.0	0.3	0.7	236	2.1	0.4	0.9	1292	7.4	1.7	7.2				
171	2.9	0.4	1.0	71	0.6	Z	X	265	1.5	0.1	0.4	18	2.3	Z	0.1
513	8.6	11.8	28.4	2030	18.3	29.0	55.1	53	0.3	1.2	4.9				
1062	17.9	3.3	8.1	2283	20.6	3.6	6.7	6519	37.5	12.4	51.7	292	37.5	1.0	36.2
496	8.3	1.9	4.6	791	7.1	1.5	2.8	920	5.3	0.5	2.0	62	8.0	0.1	2.5
												9	1.1	0.2	5.5
6	0.1	1.6	3.8												
18	0.3	0.2	0.4												
106	1.8	0.2	0.4	2053	18.6	2.3	4.3								
12	0.2	0.4	0.9	12	0.1	Z	X	5882	33.8	3.4	14.0	9	1.1	Z	0.3
2325	39.2	15.5	37.5	2153	19.5	12.3	23.3	201	1.2	1.4	5.9	186	23.9	0.4	14.1
6	0.1	Z	0.1	6	0.1	0.1	0.2								
183	3.1	0.2	0.4	218	2.0	0.4	0.8	1664	9.6	1.5	6.4	9	1.1	Z	0.3
				18	0.2	Z	X								
12	0.2	Z	X	47	0.5	0.1	0.1	18	0.1	Z	X	53	6.8	0.1	3.1
29	0.5	0.1	0.2	24	0.2	Z	X	71	0.4	Z	0.1				
6	0.1	0.1	0.3												
								6	X	Z	X	9	1.1	Z	0.3
442	7.5	1.2	3.0	578	5.2	1.4	2.7	18	0.1	0.1	0.4	9	1.1	Z	0.6
								6	X	Z	X				
				24	0.2	Z	0.1	118	0.7	0.1	0.4				
53	0.9	0.6	1.4	18	0.2	0.2	0.4	12	0.1	Z	X				
								12	0.1	Z	0.1				
								24	0.1	Z	0.1				
35	0.6	0.6	1.5											0.2	6.1
5935	100	41.3	100	11065	100	52.7	100	17401	100	24.0	100	781	100	2.9	100

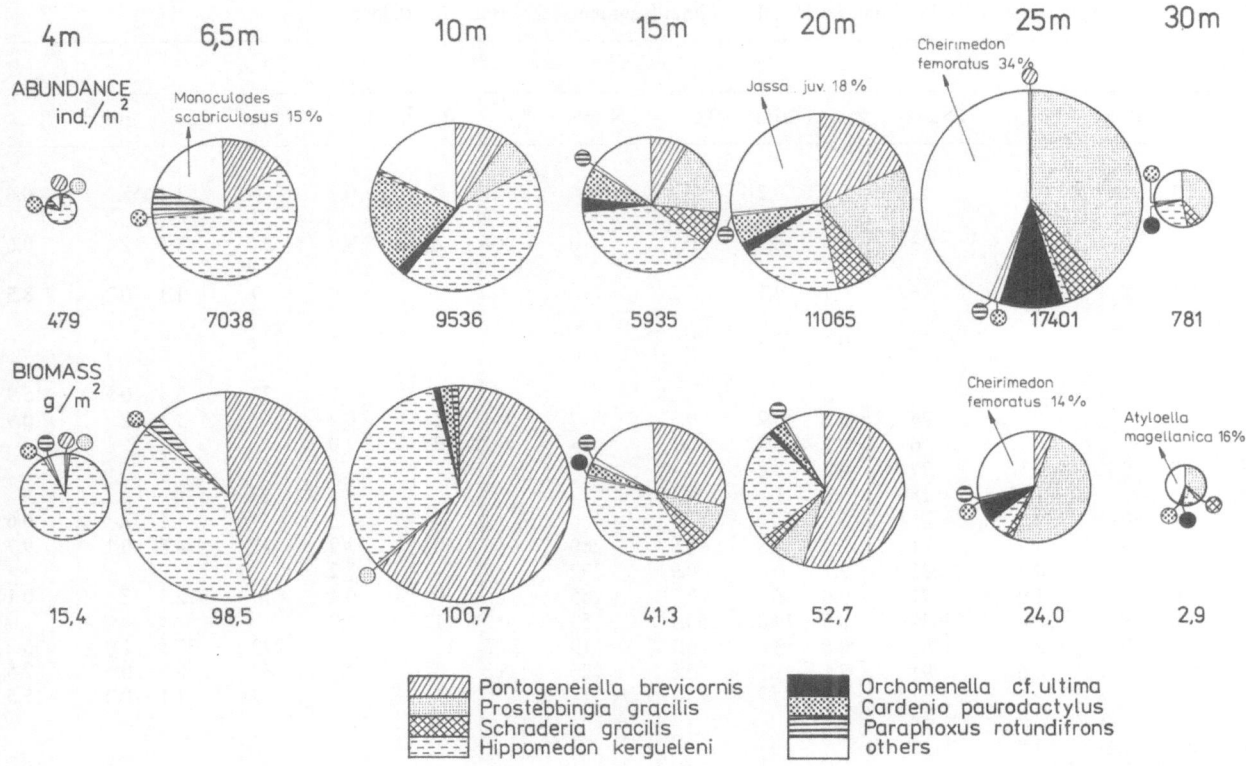

Fig. 3. The composition of Amphipoda of the soft bottom of the Admiralty Bay.

samples there were also several specimens fitting well the 'calceolata' form, but we tentatively accept the view of Thurston; further study of this problem is needed before final rejection of the idea that 'calceolata' is a good species. *S. gracilis* is also a shallow water phytophilous circuman-tarctic species. Its occurrences in large numbers in our samples can be explained in similar way as for the case of *Prostebbingia gracilis*.

Paraphoxus rotundifrons is a West-antarctic burrowing amphipod common mainly on shallow soft bottoms (Thurston, 1974a). The proper up-dated generic classification of the species will be possible after a new genus promised by J. L. Barnard & Drummond (1978) is described.

Haustorioidea of the Antarctic belong to those groups that are most inadequately studied, espe-cially in the view of recent revisions by J. L. Barnard & Drummond (1982). Our material of *Urothoe* sp. consists of only 4 specimens and at the moment we refrain from creating a new taxon.

Jassa wandeli was recognised by Lowry & Bullock (1976) as one of numerous synonyms of *Jassa falcata* (Montagu). Thurston (1974a), how-ever, distinguished in his South Orkneys material 3 taxa called *Jassa falcata* form 1, *J. falcata* form 2 and *J. falcata* form 3. The first agreed with *J. wandeli*. At least in the case of *J. falcata* form 1 (= *J. wandeli*) we fully support Thurston's sug-gestion that: 'In view of the apparent constancy of these forms they may eventually be considered good species'.

Orchomenella ultima was described by Bellan-Santini (1972) (as *Orchomene ultima*) based on the material consisting of 3 specimens found in the stomachs of nototheniid fish. This is the only orchomenid species that seems to be similar to our rather rich material, composed mainly of young specimens, but also some mature ones as well. Our *Orchomenella* is a very characteristic species, pale yellow with brown transverse stripes across each segment. Before any taxonomic decision we would like to see the typical material since several possibly important differences in the structure of 1st maxilla, gnathopods, 3rd epimeral plate and telson between our material and figures of Bellan-Santini do exist.

One of our dominant, *Cardenio paurodactylus*, a burrowing Subantarctic and West-antarctic species, was for a century wrongly placed in the family Haustoriidae (Schellenberg, 1926; K. H. Barnard, 1932; Stephensen, 1947; Thurston, 1974b; Lowry & Bullock, 1976; Bellan-Santini & Ledoyer, 1986). The first signal for its proper, we believe, systematic position – in the family Synopiidae – we noted in the paper by J. L. Barnard & Drummond (1982). In the paper by Arnaud *et al.* (1986) *C. paurodactylus* is mentioned as 'Synopiidae n. gen. n. sp.'. Quite recently J.L. Barnard & Karaman (1987) have erected for *C. paurodactylus* a new family, Cardenioidae. We consider, however, this new placement as not sufficiently justified. The paper substantiating the proper assignment of *C. paurodactylus* with remarks on its morphology and biology is published elsewhere (Jażdżewski & De Broyer, 1990). This species was hitherto rarely found and usually in few specimens with the one exception of a huge sample of about 30 000 specimens in the 'Discovery' material collected off Bouvetoya (Thurston, 1974b).

Proboloides sp. is referred to as '*Proboloides* cf. *antarcticus* Walker (?)' in the paper by Arnaud *et al.* (1986). Probably we have to do here with a new species to be described in the future.

In many papers dealing with Antarctic benthos the importance of Amphipoda, especially in abundance, has been mentioned (Gruzov *et al.*, 1967; Gallardo & Castillo 1969, 1970; Hardy, 1972; Lowry, 1975; Gallardo *et al.*, 1977; Richardson & Hedgpeth, 1977). However, this group usually ranked in third place after Polychaeta and Bivalvia. The position of Amphipoda in our materials, especially in terms of biomass, in reality is probably not so high. Our biomass values for Bivalvia can be much underestimated, since using our sampler we were obviously not able to collected the largest and rather common pelecypod species in the shallow waters of Admiralty Bay, *Laternula elliptica* (King et Broderip), that burrows in the sediments much deeper than 5 cm. The biomass of this species alone for similar depths and sediment of Borge Bay (South Orkneys) was estimated by Everson & White (1969) as reaching 1–4 kg m^{-2}.

Until now few data are available with both qualitative and quantitative analysis of the bottom fauna of the soft bottom at comparable depths (5–35 m) in the Antarctic. The best reference papers are those concerning the shores of Signy Island, South Orkney Islands (Everson & White, 1969; Bregazzi, 1972a, b; Hardy, 1972; Thurston, 1974a) and the Arthur Harbor at Anvers Island, Palmer Archipelago (Lowry, 1975; Richardson & Hedgpeth, 1977). The comparison of these data with ours suggests clear biocenotical similarity of the region of our study to the Signy Island sublittoral. In this last region also the dominant bivalves were *Mysella charcoti* and *Yoldia eightsi*, among dominant polychaetes were *Capitella* sp. and *Scoloplos marginatus* and as dominant amphipod species were mentioned *Hippomedon kergueleni*, *Cheirimedon femoratus*, *Pontogeneiella brevicornis*, *Prostebbingia gracilis* and *Paraphoxus rotundifrons*. Noteworthy is a remarkable similarity of the composition and proportional representation of the list of amphipod species found by Thurston (1974a) in his habitats D and E at the depths and substrates much resembling our study area.

On the other hand the soft bottom of Arthur Harbor at similar depths differed strongly in the composition of leading taxa. In the paper by Lowry (1975), who studied over a yearly cycle two stations at depths of ca. 20 m, the list of most abundant animals includes among the first 10 taxa only the Amphipoda *Ampelisca bouvieri* Chevreux, *Heterophoxus videns* K.H. Barnard, *Harpinia* sp. and *Megamphopus* sp. and the list of biologically most important taxa ranks these four amphipods among the top ten. Rather long lists of amphipod taxa, comparing those of the present paper and that of Lowry (1975), are highly dissimilar, with only 6 taxa in common. The data of Richardson & Hedgpeth (1977) for Arthur Harbor, but from 14 stations of much wider depth range confirmed in general Lowry's (1975) results, but the rank values evaluated for the entire generalized area lists the two first amphipods, *Heterophoxus videns* and *Ampelisca bouvieri*, at the end of the group of 15 dominant taxa. The list of nearly 50 amphipod taxa presented by these authors also strongly differed from ours with only 9 species in common. We are of the opinion, however, that these differences in the bottom faunas in general, and in particular in amphipod fauna composition between Arthur Harbor and our study area are caused more by ecological factors than zoogeographical ones. The available comparative sedimentological data (Tatur & Siciński, unpubl. data) indicate namely that the mean grain size of the sediment in the region of our study was significantly greater than that of the bottom of Arthur Harbor, where sandy silt and clayey silt predominated. Moreover, it seems that in our samples a significant role was played by broken fresh or decaying fragments of algae, as a habitat of many species connected with the phytal zone (for instance *Prostebbingia gracilis*, *Schraderia gracilis*, *Jassa* spp.). It is worthy of note that in the assemblage G of Richardson & Hedgpeth (1977), distinguished tentatively for two samples taken in Arthur Harbor from the bottom of somewhat coarser sediment and with an admixture of broken algae, the dominant species were such Amphipoda like *Cheirimedon femoratus* and *Djerboa furcipes* and among common benthic animals were also *Schraderia gracilis*, *Oradarea* sp. and *Prostebbingia gracilis*. We have to stress moreover, that not far from our study area, in the inner fjord of the Admiralty Bay – the Ezcurra Inlet, the area more isolated from the oceanic water influence, subjected to the glacier meltwater inflow and of much finer sediment, the ecological similarity of the bottom fauna to those of the Arthur Harbor becomes obvious (Jażdżewski et al., 1986).

However, it is to be noted that at the shores of the eastern continental Antarctic the amphipod assemblages of the shallow soft bottom of sand and silt seem to have different dominance structure with such species like *Heterophoxus videns*, *Ampelisca barnardi* Nicholls, *Orchomenella franklini* Walker and *Orchomenella pinguides* Walker as leading taxa (Everitt et al., 1980; Tucker & Burton, 1988; Jażdżewski, unpubl. data).

A quite interesting fact, stressing the diversity of Antarctic habitats, is the sharp difference between the amphipod fauna of the shallow sublittoral as presented in this paper and the composition of the amphipod list obtained by Voss (1988) after the preliminary elaboration of the Weddell Sea material taken from the depths from 200 to over 1000 m. In this list of more than 80 taxa one can rarely find a genus shared with our list, not to speak about the species.

In conclusion we have to repeat that Amphipoda with their species diversity, abundance and important share in biomass play a leading role in the shallow Antarctic ecosystem.

Acknowledgements

Thanks are due to Mr. J. Żychliński, M. Sc., for his inestimable help to the senior author in the underwater sampling, to Mr. E. Dibowski, the skipper of the boat for his field assistance and to Dr. C. De Broyer for many valuable comments and kind invitation to Brussels that enabled the senior author to make use of his well organized laboratory and a remarkable catalogue of world marine Amphipoda.

The paper was sponsored by Polish Academy of Sciences within the CPBP 03.03 Project.

Resumé

En 1988, en utilisant la benne du type Tvärminne et la technique du plongement autonome, dans les 7 stations successives situées aux profondeurs de 4 à 30 m, ont a prelevé les echantillons quantitatifs de benthos du fonds meubles du sublittoral de la baie d'Amirauté (île du Roi-George, Shetlands du Sud).

Du point de vue de la quantité Amphipoda, Polychaeta et Bivalvia etaient les groups dominants. Les mêmes groupes et aussi Echinoidea dominaient du point de vue de la biomasse. Les crustaces amphipodes dominaient distinctement aux profondeurs de 10 à 25 m; leur participation quantitatif surpassait 60% et la densité maximale atteignait 17 000 ind m^{-2}. A l'egard de la biomasse Amphipoda occupaient la 1-ère, 2-ème ou 3-ème places aux profondeurs particulières, avec la biomasse maximale de 100 g m^{-2} environ, pendant que la biomasse maximale du benthos total atteignait 260 g m^{-2} (10 m).

Avec les 35 espèces de 14 familles Amphipoda etaient le groupe le plus diversifié. La majorité des espèces appartenait aux Eusiridae s.l. et Lysianassidae s.l. *Prostebbingia gracilis*, *Pontogeneiella brevicornis*, *Schraderia gracilis*, *Hippomedon kergueleni*, *Orchomenella* cf. *ultima*, *Cardenio paurodactylus* et *Paraphoxus rotundifrons* etaient les formes les plus nombreuses.

References

Andres, H. G., 1981. Die Gammaridea (Crustacea: Amphipoda) der Deutschen Antarktis-Expeditionen 1975/76 und 1977/78, 1. Gammaridae, Melphidippidae und Pagetinidae. Mitt. hamb. zool. Mus. Inst. 78: 179–196.

Andres, H. G., 1982. Die Gammaridea (Crustacea: Amphipoda) der Deutschen Antarktis-Expeditionen 1975/76 und 1977/78. 2. Eusiridae. Mitt. hamb. zool. Mus. Inst. 79: 159–185.

Andres, H. G., 1983. Die Gammaridea (Crustacea: Amphipoda) der Deutschen Antarktis-Expeditionen 1975/76 und 1977/78. 3. Lysianassidae. Mitt. hamb. zool. Mus. Inst. 80: 183–220.

Andres, H. G., 1985. Die Gammaridea (Crustacea: Amphipoda) der Deutschen Antarktis-Expeditionen 1975/76 und 1977/78. 4. Acanthonotozomatidae, Paramphithoidae und Stegocephalidae. Mitt. hamb. zool. Mus. Inst. 82: 119–153.

Andres, H. G., 1986. Atylopis procerus sp. n. und Cheirimedon solidus sp. n. aus der Weddell See sowie Anmerkungen zu Orchomenella pinguides Walker, 1903 (Crustacea: Amphipoda: Gammaridae). Mitt. hamb. zool. Mus. Inst. 83: 117–130.

Andres, H. G., 1988. Zwei neue Acanthonotozomatiden aus der Bransfield Strasse, Antarktis (Crustacea: Amphipoda). Mitt. hamb. zool. Mus. Inst. 85: 111–120.

Andres, H. G. & O. Coleman, 1988. Neue Echiniphimedia-Arten aus der Antarktis (Crustacea: Amphipoda: Acanthonotozomatidae). Mitt. hamb. zool. Mus. Inst. 85: 121–140.

Arnaud, P. M., 19740 Contribution à la bionomie marine benthique des regions antarctiques et subantarctiques. Tethys 6: 465–656.

Arnaud, P. M., K. Jażdżewski, P. Presler & J. Siciński, 1986. Preliminary survey of benthic invertebrates collected by Polish Antarctic Expeditions in the Admiralty Bay (King George Island, South Shetlands, Antarctica). Pol. Polar Res. 7: 7–24.

Barnard J. L. & M. M. Drummond, 1978. Gammaridean Amphipode of Australia, Part III: The Phoxocephalidae Smithson. Contrib. Zool. 245: 1–551.

Barnard, J. L. & M. M. Drummond, 1982. Gammaridean Amphipoda of Australia, Part V: Superfamily Haustorioidea. Smithson. Contrib. Zool. 360: 1–148.

Barnard, J. L. & G. S. Karaman, 1987. Revisions in classification of Gammaridean Amphipoda (Crustacea), Part 3. Proc. Biol. Soc. Wash. 100: 856–875.

Barnard, K. H., 1932. Amphipoda. Discovery Rep. 5: 1–326.

Beliaev, G. M. & P. V. Ušakov, 1957. Nekotorye zakonomernosti količestvennogo raspredelenija donnoj fauny v vodach Antarktiki. Dokl. Akad. Nauk SSSR 112: 137–140.

Bellan-Santini, D., 1972. Amphipodes provenant des contenus stomacaux de trois espèces de poissons Nototheniidae récoltés en Terre Adélie (Antarctique). Tethys 4: 693–702.

Bellan-Santini, D. & M. Ledoyer, 1986. Gammariens (Crustacea, Amphipoda) des Iles Marion et Prince Edward. Boll. Mus. civ. St. nat. Verona 13: 349–435.

Bregazzi, P. K., 1972a. Life cycles and seasonal movements of *Cheirimedon femoratus* (Pfeffer) and *Tryphosella kergueleni* (Miers) (Crustacea: Amphipoda). Br. Antarct. Surv. Bull. 30: 1–34.

Bregazzi, P. K., 1972b. Habitat selection by *Cheirimedon femoratus* (Pfeffer) and *Tryphosella kergueleni* (Miers) (Crustacea: Amphipoda). Br. Antarct. Surv. Bull. 31: 21–33.

Chevreux, E., 1913. Amphipodes. Deux. Exp. Antarct. Franc. (1908–1910). Sci. Nat., Docum. Sci., Paris: 76–186.

Dayton, P. K. & J. S. Oliver, 1977. Antarctic soft-bottom benthos in oligotrophic and eutrophic environments. Science 197: 55–58.

De Broyer, C., 1981. *Monoculodes jazdzewskii*, une nouvelle espece antarctique (Crustacea, Amphipoda, Oedicerotidae). Bull. Acad. Pol. Sci., Cl. II 28: 381–387.

De Broyer, C., 1983. Recherches sur la systematique et evolution des crustaces amphipodes gammarides antarctiques et subantarctiques. Dissert. Univ. Cath. Louvain; 468 pp.

De Broyer, C., 1985. Amphipodes lysianassoides nécrophages des iles Kerguelen (Crustacea). 1. *Orchomenella guillei* n. sp. Bull. Mus. natn. Hist. nat., Paris, ser. 4,7, sect. A (1): 205–217.

Dell, R. K., 1972. Antarctic benthos. In: F. S. Russell and M. Yonge (eds.), Advances in Marine Biology. Academic Press, London – New York, 10: 1–216.

Dhargalkar, V. K., H. R. Burton & J. M. Kirkwood, 1988. Animal associations with the dominant species of shallow water macrophytes along the coastline of the Vestfold Hills, Antarctica. Hydrobiologia 165: 141–150.

Everitt, D. A., C. C. B. Poore & J. Pickard, 1981. Marine

benthos from Davis Station, Eastern Antarctica. Aust. J. mar. Freshwat. Res. 31: 829–836.

Everson, I. & M. G. White, 1969. Antarctic marine biological research methods involving diving. Rep. Underwat. Ass. 4: 91–95.

Furmańczyk, K. & K. Zieliński, 1982. Distribution of macroalgae groupings in shallow waters of Admiralty Bay (King George Island, South Shetland Islands, Antarctica). Pol. Polar Res. 3: 41–47.

Gallardo, V. A., 1987. The sublittoral macrofaunal benthos of Antarctic shelf. Env. Intern. 13: 71–81.

Gallardo, V. A. & J. G. Castillo, 1969. Quantitative benthic survey of the infauna of Chile Bay (Greenwich Island, South Shetland Islands). Gayana 16: 1–18.

Gallardo, V. A. & J. G. Castillo, 1970. Quantitative observations on the benthic macrofauna of Port Foster (Deception Island) and Chile Bay (Greenwich Island). In: M. W. Holdgate (ed.), Antarctic Ecology, Academic Press, London – New York 1: 242–243.

Gallardo, V. A., J. G. Castillo, M. A. Retamal, A. Yanez, H. I. Moyano & J. G. Hermosilla, 1977. Quantitative studies on the soft-bottom macrobenthic animal communities of shallow Antarctic bays. In: G. A. Llano (ed.), Adaptations within Antarctic Ecosystems. Proc. 3rd SCAR Symp. Antarct. Biol., Smith. Inst., Wash.: 361–387.

Gruzov, E. N. & A. E. Pushkin, 1970. Bottom communities of the upper sublittoral of Enderby Land and the South Shetland Islands. In: M. W. Holdgate (ed.), Antarctic Ecology, Academic Press, London – New York 1: 235–238.

Gruzov, E. N., M. V. Propp & A. E. Puškin, 1967. Biologičeskie soobščestva pribrežnych raionov morja Dejvisa (po rezul'tatam vodolaznych nabljudenij). Inf. Bjull. Sov. Ant. Eksp. 65: 124–141.

Hardy, O., 1972. Biomass estimates for some shallow-water infaunal communities at Signy Island, South Orkney Islands. Brit. Antarct. Surv. Bull. 31: 93–106.

Hedgpeth, J. W., 1969. Introduction to Antarctic Zoogeography. In: V. C. Bushnell and J. W. Hedgpeth (eds.), Antarctic Map Folio Series, Distribution of Selected Groups of Marine Invertebrates in Waters South of 35° S Latitude. Amer. Geogr. Soc., New York: 1–9.

Holman, H. & L. Watling, 1981. *Pagetina reducta* sp. n. (Crustacea: Amphipoda) with a review of the family Pagetinidae. Sarsia 66: 213–215.

Holman, H. & L. Watling, 1983a. Amphipoda from the Southern Ocean: families Colomastigidae, Dexaminidae, Leucothoidae, Liljeborgiidae and Sebidae. Biology of the Antartic Seas XIII, Antarct. Res. Ser. 38: 215–262.

Holman, H. & L. Watling, 1983b. A revision of the Stilipedidae (Amphipoda). Crustaceana 44: 27–53.

Jażdżewski, K. & C. De Broyer, 1990. The Antarctic and Sub-antarctic synopiid Morphology and systematic position of *Cardenio paurodactylus* Stebbing, 1888 (Crustacea, Amphipoda). Beaufortia 41(18): 129–133.

Jażdżewski, K., W. Jurasz, W. Kittel, E. Presler, P. Presler & J. Siciński, 1986. Abundance and biomass estimates of the benthic fauna in Admiralty Bay, King George Island, South Shetland Islands. Polar Biol. 6: 5–16.

Kangas, P., 1972. Quantitative sampling equipment for the littoral benthos.II. IBP, Norden N° 10: 9–16.

Knox, G. A. & J. K. Lowry, 1977. A comparison between the benthos of the Southern Ocean and the North Polar Ocean with special reference to the Amphipoda and the Polychaeta. In: M. J. Dunbar (ed.), Polar Oceans. Proc. Polar Oceans Conf., McGill University, Montreal: 423–462.

Lipski, M., 1987. Variations of physical conditions, nutrients and chlorophyll a contents in Admiralty Bay (King George Island, South Shetlands, 1979). Pol. Polar Res. 8: 307–332.

Lowry, J. K., 1975. Soft bottom macrobenthic community of Arthur Harbor, Antarctica. Antarctic Res. Ser. 23: 1–19.

Lowry, J. K. & S. Bullock, 1976. Catalogue of the marine gammaridean Amphipoda of the Southern Ocean. Bull. Roy. Soc. New Zealand. 16: 1–187.

Lowry, J. K. & H. E. Stoddaert, 1983. The shallow-water gammaridean Amphipoda of the Subantarctic islands of New Zealand and Australia: Lysianassoidea. J. Roy. Soc. New Zealand 13: 279–394

Mühlenhardt-Siegel, U., 1988. Some results of quantitative investigations of macro-zoobenthos in the Scotia Arc (Antarctica). Polar Biol. 8: 241–248.

Mühlenhardt-Siegel, U., 1989. Quantitative investigations of Antarctic zoobenthos communities in winter (May/June) 1986 with special reference to the sediment structure. Arch. FischWiss. 39: 123–141.

Nakajima, Y., K. Watanabe & Y. Naito, 1982. Diving observations of the marine benthos at Syowa Station, Antarctica. Proc. Vth Symp. Antarct. Biology, Mem. Nat. Inst. Pol. Res., Tokyo, Spec. Issue 23: 44–54.

Picken, G. B., 1985. Marine Habitats-Benthos. In: W. N. Bonner and D. W. H. Walton (eds.), Key Environments: Antarctica. Pergamon Press: 154–172.

Presler, P., 1980. Phenological and physiographical observations carried out during the first wintering at the Arctowski Station in 1977. Pol. Arch. Hydrobiol. 27: 245–252.

Presler, P., 1986. Necrophagous invertebrates of the Admiralty Bay of King George Island (South Shetland Islands, Antarctica). Pol. Polar Res. 7: 25–61.

Propp, M. V., 1970. The study of bottom fauna at Haswell Island by SCUBA diving. In: M. W. Holdgate (ed.), Antarctic Ecology. Academic Press, London – New York, 1: 239–241.

Pruszak, Z., 1980. Currents circulation in the waters of Admiralty Bay (region of Arctowski Station on King George Island). Pol. Polar Res. 1: 55–74.

Rakusa-Suszczewski, S., 1980. Environmental conditions and the functioning of Admiralty Bay (South Shetland Islands) as a part of the near shore Antarctic ecosystem. Pol. Polar Res. 1: 11–27.

Richardson, M. D. & J. W. Hedgpeth, 1977. Antarctic soft bottom macrobenthic community adaptations to a cold, stable, highly productive, glacially affected environment.

In: G. A. Llano (ed.), Adaptations Within Antarctic Ecosystems. proc. Third SCAR Symposium on Antarctic Biology, Smith. Inst. Wash.: 181–196.

Samp, R., 1980. Selected environmental factors in the waters of Admiralty Bay (King George Island, South Shetland Islands) December 1978 – February 1979. Pol. Polar. Res. 1: 53–66.

Schellenberg, A., 1926. Amphipoda 3. Die Gammariden. Wiss. Ergebn. dt. Tiefsee-Exped. 'Valdivia' 23: 195–243.

Siciński, J., 1986. Application of the dendrite analysis in the discussion on the biogeography of the Antarctic. Pol. Polar. Res. 7: 305–317.

Stephensen, K., 1947. Tanaidacea, Isopoda, Amphipoda and Pycnogonida. Sci. Res. Norw. Antarct. Exp. 20: 1–90.

Szafrański, Z. & M. Lipski, 1982. Characteristics of water temperature and salinity of Admiralty Bay (King George Island, South Shetland Islands, Antarctic) during the austral summer 1978/79. Pol. Polar Res. 3: 7–24.

Thurston, M. H., 1974a. The Crustacea Amphipoda of Signy Island, South Orkney Islands. Brit. Ant. Surv. Sci. Rep. 71: 1–133.

Thurston, M. H., 1974b. Crustacea Amphipoda from Graham Land and the Scotia Arc, collected by operation Tabarin and the Falkland Islands Dependencies Survey. Brit. Ant. Surv. Sci. Rep. 85: 1–89.

Tucker, M. J. & H. R. Burton, 1988. The inshore marine ecosystem off the Vestfold Hills, Antarctica. Hydrobiologia 165: 129–139.

Voss, J., 1988. Zoogeographie und Gemeinschaftsanalyse des Makro-zoobenthos des Weddellmeeres (Antarktis). Ber. Polarforsch. 45: 1–145.

Wägele, W. & H. K. Schminke, 1986. Leben in eisigen Tiefen: Benthosforschung in der Antarktis. Natur und Museum, Frankfurt a.M. 116: 184–193.

Watling, L. & H. Holman, 1980. New Amphipoda from the Southern Ocean, with partial revisions of the Acanthonotozomatidae and Paramphithoidae. Proc. Biol. Soc. Wash. 93: 609–654.

Watling, L. & H. Holman, 1981. Additional acanthonotozomatid, paramphithoid and stegocephalid Amphipoda from the Southern Ocean. Proc. Biol. Soc. Wash. 94: 181–227.

White, M. G., 1984. Marine benthos. In: R. M. Laws (ed.), Antarctic Ecology. Academic Press, London 2: 421–461.

White, M. G. & M. W. Robins, 1972. Biomass estimates from Borge Bay, Signy Island, South Orkney Islands. Brit. Ant. Surv. Bull. 31: 45–50.

Zieliński, K., 1981. Benthic macroalgae of Admiralty Bay (King George Island, South Shetland Islands) and circulation of algal matter between the water and shore. Pol. Polar. Res. 2: 71–94.

Hydrobiologia **223**: 119–126, 1991.
L. Watling (ed.), VIIth International Colloquium on Amphipoda.
© 1991 *Kluwer Academic Publishers.*

Patterns of abundance of exoedicerotid amphipods on sandy beaches near Sydney, Australia

A.R. Jones, A. Murray & R.E. Marsh
Division of Environmental Science, The Australian Museum, PO Box A 285, Sydney, N.S.W., 2000, Australia

Key words: amphipods, crustaceans, sandy beaches, population ecology, Australia east coast

Abstract

Spatial and temporal patterns of abundance of two species of exoedicerotid amphipod at several sandy beaches near Sydney, Australia are described and related to physicochemical factors. Replicate cores were taken at monthly intervals for one year from the swash zone and data were analysed by two-way (site × month), fixed-factor analysis of variance. Spatial and temporal differences in abundance were usually significant but inconsistent because of significant site × month interactions. Spatial differences in *Exoediceroides maculosus* sometimes occurred in the absence of obvious corresponding physicochemical differences although the density of stranded seagrass and algae may affect abundance. The abundance of *Exoediceros fossor* was often greater in lagoons than open beaches. Salinity, temperature and storms had no apparent effect on the temporal patterns of abundance of either species.

Introduction

Sandy beaches provide habitat for a variety of animal species and the beaches of South Africa, North America and Europe have received substantial ecological study (see McLachlan, 1983 for a review). By contrast, Australia's beaches have received scant attention (Dexter, 1983a, b; 1984; 1985; McLachlan & Hesp, 1984); this despite the large extent of Australian beaches and their enormous recreational and economic significance. Such a situation has led to recommendations for further ecological work into beaches (Fairweather, 1988) and the initiation of the present study.

Amphipods were selected for study because they can be numerical dominants in the macro-faunal communities of sheltered beaches near Sydney (Dexter, 1984, 1985). Further, some of the sampling sites chosen for this study can experience oil pollution (McGuinness, 1988) and it appears that amphipods are more sensitive to oil pollution than most other benthic species (Sanders *et al.*, 1980; Bonsdorff, 1983). Consequently, the most locally-abundant species, the exoedicerotids *Exoediceroides maculosus* (Sheard, 1936) and *Exoediceros fossor* (Stimpson, 1956), were targeted for a study investigating distributional patterns, life histories and the role of disturbance. This paper presents the first year's findings on spatial and temporal patterns of abundance and relates these to physico chemical features of the environment.

120

Methods

Beaches at five locations near Sydney were sampled: Silver Beach in Botany Bay, Clontarf and Chinaman's Beaches in Middle Harbour, and the beaches near the intermittently-open mouths of Dee Why and Curl Curl Lagoons (Fig. 1). These locations were chosen because preliminary sampling revealed the presence of abundant populations of one or both target species. Further, comparisons could be made between beaches in different bays (Botany Bay and Middle Harbour) and between lagoons and open beaches.

Silver Beach is divided by artificial rock groynes into 13 sub-beaches, three of which were sampled. These were designated Silver 3, 8 and 12. Each of these was further divided into three

shore-normal strata (hereafter termed sites) i.e. a West, Mid and East site, because initial observations indicated that prevailing winds can create an east-west wave-exposure gradient along the beach. The existence of this gradient was subsequently confirmed (see Results). The 10 metre section of beach adjacent to the rock groynes was excluded from sampling because of potential edge effects. Consequently, a total of 13 beach sites, which varied from 50 to 85 metres in length, were sampled. The Clontarf site was relocated westward about 300 metres after eight months sampling because persistent near-absence of specimens developed at the original section of beach. Sampling was monthly for 11 months (July, 1988 through May, 1989).

At each sampling time, five shore-normal

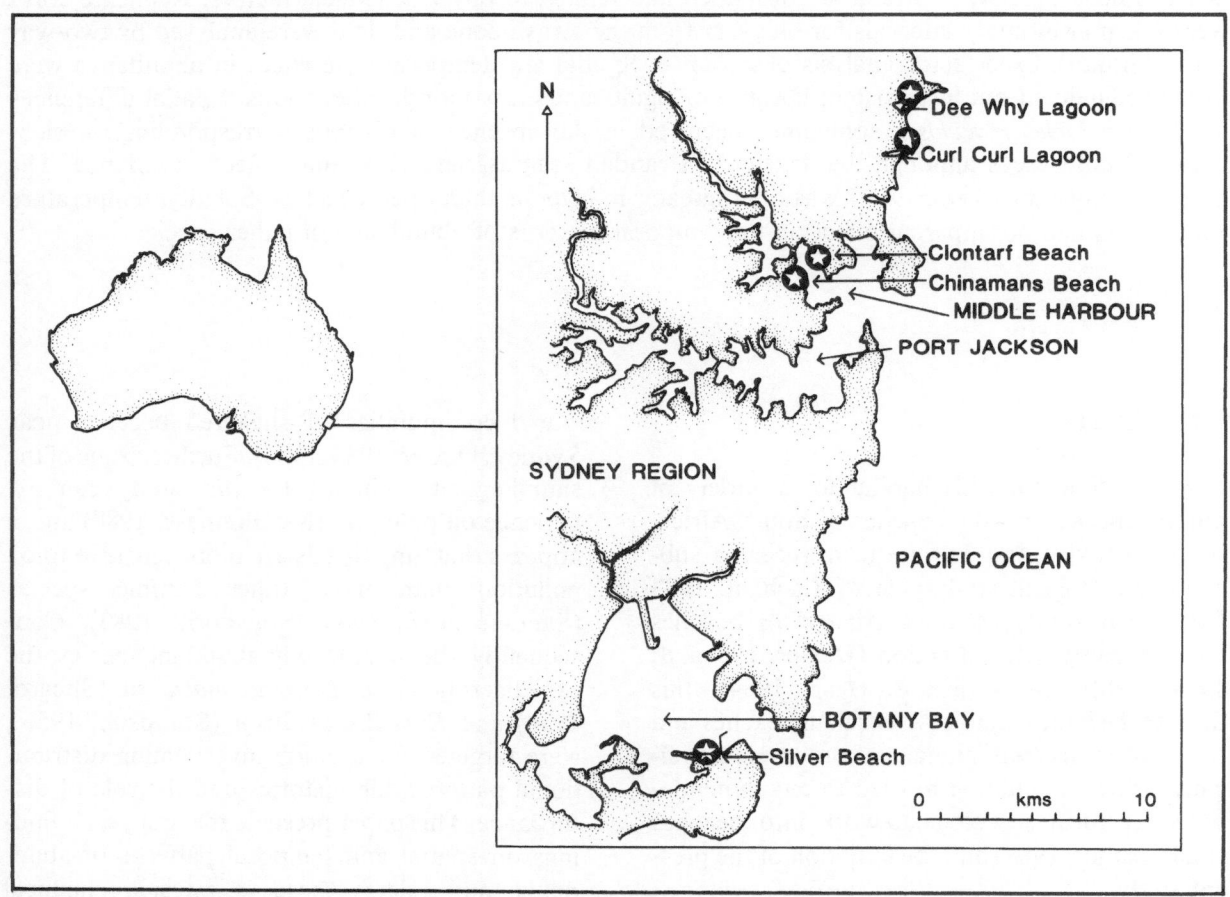

Fig. 1. Map of the Sydney region showing sampling locations. At Silver Beach, the sub-beaches 3, 8, and 12 occur progressively further west with about one km between adjacent sub-beaches.

transects were randomly located in each site. For each transect, the swash zone was divided into nine shore-parallel bands and two random 5.5 cm diameter cores were taken from each band giving 90 cores/month/site. This novel sampling scheme was devised because these species aggregate in the swash zone and migrate with the tide. A full description of the sampling method will appear elsewhere (Jones *et al.*, in prep). Coring depth was 2–3 cm after preliminary sampling showed an absence of specimens below two cm. Each core was washed through a 1 mm sieve and the specimens identified and enumerated immediately and replaced after the completion of each transect.

The salinity and temperature of the water at each beach site were measured using a Goldberg temperature-compensated refractometer and thermometer, respectively. From November, 1988 onward, the width of the swash zone for each transect was recorded as a measure of hydro-dynamic energy and the slopes of the Mid site of Silver 3, 8, 12 and at Chinaman's and Clontarf were measured on one occasion. Data on the duration and intensity of storms were obtained from a waverider buoy anchored about three kilometres offshore from Sydney and operated by the Manly Hydraulics Laboratory, Public Works Department, N.S.W.

Data analysis

A two-way (site × month), fixed-factor analysis of variance (ANOVA) was used to test for spatial and temporal differences in abundance for each species separately, and to determine the significance of interactions between space and time. To locate significant differences *a posteriori* Student-Newman-Keuls (SNK) multiple comparisons (Sokal & Rohlf, 1981) were used. Cochran's test was used to measure variance heterogeneity and, in practice, the data were transformed to ln $(x + 1)$ to reduce this heterogeneity. The monthly data for each site were pooled for this paper. Within-site analyses will be described in a later paper.

Relationships between abundance and the abiotic variables salinity and temperature were quantified using Spearman rank correlations. Partial correlations were used when salinity and temperature were themselves significantly correlated. The relationship between abundance and storm activity was assessed graphically.

Results

Environmental conditions

The means of recorded salinities at open beaches varied from 31.6 at Silver 12 Mid to 29.8‰ at Silver 3 East (Fig. 2a). Temporal variation in salinity as measured by the standard deviation differed little among these beaches and Dee Why Lagoon. The mean salinities at the lagoons were lower than at the open beaches, particularly at Curl Curl which was the most variable through time (Fig. 2a).

Both the overall mean water temperature and its temporal variation differed little among beaches with the highest and lowest means being 21.4 (Clontarf and Silver 3 East) and 20.4 °C (Silver 8 West), respectively (Table 1). The lowest

Table 1. Temperature data and width of swash zone for each beach site. Swash zones are absent from lagoons (Curl Curl, Dee Why). W, M, E = West, Mid, East, respectively.

Beach		Temperature (°C)		Swash zone Mean ± 1 s.d. (m)
		Mean ± 1 s.d.	Range	
Silver	3 W	20.6 ± 3.4	9.5	7.1 (2.5)
Silver	M	21.2 ± 3.1	9.3	6.9 (2.7)
Silver	E	21.4 ± 3.2	9.5	5.1 (2.2)
Silver	8 W	20.4 ± 3.3	9.5	7.2 (2.3)
Silver	M	20.5 ± 3.2	9.0	6.4 (1.5)
Silver	E	20.6 ± 3.1	8.3	5.2 (2.1)
Silver	12 W	21.2 ± 4.1	12.8	3.9 (1.6)
Silver	M	21.1 ± 3.3	9.5	3.7 (1.3)
Silver	E	21.1 ± 3.1	8.5	3.0 (0.9)
Chinaman's		20.9 ± 2.9	8.2	5.5 (1.1)
Clontarf		21.4 ± 2.8	7.7	2.7 (0.7)
Curl Curl		21.0 ± 3.9	11.8	–
Dee Why		20.8 ± 3.3	12.0	–

122

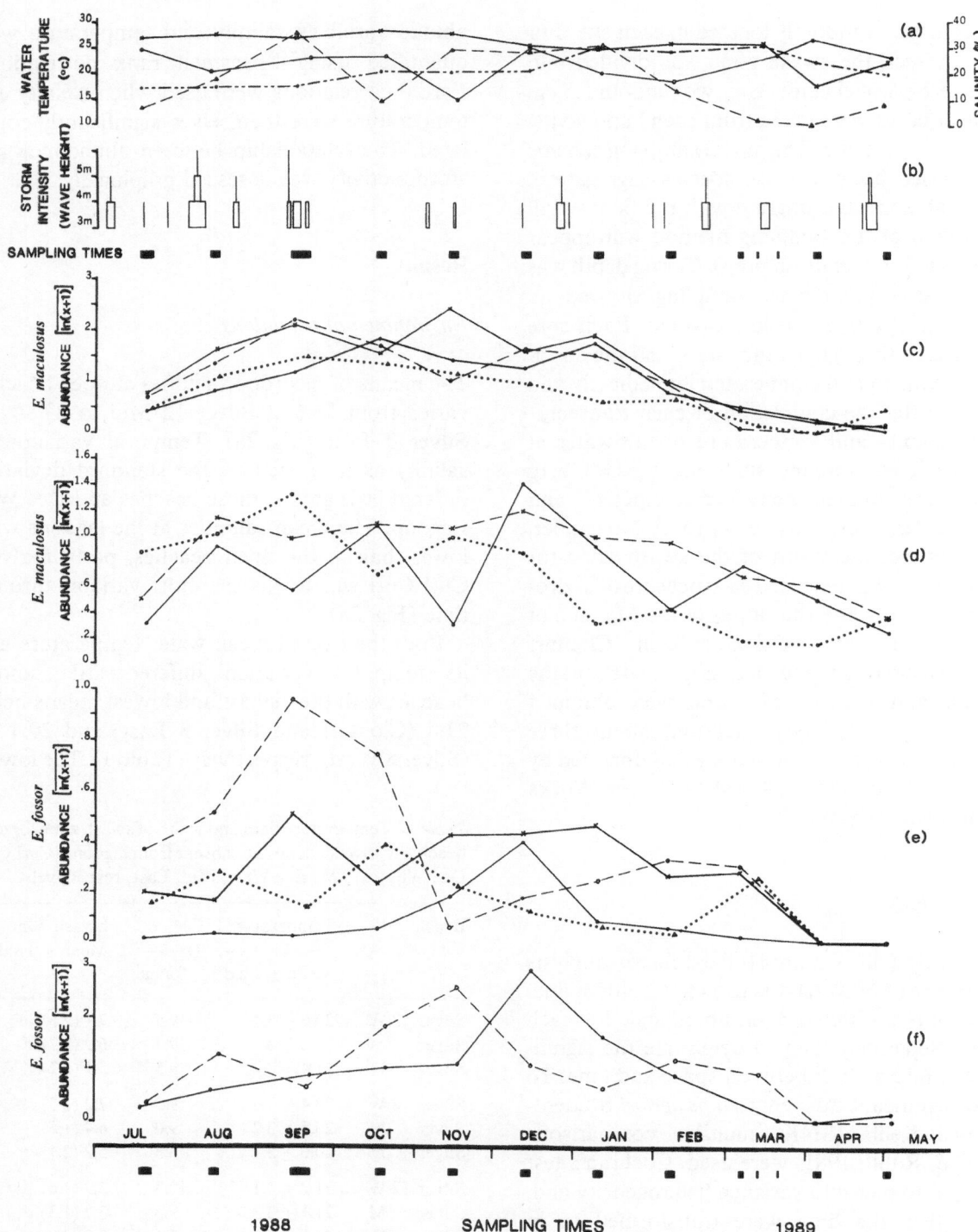

Fig. 2. Temporal distribution of physicochemical and biological variables. Standard errors were omitted for clarity but can be obtained from the senior author along with detailed (SNK) results. a) Mean salinity for Dee Why Lagoon o–o, Curl Curl Lagoon △– –△ and all open beaches x–x. Mean temperature for all sites ●– – –●. b) Duration and intensity of storms. c) Abundance of *E. maculosus* at Silver 3 West ●–●, Mid x–x, East o– –o and at Chinaman's Beach △ · · · △. d) Abundance of *E. maculosus* at Silver 8 West ●–●, Mid x–x and East o · · · o. e) Abundance of *E. fossor* at Silver 12 West ●–●, Mid x–x, East o– –o and Clontarf Beach △ · · · △. f) Abundance of *E. fossor* at Dee Why Lagoon ●–● and Curl Curl Lagoon o– –o.

recorded temperatures varied between 14.8 (Curl Curl) and 17.5 °C (Silver 12 East) and the highest between 28.8 (Silver 12 West) and 24.5 °C (Clontarf). The former occurred from July until September and the latter from January until March (Fig. 2a). The largest ranges occurred in the lagoons Curl Curl and Dee Why and at Silver 12 West. The other beaches had smaller ranges but varied little (Table 1).

The mean swash zone widths were greatest at Silver 3, Silver 8 and Chinaman's and least at Silver 12 and Clontarf (Table 1). All three Silver beaches showed a gradient of decreasing swash zone widths from west to east.

Two storms persisted for more than two days during the sampling period (Fig. 2b). These occurred in early August and late April with the former having the longer duration and greater intensity. In the two months prior to sampling, two other storms occurred in early and late May.

Patterns of abundance

The two species consistently occupied separate beaches: *Exoediceroides maculosus* on Silver 3, Silver 8 and Chinaman's Beaches and *Exoediceros fossor* on Silver 12 and Clontarf Beaches and Dee Why and Curl Curl Lagoons. Minor overlaps of distribution did occur but were rare. Consequently, the following analyses for each species exclude those beaches where that species was consistently absent or virtually so.

Exoediceroides maculosus

Significant differences in abundance occurred among beach sites (two-way ANOVA, $F_{site} = P < 0.001$; Fig. 2c, d). However, the patterns of difference were not consistent (two-way ANOVA, interaction term significant; SNK tests) and hence any general ecological importance of the main factors was obscured by complex interactions.

Differences among sites were greatest from August until January (SNK tests, Fig. 2c, d).

Although the patterns of difference varied among these times (SNK tests), the Silver 3 sites frequently yielded the most specimens and Silver 8 West and East often the least. At other times, the Silver 8 sample values often exceeded those for Silver 3 but differences were rarely significant. Silver 8 Mid and Chinaman's Beach often yielded intermediate sample abundances.

Temporal differences occurred at all sites (SNK tests, Fig. 2c, d). Abundances at most sites were large from August until January and small in July and from March until May. These amongmonth differences were greatest at the Silver 3 sites.

Exoediceros fossor

Significant differences in abundance occurred among sites (two-way ANOVA, $F_{site} = 282.2$, $P < 0.001$; Fig. 2e, f) and among months (two-way ANOVA, $F_{month} = 112.0$, $P < 0.001$; Fig. 2e, f). However, the patterns of difference were not consistent (two-way ANOVA interaction term significant; SNK tests).

Differences among sites were largest from August until February (SNK tests, Fig. 2e, f). Although the patterns of difference varied at these times (SNK tests), the two lagoon sites were frequently the richest and Silver 12 West and Clontarf often poorest. Spatial differences were rarely significant at other times with *E. fossor* being absent from all samples in April and virtually so in May.

Significant temporal differences were present at all sites but were most marked at the two lagoons (SNK tests, Fig. 2e, f). Although all sites showed least abundance at coincident times (April and May), the times of maximum abundance could occur in September (Silver 12 Mid and East), October (Clontarf), November (Curl Curl) or December (Silver 12 West and Dee Why).

Relationships with physicochemical factors

Most correlations between abundance and both temperature and salinity were not significantly

Table 2. Spearman correlation coefficients between the abundance of each species and salinity and temperature at each site. Where salinity and temperature were themselves significantly correlated (only at Silver 8 West), partial correlations with abundance were calculated. The Bonferroni procedure (Huitema, 1980) was used to adjust the type 1 error rate. * = $P < 0.05$ two sided test, $n = 11$ for all sites except Clontarf where $n = 8$. W, M, E = West, Mid, East, respectively.

Exoediceroides maculosus			*Exoediceros fossor*		
Site	Salinity	Temp.	Site	Salinity	Temp.
Silver 3 W	0.484	0.073	Silver 12 W	0.392	0.300
Silver 3 M	0.548	− 0.073	Silver 12 M	0.348	0.254
Silver 3 E	0.478	− 0.237	Silver 12 E	0.587	− 0.410
Silver 8 W	0.341	0.250	Clontarf	0.887*	− 0.386
Silver 8 M	0.466	− 0.331	Curl Curl	0.420	0.006
Silver 8 E	0.387	− 0.842*	Dee Why	0.416	0.260
Chinaman's	0.653	− 0.421			

different from zero (Table 2). The only exceptions were the negative correlation with temperature for *E. maculosus* at Silver 12 West and the positive correlation with salinity for *E. fossor* at Clontarf Beach.

Abundance was not obviously related to storm activity for either species (Fig. 2). The most persistent and intense storms, in August and April, coincided with rising and falling abundance respectively for *E. maculosus* and with rising abundance and no change, respectively for *E. fossor*. Sharp changes in abundance of both species could occur in the absence of strong storm activity (Fig. 2).

Discussion

Spatial and temporal differences in abundance of both species were not only statistically significant but also inconsistent i.e. particular spatial patterns did not persist over all months and temporal patterns varied among sites. Variability of this sort is common amongst local benthic infaunal communities and populations (Rainer, 1981; Jones *et al.*, 1986, 1988; Jones, 1987). Such variability makes it difficult to suggest controlling factors but implies that individual factors operate differently at different sites or else they operate similarly but asynchronously.

Storms could be a factor in the latter category because the sites varied in their directional aspect or degree of protection and therefore a given storm would not affect all sites equally. However, although storms have been implicated in beach faunal changes elsewhere (Scott, 1960; Brown, 1971), they had little or no apparent effect in this study, quite possibly because they were not especially severe. A second asynchronous factor affecting *E. fossor* in lagoons is the intermittent opening of lagoons to the sea. The associated fall in water level can be sufficiently rapid to cause mortality by stranding (Jones, unpubl. data).

In contrast, wind may be a factor operating differently at different sites. For example, a persistent westerly wind may move specimens from the West site on Silver Beaches 3, 8 and 12 to the East site. Some suggestions of this exist (e.g. in September). Other such factors could be the presence of large dead bivalve shells (*Anadara trapezia*) and stranded seagrass and algae whose abundance can vary among sites. The relationship between these factors and amphipod abundance will be evaluated when further data are available. Salinity and temperature explain little of these inconsistent differences because most sites experienced similar values at a given time and further, there was no evidence from correlation analysis that these factors affected abundance.

Despite such perplexing inconsistencies, some degree of pattern can be discerned for both species. For example, when *E. maculosus* was most abundant (July–January), significantly more specimens usually occurred at Silver 3 sites than at Silver 8 sites while the reverse pattern was rare throughout. These findings suggest that conditions at Silver 3 were more favourable although the beaches are similar in the following respects: salinity, temperature, slope, swash-zone width ($\approx$ exposure), directional aspect, *Anadara* shell density, and sediment grade (the last assessed subjectively) and they are only about one km apart. A probable difference between these beaches involves stranded seagrass and algae. Preliminary estimates indicate that Silver 3 has about 25% more weed than Silver 8. If stranded weed ultimately provides food for *E. maculosus*, as seems likely (Dexter, unpubl. data; Enequist, 1949), this factor could explain the observed differences. Inter-beach differences in the timing of reproductive activity and recruitment (Dexter, 1985) would also explain these differences but this is unlikely because populations at both beaches increased simultaneously following low abundance in July.

Spatial differences in the abundance of *E. fossor* also showed some degree of pattern at times of high abundance (August – February) with the two lagoon populations usually exceeding those at the open beaches. This difference is related to the degree of wave action, a factor claimed to be of major importance for beach species (Eleftheriou & Nicholson, 1975; Croker *et al.*, 1975; Withers, 1977; Dexter, 1984). This factor may be particularly influential for *E. fossor* because it is not found on highly-exposed open beaches and increases in number through semi-exposed to protected beaches (Dexter, 1983a, 1985; Jones, unpubl. data).

Lagoons also differ from open beaches by having greater ranges of salinity and temperature but this variability had no apparent negative effects. In fact, the euryhalinity of *E. fossor* is remarkable. Previously, it has been found in salinities ranging from 16.4 to 28.0‰ (Fearn-Wannan, 1968; Barnard & Drummond, 1982) and in the present study, from 2 to 35.5‰.

Temporal changes in the abundance of *E. maculosus* also showed some degree of consistency with most sites showing increases from July until September. This increase was caused by an influx of juveniles (Jones, unpubl. data) and is consistent with the reproductive peak in July–September found by Dexter (1985) in 1980. The small numbers at most sites from March until May probably mark the death of most of the July–September cohort.

The only inter-site consistent temporal pattern for *E. fossor* was the virtual absence in April and May at all sites. As no catastrophic disturbances were noted, this probably represents a failure of reproduction although the species is capable of continuous breeding, albeit with several maxima (Dexter, 1985).

In conclusion, both these exoedicerotid species display variability in their temporal and spatial patterns of abundance. Much of this variability is currently unexplained in terms of specific factors although Dexter (1984) implied that widely-fluctuating abundance in peracarids is facilitated by their being short-lived, opportunistic species. Such variability means that the assessment of the effects of disturbances such as pollution may be hampered by the absence of adequate control sites.

Acknowledgements

Keith McGuinness advised on the sampling design and statistics and Scott Markich assisted in developing the sampling methodology. Many volunteers including Kate Attwood, Vicki Tzoumis, Ross Blick, Jane Nelson, Joanne Procter, Andrew Irvine, Roger Springthorpe, Andrew West, Marilyn Luchetti, and Peggy O'Donnell generously gave their time. Greg Skilleter helped to organise the data.

We are grateful to all the above and to the Manly Hydraulics Laboratory for supplying data on wave distribution and heights. Peggy O'Donnell, Les Watling and an anonymous reviewer commented usefully on the manuscript which was typed by Sally Cowan and Kelly Havansky.

References

Barnard, J. L. & M. M. Drummond, 1984. Redescription of *Notoediceros tasmaniensis* Bousfield and a note of the synonymy of *Warreyus* Barnard and Drummond with *Exoediceroides* Bousfield (Crustacea: Amphipoda: Exoedicerotidae). Proc. r. Soc. Victoria 96: 25–32.

Bonsdorff, E., 1983. Effects of experimental oil exposure on the fauna associated with *Corallina officienalis* L. in intertidal rock pools. Sarsia 68: 149–156.

Brown, A. C., 1971. The ecology of the sandy beaches of the Cape Peninsula, South Africa Part I: Introduction. Trans. r. Soc. S. Afr. 39: 247–279.

Croker, R. A., 1968. Distribution and abundance of some interstitial sand beach amphipods accompanying the passage of two hurricanes. Chesapeake Sci. 9: 157–162.

Dexter, D. M., 1983a. Community structure of intertidal sandy beaches in New South Wales, Australia. In A. McLachlan & T. Erasmus (eds), Sandy Beaches as Ecosystems. Dr W. Junk, The Hague: 461–472.

Dexter, D. M., 1983b. A guide to sandy beach fauna of New South Wales. Wetlands 3: 94–104.

Dexter, D. M., 1984. Temporal and spatial variability in the community structure of the fauna of four sandy beaches in south-eastern New South Wales. Aust. J. mar. Freshwat. Res. 35: 663–72.

Dexter, D. M., 1985. Distribution and life histories of abundant crustaceans of four sandy beaches of south-eastern New South Wales. Aust. J. mar. Freshwat. Res. 36: 281–289.

Eleftheriou, A. & M. D. Nicholson, 1975. The effects of exposure on beach fauna. Cah. Biol. mar. 16: 695–710.

Enequist, P., 1949. Studies on the soft-bottom amphipods of the Skagerak. Zoologiska Bidrag Fran Uppsala 28: 297–492.

Fairweather, P. G., 1988. Ecological changes due to our use of the coast: research versus effort. Proc. ecol. Soc. Aust. 16: 71–77.

Fearn-Wannan, H. J., 1968. Littoral amphipods of Victoria. Part 2. Proc. r. Soc. Victoria 81: 127–135.

Huitema, B. E., 1980. The Analysis of Covariance and Alternatives. J. Wiley & Sons, N.Y., 444 pp.

Jones, A. R., 1987. Temporal patterns in the macrobenthic communities of the Hawkesbury Estuary, N.S.W. Aust. J. mar. Freshwat. Res. 38: 607–624.

Jones, A. R., C. J. Watson-Russell & A. Murray, 1986. Spatial patterns in the macrobenthic communities of the Hawkesbury Estuary, N.S.W. Aust. J. mar. Freshwat. Res. 37: 521–543.

Jones, A. R., A. Murray & G. A. Skilleter, 1988. Aspects of the life history and population biology of *Notospisula trigonella* (Bivalvia: Mactridae) from the Hawkesbury Estuary, Southeastern Australia. The Veliger 30: 267–277.

MacLachlan, A. & T. Erasmus eds., 1984. Sandy Beaches as Ecosystems. Dr W. Junk, The Hague. 757 pp.

MacLachlan, A. & P. Hesp, 1984. Faunal response to morphology and water circulation of a sandy beach with cusps. Mar. Ecol. Progr. Ser. 19: 133–144.

McGuinness, K. A., 1988. The Ecology of Botany Bay and the Effects of Man's Activities: a critical synthesis. Inst. Marine Ecology, University of Sydney. 114 pp.

Rainer, S., 1981. Temporal patterns in the structure of macrobenthic communities of an Australian estuary. Estuar. coast. shelf. Sci. 13: 597–620.

Sanders, H. L., J. R. Grassle, G. R. Hampson, L. S. Morse, S. Garner-Price & G. C. Jones, 1980. Anatomy of an oil spill: Long-term effects from the grounding of the barge 'Florida' off West Falmouth, Massachusetts. J. mar. Res. 38: 265–380.

Scott, A., 1960. The fauna of the sandy beach, Village Bay, St. Kilda: a dynamic relationship. Oikos 11: 153–160.

Sokal, R. R. & F. J. Rohlf, 1981. Biometry: The Principles and Practice of Statistics in Biological research. 2nd Edition. W. H. Freeman and Co., San Francisco. 859 pp.

Withers, R. G., 1977. Soft-shore macrobenthos along the south west coast of Wales. Estuar. coast. mar. Sci. 5: 467–484.

Hydrobiologia **223**: 127–140, 1991.
L. Watling (ed.), VIIth International Colloquium on Amphipoda.
© 1991 *Kluwer Academic Publishers.*

Local distributions of sandhoppers and landhoppers (Crustacea: Amphipoda: Talitridae) in the coastal zone of western Tasmania

A.M.M. Richardson, R. Swain & S.J. Smith*
Dept. of Zoology, University of Tasmania, GPO Box 252C, Hobart, Tasmania 7001, Australia; *Dept. of Parks, Wildlife & Heritage, GPO Box 44A, Hobart, Tasmania 7001, Australia*

Key words: Talitridae, zonation, supralittoral, maritime

Abstract

Twelve species of talitrid amphipods were recorded from pitfall transects across the supralittoral and maritime zones at three localities on the west coast of Tasmania; three were sandhoppers, the rest were landhoppers. There was a sharp demarcation between the highest range of the sandhoppers and the lowest range of the landhoppers. Organic content and sodium content of the substrate in the range of the sandhoppers were very low, but rose sharply as the sand was colonised by plants, and landhoppers replaced sandhoppers. Coastal group landhopper species were restricted to a zone about 40–70 m above the high tide mark. Cooler, wetter weather increased the activity of sandhoppers, but only affected landhopper activity slightly. These differences support the idea that landhoppers did not evolve directly from sandhopper ancestors.

Introduction

The Tasmanian landhopper fauna consists of at least 18 species (Friend, 1987a), two of which have been introduced (Richardson, unpublished data), and of the native species all but one are endemic. This diverse fauna occupies an island of about 67 000 km². The climate is temperate maritime with a strong east-west gradient in rainfall, from over 3000 mm per annum in the west to under 800 mm in the east (Bureau of Meteorology, 1979).

In his foundation study of the taxonomy and zoogeography of Tasmanian landhoppers, Friend (1987a) recognised three groups of species, based on their distributions: the western forest group (6 spp.), the eastern forest group (4 spp.) and the coastal group (3 + spp.). The members of the latter group are restricted to coastal situations and Friend speculated that they are less able to retain ions than their inland relatives and so are more dependent on ionic precipitation.

The west coast of Tasmania is directly exposed to the winds of the 'roaring forties' and consequently experiences a cool, wet climate and, presumably, a heavy precipitation of ions in rainfall and seaspray. Much of the west coast is uninhabited and although there have been repeated fires in some areas, many places remain where undisturbed native vegetation, either swamp, forest or scrub (depending on the degree of exposure), can be found to the high water mark. The litter in these situations supports coastal group landhoppers and members of the forest groups, and these may approach or overlap the upper limits of the supralittoral talitrids.

Landhoppers represent one of the few successful attempts by crustaceans to colonise land. The distribution, diversity and ecological significance of landhoppers (Friend & Richardson, 1986), in

Australia and New Zealand at least, leave no doubt that the attempt has been successful. The establishment of synanthropic species in equivalent habitats in the northern hemisphere suggests that historical biogeographic factors have been important in preventing landhoppers from achieving a greater worldwide distribution.

The physiological ecology of land invasion by talitrids has recently been reviewed by Spicer *et al.* (1987). They follow Bousfield (1984) in proposing two sources, and hence ecological routes, of the fully terrestrial species in an evolutionary radiation that began in the mid-Cretaceous as Gondwana and Laurasia separated. The sandhoppers and/or beachfleas (Bousfield's Groups II and III) gave rise to the sexually dimorphic Group IV : 2 landhoppers, presumably via sandy and rocky open-coast beaches, while the more ancient, less sexually dimorphic Group IV : 1 landhoppers arose directly from marine littoral species via the palustral talitrids (Bousfield's Group I) through salt-marshes, mangroves and perhaps coastal streams, in sheltered situations. Both beachfleas and sandhoppers are modified, physiologically and morphologically, for life in the supralittoral zone, where they experience more severe physical conditions, of water availability at least, than forest-dwelling landhoppers. No data are available to describe the adaptations of palustral talitrids. Problems of ionic and osmotic regulation appear to limit the spread of beachfleas and sandhoppers into terrestrial habitats (Spicer *et al.*, 1987), since their greatest penetration from the high water mark always occurs in islands or coastal situation where there is substantial ionic input by seaspray (e.g. Bagenal, 1957; Richardson, 1980).

The west coast of Tasmania, with its diversity of talitrids and undisturbed habitats, provides an ideal opportunity to examine the stages in the terrestrialization of landhoppers, in the form of the present adaptations of a series of species spanning both the beach/land interface and increasingly terrestrial habitats. Friend (1987b) found that South Cape Bay, on the far south coast, has a high diversity of landhoppers, including most of the members of the coastal group. This study deals with a series of species at S. Cape Bay and two other open-coast, sandy beaches.

The objectives of this study were to describe the distribution of upper supralittoral, coastal group and forest group talitrids at three localities on the west coast of Tasmania, and to correlate their distributions with environmental parameters, with a view to generating hypotheses about the stages involved in the evolution of the terrestrial habit in landhoppers.

Methods

The undisturbed parts of Tasmania's west coast are by their very nature inaccessible, so the choice of study sites was restricted to places where access was provided in one way or another. Two sites, Hibbs Bay and Mulcahy Bay, were visited twice only, by air. The third site, South Cape Bay, is two hours walk from the nearest road head, and it has been visited more often. The location of the sites is shown in Fig. 1.

At both Hibbs Bay and S. Cape Bay, two transect lines were set up in February 1989, at right angles to the high water mark, from below the extreme high water mark of spring tides (EHWS) to between 40 and 70 m inland. The entire Hibbs Bay transects were set and collected during 4 days in February 1989, but at S. Cape Bay the two transects were extended, and a second series of collections made, in July. At each meter mark along these transects a pitfall trap was set. The traps consisted of a plastic container of 5 cm diameter and 10 cm depth, let into the soil so that the lip of the trap was flush with the surface of the sand, soil or leaf litter. Saturated picric acid solution (5–10 ml) was placed in the traps as a non-volatile preservative and Petri dish lids were supported on twigs over the traps to prevent rainwater from filling them. The traps were emptied at various time intervals, depending on the time available (Table 1). Animals from each trap were preserved separately in alcohol.

At each trap site, the vegetation was noted and soil samples were collected into sealed containers.

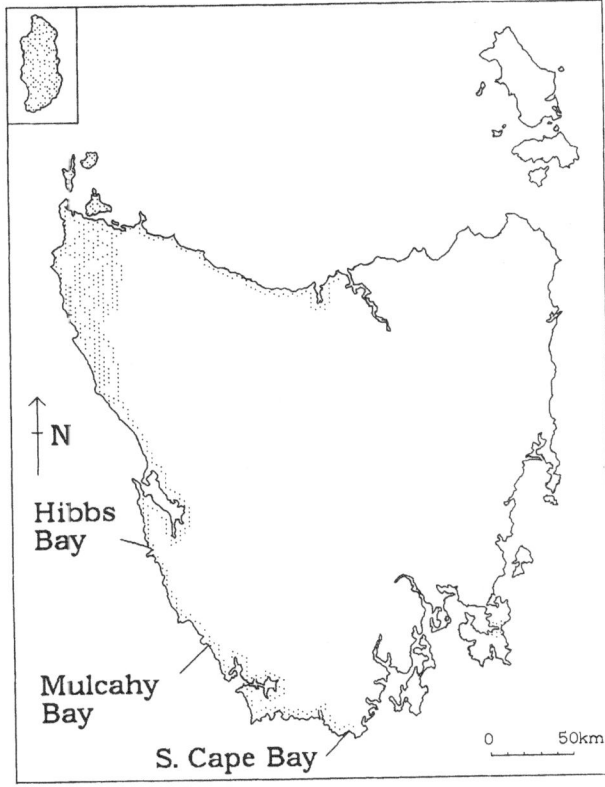

Fig. 1. Map of Tasmania showing the collecting localities. The distributions of the 3 species recorded by Friend (1987) in the coastal group are also shown.

Profiles of the transects were drawn up using simple surveying methods. In the laboratory, the soil samples were weighed, dried for 48 h at 110 °C and reweighed. Subsamples of the dried soil were ashed at 450 °C for 4 h and then reweighed. A fixed weight (1 g) of dry soil was mixed with 100 ml of distilled water and allowed to stand for 24 h; the sodium concentration of the supernatant was then measured by flame photometry.

At Mulcahy Bay, three short transects were examined by hand collecting over two days. Hand collections were also made at various times at the other two sites for comparison with the pitfall catches.

Results

A total of 12 species, three of which were supra-littoral sandhoppers, were found on the three transects; they are listed in Table 2. Two undescribed species of *Tasmanorchestia* Friend were identified, and none of the sandhoppers has been placed in a specific taxon because of the poor taxonomic knowledge of this group. Morino & Miyamoto (1988) have redefined and restricted the previously very broad genus *Talorchestia* Dana. Since the three sandhopper species found here do not fit their diagnosis of *Talorchestia* or any other sandhopper genus, we will use the generic designation 'Talorchestia'. Descriptions of the new taxa dealt with here are planned (A.M.M. Richardson, in prep.).

One of the sandhoppers, 'Talorchestia' sp.3, shows a remarkable morphology in the mature males. The hind margins of the basis, ischium and merus of peraeopod 7 have become greatly expanded, so that the extension of the basis is pedunculate; the whole limb is massive and some of the extensions apparently articulate with each

Table 1. Pitfall trapping programs at west coast sites.* the additional 15 traps were spaced at 2 m intervals. Collections were made by hand at Mulcahy Bay.

Locality	Trapline	Number of Traps	Duration (days)	Emptied
Hibbs Bay	A	71	4	daily
	B	50	4	daily
S. Cape Bay (Feb)	A	39	10	daily
	B	44	10	daily
S. Cape Bay (July)	A	65	14	once
	B	59*	14	once

Table 2. Sandhoppers and landhoppers collected at three coastal sites in western Tasmania.

Species	Eco-morphological Group*	Hibbs Bay	Mulcahy Bay	S. Cape Bay
'Talorchestia' sp. 1	Sandhopper (5-dentate)	×	×	×
'Talorchestia' sp. 2	Sandhopper (5-dentate)	×	×	×
'Talorchestia' sp. 3	Sandhopper (4-dentate)	×	×	×
Austrotroides maritimus	Landhopper (IV : 1): coastal group	×	×	×
A. longicornis	Landhopper (IV : 1): coastal group			× †
Keratroides rex	Landhopper (IV : 1): coastal group	×	×	×
Tasmanorchestia sp. 1	Landhopper (IV : 2): coastal group	×	×	×
Tasmanorchestia sp. 2	Landhopper (IV : 2): coastal group		×	
Mysticotalitrus tasmaniae	Landhopper (IV : 1): E. forest group	×		×
M. cryptus	Landhopper (IV : 1): E. forest group	×		
K. vulgaris	Landhopper (IV : 1): E. forest group	×		×
K. angulosus	Landhopper (IV : 1): E. forest group			×
Neorchestia plicibrancha	Landhopper (IV : 1): W. forest group			×

* IV : 1 – simplidactylate, weakly sexually-dimorphic landhoppers; IV : 2 – cuspidactylate, sexually-dimorphic landhoppers (Bousfield, 1984); 4-, 5-dentate sandhoppers: substrate-modifying, sand beach talitrids having 4 or 5 teeth on the left lacinia mobilis of the mandible (Bousfield, 1984); coastal, E. forest, W. forest: distributional groupings of Friend (1987).
† *Austrotroides longicornis* was collected from one site only, about 1 km inland from S. Cape Bay.

other. In addition, the dorsum of the last abdominal segment is produced into two broad, blunt projections. These features apparently develop in the last few moults before maturity and are completely absent in the females; their function is unknown.

Profiles of the transects at Hibbs and S. Cape Bay are shown in Fig. 2, with the water, organic and Na content of the soil samples from each trap. Hibbs Bay is more exposed to the westerly and south westerly winds than S. Cape Bay, and this is reflected in the gradient of the profiles. At Hibbs Bay the dunes begin with a steep face of bare sand at the top of which is the highest point of the profile, from where the transects are fairly flat. At S. Cape Bay the fore dune is very low with only a small sand cliff; there is then a series of ridges and slacks, increasing in height to the fourth ridge. The vegetation also differs; at Hibbs Bay the canopy is relatively low and the vegetation is a dense, wind-pruned coastal scrub not exceeding 5 m, while at S. Cape Bay the canopy is higher, with more tree species, and develops into a tall (> 20 m) closed eucalypt forest beyond the fourth ridge.

The levels of soil moisture, organic content and Na concentration also differ between the sites, generally being higher at Hibbs Bay (Table 3). Along each transect, levels of moisture, organic content and Na all rise sharply at the transition from the open foredune to the soil beneath the shrub canopy. Further inland there is little systematic trend, apart from a decline in Na levels at the inland end of the Hibbs B transect. There were, however, strong correlations between the levels of the three factors, particularly in the summer data (Table 4).

The July samples collected from the extensions to the S. Cape Bay transects show lower levels of Na (Transect A: $t_{56} = 4.311$, $p < 0.001$; Transect B: $t_{46} = 2.324$, $p = 0.0246$) than the summer samples, but the soil moisture and levels of organic material do not differ significantly.

The catches from the pitfall traps (Fig. 3) show

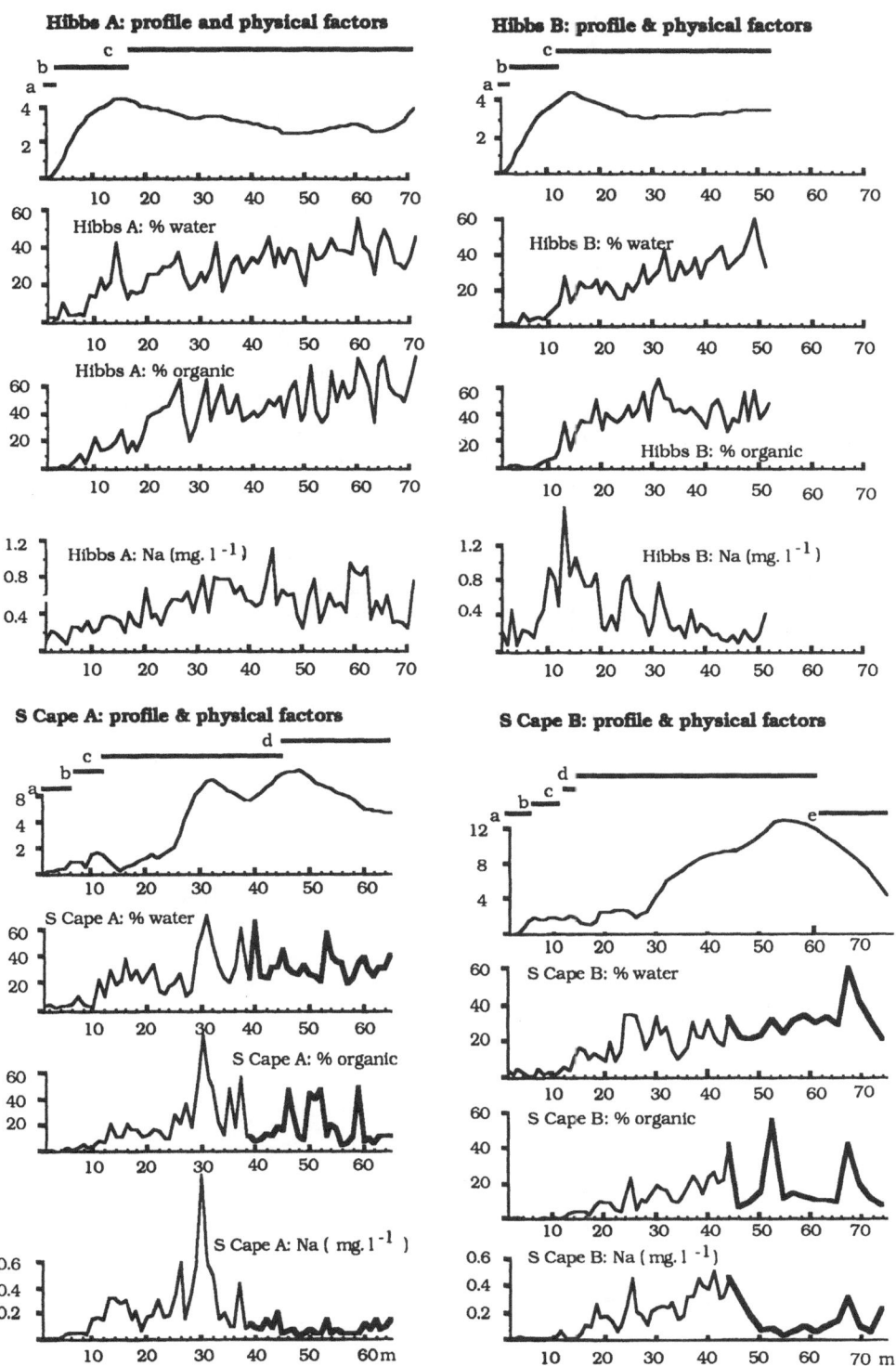

Fig. 2. Physical factors along transects at Hibbs Bay and S. Cape Bay. S. Cape A – a: open sand, b: low shrubs including *Correa* sp., c: mixed forest, d: tall eucalypt forest. S. Cape B – a: open sand, b: open sand + herbs, c: low shrubs including *Correa* sp., d: mixed forest, e: tall eucalypt forest. Hibbs A & B – a: open sand, b: low shrubs & bracken, c: wind-pruned coastal scrub.

Table 3. Mean values and 95% confidence limits for the physical factors measured at each trap site. Values for the extensions to the S. Cape Bay transects have been excluded since they were measured in July.

Factor	95% Confidence Limits		
Site	Mean	Lower	Upper
% water			
S. Cape A	29.17	23.39	34.96
S. Cape B	19.44	15.83	23.05
Hibbs A	26.07	22.88	29.26
Hibbs B	26.96	24.01	29.91
% organic			
S. Cape A	25.15	17.64	32.66
S. Cape B	11.33	8.64	14.04
Hibbs A	40.95	4.60	47.30
Hibbs B	43.17	39.28	47.09
Na concentration (mg l^{-1})			
S. Cape A	0.301	0.203	0.400
S. Cape B	0.212	0.166	0.258
Hibbs A	0.542	0.463	0.622
Hibbs B	0.468	0.364	0.573

distinct zonations in the distribution of species. Among the sandhoppers, there is a clear series, with 'Talorchestia' species 1 found closest to the sea, followed by 'Talorchestia' species 2, which in turn is replaced by 'Talorchestia' species 3. There is substantial overlap between these three in the summer samples, and this is much greater in the winter, although the series can still be distinguished.

The replacement of sandhoppers ('Talorchestia' species 3) by landhoppers (*Austrotroides maritimus* Friend, *Keratroides rex* Friend and/or *Tasmanorchestia* sp.) is very sharp (but note the single individual of 'Talorchestia' sp3 trapped at 40 m on Hibbs transect A). In the summer samples at S. Cape Bay the overlap was never more than 2 traps (= 2 m), and on several nights there was no overlap at all. During the winter series on transect A there was a complete replacement of sandhoppers by *A. maritimus* from trap 9 to trap 10, with no sandhoppers in trap 10 and no *A. maritimus* in trap 9 after 14 nights of catching.

Once beyond the landward range of the sandhoppers, the catch at S. Cape was dominated by *A. maritimus* for the next 20–30 m inland. At Hibbs Bay, two other members of the coastal

Table 4. Correlations between the physical factors measured at each trap for each sampling occasion and site. At S. Cape in July measurements were only taken from the traps on the extensions to the transects. *: $p < 0.05$.

S. Cape A (Feb.)	% water	% org.	Na conc.	S. Cape A (July)	% water	% org.	Na conc.
% water	1				1		
% org.	0.821*	1			0.269	1	
Na conc.	0.717*	0.831*	1		0.258	−0.43	1
S. Cape B (Feb.)				**S. Cape B (July)**			
% water	1				1		
% org.	0.788*	1			0.592*	1	
Na conc.	0.784*	0.902*	1		0.283	0.047	1
Hibbs A (Feb.)							
% water	1						
% org.	0.865*	1					
Na conc.	0.578*	0.628*	1				
Hibbs B (Feb.)							
% water	1						
% org.	0.792*	1					
Na conc.	−0.172	0.032	1				

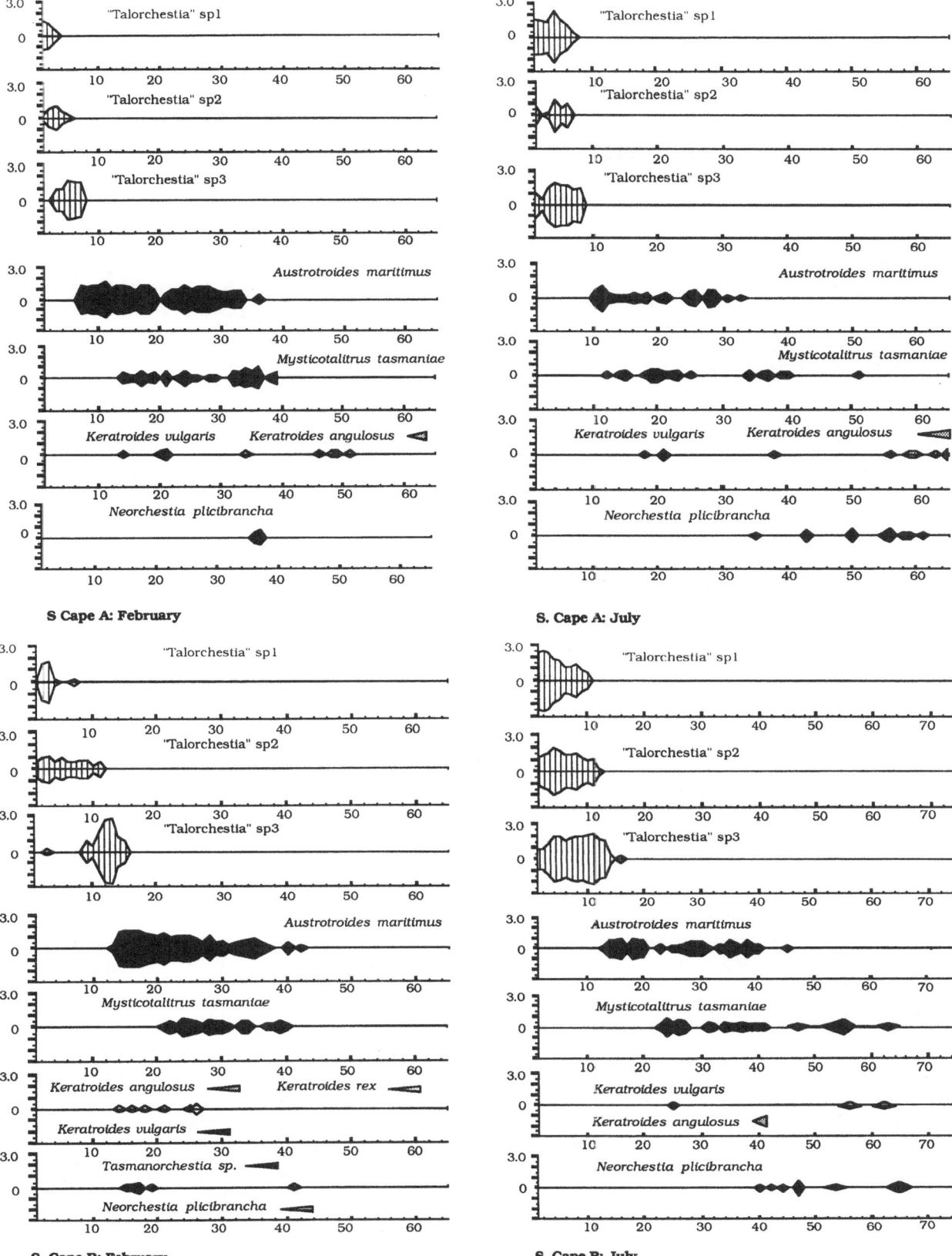

Fig. 3.

134

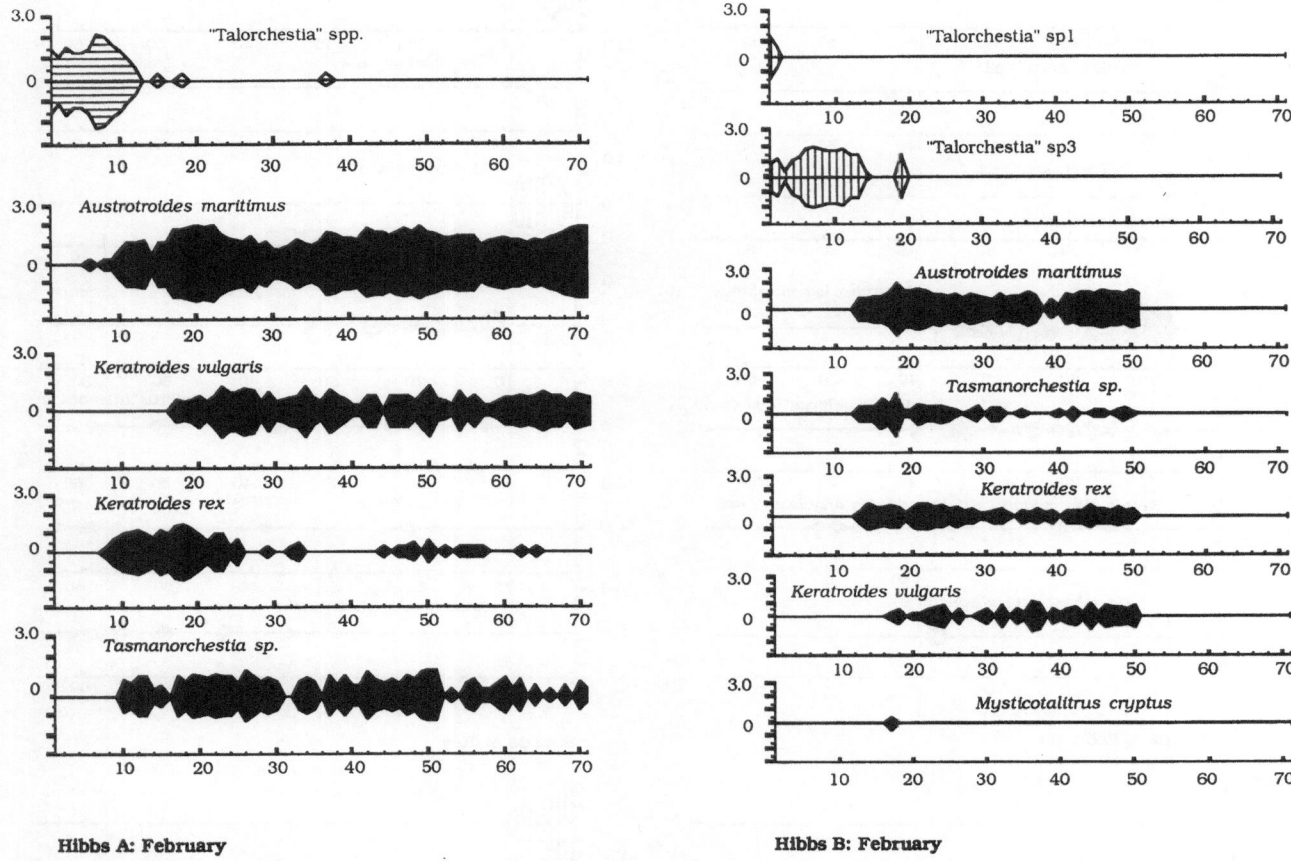

Hibbs A: February

Hibbs B: February

Fig. 3. Distribution and abundance of sandhoppers and landhoppers along pitfall transects at Hibbs Bay and S. Cape Bay. Horizontal axis is in metres from trap 1; vertical axis is $\log_{10}(X + 1)$, where X is the daily catch of each species in each pitfall.

group, *Keratroides rex* and *Tasmanorchestia* sp., were also common over the full length of the transect, although both showed a tendency to decline in numbers further from the sea. *Tasmanorchestia* sp. and *Keratroides rex* also appeared at S. Cape Bay, but in very low numbers, and only on transect B in the summer series.

At Hibbs Bay, non-coastal group species were represented only by *Keratroides vulgaris* and *Mysticotalitrus cryptus* Friend, but at S. Cape Bay five species of forest amphipods were present: *Keratroides vulgaris* Friend, *Keratroides angulosus* Friend, *Mysticotalitrus tasmaniae* (Ruffo), *M. cryptus* and *Neorchestia plicibrancha* Friend. Of these, only *M. tasmaniae* was caught frequently, appearing in the middle range and tending to disappear by the most landward traps. *Keratroides*

vulgaris appeared spasmodically throughout the vegetated section of the transects, while *Neorchestia plicibrancha* and *K. angulosus* were most frequently caught at the landward end.

The catch data at both sites were examined for any positive or negative associations between landhopper species. At Hibbs Bay there were positive correlations between the three coastal group species, but no strong relationships emerged in the S. Cape Bay catches. With the exception of *A. maritimus* and *M. tasmaniae*, the numbers of landhoppers caught there were too low for a meaningful analysis.

Samples collected by hand, during the day, (Table 5) from the 4 slacks of S. Cape Bay transect B did not contain the same proportions of species as the pitfall traps at all sites. At the

Table 5. Hand collections compared with pitfall catches at S. Cape Bay transect B, and various other hand collections from the area.

Locality	Collection type	
Species	Hand	Pitfall
First Slack		
'Talorchestia' sp. 2	0	1
'Talorchestia' sp. 3	3	96
A. maritimus	125	1
K. vulgaris	20	0
2nd Slack		
'Talorchestia' sp. 3	0	3
A. maritimus	125	24
K. vulgaris	78	0
M. tasmaniae	30	0
K. angulosus	6	0
N. plicibrancha	3	0
3rd Slack		
A. maritimus	2	4
K. vulgaris	128	1
M. tasmaniae	9	8
K. angulosus	11	0
N. plicibrancha	8	0
4th Slack		
K. vulgaris	40	0
M. tasmaniae	4	4
K. angulosus	75	2
N. plicibrancha	35	6
70 m inland from end of Transect B		
K. vulgaris	1	–
K. angulosus	14	–
N. plicibrancha	6	–
60 m inland from end of Transect A		
M. cryptus	5	–
N. plicibrancha	11	–
1 km inland, *Melaleuca*/eucalypt regrowth		
M. cryptus	14	–
N. plicibrancha	14	–
Austrotroides longicornis	6	–
5 km inland, rainforest		
N. plicibrancha	8	–
S. Cape Bay Lagoon, marginal reeds		
Tasmanorchestia sp.	6	–
A. maritimus	1	–

seaward end, hand samples failed to catch sandhoppers but caught a much higher number of *A. maritimus* and *K. vulgaris*. Further landward,

most species were much less likely to be trapped than caught by hand, and only *A. maritimus* and *M. tasmaniae* were over-represented in the traps.

Further inland, beyond the traplines at S. Cape Bay, the landhopper community in forest litter changes, with no trace of the coastal group species or *Mysticotalitrus tasmaniae*. *K. angulosus* and *Neorchestia plicibrancha* become the dominants, with *Austrotroides longicornis* Friend appearing at a drier site about 1 km from the beach. Further inland still, *N. plicibrancha* was the only species collected in rainforest several kilometers from the coast. The collection of *Tasmanorchestia* sp. in some numbers from near a sheltered lagoon behind the main beach is interesting in view of its rarity in the pitfall catches.

The summer and winter catches as S. Cape Bay differ only slightly in the distribution of species, the biggest differences being the blurring of the zonation of sandhoppers mentioned above. A comparison of the numbers of each species caught per day in summer and winter (Table 6), shows that while the catch of landhoppers (*A. maritimus* and *M. tasmaniae*) was higher in February, the catch of two of the sandhoppers was higher in the winter.

The hand collections at Mulcahy Bay (Fig. 4) do not extend far inland, but show a similar sharp transition between the sandhoppers and the landhoppers to that shown at the other sites. The most seaward landhoppers species are *Keratroides rex* and two species of *Tasmanorchestia*. It is not known whether *Austrotroides maritimus* also occurs at this site.

The 10 day series of daily catches at S. Cape

Table 6. Comparison between summer and winter catches of the five most numerous species at S. Cape Bay, using paired t-tests. ***: $P > 0.001$, **: $0.001 < P < 0.01$, *: $0.01 < P < 0.05$, ns: $P > 0.05$.

Species	Transect A	Transect B
'Talorchestia' sp. 1	-1.263^{ns}(J > F)	-2.684^{*}(J > F)
'Talorchestia' sp. 2	-0.763^{ns}(J > F)	-3.143^{**}(J > F)
'Talorchestia' sp. 3	-0.956^{ns}(J > F)	-0.968^{ns}(J > F)
A. maritimus	5.970^{***}(F > J)	4.306^{***}(F > J)
M. tasmaniae	2.745^{***}(F > J)	2.359^{*}(F > J)

136

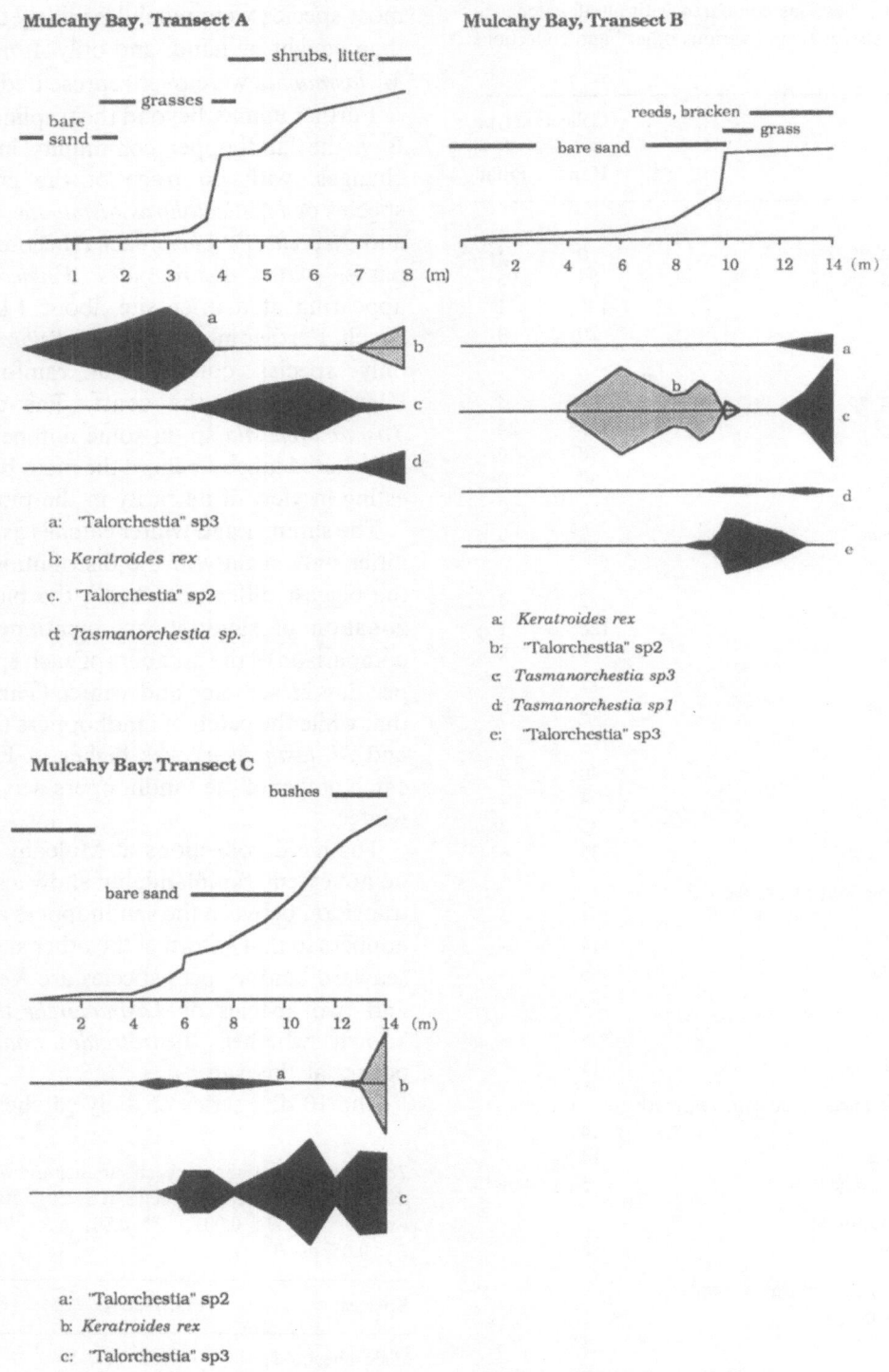

Fig. 4. Transect profiles, and the distribution and abundance of sandhoppers and landhoppers collected by hand, on three transects at Mulcahy Bay. Horizontal axis is in metres from trap 1; vertical axis is catch of each species at each station.

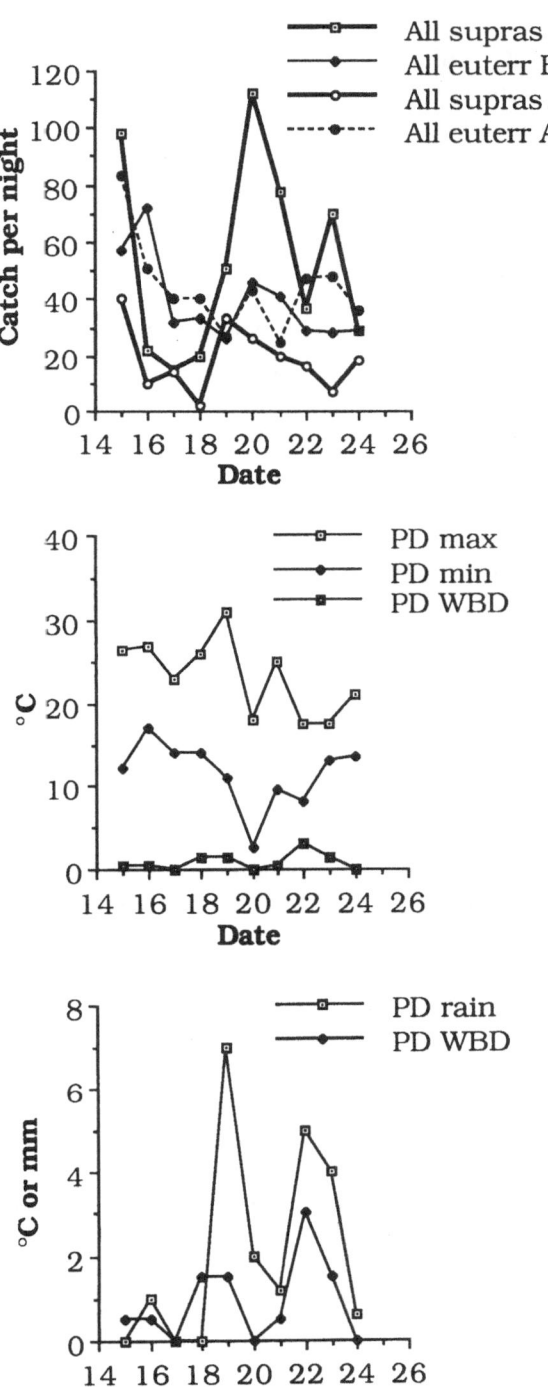

Fig. 5. Daily pitfall catches of sandhoppers and landhoppers and weather parameters over 10 days at S. Cape Bay in February, 1989. PDmax: daily maximum temperature at Port Davey (°C); PDmin: daily minimum temperature at Port Davey (°C); PD WBD: daily wet bulb depression at Port Davey (°C); PD rain: daily rainfall at Port Davey (mm).

Bay allows some examination of how well the various species were trapped in comparison to the weather. Figure 5 shows the daily catches of sandhoppers and landhoppers for both transects alongside the daily maximum and minimum temperatures, wet bulb depression and rainfall at Port Davey, the nearest suitable weather station, about 50 km NW of S. Cape Bay. Rain fell on 13 February, just before the trapping began, and there followed a short spell of warm fine weather which was broken during the middle of the trapping series by the passage of a cold front on 19 February. The daily maximum temperature dropped by over 10 °C, the minimum temperature dropped to 2 °C, 7 mm of rain fell and the wet bulb depression rose by 1.5 °C. Over the next few days temperatures rose a little and there was further rain.

The responses of the two groups of amphipods to the passage of the front differ. The sandhoppers' activity had been dropping since the start of the trapping run, but rose sharply during the passage of the front and remained high over the remaining days. In contrast, the landhoppers, whose numbers were dominated by *A. maritimus*, showed little or no change in their activity following the change in the weather.

While there is a clear relationship between the sandhoppers/landhopper interface and a sharp change in soil moisture, soil organic content and Na concentration, no consistent correlations were found between the abundances of any individual species and the levels of these soil parameters measured at each trap.

Discussion

The collections described here emphasize the remarkable and perhaps unique diversity of talitrid amphipods in coastal regions of western Tasmania (Friend, 1987a). While New Zealand shores should provide similar conditions, and the supralittoral talitrids and landhoppers are diverse (Hurley, 1956, 1957; Duncan, 1984), the New Zealand landhopper fauna apparently lacks simplidactylate landhoppers, perhaps because of its early separation from Gondwana (Bousfield, 1984). Macintyre (1963) examined the supralittoral fringe of beaches in the Canterbury region of New Zealand's South Island and recorded 7 species of sandhoppers and beachfleas, but did not mention landhoppers. Several New Zealand supralittoral species (*Talorchestia spadix* Hurley, *T. chathamensis* Hurley, *T. telluris* (Bate), *Orchestia chiliensis* Milne-Edwards, *O. miranda* Chilton and *O. parva* (Chilton)), and some from the African South Atlantic coast, show modification of the last walking leg of mature males similar to 'Talorchestia' sp3, but none of them, except perhaps *Talorchestia telluris* (in which the extension is to the carpus of peraeopod 7), shows such extreme enlargement, or has a pedunculate extension to the basis of peraeopod 7. Hurley (1956) suggests that *T. telluris* is modified for burrowing, but makes no suggestions why the modifications should be confined to mature males.

All of the landhoppers species collected here were also found in the S. Cape Bay area by Friend (1987b) in a study of habitat use by landhoppers. He also found *Austrotroides leptomerus* Friend, but almost always in above-ground accumulations of litter. Friend notes the disappearance of coastal group species within 100 m of the high water mark.

The species we collected include representatives of three of Bousfield's talitrid groups: sandhoppers, simplidactylate landhoppers and cuspidactylate landhoppers. The absence of beachfleas reflects the relatively high-energy nature of the sand beaches, but members of this group have been collected nearby (AMMR, unpublished data). Palustral talitrids also occur in Tasmania (*Eorchestia* sp.: Bousfield, 1984; A.M.M. Richardson, unpublished records), but they were not found at any of the sites studied here. If palustral talitrids are found in the salt-marshes and swamps associated with inlets and lagoons on Tasmania's west coast, a comparison of their ecological range and physiological adaptations with those of the cuspidactylate and simplidactylate landhoppers, with which they should be contiguous, will be of great interest.

Pitfall traps, as used here, catch animals which

are walking on the surface during their nocturnal activity. They may fail to give a true indication of the range of species if a) the nocturnal range differs greatly from the range of daytime refuges, b) if the activity of species is differentially affected by weather conditions, or c) if some species are less active on the surface than others. It is likely that all of these effects have influenced the results obtained here.

Sandhoppers were not collected by hand during the day much further landward than the sand cliffs at the top of the beaches. At night they apparently moved landward (especially 'Talorchestia' sp3) to feed on the vegetation (A. Richardson, unpublished observations), but no further than the seaward end of the range of coastal group landhoppers. The sharp interface between 'Talorchestia' Sp3 and *Austrotroides maritimus* at S. Cape Bay is particularly noteworthy. The fact that neither species apparently crossed the 1 meter gap between traps 9A and 10A over the 14 nights of the July trapping series suggests that there are very strong behavioural cues which regulate their movement. These could have to do with their response to light intensity and canopy cover (Ugolini *et al.*, 1986), soil chemistry or even a direct antagonistic interaction between individuals of the two species. Further work is required to discover how this sharp species replacement is controlled.

The sandhoppers responded much more strongly to the cooler wetter weather in the middle of the February trapping series at S. Cape Bay than the landhoppers, whose activity levels remained almost unchanged. Conditions on the supralittoral fringe are harsh and unpredictable (Morritt, 1987) compared with those beneath the coastal vegetation. Many dead sandhoppers were seen on the sand at S. Cape Bay when the February trapping series was started. It seems likely that the sandhoppers have developed more opportunistic behavioural strategies than the landhoppers.

The tendency for some landhoppers to be restricted to the lower layers of the litter, and so to be underrepresented in surface pitfalls, has been documented previously (Friend & Richardson,

1977; Richardson & Devitt, 1984). In both cases, species present on the transects studied here were involved (*Keratroides angulosus* occurring below *K. vulgaris*, and *K. vulgaris* below *Mysticotalitrus tasmaniae* at another site). Comparison of the pitfall catches with daytime hand collections from S. Cape Bay suggests that *Austrotroides maritimus* dominates the surface layer at night at the seaward end of the terrestrial part of the transects, to the exclusion of some of the other landhoppers. At the landward end, *K. angulosus* was trapped less frequently than other species such as *Neorchestia plicibrancha*.

There is clearly a sharp boundary between the ranges of sandhoppers and landhoppers on these transects, and the soil conditions change at the same point. It has been generally assumed that supralittoral species live in an environment where ions are in abundant supply (e.g. Spicer *et al.*, 1987). Indeed Macintyre (1963) recorded high salinity in the New Zealand supralittoral on a transect across a sand spit, the salinity levels declining sharply across the dunes. The levels of Na recorded here show an opposite trend: low levels in the supralittoral increasing as the transects proceed into terrestrial vegetation. These observations suggest that surface beach sand, although experiencing periodic abundance of ions, can also be readily leached. Leaching will occur frequently in areas of high rainfall such as Tasmania's west coast. This frequently-leached zone will be broader on exposed sandy beaches with low topography, where the sandhopper habitat at the top of the beach may only be immersed during storm surges. Thus the supralittoral fringe presents its inhabitants with problems not only of temperature fluctuation and water stress, but also of low ion availability, at least periodically.

The measured levels of Na rise as soon as soil organic content rises, coincident with the start of the vegetation cover. It is likely that Na and other ions are held in complexes in decaying plant material which resist leaching. This explains the decline in Na concentration on Macintyre's (1963) transect across unforested dunes which would have had little organic matter in the soil. With the exception of Hibbs transect B, there was no evidence that Na levels fall at the landward ends of the transects. Buckney & Tyler (1973) recorded clear maritime influence in the ionic dominance of lake waters many kilometers inland from Tasmania's west coast, and although this gives no indication of the quantity of input, it shows that sea spray can be carried considerable distances inland. Further work is required to discover the rates of ion input on sand beaches, coastal habitats and inland forests, and this should be combined with studies of the ionic requirements of talitrids from these habitats.

The sharp change in soil conditions between the habitat of sandhoppers and landhoppers further supports Bousfield's (1984) suggestion that the exposed sand beach/forest interface is not likely to have been the evolutionary route by which talitrid amphipods reached fully terrestrial habitats. The sharp boundary in species ranges found here contrasts with observations of beachfleas, such as *Orchestia gammarellus* and *Traskorchestia traskiana* (Richardson, 1980; Bousfield, 1982; Spicer *et al.*, 1987), cohabiting with landhoppers and oniscoid isopods in terrestrial habitats well above the high water mark, or *Orchestia cavimana* penetrating long distances inland at the edges of streams (Morritt, 1988). A study of the osmoregulatory abilities of a landhopper, *Arcitalitrus dorrieni*, compared to those of beachfleas (including *O. cavimana*) and sandhoppers (Morritt, 1988) suggests that the ancestors of simplidactylate landhoppers did not pass through a prolonged freshwater stage in their evolutionary history. Thus the saltmarsh or mangrove route is a more likely setting for the terrestrialization of talitrid amphipods.

Sheltered shores, with a compressed sea/land boundary, can be found in the sheltered inlets on Tasmania's west coast such as Macquarie Harbour or the Port Davey/Bathurst Harbour complex. Such localities will provide another interface where landhoppers may be found contiguously with beachfleas and perhaps with palustral talitrids. Since the latter groups are more closely-related to the ancestral stock of landhoppers (Bousfield, 1984) these localities may provide further interesting ecological information.

140

Acknowledgements

We are grateful to the Tasmanian Department of Parks, Wildlife & Heritage, who provided transport to the remote sites through their scheme of Directed Wildlife Research Projects to study the biota of the Western Tasmania World Heritage Area. We thank Jerry Tupacz, Mary Mulcahy, Caron Summers, Maria Moore, Stefan Eberhard and Mark Nelson for assistance in the field, and Mary Mulcahy for assistance in the laboratory. An anonymous reviewer made helpful comments on the MS.

References

Bagenal, T. B., 1957. The vertical range of some littoral animals on St. Kilda. Scot. Nat. 69: 50–51.

Bousfield, E. L., 1982. The amphipod superfamily Talitroidea in the northeastern Pacific region. I. Family Talitridae: systematics and distributional ecology. National Museum of Natural Sciences, Ottawa. Publications in Biological Oceanography 11: 1–75.

Bousfield, E. L., 1984. Recent advances in the systematics and biogeography of landhoppers (Amphipoda: Talitridae) of the Indo-pacific region. In F. J. Radovsky, P. H. Raven & S. H. Somer (eds), Biogeography of the tropical Pacific. Bishop Museum Special Publication 72: 171–210.

Bureau of Meteorology, 1979. Climate of Tasmania. Tasmanian Year Book 13: 27–38. Commonwealth Bureau of Statistics, Tasmanian Office, Hobart.

Buckney, R. T. & P. A. Tyler, 1973. Chemistry of Tasmanian inland waters. Int. Revue ges. Hydrobiol. Hydrogr. 58: 61–78.

Duncan, K. W., 1984. The systematics and physiology of New Zealand landhoppers. Vols I & II. Unpublished Ph.D. thesis, University of Canterbury, Christchurch, New Zealand.

Friend, J. A., 1987a. The terrestrial amphipods (Amphipoda: Talitridae) of Tasmania: systematics and zoogeography. Rec. aust. Mus., Suppl. 7: 1–85.

Friend, J. A., 1987b. A survey of the terrestrial amphipod fauna of the Western Tasmania World Heritage Area. National Parks and Wildlife Service, Tasmania. 18 pp, 30 figs + appendix.

Friend, J. A. & A. M. M. Richardson, 1977. A preliminary study of niche partition in two Tasmanian terrestrial amphipod species. In U. Lohm & T. Persson (eds), Soil Organisms as Components of Ecosystems, Ecol. Bull. (Stockholm) 25: 24–35.

Friend, J. A. & A. M. M. Richardson, 1986. Biology of terrestrial amphipods. Ann. Rev. Ent. 31: 25–48.

Hurley, D. E., 1956. Studies on the New Zealand amphipod fauna. No. 13. Sandhoppers of the genus *Talorchestia*. Trans. r. Soc. N.Z. 84: 359–389.

Hurley, D. E., 1957. Terrestrial and littoral amphipods of the genus *Orchestia*, Family Talitridae. Trans. r. Soc. N.Z. 85: 149–199.

Macintyre, R. J., 1963. The supralittoral fringe of New Zealand sand beaches. Trans. r. Soc. N.Z. (Gen.) 1: 89–103.

Morino, H. & H. Miyamoto, 1988. Redefinition of *Talorchestia* (Amphipoda: Talitridae) with description of a new species from the tropical west Pacific. J. crust. Biol. 8: 91–98.

Morritt, D., 1987. Evaporative water loss under desiccation stress in semiterrestrial and terrestrial amphipods (Crustacea: Amphipoda: Talitridae). J. exp. mar. Biol. Ecol. 111: 145–157.

Morritt, D., 1988. Osmoregulation in littoral and terrestrial talitroidean amphipods (Crustacea) from Britain. J. exp. mar. Biol. Ecol. 123: 77–94.

Ugolini, A., F. Scapini & L. Pardi, 1986. Interaction between solar orientation and landscape visibility in *Talitrus saltator* (Crustacea: Amphipoda). Mar. Biol. 90: 449–460.

Richardson, A. M. M., 1980. Notes on the occurrence of *Talitrus dorrieni* Hunt (Crustacea: Amphipoda: Talitridae) in south-west England. J. nat. Hist., Lond. 14: 751–757.

Richardson, A. M. M. & D. M. Devitt, 1984. The distribution of four species of terrestrial amphipods (Crustacea, Amphipoda: Talitridae) on Mt. Wellington, Tasmania. Aust. Zool. 21: 145–156.

Spicer, J. I., P. G. Moore & A. C. Taylor, 1987. The physiological ecology of land invasion by the Talitridae (Crustacea: Amphipoda). Proc. r. Soc. Lond. (B) 232: 95–124.

Hydrobiologia **223**: 141–148, 1991.
L. Watling (ed.), VIIth International Colloquium on Amphipoda.
© 1991 *Kluwer Academic Publishers.*

Ingestion of live filamentous diatoms by the Great Lakes amphipod, *Diporeia* sp.: a case study of the limited value of gut contents analysis

Michael A. Quigley & Henry A. Vanderploeg
U.S. Department of Commerce, National Oceanic and Atmospheric Administration, Great Lakes Environmental Research Laboratory, 2205 Commonwealth Blvd., Ann Arbor, MI 48105, USA

Key words: Amphipod, *Diporeia*, diatoms, feeding

Abstract

Individuals of the Great Lakes amphipod, *Diporeia* sp. (formerly named *Pontoporeia hoyi*) were collected from a 45-m deep station in southeastern Lake Michigan and isolated in small laboratory feeding vessels at 4 °C, after the animals had voided their guts over a 24-hour period. Over a 20-day period, following introduction of a single ration of live cells of the filamentous diatom, *Melosira varians*, 9 of 10 animals had ingested this material, and 7 of these 9 individuals had deposited fecal pellets. Subsequent examination of gut contents and fecal pellets showed that although animals had ingested whole algal cells/filaments, little of the material in gut contents or fecal pellets bore any identifiable structural similarity to cells/filaments prior to ingestion. The results suggest that earlier studies of pontoporeiid gut contents may seriously underestimate the importance of algal components in the amphipod's diet and imply that *Diporeia* sp. growth and production may be more closely linked to primary production than previously thought.

Introduction

Diporeia sp. (formerly named *Pontoporeia hoyi*, prior to a revision by Bousfield, 1989) is a prominent benthic macroinvertebrate of North America's Laurentian Great Lakes and a prey item that supports the food requirements of a wide variety of valuable commercial and sportfish species (Wells, 1980). In Lake Michigan, *Diporeia* sp. constitutes up to 60% of the resident macroinvertebrate biomass (Nalepa *et al.*, 1985). The amphipod commonly inhabits a thin (1–2 cm) surficial sediment layer (Nalepa & Robertson, 1981), where it presumably feeds on recently-

deposited detritus. To date, however, a definitive study and documentation of the nutritional sources supporting *Diporeia* sp. production has not been conducted. Laboratory sediment preference studies performed in the 1960's (Marzolf, 1965) indicated that when quantities of *Diporeia* sp. were introduced into aquaria containing a variety of sediment types, significantly more animals colonized fine, bacteria-rich sediments, compared to coarser or sterile substrates. Although subsequent gut analysis studies of both *Diporeia* sp. (formerly named *Pontoporeia affinis* (= *P. hoyi*) (Great Slave Lake, N.W. Territories, Canada; Moore, 1977) and *Monoporeia affinis* (formerly named *Pontoporeia affinis* prior to revision by Bousfield, 1989) (Baltic Sea; Ankar, 1977) indicated that particles found in the gut were $\leq 10 \mu m$), the nutritional importance of microflora associated with particles in this size range is unknown. *Diporeia* sp. collected from Great Slave Lake contained algae, detritus, sand grains and oligochaete setae in their guts (Moore, 1977). While identifiable algae composed $\leq 1.5\%$ of gut contents, most species ingested were planktonic forms. In all ingested diatom species, intact chloroplasts were not observed, suggesting that cells were partially digested during passage through the gut, or that the cells were dead prior to ingestion. Green algae, (particularly *Scenedesmus quadricauda*) were seldom broken down and cell walls of most ingested species remained intact during gut passage. While ruptured chloroplasts indicated partial digestion and possible nutritional importance of algae, the relatively small fraction of algae in overall gut volume indicated that they constituted only a minor energy source.

Examination of guts of *Diporeia* sp. collected from a 97-m deep southern Lake Michigan site showed that 1% of the gut volume was composed of identifiable biological remains, including *Cyclotella* and *Melosira* fragments (Evans *et al.*, 1990).

Similarly, gut analysis of *M. affinis* populations in Lake Erken, Sweden indicated that algae composed only 2.3% of gut volume with detritus (97.6%) and animal remains (0.1%) constituting the remaining volume (Johnson, 1987). Although diatoms represented less than half of the algal component (and 1% of the total gut volume), highest *M. affinis* growth and production occurred soon after diatom blooms which provided a rich supply of organic matter to profundal sediments (Johnson, 1987). Though blooms of other algae (primarily *Cladophora*) occurred in October and November, corresponding *M. affinis* production was unusually low and implied that nutritional value of *Cladophora* and associated organic matter was minimal.

Elsewhere, the close association between diatom production in the water column and subsequent pontoporeiid production suggests that diatoms or fresh diatom-derived detritus may be far more nutritionally important than previous gut contents studies have revealed. In Lake Ontario,

Diporeia sp. instantaneous growth rates were highest in spring and closely followed spring peaks in phytoplankton productivity (Johnson & Brinkhurst, 1971). In the Baltic Sea, most of the annual growth of *M. affinis* and *Pontoporeia femorata* populations was restricted to spring, when peak phytoplankton production occurred (Hill & Elmgren, 1987). Although direct linkage between *Diporeia* sp. production and phytoplankton production has not been documented in Lake Michigan, there is evidence that the amphipod's reproduction is synchronized closely with peaks in phytoplankton productivity to ensure that the release of young occurs at times when phytoplankton-generated organic matter is most abundant (Lubner, 1979). In addition, the rapid increase in *Diporeia* sp. lipid content following spring diatom blooms (Gardner *et al.*, 1985) suggests that primary production in overlying water may subsequently provide a rich food source for the amphipod's growth and production (Gardner *et al.*, 1990). Lipid content of Lake Michigan *Diporeia* sp. may reach as much as 50% (dry weight) (Gardner *et al.*, 1985) and recent work (Quigley *et al.*, 1989) has indicated that most of this lipid fraction consists of storage lipids (triacylglycerols) which play a major energetic role throughout the animal's life history.

Recent gut fullness studies (Dermott & Corning, 1988; Quigley, 1988; Evans *et al.*, 1990) have demonstrated that *Diporeia* sp. feeds intermittently, unlike most amphipod species, which feed continuously (e.g. *Hyalella azteca*, Hargrave, 1970; *Gammarus pulex*, Welton *et al.*, 1983). Given the ability of *Diporeia* sp. to store large quantities of lipid, intermittent feeding may represent a strategy that allows the amphipod to fully exploit high-quality food during the brief period that it is available (i.e. following the deposition of organic matter from spring diatom blooms), and to feed sparingly at other times, while relying on lipid stores to support metabolic needs (Gauvin *et al.*, 1989). In fact, *Diporeia* sp. percent gut fullness data from Lake Michigan (Quigley, 1988; Evans *et al.*, 1990) and from Lake Ontario (Dermott & Corning, 1988) show that feeding is intensified following spring diatom blooms. Thus, intensified feeding and associated increases in lipid storage at this time suggest that *Diporeia* sp. growth and production is closely linked with, and highly dependent on, inputs of organic matter derived from spring blooms. At present, little is known about the viability or decompositional stage of algal cells deposited at the sediment-water interface following spring blooms. However, silica flux measurements from Lake Michigan sediments (Quigley & Robbins, 1984; Conley & Schelske, 1989) have demonstrated that most of the silica released into overlying water is derived from a surficial floc layer of biogenic silica (primarily diatom frustules). Since mineralization of frustules can only proceed after disintegration of the outer diatom cell membrane, this floc layer should have considerable concentrations of senescent and dying diatoms. In fact, the introduction of moribund diatoms to the surface of estuarine Chesapeake Bay cores produced linear increases in the release of dissolved silica and implied that most of the upward flux of dissolved silica originated from a surficial floc layer that was rich in biogenic silica (primarily diatom frustules) (Yamada & D'Elia, 1984). Data on the seasonal concentration of photosynthetic pigments in surficial nearshore Lake Michigan sediments also demonstrated that pigment abundance is greatest in spring (Nalepa & Quigley, 1987). Senescent diatoms, in fact, might prove to be of particular nutritional value to *Diporeia* sp. since they may contain large amounts of lipids that are known to accumulate during low-growth periods (Parrish, 1986). The potentially high nutritional value of Baltic Sea surficial sediments, composed extensively of freshly-deposited phytoplankton, has also been cited in explaining high carbon assimilation efficiencies noted in *M. affinis* ($\sim 40\%$) and *P. femorata* (28%) (Lopez & Elmgren, 1989). Moreover, the sediment preference tests of Marzolf (1965) showed that *Diporeia* sp. selected sandy sediments overlain by a thin layer of fine detritus compared to sands into which this same detritus layer had been mixed. Although previous gut contents studies have tended to discount the direct ingestion of diatoms and other algal cells, such conclusions are limited since mechanical and digestive breakdown might greatly obscure the original identity and composition of the ingested material, and lead to an underestimate of the significance of algal components in the overall nutrition of *Diporeia* sp. If this was the case, we hypothesized that trituration and digestion of algal cells by the amphipod is extensive and that utilization of such food sources is far more important than previously believed.

To test this hypothesis, we introduced quantities of the diatom *Melosira varians* to individual *Diporeia* sp., and evaluated the degree of ingestion and breakdown of these live cells by the amphipod. *Melosira varians* is a large filamentous diatom that is typical in size and morphology among *Melosira* species of the Great Lakes. *Melosira spp.* are dominants of the spring bloom and are known to rapidly sink out of the water column at the onset of stratification (Holland & Beeton, 1972; Fahnenstiel & Scavia, 1987). Moreover, the large size and morphology of *M. varians* render it resistant to ingestion by some zooplankton during its pelagic existence, although *Diaptomus spp.* can ingest if after orienting the filaments normal to the long axis of the body (Vanderploeg *et al.*, 1988). As such, this diatom represented a particularly appropriate food item and a rigorous test of the digestive capabilities of *Diporeia sp.*

Materials and methods

Diporeia sp. obtained from a 45 m deep site in southeastern Lake Michigan were held in the laboratory at 4 °C. in sandy-silt sediments collected from the site. Prior to feeding tests, approximately 30 juveniles (5–6 mm total length) were isolated for 48 hours in lake water to allow the animals to void their guts. Following this isolation interval, 10 test animals with completely empty guts were individually isolated in feeding test vessels consisting of 20 ml high-density linear polyethylene liquid scintillation vials containing a 1 cm layer of 500–550 µm ignited (550 °C. for 4 hours) sand overlain by filtered (2 µm) lake water. The large particle size provided a highly porous, well-oxygenated sediment environment and animals readily burrowed into the substrate following introduction. The coarse ignited sand also provided an inert substrate of particles that were too large to be ingested by *Diporeia* sp. and ensured that only introduced food could be potentially ingested. The substrate also permitted rapid location and retrieval of any fecal pellets produced during feeding experiments.

The tops of test vials consisted of open-ended polyethylene screw caps that were covered with 500 μm nylon screen. Test vials were held in an acrylic rack, submerged in a shallow tray, where filtered lake water was continuously aerated. The system also allowed individual vials to be periodically removed from the tray to inspect animals and sediments. This was accomplished by replacing the vial's screw cap with a double-width (5 × 7.5 cm) microscope slide and then placing the vial on the stage of a dissecting microscope where animals and sediments could be readily observed.

Following introduction to test vials, individual animals were held in the containers for 24 hours to adjust to conditions. Rations of laboratory-cultured *Melosira varians* (Vanderploeg *et al.*, 1988) were then introduced to individuals as follows: A 5 ml quantity of *Melosira* and culture water was drawn from the bottom of the culture flask and transferred to a 100 ml beaker where it was combined with 75 ml of filtered lake water. This suspension was then vigorously agitated by magnetic stirrer to ensure uniform distribution of *Melosira* filaments. Each test vial was then removed from the holding tray, opened, and 5 ml of the vial's water removed. A 2 ml volume of water from the *Melosira* suspension was then added and the cells allowed to settle to the sand substrate before the test vial was topped off with lake water and returned to the holding tray. To estimate amount of *Melosira* rations introduced to individuals, we also transferred 2 ml volumes of the suspension to pre-weighed aluminum dishes and these were dried for 24 hours at 60 °C and reweighed. Mean ration weight was 289 μg (dry weight) $\pm$ (SE) 16 μg ($n = 5$). Following food introduction, all 10 test vials were incubated in

darkness at 4 °C. for a 20-day period during which vials were periodically removed from trays to examine conditions of test animals, the food ration, and to retrieve any recently-deposited fecal pellets. Any fecal pellets recovered were immediately preserved in 5% formalin solution.

Results and discussion

From the very outset of the feeding experiment, it was apparent that *Diporeia* sp. would readily ingest introduced *Melosira* cells/filaments. In one instance during the initial introduction of the *Melosira* ration, we observed one animal which emerged from its partially-buried position, to move directly to a large clump of *Melosira* filaments. While lying on its side at the sediment surface, the animal could be clearly observed guiding tangled masses of *Melosira* to its mouth where ingestion occurred.

Over the first 10 days of the test period, 6 of 10 test animals had partially or totally full guts of *Melosira* and 4 of these 6 animals had deposited fecal pellets. By 20 days, 9 animals had *Melosira* gut contents and 7 of these 9 had produced from 1 to 15 fecal pellets.

Fecal pellets produced by test animals fed *Melosira* were visibly different than *Diporeia* sp. fecal pellets we have previously collected from Lake Michigan sediments, and from sediment-fed animals in the laboratory. These fecal pellets typically consisted of cylindrically-shaped, highly compacted fine material that was tightly enclosed within a durable peritrophic membrane. By contrast, *Melosira*-fed *Diporeia* sp. released fecal pellets that were more amorphous, had less compacted material and more loosely fitting peritrophic membranes. Gut turnover rates (as defined by fecal pellet (F.P.) output per unit time) of *Melosira*-fed animals were also far lower than rates we have previously observed in sediment-fed *Diporeia* sp. (Quigley, unpublished data). The maximum of 15 fecal pellets produced over 20 days by *Melosira*-fed animals (0.75 F.P. d^{-1}) was far lower than the 100 or more fecal pellets we have observed produced by sediment-fed animals over 10 days (10 F.P. d⁻¹ or a daily gut turnover rate of 1.5–2.0). Similar gut turnover rates have also been noted for sediment-fed Lake Ontario *Diporeia* sp. (Dermott & Corning, 1988) which took a mean of 6.7 hours to completely fill guts at 5 °C. Since the animals would require some time to void their guts, the expected daily gut turnover rate will be less than a maximum of 3.6 (assumes instantaneous emptying of gut) and is probably closer to the range of 1.5–2, we observed in Lake Michigan *Diporeia* sp.

The markedly lower gut turnover rate of *Melosira*-fed animals might, in part, be due to the relatively higher nutritive quality of this food item and a corresponding response by *Diporeia* sp. that extended the residence time of this material in the gut, thereby allowing more time for the full assimilation of this food. Alternatively, the relatively long gut passage time of ingested *Melosira* might represent an inability of *Diporeia* sp. to efficiently process the ration in the complete absence of fine sediment particles which may be required for rapid, efficient passage of gut contents.

Inspection of both gut contents and fecal pellets (under a compound microscope, Figs. 1 and 2) revealed that *Diporeia* sp. was extremely efficient in breaking down live *Melosira* cells. Whole intact cells were rarely observed in the gut contents or in fecal pellets. While some cell wall structures were occasionally observed, most of the remaining gut and fecal pellet material bore no identifiable similarity to the *Melosira* cells that had been originally ingested.

The extensive breakdown of *Melosira* by *Diporeia* sp. reflects an ability to crush and grind ingested material and is similar to the behavior exhibited by *M. affinis* in the Baltic Sea, where this species is capable of crushing the 220–250 μm shells of the spat of the bivalve *Macoma balthica*, presumably ingesting the remains (Elmgren *et al.*, 1986). We examined *Diporeia* sp. mandibles and found that they con-

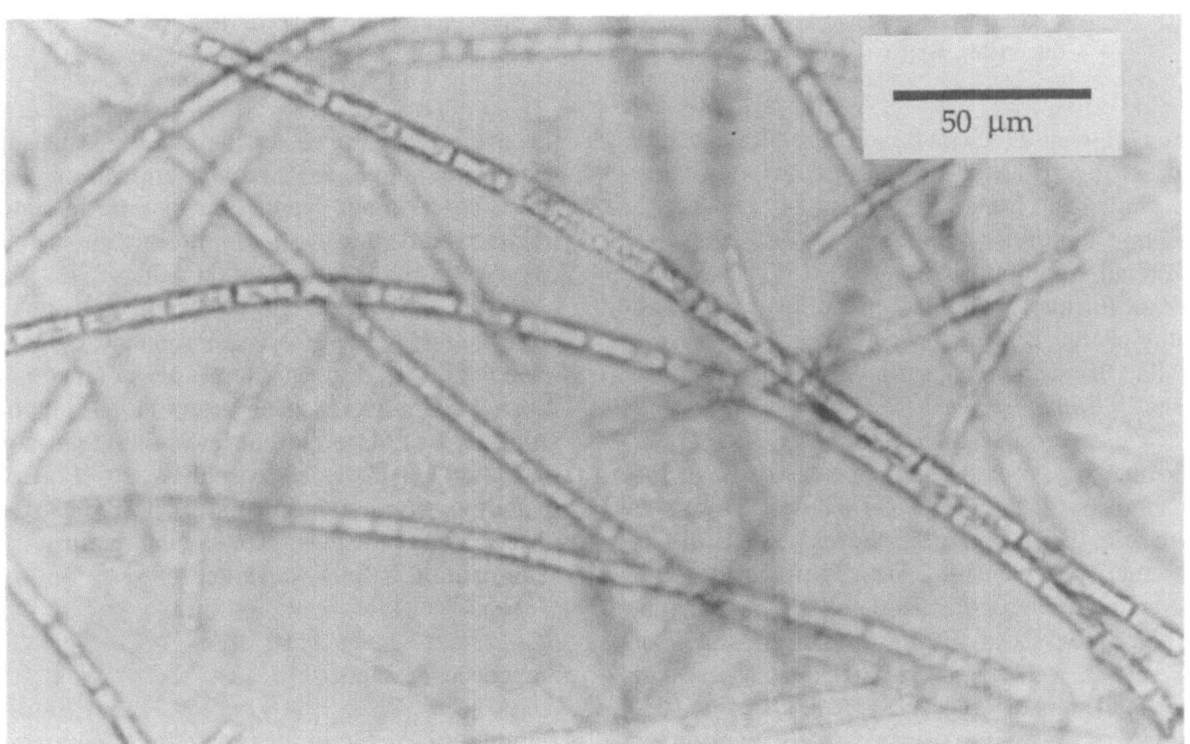

Fig. 1. Live *Melosira* varians cells/filaments prior to ingestion by *Diporeia*. Black bar (upper right) denotes scale (50 μm).

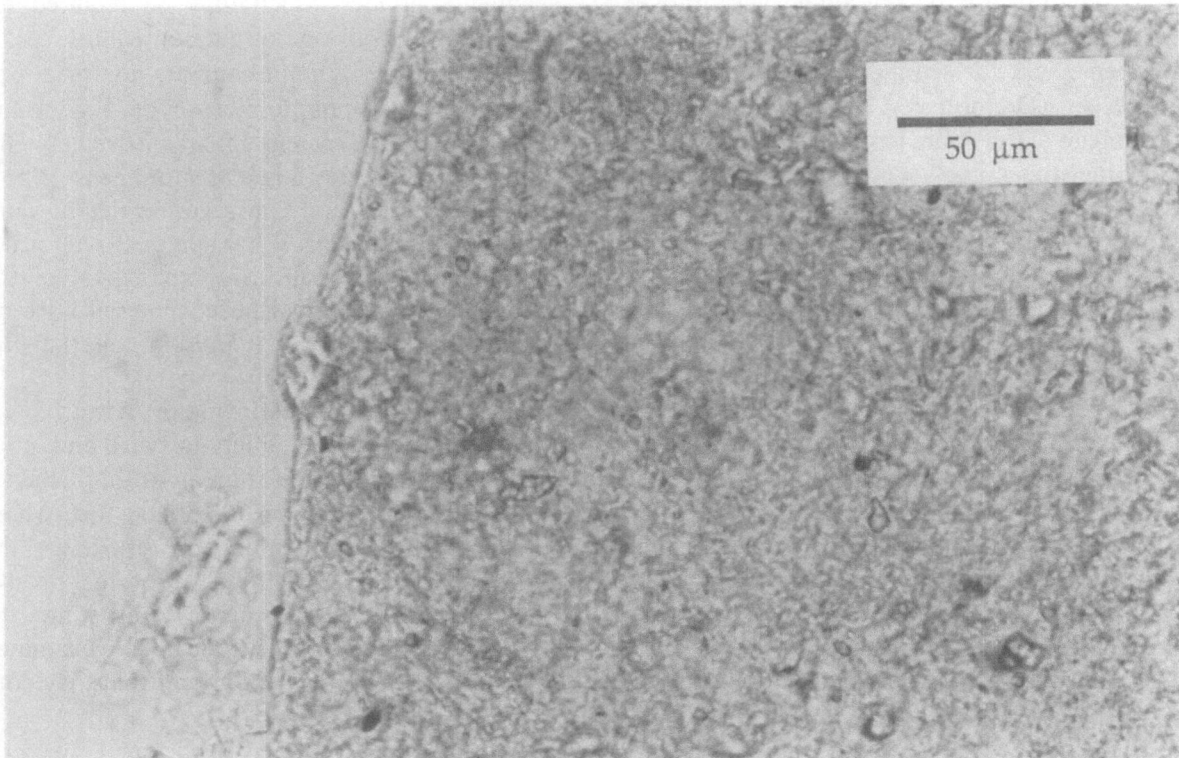

Fig. 2. Portion of fecal pellet collected from a *Melosira*-fed *Diporeia*. Note peritrophic membrane (left) and absence of any discernible remaining *Melosira* cellular structure. Black bar (upper right) denotes scale (50 μm).

sisted of well-developed multi-toothed incisors and strong triturative molar processes. These morphological traits are typical of the biting-rasping mandible commonly found in gammaridean amphipods (Bousfield, 1973; Barnard, 1969). Further mechanical breakdown may also occur in the foregut if *Diporeia* sp. possesses a cardiac stomach (and gastric mill) similar to those noted in *Gammarus lacustris* (Schmitz, 1967) and *Hyalella azteca* (Schmitz & Scherrey, 1983). At present, there is little detailed knowledge of the morphology of the gut of *Diporeia* sp. However, Bousfield (pers. comm.) believes that its gut is structurally typical of most gammarids and should include a cardiac stomach and gastric mill. In summary, given the demonstrated ability of *Diporeia* sp. to break down and digest *Melosira* to the extent that few recognizable features remain, it appears that previous gut content analysis of field-sampled animals may grossly underestimate the overall importance of algal material in supporting the amphipod's growth and production. Thus, *Diporeia* sp. production may be far more intimately linked with primary production in overlying water than previously thought.

Finally, we believe that our results illustrate the often limited value of extrapolating gut contents observations to draw inferences about an aquatic invertebrate's overall diet. Peters (1984) emphasized the use of caution in interpreting gut contents data and provided a useful review of factors that often confound or limit such studies. Our observations of *Diporeia* sp. feeding reaffirm that such caution is fully warranted.

Acknowledgements

We thank L.S. Goad for providing *Melosira varians* and J.R. Liebig for his help in maintaining

M. varians cultures. W.S. Gardner, J.F. Cavaletto and T.F. Nalepa reviewed the manuscript and we appreciate their comments.

References

Ankar, S., 1977. The soft bottom ecosystem of the northern Baltic proper with special reference to the macrofauna. Contrib. Asko Lab., Univ. Stockholm 19: 1–62.

Barnard, J. L., 1969. The families and genera of marine gammaridean amphipods. U.S. National Museum Bull. 271, Smithsonian Institution Press, Washington, D.C. 535 pp.

Bousfield, E. L., 1973. Gammaridean Amphipoda of New England. Cornell Univ. Press., Ithaca N.Y., 312 pp.

Bousfield, E. L., 1989. Revised morphological relationships within the amphipod genera *Pontoporeia* and *Gammaracanthus* and the 'glacial relict' significance of their postglacial distributions. Can. J. Fish. aquat. Sci. 46: 1714–1725.

Conley, D. J. & C. L. Schelske, 1989. Processes controlling the benthic regeneration and sedimentary accumulation of biogenic silica in Lake Michigan. Arch. Hydrobiol. 116: 23–43.

Dermott, R. M. & K. Corning, 1988. Seasonal ingestion rates of *Pontoporeia hoyi* (Amphipoda) in Lake Ontario. Can. J. Fish. aquat. Sci. 45: 1886–1895.

Elmgren, R., S. Ankar, B. Marteleur & G. Ejdung, 1986. Adult interference with postlarvae in soft sediments: the *Pontoporeia-Macoma* example. Ecology 67: 827–836.

Evans, M. S., M. A. Quigley & J. A. Wojcik, 1990. Comparative ecology of *Pontoporeia hoyi* populations in southern Lake Michigan: the profundal region versus the slope and shelf regions. J. Great Lakes Res. 16: 27–40.

Fahnenstiel, G. L. & D. Scavia, 1987. Dynamics of Lake Michigan phytoplankton: primary production and growth. Can. J. Fish. aquat. Sci. 44: 499–508.

Gardner, W. S., T. F. Nalepa, W. A. Frez, E. A. Cichocki & P. F. Landrum, 1985. Seasonal patterns in lipid content of Lake Michigan macroinvertebrates. Can. J. Fish. aquat. Sci. 42: 1827–1832.

Gardner, W. S., M. A. Quigley, G. L. Fahnenstiel, D. Scavia & W. A. Frez, 1990. *Pontoporeia hoyi*, an apparent direct trophic link between spring diatoms and fish in Lake Michigan. In *Structural and Functional Properties of Lakes*. (M. Tilzer, Ed.), Springer Verlag, Heidelberg. pp. 632–644.

Gauvin, J. M., W. S. Gardner & M. A. Quigley, 1989. Effects of food removal on nutrient release and lipid content of Lake Michigan *Pontoporeia hoyi*. Can. J. Fish. aquat. Sci. 46: 1125–1130.

Hargrave, B. T., 1970. The utilization of benthic microflora by *Hyalella azteca* (Amphipoda). J. anim. Ecol. 39: 427–437.

Hill, C. & R. Elmgren, 1987. Vertical distribution in the sediment in the co-occurring benthic amphipods *Pontoporeia affinis* and *P. femorata*. Oikos 49: 221–229.

Holland, R. E. & A. M. Beeton, 1972. Significance to eutrophication differences in nutrients and diatoms in Lake Michigan. Limnol. Oceanogr. 17: 88–96.

Johnson, M. G. & R. O. Brinkhurst, 1971. Production of benthic macroinvertebrates of the Bay of Quinte and Lake Ontario. J. Fish. Res. Bd. Can. 28: 1699–1717.

Johnson, R. K., 1987. The life history, production and food habits of *Pontoporeia affinis* Lindstrom (Crustacea: Amphipoda) in mesotrophic Lake Erken. Hydrobiologia 144: 277–283.

Lubner, J. F., 1979. Population dynamics and production of the relict amphipod *Pontoporeia hoyi* at several Lake Michigan stations. Ph.D. Thesis, Univ. Wisconsin, Milwaukee, WI. 98 pp.

Lopez, G. & R. Elmgren, 1989. Feeding depths and organic absorption for the deposit-feeding benthic amphipods *Pontoporeia affinis* and *Pontoporeia femorata*. Limnol. Oceanogr. 34: 982–991.

Marzolf, G. R., 1965. Substrate relations of the burrowing amphipod *Pontoporeia affinis* in Lake Michigan. Ecology 46: 579–592.

Moore, J. W., 1977. The importance of algae in the diet of *Gammarus lacustris* and *Pontoporeia affinis*. Can. J. Zool. 55: 637–641.

Nalepa, T. F. & M. A. Quigley, 1987. Distribution of photosynthetic pigments in nearshore sediments of Lake Michigan. J. Great Lakes Res. 13: 37–42.

Nalepa, T. F., M. A. Quigley, K. Childs, J. M. Gauvin, T. S. Heatlie, M. Parker & L. Vanover, 1985. The macrobenthos of southern Lake Michigan. NOAA Data Report, ERL GLERL-28, Great Lakes Environmental Research Laboratory, Ann Arbor, MI.

Nalepa, T. F. & A. Robertson, 1981. Vertical distribution of the zoobenthos in southeastern Lake Michigan with evidence of seasonal variation. Freshwat. Biol. 11: 87–96.

Parrish, C. C., 1986. Dissolved and particulate lipid classes in the aquatic environment. Ph.D. Thesis, Dalhousie Univ., Halifax, N.S. 259 pp.

Peters, R. H., 1984. Methods for the study of feeding, filtering and assimilation by zooplankton. In Downing, J. A. & F. H. Rigler, A Manual on Methods for the Assessment of Secondary Productivity in Fresh Waters. I.B.P. Handbook No. 17, Blackwell Sci. Pubs. pp. 336–412.

Quigley, M. A., 1988. Gut fullness of the deposit-feeding amphipod *Pontoporeia hoyi* in southeastern Lake Michigan. J. Great Lakes Res. 14: 178–187.

Quigley, M. A., J. F. Cavaletto & W. S. Gardner 1989. Lipid composition related to size and maturity of the amphipod *P. hoyi*. J. Great Lakes Res. 15: 601–610.

Quigley, M. A. & J. A. Robbins, 1984. Silica regeneration processes in nearshore southern Lake Michigan. J. Great Lakes Res. 10: 383–392.

Schmitz, E. H., 1967. Visceral anatomy of *Gammarus lacustris lacustris* Sars (Crustacea: Amphipoda). Am. Midl. Nat. 78: 1–54.

Schmitz, E. H. & P. M. Scherrey, 1983. Digestive anatomy

of *Hyalella azteca* (Crustacea: Amphipoda). J. Morph. 175: 91–100.

Vanderploeg, H. A., G. Paffenhoffer & J. R. Liebig, 1988. *Diaptomus* vs. net phytoplankton: Effects of algal size and morphology on selectivity of a behaviorally flexible, omnivorous copepod. Bull. mar. Sci. 43: 377–394.

Wells, L., 1980. Food of alewives, yellow perch, spottail shiners, trout-perch and slimy and fourhorn sculpin in southeastern Lake Michigan. U.S. Fish & Wildlife Serv. Tech. Pap. 98.

Welton, J. S., M. Ladle, J. A. B. Bass & I. R. John, 1983. Estimation of gut throughput time in *Gammarus pulex* under laboratory and field conditions with a note on the feeding of young in the brood pouch. Oikos 41: 133–138.

Yamada, S. S. & C. F. D'Elia, 1984. Silicic acid regeneration from estuarine sediment cores. Mar. Ecol. Prog. Ser. 18: 113–118.

Hydrobiologia **223**: 149–158, 1991.
L. Watling (ed.), VIIth International Colloquium on Amphipoda.
© 1991 *Kluwer Academic Publishers.*

A comparison of water loss and gill areas in two supralittoral amphipods from New Zealand

I.D. Marsden
Zoology Department, University of Canterbury, Christchurch, New Zealand

Key words: Amphipod, gill area, desiccation tolerance, water loss, oxygen consumption

Abstract

Total gill area and gill distribution were measured for the sandhopper *Talorchestia quoyana* (Milne-Edwards) and the beach flea *Transorchestia chiliensis* (Milne-Edwards). For both species the gill structure and proportional area contributed by individual gills was similar. Gill 6 (G6) was the largest, providing 36% of the gill area in *Tal. quoyana* and 30% in *Tr. chiliensis*. The gill area/total dry weight relationships were similar, $Y = 1.3 X^{0.79}$ for *Tal. quoyana* and $1.4 X^{0.78}$ for *Tr. chiliensis*. Small, medium and large amphipods survived > 24 h in aerial conditions close to 100% RH at 15 °C. Rates of water loss in desiccating conditions increased with decreasing RH. Lethal water loss exceeded 30% weight loss for both species. Rate of water loss, (R) mg water loss. mg wet wt tissue. h^{-1} exposed to 75% RH for *Tr. chiliensis* was 0.21, resulting in total mortality within 2 h. Medium *Tal. quoyana* were the most resistant group surviving 4 h exposure to 75% RH with R = 0.08. Differences in desiccation tolerances of the two amphipods are not explained by body water content, gill area relationships or the larger maximal size of *T. quoyana*. Results were combined with those from other talitrids to examine the relationship between gill area, water content, desiccation habitat and oxygen consumption in aerial and aquatic conditions. There were no consistent relationship between gill area, O_2 uptake and desiccation resistance. Amphipods show compensatory respiratory adaptation with individuals from all habitats, showing similar rates of oxygen uptake, either in air or in water, whichever was their most usual respiratory medium. Q^{10} values close to 2.0 were found in all ecomorphological groups. Sandhoppers, including *Tal. quoyana*, are best able to survive terrestrial conditions associated with a low humidity environment. It is concluded that the water loss characteristics of *Tr. chiliensis* limit its distribution on sand beaches to areas of high relative humidities.

Introduction

Organisms which occupy high shore levels on sand beaches are exposed to temperature and desiccation stress exceeding those of other transitional habitats (MacIntyre, 1963; Newell, 1979). Crustaceans found in the supralittoral zone have evolved mechanisms of aerial gas exchange which typically, in the Isopoda and Decapoda, are associated with minimising water loss through a less permeable exoskeleton (Edney, 1960) a reduction in gill area, and the development of highly modified cutaneous lungs or respiratory surfaces (Innes & Taylor, 1986). Terrestrial Amphipoda lack many of the adaptations listed above. They have retained the simple sac-like gills found in aquatic species (Moore & Taylor, 1984; Spice & Taylor, 1986), have reduced resistance to desiccation (Friend & Richardson, 1977; Richardson & Devitt, 1984; Lazo-Wasem, 1984) and are restricted to areas of high, year round rainfall (Friend & Richardson, 1986). In contrast, beach hoppers generally have low weight specific gill areas (Moore & Taylor, 1984; Spicer & Taylor, 1987a) and rates of weight loss in desiccating conditions for the beach hopper *Talitrus saltator* are half those recorded for landhoppers (Morritt, 1987). Studies by Williamson (1951), Platzman (1960) and Morritt (1987) suggest that beachhoppers are more tolerant of desiccation than beachfleas, a feature which appears consistent with their ecomorphological groupings (Bousfield, 1982). To date, however, the relationship between relative gill area, desiccation tolerance and ability to respire in aerobic conditions has not been explored fully.

The present study compares the gill areas and water loss in a beach flea and sandhopper which cooccur in the supralittoral zone on sand beaches. The beach flea, *Transorchestia chiliensis*, occurs in estuaries, rocky shores and extends onto sand beaches where it is found amongst drift seaweed. *Talorchestia quoyana* is typically found on exposed oceanic beaches (MacIntyre, 1963; Fincham, 1977) but it is also found at hightide levels on rocky shores where sand substratum is available. Distribution patterns may overlap and both species emerge during darkness to feed on the freshly deposited driftweed at the most recent high tide mark. The oxygen consumption and salinity tolerances of the two species have been investigated previously (Marsden, 1980; Marsden, 1984; Marsden, 1989) and it is therefore possible to relate gill area, desiccation tolerance, bimodal oxygen uptake and the ability of amphipods to survive in low salinity habitats. Friend & Richardson (1986) and Spicer *et al.* (1987) suggest that the ability of amphipods to regulate their internal osmotic pressure is the principal factor which has allowed certain groups to colonise fully terrestrial habitats.

Materials and methods

Talorchestia quoyana was collected from the ocean sand beach at Brighton, Christchurch and *Transorchestia chiliensis* from the adjacent Avon Heathcote Estuary. Amphipods were stored at 15 °C in constant darkness with substratum and drift wrack collected from the habitats, maintaining 100 percent Relative Humidity (% RH). All experiments were undertaken during the summer (December to January), and amphipods were held for 3 to 4 days prior to experimentation. Collections for the gill area measurements were made on the same day for both species to reduce possible differences due to seasonal variations.

Weight loss experiments were undertaken at 15° using individual amphipods within 200 ml experimental chambers, where a fine wire mesh platform separated the amphipod from the saturated salt solutions used to produce, 100%, 75% and 20% RH (Winston & Bates, 1963). Chambers were allowed to equilibrate for 24 h at 15 °C before the addition of a preweighed amphipod. Individuals were isolated from the habitat collections onto filter paper wetted with seawater and weighed using a Cahn microbalance accurate to 0.001 mg. Forceps were used to hold the amphipod by its 7th peraeopod and transfer it between the weighing unit and the experimental chamber. At least 20 individuals of the size range present in the habitat were included in each experiment, including males and females, but excluding gravid females. Mortality was followed at 1 h intervals and weight loss measured. A sample of at least 5 individuals was weighed for each of the three size classes for *Tal. quoyana* exposed to 75% RH. At the end of each experiment the dry weight of each individual was measured to enable initial water content to be calculated.

The water content of *Tal. quoyana* and *Tr. chiliensis* was measured for a sample of 36 individuals of each species collected from the habitat on the same day and stored overnight at 15 °C.

Gill areas of *Tal. quoyana* and *Tr. chiliensis* were calculated following the procedures outlined

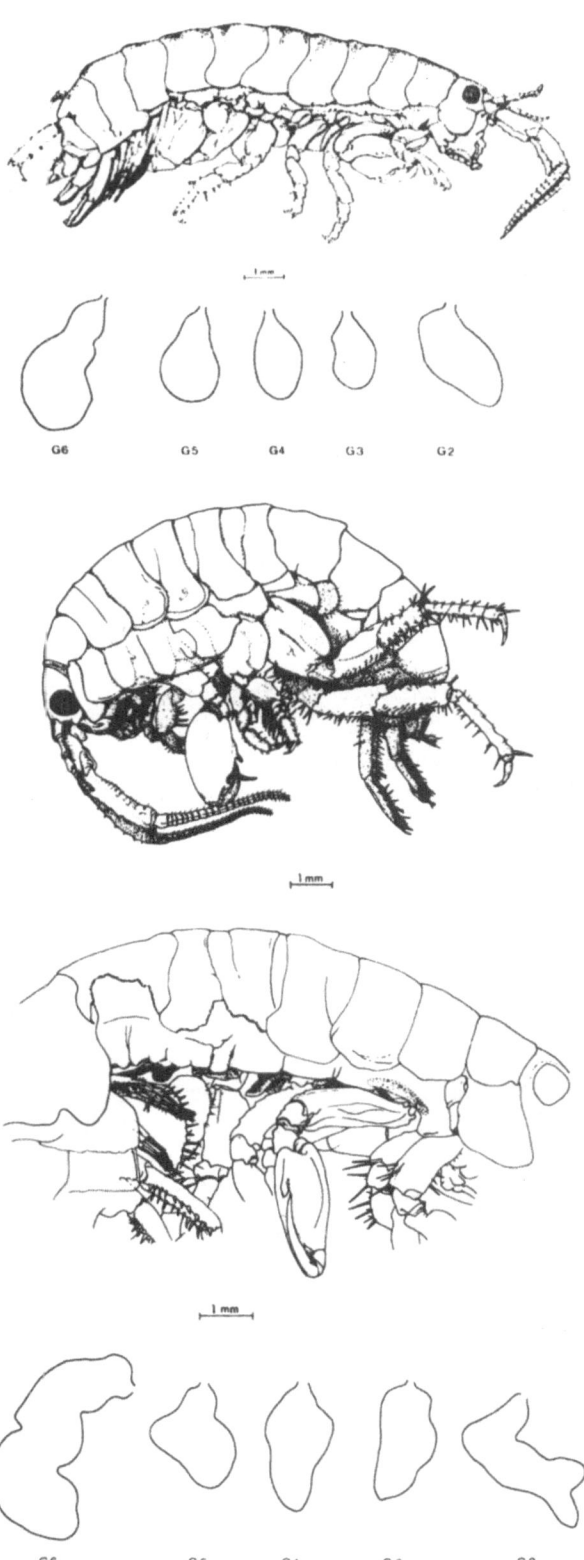

in Moore & Taylor (1984) and Marsden (1985). For individuals of known body length the area of each coxal gill was measured and the total gill area plotted against dry body weight calculated from a regression equation relating body length (mm) to dry body weight (mg).

Results

Gill structure and area

In both *Tr. chiliensis* and *Tal. quoyana* the gills are robust, sac like structures, attached to peraeopods 2 to 6 (Fig. 1). Gill 2 (G2) in both species is large, however, in *Tal. quoyana*, the end is bilobed projecting anteriorly towards peraeopod 1. G3, 4 and 5 are the smallest gills in both species with G6 the largest, projected posteriorly. In *Tal. quoyana* G6 is W shaped curving around the base of peraeopod 7. Figure 2 shows the percentage contribution to the total gill area of gills for each species. G6 contributed 30% of the total gill area in *Tr. chiliensis* and 36% of the total gill area in *Tal. quoyana*.

The relationship between total gill area and total dry body weight is shown in Fig. 2 for each species, which also includes the equations of the regression lines (significant at the 1% level).

Despite the overall larger maximum size for *Tal. quoyana* the gill area relationships of the two species were similar (analysis of covariance F (slope) = 0.87 and F (intercept) = 1.76). Male and female amphipods of both species appear to have similar gill areas and length/dry weight relationships regardless of difference in external morphology.

Water content

Figure 3 shows the water content of male, female and juvenile individuals of the two species plotted

Fig. 1. Top – *Transorchestia chiliensis* with limbs removed to show location of coxal gills G2–G6. Bottom – *Talorchestia quoyana*, gill locations and coxal gills G2–G6.

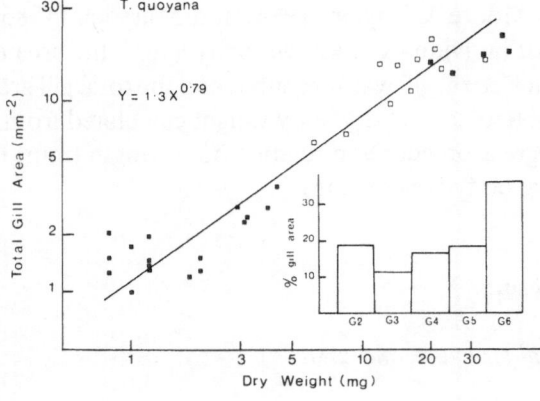

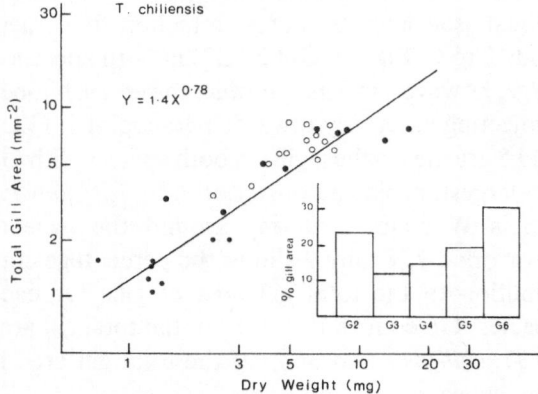

Fig. 2. Relationship between total gill area and dry weight for *Tr. chiliensis* and *Tal. quoyana*. Insets show % contribution by each coxal gill. Small symbols, juveniles; open symbols, females; closed symbols, males.

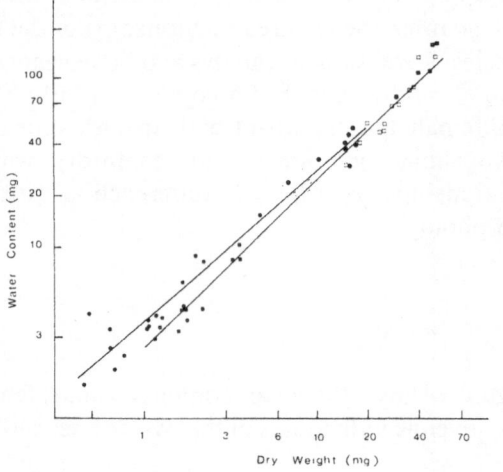

Fig. 3. Water content of *Tr. chiliensis* (circles) and *Tal. quoyana* (squares). Small symbols juveniles; open symbols, females; closed symbols, males.

against dry weight. Although a direct linear relationship was significant for both species, it explained less variation for *Tal. quoyana* than when a logarithmic transformation was applied to both axes. Water content for *Tr. chiliensis* follows the equation 3.7 dry wt$^{0.87}$ ($r = 0.97$, % variance explained = 95.9) and for *Tal. quoyana* 2.5 dry wt$^{0.99}$ ($r = 0.99$, % variance explained = 98.4). These differences in water content with size suggest that small *Tal. quoyana* have a lower water content than *Tr. chiliensis* of a similar size but these differences become insignificant in larger individuals. Water content values reported here correspond to between 68% and 73% of the wet weight for *Tal. quoyana* and between 72% and 76% for *Tr. chiliensis*. No significant differences were found between water content of adult male and female amphipods in either species.

Desiccation tolerance and weight loss

In experimental conditions where relative humidities were maintained close to saturation, both species survived 24 h exposure at 15 °C (Table 1). This result is consistent with all other talitrids examined to date including aquatic species (Spicer *et al.*, 1987).

Most *Tal. quoyana* survived more than 4 h exposure, at 75% RH whereas all *Tr. chiliensis* died within 2 h irrespective of their body size. On exposure to 20% RH, size effects were obvious with small individuals of both species being less tolerant of water loss than larger individuals.

The effect of body size on weight loss of *Tal. quoyana* exposed to 20% RH for 1 h is shown in Fig. 4 together with measurements made on the same animals 4 h after exposure. All juvenile *Tal. quoyana* died as a result of lethal water loss with rates of water loss (R) greater than 0.6 mg water. mg wet wt. h^{-1}. R values for larger amphipods which survived 1 h exposure were 0.25. Percent weight loss decreased with increasing body size although the exact relationship remains unclear. The relationship could be represented by a linear or an exponential relationship, both regression analyses were significant at

Table 1. Percent mortalities for *Tr. chiliensis* and *Tal. quoyana* exposed to aerial conditions at 15 °C.

Tal. quoyana	Exposure time (h)				
	1	2	3	4	24
Sand	0	0	0	0	0
Wet paper	0	0	0	0	0
100% RH	0	0	0	0	0
75% RH	0	0	0	40	100
20% RH – small	100	100	100		
– medium	0	0	60	85	100
– large	0	0	70	90	100
Tr. chiliensis	1	2	3	4	24
Stones	0	0	0	0	0
Wet paper	0	0	0	0	0
100% RH	0	0	0	0	0
75% RH	35	100	100		
20% RH – small	100	100	100		
– medium	0	80	100		
– large	0	100	100		

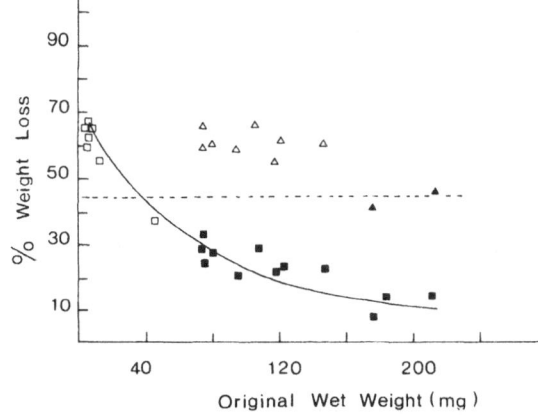

Fig. 4. Weight loss of *Tal. quoyana* exposed to RH 20% for 1 hr (squares) and 4 h (triangles). Closed symbols indicate live amphipods, dotted line, the estimated lethal water loss.

the 5% level. However it was clear that small amphipods lost water at rates greater than expected from a simple surface area to weight relationship, and for this reason, the curve fitted by eye was retained. Weights of *Tal. quoyana* recorded after 4 h exposure appeared to follow the same weight loss pattern, supporting previous studies which suggest that weight loss in amphipods is a passive rather than active process (Moore & Francis, 1985). The estimated lethal water loss for *Tal. quoyana* at 20% RH was between 30 and 40% of the original body weight or a loss of more than 50% of the total body moisture.

Figure 5 shows the weight loss of a size range of *Tr. chiliensis* following 1 h exposure to 20% RH. During this time all individuals less than 15 mg original weight had died, with the remainder becoming increasingly immobile. Rates of water loss (R) varied between 0.6 for small individuals to 0.4 for larger size groups. Percent weight loss was similar for all individuals above 30 mg original weight, but generally weight loss decreased regularly with increasing body size. Lethal water loss in *Tr. chiliensis* was estimated between 45 and 50% of the original weight which is equivalent to a 70% loss in total body water.

Weight loss for three size groups of *Tal. quoyana* exposed to TH 75% are shown in Figure 6. Initial weight loss over the first hour of exposure was higher than during the remaining period over which the weight loss was linear. Medium *Tal. quoyana* were the most resistant group with weight losses of 0.1 mg. mg h^{-1} for the first hour followed by an average rate of 0.008 mg. There was one mortality recorded in the sample

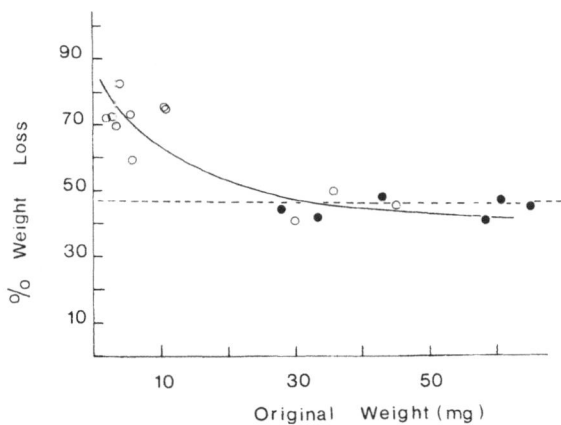

Fig. 5. Weight loss for *Tr. chiliensis* exposed to RH 20% for 1 h. Closed symbols indicate live amphipods, dotted line, the estimated water loss.

154

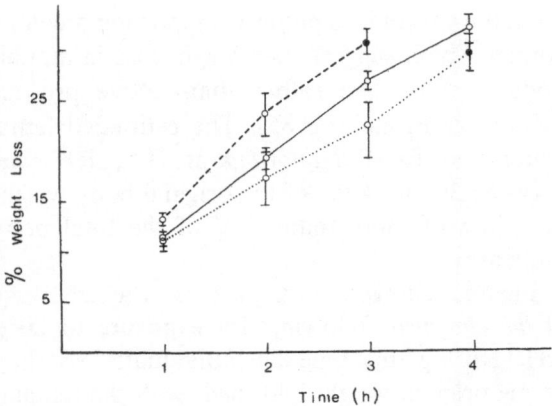

Fig. 6. Weight loss for *Tal. quoyana* exposed to RH 75% for up to 4 h for small < 50 mg wet wt. (-----), medium 50–100 mg (——), and large > 100 mg (·····) individuals; closed symbols indicate samples where some mortality occurred during the time interval.

of small individuals following 3 h exposure and two mortalities in the sample of large individuals after 4 h. Lethal water loss for *Tal. quoyana* was estimated at approximately 30% of the original weight although this is probably an underestimate for medium amphipods.

Similar experiments on *Tr. chiliensis* were terminated when amphipods failed to survive 2 h exposure to 75% RH. During the first hour, however, rates of water loss were similar for the complete size range of amphipods R = 0.21 (± 0.039) which is approximately twice the initial weight loss for *Tal. quoyana*. In control experiments for both species, exposed to humidity conditions approximating 100% RH, little weight loss was recorded and some individuals gained weight slightly. Changes in weight (± 2 SE) expressed as mg weight loss per mg wet weight per hour were 0.003 (± .0037) for *Tal. quoyana* and 0.054 (± 0.049) for *Tr. chiliensis*.

Discussion

The total gill area and arrangement of the gills in the beach flea *Tr. chiliensis* and the sandhopper *Tal. quoyana* are similar yet the two species differ considerably in their resistance to desiccation. Previous studies suggest such differences might relate to two main features, either restrictions due to phylogenetic constraints or their location within an evolutionary sequence onto land.

The gills of the two species under investigation are simple sac-like structures similar to the talitrid species described by Moore & Taylor (1984) and Spicer & Taylor (1986). They investigated three species of '*Orchestia*' in which the gill attached to peraeopod 2 provided the greatest contribution to total gill area. *Tr. chiliensis*, however, like the subtropical species, *Chroestia lota* (Marsden, 1985) and all *Talorchestia* species studied to date, has two enlarged gills. Gill G6 contributes up to 32% of the overall gill area and G2 contributes 24% of the gill area. As this value is close to literature values for other temperate '*Orchestia*' it seems that compensatory reduction must have occurred in the relative gill area of G3 and G4 of this species.

Although *Talorchestia* species typically have an enlarged 6th coxal gill (Spicer & Taylor, 1986), *Tal. quoyana* has the most enlarged gill measured to date contributing 36% of the total gill area. In this respect it is most similar to *T. deshayesii* which is also a sand dwelling species. Large gill areas might represent a significant potential for water loss in the posterior part of the body, especially in *Tal. quoyana* where the combined gill areas for G5 and G6 are 54% of the total gill area. The posterior gills in *Tal. quoyana* are, however, protected by the flexed abdomen when the amphipod is at rest or walking and it seems likely this behaviour actively reduces potential water loss from the gills and perhaps increases the possibility of O_2 uptake due to the proximity to pleopods.

Although *Tal. quoyana* achieves a larger maximum size than *Tr. chiliensis* the weight specific gill area and the size relationships are similar. The total gill areas are within the range given for other talitrid amphipods (Spicer & Taylor, 1986), however the weight exponents are closer to the sandhoppers than beachfleas. These results suggest that the two ecological groups cannot be characterised simply by the different slopes of their gill area/dry body weight regression plots.

Previous studies have highlighted the role of habitat as a general determinant of relative gill area but its importance in invasion into low salinity environments has remained unanswered. *Tr. chiliensis* is euryhaline (Marsden, 1980) and like *O. gammarellus* and *Tal. saltator* (Moore & Francis, 1985; Spicer & Taylor, 1987b; Morritt, 1988) may be expected to regulate its haemolymph osmolality under conditions of reduced salinity. Gills may be implicated in such control mechanisms and this may provide one explanation for increased gill areas in the above species compared with their stenohaline counterparts.

The ability of amphipods to survive water loss depends on many factors, including rate of water loss, relative body size and total water content. Relatively few studies have measured water loss in amphipods and, due to different methodologies, it is often difficult to compare results in a realistic ecological framework. In addition amphipod weight relationships may be affected greatly by their internal water balance, influenced for example by the habitat, the availability of moist food, or internal factors such as the time since last feeding or the amount of food in the gut (Moore & Francis, 1984). Recently Morritt (1987) investigated water loss under desiccating stress in three species of talitrid amphipods and confirmed earlier results by Williamson (1951) and Moore & Francis (1985) showing lower transpiration rates in the beach hopper, *Talitrus saltator*, than in the beach flea, *Orchestia gammarellus*, and the landhopper, *Arcitalitrus dorrieni*. As in the present study, juveniles had higher weight-specific rate of water loss in still air than adults and elevated rates of water loss were recorded under conditions of high desiccation stress. Weight loss of 0.21 mg. mg wet wt. h^{-1} for *Tr. chiliensis* is similar to that recorded for the landhoppers *A. sylvaticus* and *A. dorrieni* (Lazo-Wasem, 1984; Morritt, 1987). Adult *Tal. quoyana*, however, show similar weight loss value to *Talitrus saltator* with juveniles exhibiting lower transpiration rates (0.09 mg water. mg wet wt. h^{-1}) on exposure to 75% RH at 15 °C.

In most invertebrates, both the rate of water loss and the lethal water loss is related to the initial body water content, a feature which often explains differences in water loss between species. The usual water content for *Tal. quoyana* was between 68% and 73%, values which are similar to *O. gammarellus* (Moore & Francis, 1985). In contrast with this, water content or *Tr. chiliensis* was closer to the aquatic gammarids like *G. locusta* and *G. duebeni* (Sutcliffe, 1971; Lockwood & Inman, 1973). For the two species examined it seems unlikely that differences in water content of up to 5% original weight could account for the observed differences in rates of water loss, although, it might explain the higher lethal weight loss value for *Tr. chiliensis*.

Examination of the amphipod literature provided only a single value for lethal weight loss in semiterrestrial amphipods. Lethal weight loss for *O. gammarellus* was between 17 and 20% initial weight or a 25% loss of body water (Moore & Francis, 1985). This value is similar to the intertidal mud crab *Helice crassa* (22%) but below that for *Macrophthalmus hirtipes* (30%) which occurs at slightly lower tidal levels (Jones and Simons, 1982). Some high shore crabs, such as *Cyclograpsus lavauxi* (Innes et al., 1986) survive losses in body weight between 31 and 36% corresponding to 45% of their total body moisture. When exposed to 75% RH, both *Tal. quoyana* and *Tr. chiliensis* survived weight loss corresponding to approximately 60% body moisture. This remarkable ability may indicate a potential for utilising water loss as a mechanism for evaporative cooling. Such a phenomenon has been demonstrated for isopods and burrowing mud crabs and would have an adaptive function in species inhabiting surface wrack deposits.

In the Amphipoda, if the gills are the major site of water loss, species with reduced gill areas should have low rates of water loss in desiccating conditions. Table 2 attempts to relate gill area to desiccation tolerance, habitat, and oxygen consumption ($\dot{V}O_2$) for species where sufficient information is available. All estimates were based on a standard amphipod (5 mg dry weight) and oxygen uptake h^{-1} at 15 °C using a Q^{10} value of 2. The species were arranged according to their gill areas, a sequence which corresponds almost exactly to the temperature/humidity characteris-

Table 2. Relating gill area, oxygen uptake and habitat in an ecological sequence of talitrid amphipods – calculated for an individual 5 mg dry weight.

Species	Habitat	% water content (R)	Total gill area	Aq VO$_2$	Air VO$_2$	Q10 (W)Aquatic (A)Air
Tatitrus saltator	supralittoral, sand, marsh	? (0.11)$_M$	2$_B$		7.5$_K$	(A) 2.5(5–15)
Chroestia lota	supralittoral, sand, brackish, shingle	72–75	3$_G$	6$_G$	6$_G$	(W) 2.3(20–30) (A) 2.3(20–30)
Orchestia gammarellus	supralittoral, wrack, brackish, rocks saltmarsh	68.5$_{C,L}$ (0.11)$_M$	4.2$_{B,O}$	5.8$_J$	6$_J$	(W) 1.8(10–20) (A) 1.6(10–20)
Talorchestia quoyana	supralittoral, wrack, sand beach, oceanic	68–73 (0.08)	4.5	3$_I$	3.5$_I$	(W) 1.6(5–15) (A) 2.0(5–15)
Transorchestia chiliensis	supralittoral, brackish, rocks, shingle, sand	72–76 (0.21)	4.5	4$_H$	9$_H$	(W) 1.3(5–15) (A) 2.1(5–15)
Orchestia mediterranea	supralittoral wrack, stones	?	6.2$_B$	6.8$_J$	6.4$_J$	(W) 1.9(10–20) (A) 2.4(10–20)
Orchestia cavimana	freshwater, brackish highwater	?	6.5$_B$	4.0$_J$	6.5$_J$	(W) 3.0(10–20) (A) 2.3(10–20)
Arcitalitrus dorrieni	terrestrial, leaf litter	? (0.18)$_M$	11$_B$	2.5$_J$	8.0$_J$	(W) 1.4(10–20) (A) 2.0(10–20)
Gammarus duebeni	marine, highwater, brackish	73–76$_E$	13$_O$	2.0$_N$		(W) 2.0(5–15)
Gammarus locusta	marine, low tide	70–74$_A$	19$_O$	5.4$_N$		(W) 2.3(5–15)
Gammarus pulex	freshwater	80$_F$	21$_O$	9.5$_D$		

Reference material:
A. Sutcliffe, 1971
B. Spicer and Taylor, 1986
C. Spicer and Taylor, 1987a
D. Wright and Wright, 1976
E. Lockwood and Inman, 1973
F. Butterworth, 1968
G. Marsden, 1985
H. Marsden, 1984
I. Marsden, 1989
J. Spicer and Taylor, 1987b
K. Williams, 1981
L. Moore and Francis, 1985
M. Morritt, 1987
N. Bulnheim, 1979
O. Moore and Taylor, 1984

tics of the habitats. Supralittoral species occur at the op of the list with fully aquatic species at the bottom and the euterrestrial species at inter-mediate levels. Although the table illustrates some correlation between water content, desiccation and gill area, rates of oxygen uptake appear highly variable with similar oxygen uptake in species with both a low and high gill area (e.g., *G. locusta* and *C. lota*). As expected, fully aquatic species show increased rates of oxygen consumption in aquatic conditions with the reverse for those species normally found in air. These differences are further reflected in the Q^{10} values with values greater than 2 associated with the usual respiratory medium. This was also a feature in *G. oceanicus* (Halcrow and Boyd, 1967) and *Talorchestia margaritae* (Venables, 1981). Interestingly, those supralittoral amphipods found in brackish water appear to have relatively high Q^{10} values both in air and water.

Amphipods which coexist in particular habitats or environments are usually distributed in relation to their aquatic requirements, for example in a series of *Gammarus* species from marine, brackish and freshwater habitats (Bulnheim, 1979) and landhoppers occupying grasslands (Duncan, 1969). Although *Tal. quoyana* is more tolerant of desiccation than *Tr. chiliensis* they cooccur on rocky shores possessing sand at upper levels. *Tal. quoyana*, however, is a sand burrowing beach hopper; larger individuals can remain within burrows thus evading potential thermal and desiccation stress. In contrast, the beach flea *Tr. chiliensis*, shelters below wrack or stones and is exposed to a widely fluctuating environmental regime. Interestingly juveniles of both species are found in surface sand or under drift weed and appear especially sensitive to water loss. Given the differences in the desiccation tolerances of the two species it seems likely that the distribution of *Tr. chiliensis* would be detrimentally affected by desiccation stress. This assertion is supported by field observations which show that *Tr. chiliensis* is generally absent from sand beaches without sufficient drift seaweed to provide shelter and conditions of high humidity.

This study was unable to ascertain why the beach flea is more sensitive to water loss than the beach hopper. The gill areas of the two species are similar and Dahl (1977) suggests the branchial cuticle is likely to be of a similar thickness in most species. Another feature likely to be important in water loss is the permeability of the general body surface. Duncan (1985), Moore & Francis (1985) and Halcrow & Bousfield (1987) have shown high densities of cuticular pores over the surface of many amphipods including supralittoral and terrestrial species. They are present also in the two species under investigation. Initial examination of SEM photographs of the body surface show few obvious differences between the two species, suggesting that microcuticular structures may be unimportant in determining rates of water loss. It is suggested here that future studies directed at determining mechanisms of water loss of amphipods need to investigate cuticle thickness and surface secretions as well as other sites of water loss including the antennal gland, mouth and anus.

Acknowledgements

I would like to thank Clinton Duffy for providing the whole animal illustrations and Paul Creswell for assistance with gill areas. Kelvin Duncan contributed greatly with discussions on amphipod water loss mechanisms. Thanks also to the secretaries of the Zoology Department for their help in preparing the final manuscript.

References

Bousfield, E. L., 1982. The amphipod family Talitroidea in north-eastern Pacific regions. 1. Family Talitridae; systematics and distributional ecology. Publications in Biological Oceanography, National Museum of Canada, no. 11, 73 pp.

Bulnheim, H. P., 1979. Comparative studies on the physiological ecology of five euryhaline *Gammarus* species. Oecologia 44: 80–86.

Butterworth, P. E., 1968. An estimation of the haemolymph volume in *Gammarus pulex*. Comp. Biochem. Physiol. 26A: 1123–1125.

Dahl, E., 1977. The amphipod functional model and its bearing upon systematics and phylogeny. Zoological Scripta 6: 221–228.

Duncan, K. W., 1969. The ecology of two species of terrestrial Amphipoda (Crustacea: Family Talitridae) living in waste grassland. Pedobiologia 9: 323–341.

Duncan, K. W., 1985. Cuticular microstructures of terrestrial Amphipoda (Crustacea: Family Talitridae). Zool. Anz 215: 140–146.

158

Edney, E. B., 1960. Terrestrial adaptations. In: The Physiology of Crustacea (T.H. Waterman, Ed.) New York: Academic Press, vol. 1 pp 367–393.

Fincham, A. A., 1977. Intertidal sand-dwelling peracarid fauna of North Island, New Zealand. N.Z. J. Mar. Freshwat. Res. 11: 677–696.

Friend, J. A. & A. M. M. Richardson, 1977. A preliminary study of niche partition in two Tasmanian terrestrial amphipod species. Ecological Bulletin (Stockholm) 25: 24–35.

Friend, J. A. & A. M. M. Richardson, 1986. Biology of terrestrial amphipods. Ann. Rev. Ent. 31: 25–48.

Halcrow, K. & E. L. Bousfield, 1987. Scanning electron microscopy of surface microstructures of some gammaridean amphipod crustaceans. J. Crust. Biol. 7: 274–287.

Halcrow, K. & C. M. Boyd, 1967. The oxygen consumption and swimming activity of the amphipod Gammarus oceanicus at different temperatures. Comp. Biochem. Physiol. 23A: 233–242.

Hurley, D. E., 1968. Transition from water to land in amphipod crustaceans. Am. Zoo. 8: 327–353.

Innes, A. J., M. E. Forster, M. B. Jones, I. D. Marsden & H. H. Taylor, 1986. Bimodal respiration, water balance and acid base regulation in a high shore crab, Cyclograpsus lavauxi H. Milne Edwards. J. exp. mar. Biol. Ecol. 100: 127–145.

Innes, A. J. & E. W. Taylor, 1986. The evolution of air-breathing in crustaceans: a functional analysis of branchial, cutaneous and pulmonary gas exchange. Comp. Biochem. Physiol. 85A: 621–637.

Jones, M. B. & M. J. Simons, 1982. Habitat preferences of two estuarine burrowing crabs Helice crassa Dana (Grapsidae) and Macrophthalmus hirtipes (Jacquinot) (Ocypodidae). J. exp. mar. Biol. Ecol. 56: 49–62.

Lazo-Wasem, E. A., 1984. Physiological and behavioral ecology of the terrestrial amphipod Arcitalitrus sylvaticus (Haswell, 1880). J. Crust. Biol. 4: 343–355.

Lockwood, A. M. P. & C. B. E. Inman, 1973. Water uptake and loss in relation to salinity of the medium in the amphipod crustacean Gammarus duebeni. J. exp. Biol. 58: 149–163.

MacIntyre, R. J., 1963. The supralittoral fringe of New Zealand sand beaches. Trans. r. Soc. N.Z. 88: 89–103.

Marsden, I. D., 1980. Effects of constant and cyclic temperatures on the salinity tolerance of the estuarine sandhoppers, Orchestia chiliensis. Mar. Biol. 59: 211–218.

Marsden, I. D., 1984. Effects of submersion on the oxygen consumption of the estuarine sandhopper Transorchestia chiliensis (Milne Edwards, 1840). J. exp. mar. Biol. Ecol. 79: 263–276.

Marsden, I. D., 1985. Some factors affecting survival and oxygen uptake in a subtropical beach flea. J. exp. mar. Biol. Ecol. 88: 213–225.

Marsden, I. D., 1989. An assessment of seasonal adaptation in the beach hopper Talorchestia quoyana (Milne Edwards). J. exp. mar. Biol. Ecol. 128: 203–218.

Moore, P. G. & C. H. Francis, 1985. On the water relations and osmoregulation of the beach-hopper Orchestia gammarellus (Pallas) (Crustacea: Amphipoda). J. exp. mar. Biol. Ecol. 94: 131–150.

Moore, P. G. & A. C. Taylor, 1984. Gill area relationships in an ecological series of gammaridean amphipods (Crustacea). J. exp. mar. Biol. Ecol. 74: 179–186.

Morritt, D., 1987. Evaporative water loss under desiccation stress in semiterrestrial and terrestrial amphipods (Crustacea: Amphipoda: Talitridae). J. exp. mar. Biol. Ecol. 111: 145–157.

Morritt, D., 1988. Osmoregulation in littoral and terrestrial talitroidean amphipods (Crustacea) from Britain. J. exp. mar. Biol. Ecol. 123: 77–94.

Newell, R. C., 1979. Biology of Intertidal animals, 3rd Edition, Faversham, Kent Marine Ecological Surveys Ltd, 781 pp.

Platzman, S. J., 1960. Comparative ecology of two species of intertidal amphipods: Talorchestia megalophthalma and orchestia agilis. Biol. Bull. (Woods Hole, Mass.) 119: 333.

Richardson, A. M. M. & D. M. Devitt, 1984. The distribution of four species of terrestrial amphipods (Crustacea, Amphipoda: Talitridae) on Mount Wellington, Tasmania. Aust. Zool. 21: 143–156.

Spicer, J. I., P. G. Moore & A. C. Taylor, 1987. The physiological ecology of land invasion by the Talitridae (Crustacea: Amphipoda). Proc. r. Soc. Lond. B 232: 95–124.

Spicer, J. I. & A. C. Taylor, 1986. A comparative study of the gill area relationships in some taltrid amphipods. J. Nat. Hist. 20: 935–947.

Spicer, J. I. & A. C. Taylor, 1987a. Respiration in air and water of some semi- and fully terrestrial talitrids (Crustacea: Amphipoda: Taltridae). J. exp. mar. Biol. Ecol. 106: 265–277.

Spicer, J. I. & A. C. Taylor, 1987b. Ionic regulation and salinity related changes in haemolymph protein in the semi-terrestrial beech-flea Orchestia gammarellus Pallas (Crustacea: Amphipoda). Comp. Biochem. Physiol. 88A: 243–426.

Sutcliffe, D. W., 1971. Regulation of water and some ions in gammarids. 1. Gammarus duebeni. J. exp. Biol. 55: 324–344.

Taylor, A. C. & J. I. Spicer, 1986. Oxygen transporting properties of the blood of two semi-terrestrial amphipods. Orchestia gammarellus (Pallas) and O. mediterranea (Costa). J. exp. mar. Biol. Ecol. 97: 135–150.

Venables, B. J., 1981. Oxygen consumption in a tropical beach amphipod Talorchestia margaritae Stephenson: effects of size and temperature. Crustaceana (Leiden) 41: 89–94.

Williams, J. A., 1981. The respiratory rhythm and respiratory quotient of Talitrus saltator. Comp. Biochem. Physiol. 70A: 639–641.

Williams, D. I., 1951. Studies on the biology of Talitridae (Crustacea: Amphipoda): effects of atmospheric humidity. J. mar. biol. Ass. U.K. 30: 73–90.

Winston, P. W. & D. H. Bates, 1960. Saturated solutions for the control of humidity in biological research. Ecology 41: 232–237.

Wright, C. A. & A. A. Wright, 1976. The respiratory quotient of Gammarus pulex L. Comp. Biochem. Physiol. 53A: 45–46.

Hydrobiologia **223**: 159–169, 1991.
L. Watling (ed.), VIIth International Colloquium on Amphipoda.
© 1991 *Kluwer Academic Publishers.*

Lack of oxygen and low pH as limiting factors for *Gammarus* in Hessian brooks and rivers

Meertinus P.D. Meijering
Fachgebiet Fließgewässerkunde, Fachbereich Landwirtschaft, Gesamthochschule Kassel,
Nordbahnhofstraße 1a, D-3430 Witzenhausen, Germany

Key words: oxygen, pH, freshwater, Gammarus, brooks, rivers

Abstract

During two decades (1968–1988) the occurrence of *Gammarus* was studied in Hessian running waters draining some 7700 km² of the Fulda-Eder-basin and two adjacent areas. About 1530 sites were sampled, most of them in 1968, 1978, 1981/82 and 1985, which made comparisons possible, primarily with respect to space, but partly also to time. After consideration of the relief of the landscape, geological conditions, and water-quality in the neighbourhood of human settlements, lack of oxygen (due to organic pollution) and low pH (souring) can be recognized as the most important factors which influence and alter natural distribution patterns of *Gammarus fossarum*, *G. pulex* and *G. roeseli*. Both vanishing and recovering populations were observed in the course of twenty years.

Introduction

There are physiological relationships between influences of both souring and diminishing concentration of oxygen, since the increase of free hydrogen ions in waterbodies affects respiration rates in cladocerans and gammarids (Ivanova & Kleckowski, 1972; Alibone & Fair, 1981; Brehm & Meijering, 1982). Accordingly *G. fossarum*, suffering more from oxygen depletion than *G. pulex*, is also first to vanish in the case of water souring (Meijering, 1982). Another very sensitive *Gammarus* is *G. lacustris* (Vangenechten, 1979), in which, however, long term adaptations to lower pH may improve the stability of a population (Økland & Økland, 1985). Differences in pH tolerance were also found in various populations of *Hyalella azteca* (France, 1987).

Factors other than oxygen content and pH were also found to play important roles in the distribution of *G. fossarum* and *G. pulex*, such as

temperature and current (Roux, 1971; Meijering, 1972), different migration habits and intensities (Meijering, 1971, 1972) and food-preferences (Koch, 1990). In an excellent series of papers comparing population dynamics of *G. fossarum*, *G. pulex* and *Echinogammarus berilloni*, a large complex of biotic and abiotic factors was taken into consideration (Goedmakers, 1980, 1981a, 1981b; Goedmakers & Pinkster, 1981). These results should serve as a warning about the danger of oversimplification which can result from a reduction of our question to just one or another factor.

A large number of sites were sampled in the Fulda-Eder-watershed in Hesse, and the occurrence of *Gammarus* as an important indicator for souring and water pollution was registered (Jahr *et al.*, 1980; Pieper & Meijering, 1981, 1982a, 1982b, 1983; Teichmann & Meijering, 1981; Meijering & Pieper, 1985). Additionally long termed comparisons became possible in running

160

waters of the Schlitzerland, whose *Gammarus* populations have been known for about 20 years (Meijering, 1971, 1984). All these results will be discussed in context and applied to a brook system in the Kaufungen Forest, where a large number of limnological studies were recently conducted.

Results

In a special program on the distribution of *Gammarus*, which was sponsored by the Hessian Foundation for Nature Preservation (Stiftung Hessischer Naturschutz), all running waters of the Fulda-Eder-watershed shown in Fig. 1 were

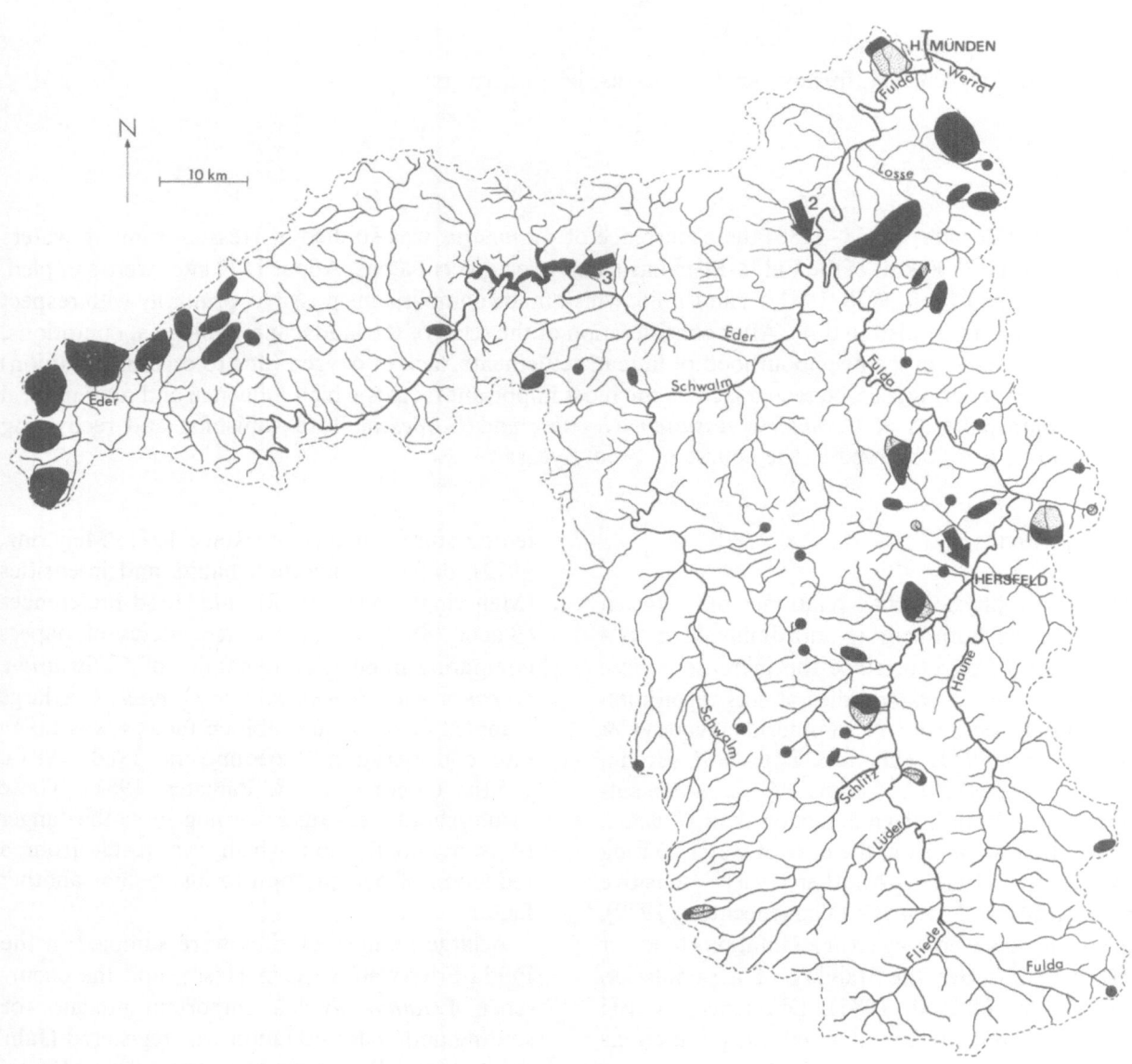

Fig. 1. Souring of upper courses in the Fulda-Eder-watershed as indicated by the distribution of *Gammarus*. Black areas are free of *G. fossarum*, while in gray areas this species is accompanied by *G. pulex*. Arrow 1 points to the town of Hersfeld, arrow 2 to the mouth of the Eder. Arrow 3 marks a dam in the Eder.

sampled at defined places: The crossings between these waters and the 50 m contours from 150 to 750 m above sea level. The number of sites resulting from this system was 1167, which were nearly all situated at altitudes between 200 and 650 m according to the general level of the country.

Early results of the program were presented at the Vth International Colloquium on *Gammarus* and *Niphargus* in Łódź in 1981. So the methods of sampling were published already and can be found in the proceedings of that meeting (Pieper & Meijering, 1982a), were a map shows all sampling sites in part of the Fulda-watershed, south of Hersfeld. Maps of other parts of the Fulda-Eder-watershed are available in preliminary publications (Pieper & Meijering, 1982b, 1983).

Although we should expect *Gammarus* nearly everywhere in the mentioned region, which was originally covered with deciduous woods, this is far from reality now as can be seen in Fig. 2. In the lower parts and especially on elevations of 200 and 250 m, where running waters are developed at least as strong brooks or small rivers, the occurrence of *Gammarus* was merely sporadic. *G. roeseli*, which usually inhabits larger running waters, is completely subject to the influence of water pollution and is restricted to less than 10% of the potential dwelling places. The species was not found higher than 350 m, which, however,

could be a natural limit of distribution within the relief of this watershed.

At first glance the general situation of *G. pulex* is very similar. This species was not found in more than about 20% of the sites at lowest levels, i.e., in larger running waters, but it is found at elevations up to 450 m above sea level, where most sites are in smaller brooks or even near springs, (see Fig. 1). Also, the level of 450 m is close to the natural upper limit of distribution of *G. pulex*, so that this species, which predominantly settles on low levels, is also very much exposed to water pollution. *G. fossarum*, on the other hand, is much more confined to small running water, which, however, may be situated on all levels of the considered watershed, since in Central Europe *G. fossarum* goes up to more than 1000 m. It may be, that *G. fossarum* on lower levels is limited by the competition from the two other gammarids, but without doubt water pollution is also responsible for its retreat from most sites lower than 200 m. On higher levels up to 60% of the sites are occupied by *G. fossarum*, and there is a gradient towards higher altitudes, although with an interruption at 550 and 600 m above sea level (see Fig. 2).

In order to explain the course of the *G.-fossarum*-curve in Fig. 2 it is necessary to analyze the distribution of this species in different parts of the watershed separately. In Fig. 3 the

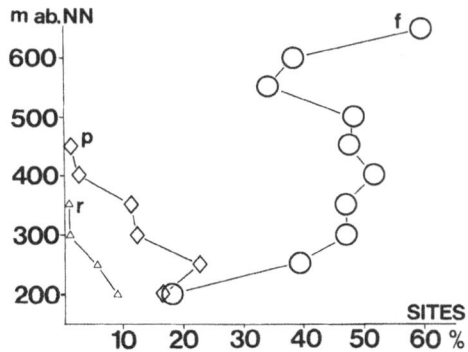

Fig. 2. Occurrence of *G. roeseli* (*r*), *G. pulex* (*p*) and *G. fossarum* (*f*) in percentage of sampling sites on different elevations above sea level (NN) in the entire Fulda-Eder-river system.

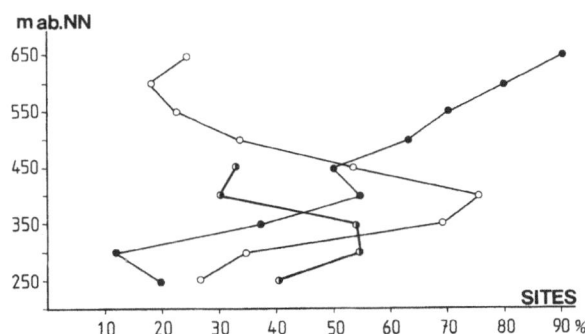

Fig. 3. Occurrence of *G. fossarum* in three parts of the Fulda-Eder-watershed (see Fig. 1). ● = area south of Hersfeld, ◐ = area between Hersfeld and Hannoversch Münden (but without Eder-system) and ○ = area west of Eder-dam.

162

results are presented for the watershed of the upper Fulda south of Hersfeld (see Fig. 1 and Pieper & Meijering, 1982a) with the mountainous areas of the Rhön and the Vogelsberg, which are not so densely populated by humans than the lower parts of the country. Here we found a regular gradient running up to 90% at a level of 650 m. Two curves, of the watersheds of the lower Fulda north of Hersfeld, and of the upper Eder west of the place where it is dammed, show a completely different picture. At lower levels the gradient goes up steadily, but at 350 and 400 m respectively they both decrease dramatically, and not more than 30 or even 20% of the sites are settled by *G. fossarum* anymore. And since there are only few human settlements in higher regions of both the Fulda- and the Eder-watersheds, they cannot be responsible for the retreat of *G. fossarum* from all the places marked in Fig. 1. And indeed, there are other reasons why gammarids vanished from the top levels, which lie at 400 to 450 m on both sides of the lower Fulda and at 500 to 700 m in the Rothaar-mountains, from where the Eder originates.

All sites at high elevations, which were free of *G. fossarum* or in which *G. fossarum* was accompanied by *G. pulex* (black or gray in Fig. 1) and which additionally were free of polluted water (water quality group I or I–II according to the German classification system used in the state of Hesse), altogether 192 sites, were collected in Fig. 4, grouped on geological criteria. The sites, where *G. fossarum* is lacking or where it is accompanied by *G. pulex*, are concentrated on two types of geological basements. Half of the sites, 95, are in brooks whose watersheds are dominated by red sandstone; these are the groups 1 and 4 in Fig. 4. Another 70 sites lie in Devonian greywacke and sandstone, which are the groups 2, 3, 5, 6, 7, and 8. Furthermore 8 sites from the rest are in greywacke-basements and 7 in red sandstone, the latter mainly accompanied by basaltic intrusions. All these basements are well known for their very poor pH-buffer capacity, which makes them vulnerable to souring. Red sandstone in Hesse is mainly situated along both sides of the lower River Fulda, more or less in the north-eastern part of the

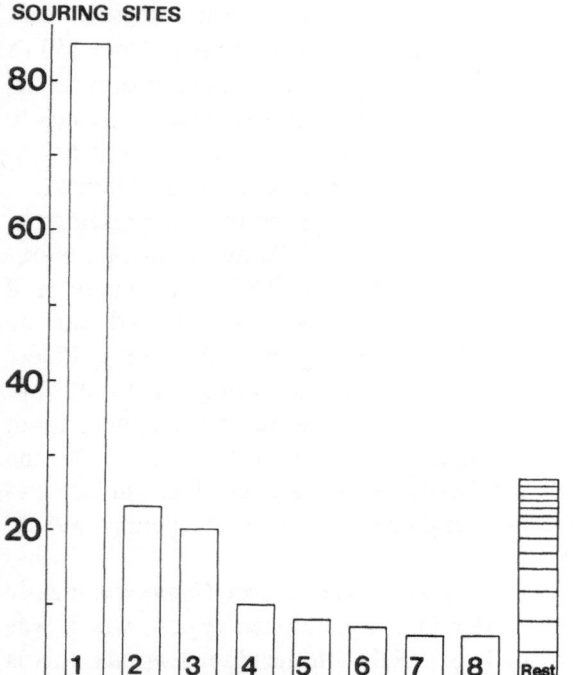

Fig. 4. Geological conditions in the watersheds of 192 sampling sites, where souring was indicated by *Gammarus*. Lower and medium Red sandstone (1); Devonian (Eifel) greywacke (2); Devonian greywacke (Oberems) (3); lower and medium Red sandstone + Miocaenian deposits (4); Devonian (Siegen) greywacke-sandstone (5); Devonian (Unterems) greywacke – sandstone + some Pleistocaenian deposits (6); Devonian (Oberems) greywacke and sandstone + Devonian (Eifel) greywacke (7); Devonian (Oberems) sandstone and greywacke + Devonian (Givet) sandstone + Devonian (Eifel) greywacke and sandstone (8).

Fulda-Eder-watershed (Fig. 1), while Devonian greywacke and sandstone lie on high levels of the Rothaar-mountains, in the western part of the Eder-watershed.

In Fig. 5 the percentage of sampling sites settled by *G. fossarum* is given for several geological conditions in the watersheds; only sites in unpolluted water (quality class I) were taken into consideration. On the whole 32 sites were found on tertiary basalt (a), 36 in a region of carbonic greywacke and limestone (b) and 70 of lower and medium red sandstone (c), all according to the geological map of the state of Hesse on the scale 1 : 300 000. All the basaltic sites were settled by *G. fossarum*, little more than 90% in the region of

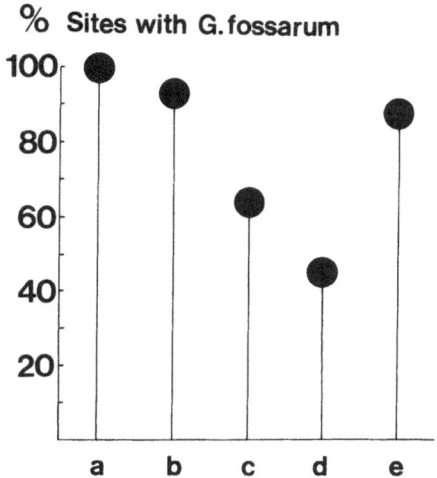

Fig. 5. Percentage of sampling sites settled by *G. fossarum* under different geological conditions. Tertiary basaltic basements (a); Carbonic (Kulm) greywacke + limestone (b); lower and medium Red sandstone (according to a map 1 : 300 000) (c); lower and medium Red sandstone (map 1 : 25 000) (d); lower and medium Red sandstone + various other deposits (map 1 : 25 000) (e).

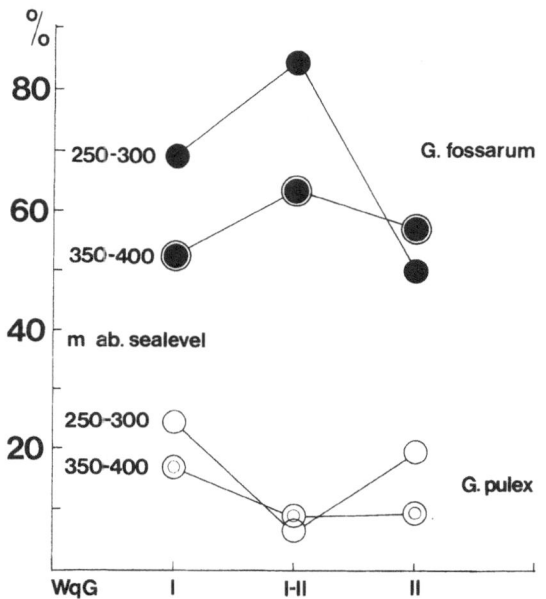

Fig. 6. Percentage of sampling sites settled by *G. fossarum* and *G. pulex* in Red Sandstone watersheds at different elevations above sea level and in various water qualities. Number of sites in WqG I: 69, in WqG I–II: 108, and in WqG II: 47.

greywacke and limestone, but not more than about 60% in the red sandstone area. This is another confirmation, that areas with poor pH-buffer capacity are difficult places for *G. fossarum*. There was an opportunity to look closer into the distribution of *G. fossarum* in just a small part of the Fulda-watershed, the Schlitzerland, from where a geological map on the scale 1 : 25 000 is available. Here it was possible to sort out watersheds with pure red sandstone basements from those with small additions, which are not visible on a map 1 : 300 000. The result is striking. From 20 sites with pure red sandstone watersheds only 9 were settled by *G. fossarum* (d), but from 23 sites, which proved to have other geological basements within their watersheds as well, not less than 20 were found to be settled (e)! So the result of group c can be looked upon as a mixture of sites with more or less homogeneous red sandstone watersheds.

In Fig. 6 the occurrence of *G. pulex* and *G. fossarum* were compared with respect to water-pollution, this on elevations of 250, 300, 350, and 400 m above sea level and entirely in red sandstone watersheds. It is obvious that *G. fossarum*

is most frequent in just slightly polluted water (water quality group I–II), and that this species is less frequently found in unpolluted water (WqG I) on one hand and in moderately polluted water (WqG II) on the other. Another effect is, that in better water qualities *G. fossarum* is less frequent on higher levels, which reflects the fact, that on levels of 350 or 400 m a lot of sites are in the vicinity of the tops of hills, especially in the watershed of the lower Fulda north of Hersfeld, where red sandstone basements mainly occur (compare Figs. 1 and 3). In moderately polluted water (WqG II), however, higher altitudes obviously give better chances to *G. fossarum*, but primarily a certain degree of suppression caused by waste water is visible. It should be stressed that the percentages of sites settled by one species or the other must not be added, since they both give figures for the specific settlements within all the sites sampled. So it is remarkable that the curves of *G. pulex* run just the opposite way to those of *G. fossarum*, and it seems likely that competition is active at 250 and 300 m, where *G. pulex* gets chances in places where *G. fossarum* is more or less suppressed. At 350 and 450 m, where *G. pulex* is pretty close to its natural distribution limit, this still is clear with respect to WqG I (souring sites) and, typically, to a less extend with respect to water pollution (moderately wasted sites), where *G. fossarum* was more frequent. In Fig. 6 both agents of low pH and O_2-losses are visible for *G. fossarum* and *G. pulex* and can be compared.

Since O_2-concentration and acidity were recognized to be of big influence on the distribution patterns of *Gammarus* species, it was very likely that current changes in *Gammarus* distributions could be derived from changes in either the amount of O_2 or H-ions in the water. As far as water pollution is concerned, circumstances in the rivers Lauter and Schlitz (Fig. 7 and 8) can serve as an example. Four sampling stations were under observation there for about 20 years. In 1970 the purification plant of Lauterbach was already overloaded, so that *G. fossarum* and *G. pulex* were found not more than sporadically and only *G. roeseli*, the most stable species (Scholz & Meijering, 1975), was abundant. In lower parts of the Lauter river, where decomposition processes were fully developed and O_2 usually was scarce, no more than a few specimens of *G. roeseli* were found at that time. This bad water was mixed with that coming from the river Altfell, which had good

quality (I–II). And although there were further pollutions from Bad Salzschlirf, water qualities in the river Schlitz were reasonable, and in the middle section all three *Gammarus* species were found in big numbers. But after that recovery the old purification plant of the town of Schlitz, being absolutely overloaded, destroyed *Gammarus* and left just a very few specimens of *G. roeseli*.

Up to 1976 several villages surrounding Lauterbach were linked to the drains of that town, and since a planned new purification plant was not built before 1985, the situation became critical. No gammarids were found anymore at site 1; *G. roeseli* was exterminated from site 2, where just some sporadic intruders from a nearby brooklet were seen, a few *G. fossarum* and one *G. pulex*; *G. fossarum* vanished from site 3, in spite of the fact that the community of Bad Salzschlirf had built a new purification plant in 1975. In 1981 the situation of site 2 had been spreaded down to site 3, and so that year can be looked upon as to be the worst in all the history of *Gammarus*-populations in Lauter and Schlitz. A bad example of river restauration: As soon as people are

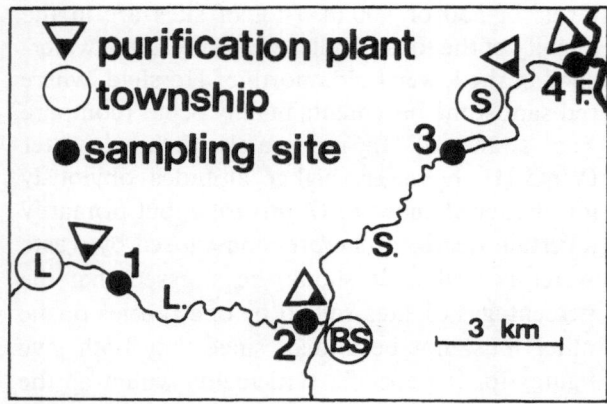

Fig. 7. Sampling sites in the rivers Lauter (L.) and Schlitz (S.), which run into the Fulda (F.). These are influenced by waste-water of the townships of Lauterbach (L), Bad Salzschlirf (BS) and Schlitz (S).

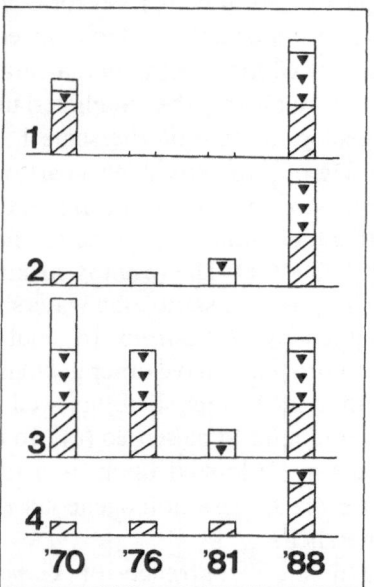

Fig. 8. Occurrence of *G. roeseli* (barred), *G. pulex* (triangles) and *G. fossarum* (white) during 18 years of increasing and after that decreasing water pollution. High columns: *Gammarus* abundant. Low columns: *Gammarus* sporadic.

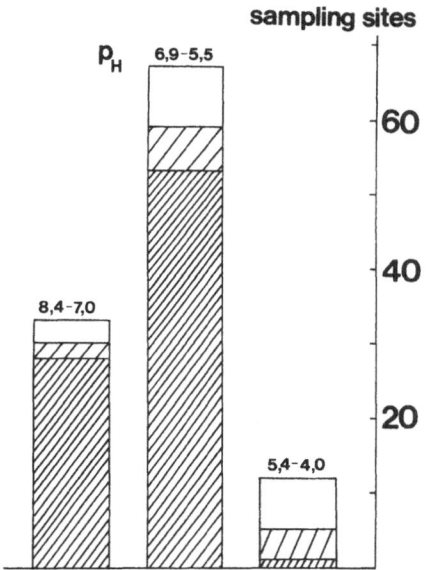

sampling sites

Fig. 9. Retreat of *Gammarus* in brooks of the Schlitzerland with different pH-conditions. Bars: Settled in 1968; Narrow bars: Settled in 1982.

connected with a purification plant, regardless whether it is overloaded or not, they have to pay duties. The building of a canal system had priority before the building of a new purification plant, and this was the reason, why *Gammarus* and the rest of the fluviatile biocoenosis had to pass such 'a steep valley' as can be seen from Figs. 7 and 8.

In 1980 the town of Schlitz opened a modern purification plant which drained to the Fulda river, but it was not before 1988 that *Gammarus* began to recover at site 4, since up to 1985 that place was still under the influence of the Lauterbach discharge of waste water. With the clean-up of the river complete, now *Gammarus* is recovering, firstly *G. roeseli*, secondly *G. pulex* and finally *G. fossarum*, which is the sequence opposite that of their decline.

Rivers got priority in sanitation programs, while streams and brooklets were taken into closer consideration only in very recent years. So a comparison between 1968 and 1982 in small running waters of the Schlitzerland shows a decline of *G. fossarum* and *G. pulex* in the course of 14 years. It should be noted, that this was an

additional program to the one described before as a Fulda-Eder-watershed program, although the Schlitzerland is part of that region.

Figure 9 contains the results from 112 sampling sites (see Meijering, 1984), which were grouped according to acidity in 1982. It is obvious that the number of sites without *Gammarus* was increased in the course of 14 years, mainly in connection with souring upper courses of small brooks (see Meijering, 1989), but in some cases also by the increase of waste water discharges from single farms in lower courses, where pH values were more or less basic. On the whole the situation in small tributaries remained better than in rivers, which is due to the fact that there is such a variety of conditions that provides niches and shelter places to widespread, numerous and mobile species such as the gammarids.

Another stream system, which is a tributary to the artificially salted Werra river, was studied during the past 5 years. This Wilhelmshäuser brook-system is settled by *G. fossarum* and *G. pulex*, but in the lowest sampling site close to the Werra river there are some *G. tigrinus* as well. This species was introduced into the Werra to replace *G. pulex* and *G. roeseli*, which formerly were common there (Schmitz, 1960; Meijering, 1980). *G. tigrinus* was not able to migrate upstream in the Wilhelmshäuser brook, in which the conductivity always lies around some 600 μS, while in the Werra we can find figures between 4000 and 11 000 μS, which means that the water is brackish (mixohaline). It may be, that in former years there also were some *G. roeseli* in the lowest part of the brook, but since currents are string there it is not likely that they were able to migrate upstream. Now-a-days there is no *G. roeseli* at all either in the Werra or in the Wilhelmshäuser brook.

The distribution patterns of *G. pulex* and *G. fossarum* are quite different. *G. pulex* (Fig. 10) can be found everywhere with just a few exceptions. They are lacking in six springs, three of them in headwater positions and three others as small tributaries coming in from aside. *G. fossarum*, on the other hand, is clearly restricted to the southern headwaters, where, however, it is lacking in several springs. In all sites where both

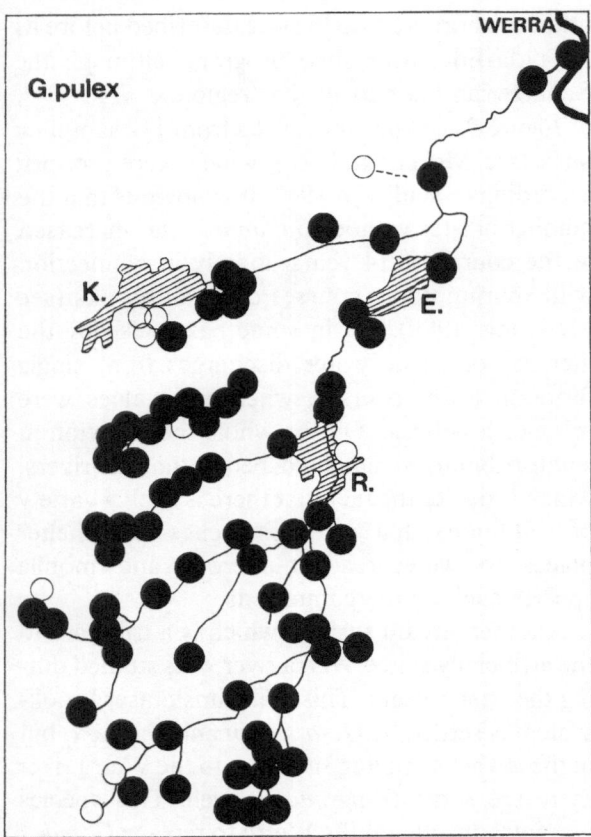

Fig. 10. Distribution of *G. pulex* in the Wilhelmshäuser brook-system, a tributary of the river Werra. K. = Kleinalmerode, E. = Ellingerode, R. = Roßbach.

ing in from SW was judged to belong to the WqG II in its upper course and to IV, III–IV and finally II east of Kleinalmerode. Summarizing we could give as an interpretation, that *G. fossarum* can be found in WqG I and I–II, while *G. pulex* was found up to WqG III.

In our studies, however, 75 sampling sites all over the Wilhelmshäuser brook-system were judged. From the results (Fig. 12) we can derive a more differentiated picture. *G. fossarum* was found in WqG I, I–II and II, in the latter group only when there are better conditions in the vicinity. If, however, short sections with water qualities I–II are limited by others which are worse, then obviously *G. fossarum* has difficulties to maintain itself, at least in case of competition from *G. pulex*. So there is a barrier of polluted water on the northern and western limit of species live together, *G. pulex* was found to be dominant (Meijering, 1988).

Explanations for these distributions can lean greatly on pH-conditions and water pollution. Water-quality measurements were done officially by the state of Hesse for all the country according to the LAWA-recommendations for biological ratings of running waters. This was also done for the Wilhelmshäuser brook-system, but just roughly because of the numerous waters in Hesse which had to be considered. The result as it can be taken from the official map was as follows. All the tributaries flowing together just south of the Roßbach-village were WqG I; the brook within that village WqG I–II; the rest of the Wilhelmshäuser brook down to the Werra was considered to be WqG II. The last tributary com-

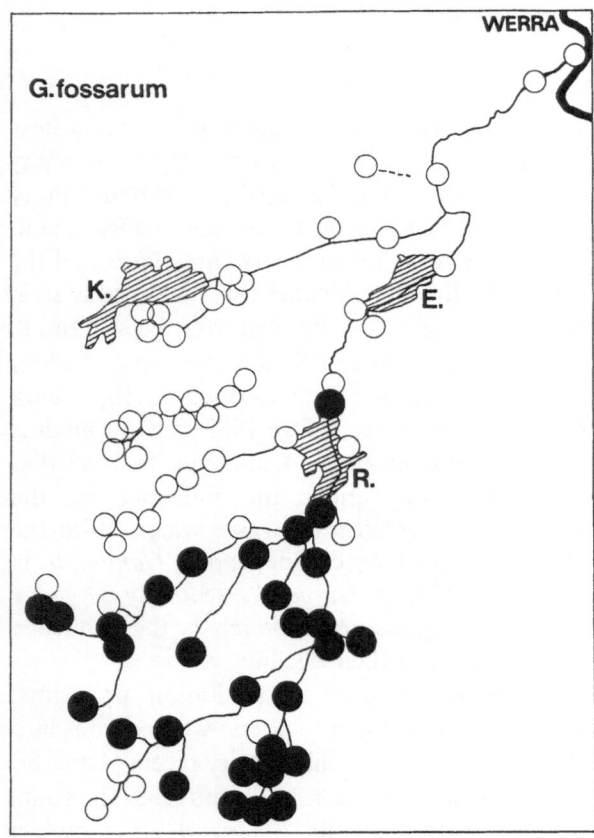

Fig. 11. Distribution of *G. fossarum* in the Wilhelmshäuser brook-system.

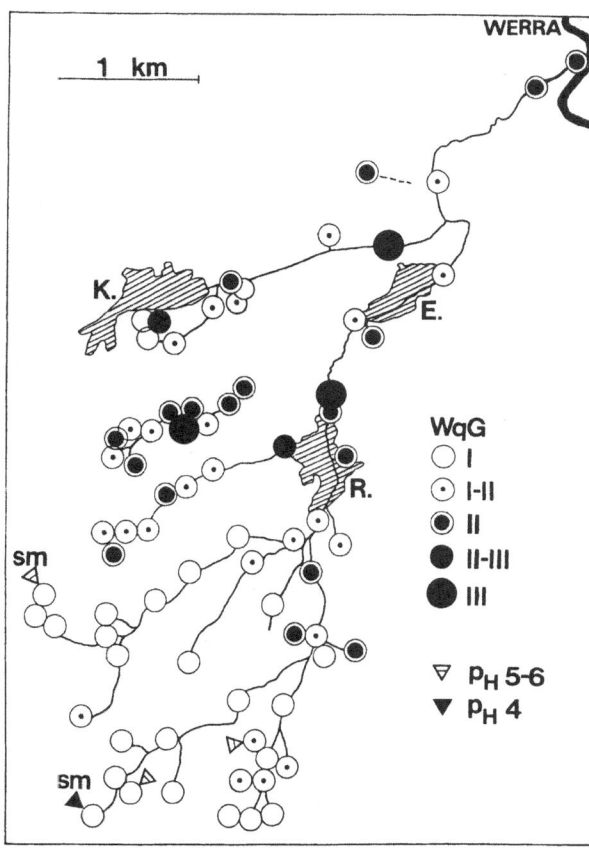

Fig. 12. Water quality and souring of the Wilhelmshäuser brook-system. WqG = Water quality Group (see text); sm = medium Red sandstone.

system. It is also lacking in a spring at Roßbach with a pretty high conductivity (some 1500 μS) and in 2 springs near the watershed in the SW part of the system, where we have a special geological situation. On the whole the Wilhelmshäuser brook-system drains the north-eastern slope of the Kaufungen Forest, a group of hills going up to about 600 m. On the south-western slopes of the Kaufungen Forest there is a basement of medium red sandstone with extremely souring brooks running off to the Fulda river (see Fig. 1 just south of the Werra). The north-eastern slopes, however, lie on lower red sandstone, which has a better pH-buffer capacity. In the watershed of the Wilhelmshäuser brook-system there are some springs lying just in medium red sandstone (sm) and they are free of *G. pulex* and, of course, *G. fossarum*. Two other springs in the neighbourhood are also relatively sour, and these are settled by *G. pulex*, but not by *G. fossarum*.

Conclusions

Woodland streams usually are characterized by factors which help to guarantee a good respiring climate for stream dwellers. There are the current, moderate temperatures during the summer months, and a sufficient supply of oxygen. Additionally, it is necessary, that the concentration of hydrogen ions is not too high, since otherwise respiration can be hindered.

G. fossarum can be considered to be a typical woodland-brook-element, not only from its resistance to currents, which is better than in *G. pulex* and *G. roeseli* (Roux, 1971; Meijering, 1972), but also with respect to its feeding habits (Koch, 1990). *G. roeseli*, on the contrary is much more a dweller of rivers, where temperatures go higher during the summer months, where currents can be slow or even absent in a variety of places within the river bed, and where biogenic oxygen may bring about oscillations of the O_2-content with high peaks in the early afternoon and minima in the early morning. *G. pulex* is more generalized as a dweller of streams and rivers, and often it is also found in springs with pretty low oxygen content.

Roßbach, another one in the brook coming down from Kleinalmerode and a third at the mouth into the salted Werra river. One tributary, the Lost Stream west of Roßbach is isolated anyway. So *G. fossarum* was restricted to just a part of the system, which is not so much polluted, more influenced by forests, and in its headwaters on higher elevations than the other tributaries. Here we find the species together with *G. pulex*, which, however, is more abundant.

G. pulex obviously is able to overcome these barriers and settles nearly everywhere, even in the Lost Stream, where it only avoids one dirty spring area, which is polluted from cattle. It was not found in another temporary and polluted spring south of Kleinalmerode and not in an isolated spring beside the lower course of the brook-

Additionally *G. pulex* is the most active migrator of the *Gammarus* species of running waters (see Meijering, 1971).

Keeping in mind these basic figures we can understand the distribution of *G. fossarum* in lotic areas with good breathing conditions, which are greatly endangered by souring and water pollution. So *G. fossarum* must be the first to be harmed, when either souring or water pollution occurs, while *G. pulex* is able to resist much better. Another thing, *G. pulex* is more a lowland-dweller and so also adapted to higher temperatures (Roux, 1971), which usually are found in lowland-brooks and -rivers during the summer. This obviously is one reason, why *G. pulex* proved to be pretty robust as soon as water pollution occurs; very much the same can be said in the case of *G. roeseli* (Meijering *et al.*, 1974; Scholz & Meijering, 1975; Meijering, 1988).

In Hesse *G. fossarum* settles in higher and highest regions, where currents are stronger, temperatures lower and the water better oxygenated, especially where brooks are situated beyond rural and human settlements. However, in several parts of Hesse, *G. fossarum* is endangered by souring, which ruins the breathing climate in running waters. *G. pulex* as a more stable species dwells in brooks and rivers, often together with *G. fossarum* on one hand or *G. roeseli* on the other. Especially when *G. fossarum* is more or less suppressed, probably because of souring or water pollution, *G. pulex* can coexist in the same places, can become dominant or even the only *Gammarus*, as in springs with pretty low O_2-contents or low pH.

G. pulex is confined to regions under 450 m, at least in the relief of Hessian hills. Exceptionally they were found at 500 m, as in the headwaters of the Wilhelmshäuser brook-system. But this means that *G. pulex* very often suffers from severe waste water, just like *G. roeseli*, which is confined to rivers or big brooks. The sanitation of rivers, however, will improve their chances in Hessian running waters now, since efforts to purify water discharges become more and more successful, firstly in rivers, but in the meantime also in streams.

References

Alibone, M. R. & P. Fair, 1981. The effects of low pH on the respiration of *Daphnia magna* Straus. Hydrobiologia 85: 185–188.

Brehm, J. & M. P. D. Meijering, 1982. Zur Säure-Empfindlichkeit ausgewählter Süßwasser-Krebse (*Daphnia* und *Gammarus*, Crustacea). Arch. Hydrobiol. 95: 17–27.

France, R. L., 1987. Differences in H-ion sensitivity among *Hyalella azteca* populations: An illative hypothesis invoking natural selection. Ann. Kon. Belg. Vereniging Dierkde. 117: 129–137.

Goedmakers, A., 1980. Population dynamics of three Gammarid species (Crustacea, Amphipoda) in a French chalk stream. Part I. General aspects and environmental factors. Bijdr. Dierk. 50: 1–34.

Goedmakers, A., 1981a. dto. Part II. Standing crop. Bijdr. Dierk. 51: 31–69.

Goedmakers, A., 1981b. dto. Part IV. Review and implications. Bijdr. Dierk. 51: 181–190.

Goedmakers, A. & S. Pinkster, 1981. dto. Part III. Migration. Bijdr. Dierk. 51: 145–180.

Hessischer Minister für Umwelt und Reaktorsicherheit (ed.), 1986. Biologischer Gewässerzustand 1986. Map 1 : 200000, Wiesbaden.

Ivanova, M. B. & R. Z. Kleckowski, 1972. Respiratory and filtration rates in *Simocephalus vetulus* (O. F. Müller) (Cladocera) at different pH. Pol. Arch. Hydrobiol. 19: 303–318.

Jahr, W., M. P. D. Meijering & W. Wüstendörfer, 1980. Zur Situation der Gattung *Gammarus* (Flohkrebse) im Vogelsberg. Beitr. Naturkde. Osthessen 16: 3–12.

Koch, K. D., 1990. Ernährungsökologische Untersuchungen an *Gammarus pulex* (L.) und *Gammarus fossarum* Koch, 1835 (Crustacea, Amphipoda) in einem Wiesenbach und einem Waldbach. Beitr. Naturkde. Osthessen 26: 3–126.

Meijering, M. P. D., 1971. Die *Gammarus*-Fauna der Schlitzerländer Fließgewässer. Arch. Hydrobiol. 68: 575–608.

Meijering, M. P. D., 1972. Experimentelle Untersuchungen zur Drift und Aufwanderung von Gammariden in Fließgewässern. Arch. Hydrobiol. 70: 133–205.

Meijering, M. P. D., 1980. Die Werra–Lebensraum für Meerestiere? Die Weser 54: 7–10.

Meijering, M. P. D., 1982. Zum Lebensformtypus *Gammarus* und dessen Indikationswert für Fließgewässerschäden. Natur und Mensch, Jahresmitt. 1982: 133–138.

Meijering, M. P. D., 1984. Die Verbreitung von Indikatorarten der Gattung *Gammarus* im Schlitzerland (Osthessen) in 1968 und 1982. In: Gewässerversauerungen in der Bundesrepublik Deutschland (Ed. Unweltbundesamt). Materialien 1/84: 96–105.

Meijering, M. P. D., 1987. Die *Gammarus*-Fauna im Pfuhlgraben-Bachsystem bei Wehrda – ein längerfristiger Vergleich. Beitr. Naturkde. Osthessen 23: 71–79.

Meijering, M. P. D., 1988. Emissionsbedingte Gewässerveränderungen und ihre Wirkung auf Wassertiere. Tagungsber. Fachgruppe 'Fischkrankheiten' d. Dtsch. Veterinärmed. Ges. 1988: 129–134, Gießen.

Meijering, M. P. D., 1989. Flohkrebse (*Gammarus*) als Indikatoren für Sauerstoffschwund und Versauerung in Fließgewässern. Mitt. DVWK, 17: 369–381.

Meijering, M. P. D., A. G. L. Hagemann & H. E. F. Schröer, 1974. Der Einfluß häuslicher Abwässer auf die Verteilung von *Gammarus pulex* L. und *Gammarus fossarum* Koch in einem hessischen Mittelgebirgsbach. Limnologica (Berlin) 9: 247–259.

Meijering, M. P. D. & H. G. Pieper, 1985. Zur Verbreitung von *Gammarus* (Crustacea: Amphipoda) im Fulda-Eder-Abflußgebiet, mit besonderer Berücksichtigung der Bachversauerung. Mittlg. Erg. Stud. Ökol. Umwelts. 10: 91–123.

Økland, K. A. & J. Økland, 1985. Factor interaction influencing the distribution of the freshwater 'shrimp' *Gammarus*. Oecologia (Berlin) 66: 364–367.

Pieper, H. G. & M. P. D. Meijering, 1981. Zur Situation der Gattung *Gammarus* im Abflußgebiet der oberen Fulda. Beitr. Naturkde. Osthessen 17: 61–69.

Pieper, H. G. & M. P. D. Meijering, 1982a. *Gammarus* occurrence as an indication for stable conditions in Hessian woodland brooks and rivers. Pol. Arch. Hydrobiol. 29: 283–288.

Pieper, H. G. & M. P. D. Meijering, 1982b. Zur Situation der Gattung *Gammarus* im Abflußgebiet der unteren Fulda. Beitr. Naturkde. Osthessen 18: 17–24.

Pieper, H. G. & M. P. D. Meijering, 1983. Zur Situation der Gattung *Gammarus* im Abflußgebiet von Eder und Diemel. Beitr. Naturkde. Osthessen 19: 75–84.

Roux, A. L., 1971. Les Gammares du groupe *pulex*. Essai de systématique biologique. II. Quelques caractéristiques écologiques et physiologiques. Arch. Zool. exp. gén. 112: 471–503.

Schmitz, W., 1960. Die Einbürgerung von *Gammarus tigrinus* Sexton auf dem europäischen Kontinent. Arch. Hydrobiol. 57: 223–225.

Scholz, E. & M. P. D. Meijering, 1975. Vergleichende Untersuchungen zur Abwasserresistenz von *Gammarus pulex* L. und *Gammarus roeseli* Gervais in osthessischen Fließgewässern. Beitr. Naturkde. Osthessen 9/10: 81–85.

Teichmann, W. & M. P. D., Meijering, 1981. Zur Situation der Gattung *Gammarus* im Kaufunger Wald. Beitr. Naturkde. Osthessen 17: 71–84.

Vangenechten, J., 1979. Biologische gevolgen van de verzuring van de milieu door verbranding van fossiele brandstoffen voor de energievoorziening. Consensus 4: 117–126.

Hydrobiologia **223**: 171–176, 1991.
L. Watling (ed.), VIIth International Colloquium on Amphipoda.
© 1991 *Kluwer Academic Publishers.*

Volumetric growth in gammaridean Amphipoda

D. J. Wildish & B. Frost
Department of Fisheries and Oceans, Biological Station, St. Andrews, N.B. E0G 2XO, Canada

Abstract

A non-destructive, direct volumetric method is described for measuring absolute growth rates throughout embryonic and postembryonic life of gammaridean Amphipoda. The method is demonstrated by following embryonic and early postembryonic growth of individuals from a population of *Platorchestia platensis* Krøyer 1844 living on a fixed shingle shore in the Bay of Fundy, Canada.

Introduction

Growth measures in gammaridean Amphipoda, as in other animals, depend on an accurate measure of size. Size may be measured either as biomass, length or volume at frequent but known ages throughout the life history. Unfortunately, no morphological features of amphipods presently known can be used to indicate physiological age equivalent to a teleostan scale or lamellibranch growth ring. Some claims to the contrary, e.g. for talitrids by Amanieu (1969), Duncan (1969), Tamura & Koseki (1974) and Louis (1977), which suggest that the number of second antennal segments indicates age, are only approximately correct. Thus, the second antennal segment number is variable between species; for example, in juvenile *Orchestia gammarellus* an average of 1 and in *O. mediterranea* an average of 1.5 second antennal segments are added at each molt (Wildish, 1969). In juvenile *Talitrus saltator*, approximately 2 segments are added to the second antenna at each molt (David, 1936). In laboratory cultured juvenile *Traskorchestia traskiana*, Page (1979) showed that an average of 1 second antennal flagellum segment was added per molt, although individual variation was from 0–2 segments. It was also found by this author that the rate of addition of segments was slowed by the onset of sexualization, was different between males and females, and the allometric relationship between body length and number of second antennal flagellum segments was seasonally variable. For these reasons, antennal segment numbers are not considered to be reliable absolute indicators of age or molt stage when individual growth is to be measured throughout postembryonic life.

Since microtags capable of remaining in place over a number of molts are not yet available for amphipods, it is necessary to use isolated, cultured individuals or single broods which leave the marsupium or hatch at a known time in order to measure growth at age accurately. Size measures used to determine postembryonic growth include body, head and limb lengths, all of which are destructive because they involve limb removal (Charniaux-LeGrand, 1952) or that the subjects are killed (Wildish, 1972). Non-destructive measurements may be made when molted skins or anaesthetized amphipods are used (Wilder, 1940). Because talitrids such as *Orchestia* rapidly consume their cast molt skins in laboratory culture (Wildish, unpublished) and anaesthetics may interfere with metabolic functions, the latter methods are not generally appropriate. Biomass measures may also be destructive if the dry weight is used, but not if wet weight is the measure. Venables (1981) has used the latter method weighing groups of live *Talorchestia margaritae* in tared test tubes to the nearest 0.1 mg. Live talitrids rapidly loose water and hence weight in dry air (Wildish, 1969) and this method may therefore have drawbacks during transfer operations. Embryonic dimensions in one to three planes have been used to estimate volume (Williams, 1978; Wildish, 1982), assuming that embryos are spheroidal or ellipsoidal in shape. We could find no studies in which the volume of amphipod embryos or post-embryos was measured directly.

Crustacea are notable because of the step-like nature of their postembryonic growth due to periodic ecdyses followed by rapid size expansion and a prolonged anecdysis or intermolt when growth is practically static. Observations which establish such a pattern have been made chiefly in decapods (Hartnoll, 1982) and only rarely in Amphipoda. An exception is provided by the work of Kinne (1959) on *Gammarus duebeni* who reported the body length for each molt throughout the life history. The aim of this presentation is to introduce volume measurement as a means of determining size in both embryonic and post-embryonic stages and to describe a non-destructive method to measure growth of an individual throughout most of its life. The method is demonstrated with data from the cosmopolitan shorehopper *Platorchestia platensis* Krøyer 1844 (Amphipoda, Gammaridea, Talitroidea).

Methods

The volumeter consisted of a 0.25 ml Gilmont micrometer burette which was sealed with stopcock grease (Fig. 1) to a 1 ml graduated Hamilton barrel syringe. Flexible plastic tubing of 3 mm internal diameter was connected to the syringe tip and a 0.1 ml pipet with 100 divisions each = 1 μl or 1 mm^3. The joints were sealed with stopcock grease. The syringe was filled with 0.45 μm Millipore filtered, locally available seawater of salinity = 30‰ to which a small amount of surfactant was added to decrease surface tension.

Volume measurements were made by noting the initial level of the pipet to the nearest one-half division. A thermometer magnifying glass was

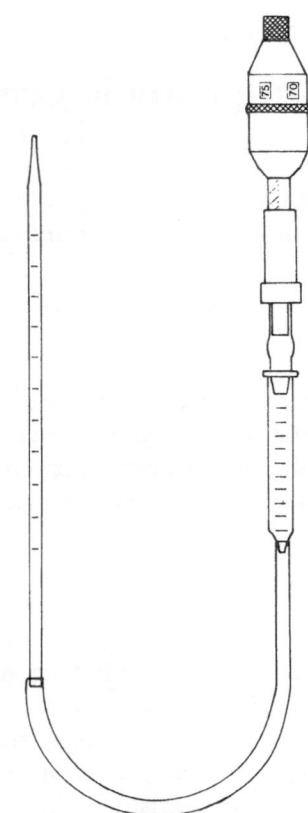

Fig. 1. Diagram showing the volumeter which is supported by a stand and supporting arm during use.

found to be helpful to view the base of the meniscus. After the introduction of the animal to the seawater at the top of the syringe barrel on a camel hair brush or a fine needle, the setting on the pipet was readjusted to its initial level and the difference in volume on the micrometer burette recorded.

Embryos were taken from ovigerous females freshly collected from a fixed shingle shore near St. Andrews, N.B., Canada. All embryos from a single brood were preserved in 5% formalin in seawater and the brood volume was measured on the day of collection. Embryos were chosen to represent the developmental stages recognized by Dorsman (1935). It was not possible to measure individual embryo volume because the lower limit of determination of the volumeter was 0.5 mm^3. Therefore, individual embryo volume was estimated as the ratio of brood volume to number of embryos. Reproducibility of the volume determination using two small lead spheres was 6.83 mm^3 $\pm$ 15.0% (mean $\pm$ standard error). Repeated volume determinations on live animals led to drying out, loss of volume and death of embryos and the juveniles tested. Individual embryos were also viewed with a Zeiss photomicroscope and linear measurements of half the length, r_1, half the width, r_2, and half the height, r_3, were made. Embryo volume was estimated as the sphere $4/3\ \pi r_1^3$ or ellipsoid $4/3\ \pi r_1 r_2 r_3$.

Live postembryonic individuals of *P. platensis* were obtained in the following way: talitrids collected near St. Andrews were placed in a large covered aquarium lined with seawater-dampened cotton wool to which fresh *Fucus* fronds were added. Subcultures were started by removing pairs in pre-copula and placing them in Petri dishes as in Wildish (1972). The cotton wool was dampened with locally available seawater and small pieces of *Fucus* for food and shelter were replaced weekly. Due to the small size of the earliest postembryonic juveniles, it was necessary to measure the whole brood and hence the volume of the first two-three molt stages of postembryonic juveniles is represented as brood means. In later molt stages, volume was determined for individuals singly. Day one was recorded as the day the juveniles left the marsupium. Observations were made thereafter every 2–3 d for as long as the juveniles survived. Problems in culturing juveniles were the main cause of mortalities. The laboratory cultures were maintained at laboratory temperatures of ca. 25 °C and at ambient summer photoperiods. Several of the cultures were kept outside for the first 15 d before they too were moved to the laboratory.

Results

Embryos

Direct volumetric measurements of embryo size are shown in Table 1. For each developmental stage, the overall mean volume of a single embryo from 3–5 broods and 30–56 actually measured

Table 1. Comparison of methods for determining the individual embryo volume in broods of *Platorchestia platensis* for the developmental stages recognized by Dorsman (1935). Units are mm^3, n_0 = number of broods and n = number of embryos measured.

Method	Embryo developmental stage				
	I	II	III	IV	V
Direct x		0.14	0.21	0.25	0.26
volumetry n_0		3	2	3	2
Ellipsoidal x		0.13	0.20	0.23	
estimation SD		0.009	0.011	0.011	
n		20	10	20	
Spherical x		0.19	0.38	0.48	
estimation SD		0.024	0.032	0.032	
n		20	10	20	

embryos is given. These results are compared with estimated volumes based on microscopic linear measurements. A Kruskal-Wallis nonparametric test upholds the hypothesis that there is no difference between the means determined by direct volumetry and ellipsoidal estimation for embryos of stages II, III and IV at $p = 0.05$. A similar test shows that the spherical measurement overestimates embryo volume. Stage 1 embryos which were within 48 h of laying were difficult to work with because the outer membranes were so delicate and frequently ruptured due to handling stresses. As a result, we were unable to measure either direct or linear estimates with this developmental stage.

Postembryonic juveniles

Newly hatched juveniles ranged from 0.2–0.5 mm^3 in volume. Juvenile growth was followed at subsequent molts and produced the step-like growth shown in Fig. 2. The timing of ecdysis was not always observed and the position of the step indicating ecdyses was interpolated between consecutive volume measurements; it is hence correct to within 3 d. Animal 1 reached molt stage 8 and 2a was in molt stage 7 by the end of the observation period.

In all, 10 individuals were followed for up to

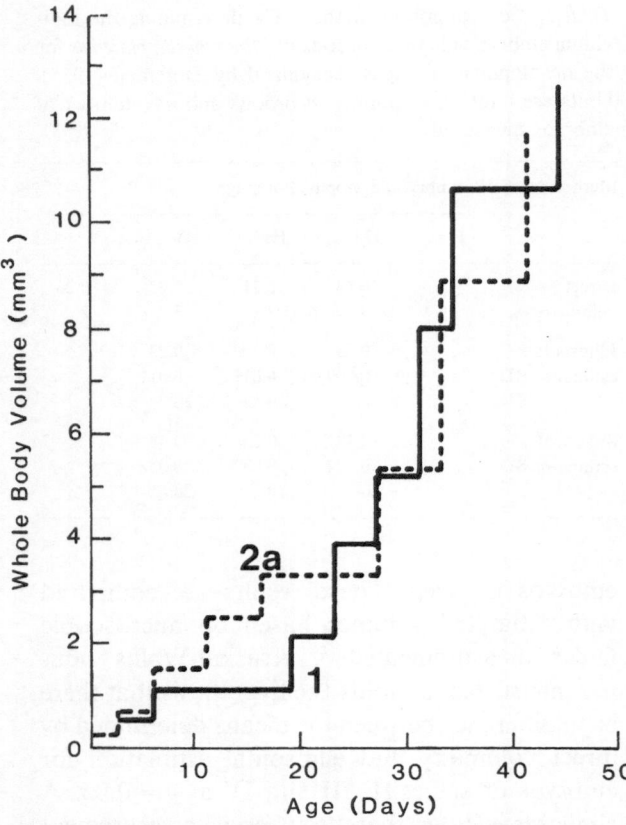

Fig. 2. Volume growth pattern for two individual *Platorchestia platensis*. Solid line brood 1 – outside for first 15 d, broken line brood 2a – laboratory reared.

46 d which, in two cases, was the ninth molt stage (Table 2). In most cases, the time of hatching was not observed, but it is thought that 1–3 d were spent in the marsupium following hatching. Variations in volume at the same molt stage were apparent with juveniles from brood 5b (Table 2) being relatively large at a given stage.

Considering group data for age in days, x, on body volume $\times 10$, y, shows that the best fit between these variables is described by a linear log/log relationship.

$$\ln(y) = 1.484 \ln(x) - 1.16, \qquad R^2 = 0.75,$$
$$N = 48$$

Growth in volume shown in Fig. 2 must slow before the adult individuals (approximately 50 mm^3, and maximum of 24 mo age) appear since the data shown are for a maximum of 13 mm^3 and 2 mo age.

The juvenile group data show considerable variations in the length of an intermolt and the volume increment at each molt, although in general the intermolt period increases with age and the volume increment at successive molts decreases. Expressed similarly to the Hiatt growth diagram (Mauchline 1976), that is the premolt body volume in mm$^3 \times 10$, x, on the percentage volume increment at a molt, y, the data are inversely related by the equation:

$$\ln(y) = -0.209 \ln(x) + 5.78, \qquad R^2 = 0.76,$$
$$N = 48.$$

Table 2. Summary of individual juvenile volume, mm^3, from different broods of *Platorchestia platensis*.

Molt stage	Brood number									
	1	2a	2b	3a	3b	4a	4b	5a	5b	6
0		0.2	0.2	0.3	0.3	0.3	0.3			0.5
1	0.5	0.6	0.6	0.6	0.6					1.6
2	1.1	1.5	1.5	1.3	1.3					2.4
3	2.1	2.5	2.5	2.4	2.4					5.0
4	3.9	3.3	3.4	3.8	3.8					
5	5.2	5.3	4.6	5.2	5.3	4.9	4.4	6.3	6.6	
6	8.0	8.9	6.4	8.3	7.9	7.3	7.1	8.5	10.3	
7	10.6	11.7	8.8	11.0	11.3	10.3	10.2	10.3	13.1	
8	12.6		10.2			13.7	12.9	12.8	15.2	
9								15.6	18.2	

Discussion

Because the minimum detection limits of the volumeter were 0.5 mm³, it was not possible to follow individual embryonic development or early juvenile life with the described apparatus. If sufficient numbers were present in a brood to exceed the minimum detection limit, an overall brood mean could be determined. In later models, it may be possible to further miniaturize the volumeter as in Cibarowski (1983) so that individual embryos and early juveniles can be measured volumetrically.

Two difficulties with the method are that of carryover of fluids or air bubbles on the animals and physical damage to them which occurs during transfer between culture containers and the volumeter. The former was not a serious problem if the cultures were kept damp but not wet. For the latter, repeated volume determinations on the same live animal led to a decline in volume attributed to drying out and physical damage occurring during transfer operations. The transfers were made as rapidly as possible to minimize this problem.

Our results with embryos suggest that microscopic estimation as the ellipsoid gave results in close agreement with those obtained by direct volumetric measurement, despite the irregularity of their shape.

One advantage of using volume as the measure of growth is that some recently described field methods for determining benthic production on a geographic scale utilize volume for this purpose. Biomass determined from volume measurements can be used to calculate crude production using P : B ratios derived by some variant of the Banse and Mosher equation (Banse & Mosher, 1980; Schwinghamer et al., 1986). Although these methods produce crude estimates, they may be justified where detailed cohort analysis of populations of each species is unavailable and the aim is to estimate the geographic distribution of production. Other possible ways of measuring amphipod volume include by particle counter or flow cytometry. Both would be considerably more expensive to set up than the method described here.

Notable results of the growth observations with *P. platensis* are that both intermolt duration and growth increment at a molt are variable, contrary to the suggestion in Wildish (1972) with interpolated linear measurements of related talitrids. The general volumetric relationship between the body volume increment at a molt as a percentage of total body volume is a full logarithmic inverse relationship. This contrasts with the semi-logarithmic inverse relationship found by Mauchline (1976) using linear measurements of other crustaceans. This relationship may not always follow a single straight line as found by Sheader (1981) in *Parathemisto gaudichaudi* with differences between juvenile and more mature stages.

Hartnoll (1982) has cautioned that laboratory culture conditions may adversely affect growth rates of crustaceans, and this could apply to the growth measures of *P. platensis* we present here. Although we can give no direct evidence for *P. platensis*, Cooper (1965), working with the freshwater talitroid, *Hyalella azteca*, showed that laboratory growth was equal to, or greater than, that observed in wild populations using cohort analysis. We believe that the culture conditions used do not adversely affect the growth rates of talitroid amphipods.

References

Amanieu, M., 1969. Cycle reproducteur à Archachon d'une population de *Orchestia gammarella* (Pallas) (Amphipode, Talitridae). Bull. Inst. Océanogra. Monaco 68: 1–24.

Banse, M., S. Mosher, 1980. Adult body mass and annual production/biomass relationships of field populations. Ecol. Monogr. 50: 355–379.

Charniaux-LeGrand, H., 1952. Le cycle d'intermue des amphipodes et ses particularités chez les formes terrestres (Talitridae). Travaux de la Station Biologique de Roscoff VII: 178–203.

Cibarowskii, J. J. H., 1983. A simple volumetric instrument to estimate biomass of fluid-preserved invertebrates. Can. Entomol. 115: 427–430.

Cooper, W. E., 1965. Dynamics and production of a natural population of a freshwater amphipod, *Hyalella azteca*. Ecol. Mon. 35: 377–394.

David, R., 1936. Recherches sur la biologie et l'intersexualité de *Talitrus saltator* Mont. Bull. Biol. Fr. Belg. 70: 332–357.

Dorsman, B. A., 1935. Notes on the life history of *Orchestia bottae* Milne Edwards. Thesis, Leiden, p. 1–58.

Duncan, K. W., 1969. The ecology of two species of terrestrial Amphipoda (Crustacea, Family Talitridae) living in waste grassland. Pedobiologia 9: 323–341.

Hartnoll, R. G., 1982. Growth. In: Abele, L. G. (ed.) The biology of Crustacea, Vol. 2. Academic Press, New York and London, p. 111–196.

Kinne, O., 1959. Ecological data on the amphipod *Gammarus duebeni*. A monograph. Veroeff. Inst. Meeresforsch. Bremerhaven 6: 177–202.

Louis, M., 1977. Étude des populations de Talitridae des étangs littoraux Méditerranéens. II. Identification des cohortes, cycles et fécondité. Bull. Ecol. 8: 75–86.

Mauchline, J., 1976. The Hiatt growth diagram for Crustacea. Mar. Biol. 35: 79–84.

Page, H. M., 1979. Relationship between growth, size, molting, and number of antennal segments in *Orchestia traskiana* Stimpson (Amphipoda, Talitridae). Crustaceana 37: 247–252.

Schwinghamer, P., B. T. Hargrave, D. Peer, C. M. Hawkins, 1986. Partitioning of production and respiration among size groups of organisms in an intertidal benthic community. Mar. Ecol. Prog. Ser. 31: 131–142.

Sheader, M., 1981. Development and growth in laboratory-maintained and field populations of *Parathemisto gaudichaudi* (Hyperiidea: Amphipoda). J. mar. biol. Assoc. U.K. 61: 769–787.

Tamura, H., K. Koseki, 1974. Population study on a terrestrial amphipod, *Orchestia platensis japonica* (Tattersall) in a temperate forest. Jap. J. Ecol. 24: 123–139.

Venables, B. J., 1981. Energy allocation for growth and metabolism in *Talorchestia margaritae* (Amphipoda, Talitridae). Crustaceana 41: 182–189.

Wilder, J., 1940. The effects of population density upon growth, reproduction and survival of *Hyalella azteca*. Physiol. Zool. 13: 439–461.

Wildish, D. J., 1969. Studies on taxonomy, ecology and behavior of *Orchestia gammarella* (Pallas), *O. mediterranea* A. Costa and *O. cavimana* Heller in the Medway Estuary. Ph.D. Thesis, London, 252 p.

Wildish, D. J., 1972. Postembryonic growth and age in some littoral *Orchestia* (Amphipoda, Talitridae). Crustaceana Suppl. III: 267–274.

Hydrobiologia **223**: 177–180, 1991.
L. Watling (ed.), VIIth International Colloquium on Amphipoda.
© 1991 *Kluwer Academic Publishers.*

Respiration of *Orchomene plebs* (Hurley, 1965) and *Waldeckia obesa* (Chevreux, 1905) from Admiralty Bay (South Shetlands Islands, Antarctic)

S. Rakusa-Suszczewski & A. Lach
Department of Polar Research, Institute of Ecology, 05-092 Lomianki, Dziekanow Lesny, Poland

Abstract

Metabolism of *Orchomene plebs* from Admirably Bay is higher than that of *Waldeckia obesa*, and similar to that of *P. plebs* from McMurdo Sound. The range of variation in respiration is highest below freezing which means that these are optimum temperatures for the Amphipoda species under study.

Introduction

The Amphipoda *Orchomene plebs* (Hurley, 1965) and *Waldeckia obesa* (Chrevreux, 1905) have a circumantarctic distribution (Bellan-Santini, 1972) at a depth from the low tide level to 800 m (Lowry & Bullock, 1976). In Admiralty Bay, these two species represent an important component of the community of necrophagous animals (Arnaud *et al.*, 1986; Presler, 1986). *O. plebs* is smaller (up to 24 mm) than *W. obesa* (up to 28 mm) (Bellan-Santini, 1972). A relationship between the wet body weight and body length in *O. plebs* can be described by the formula $W = 0.0544 L^{2.71}$, where L is the body length in mm (Rakusa-Suszczewski, 1982). A detailed study of *O. plebs* has been conducted at McMurdo Station (Rakusa-Suszczewski & McWhinnie, 1976). Measurements of metabolism in *O. plebs* (Rakusa-Suszczewski, 1982) showed that oxygen consumption varied depending on whether the animals were fed or starved. The initial conditions also determined the response of animals to increasing temperatures. Oxygen consumption depended on the concentration of salt and on the time of the acclimation of animals to experimental conditions. At $T = -1.8 °C$ and $S = 35‰$ the rate of respiration in *O. plebs* declined to one-third after six hours of the experiment (Rakusa-Suszczewski & McWhinnie, 1976). In McMurdo Sound, water temperatures are -1.86 or $-1.9 °C$, and annual fluctuations do not exceed $\pm 0.2 °C$ (Littlepage, 1965).

The objective of this study was to assess the respiration rate in *O. plebs* and *W. obesa* from Admiralty Bay, where mean annual water temperature is $-0.4 °C$, and at a depth of 50 m it may reach $-1.73 °C$ (Lipski, 1987).

Methods

Respiration was measured in a Gilson respirometer consisting of 14 vessels submerged in a bath in which the temperature was maintained thermostatically to an accuracy of 0.1 °C. Each animal was placed in a separate respirometric vessel containing from 8 to 15 ml of filtered sea water ($\pm 34‰$). A 25% solution of KOH or NaOH was used for absorption of CO_2. During the experiments, which took up to 4.5 hours, the vessels with animals were shaken automatically. *O. plebs* and *W. obesa* were captured using traps baited with meat. Twelve hours prior to the experiment, the animals were acclimated to the experimental temperature, and oxygen was added to the water with the animals. Wet weights were determined after the animals were carefully dried with filter paper.

Results and discussion

Oxygen consumption in *O. plebs* and *W. obesa* depended on their body weights and the ambient temperature (Tables 1 and 2, Fig. 1). In *O. plebs*, respiration increased in proportion to wet weight raised to powers from 0.66 to 0.84 (Table 1) and in *W. obesa* (Table 2), the coefficient *b* was more variable, this being due to a narrower weight range of the animals used, rather than to species-specific differences. The values of $b = 0.79$

obtained for all amphipods for which respiration was measured at temperatures below zero (Table 3) do not contradict the statement (cf. Rakusa-Suszczewski, 1975) that in crustaceans living at temperatures above freezing the value of *b* is the same as at temperatures above freezing. This was also confirmed by Maxwell & Ralph (1985) for temperatures close to 0 °C and Suscenja at 20 °C ($b = 0.79$ for all Amphipods).

At -1.8 °C the respiration of *O. plebs* was higher than the respiration of *W. obesa* with similar wet weight (Fig. 1). Assuming that metabolism is proportional to the wet body weight raised to power 0.79 (cf. Table 3), the value of a specific metabolism ($a = R/W^{0.79}$) was higher in *O. plebs* than in *W. obesa* at all experimental temperatures

Table 1. Respirtion of Orchomene plebs at different temperature.

T °C	Month year	Number of animals	Mean wet weight mg ± S.D.	Mean respiration $\mu 10_2$ ind^{-1} h^{-1} ± S.D.	Correlation coeffic $/r/$	Regression $R = aW_w^b$	$a = \dfrac{R}{W_w^{0.79}}$
-1.8	Nov. 88	39	194 ± 101	13 ± 7.1	0.89	$a = 0.156$ $b = 0.84$	0.20
-1.0	Nov., Dec. 88	38	148 ± 86	9.9 ± 6.6	0.93	$a = 0.156$ $b = 0.82$	0.19
0.0	Dec. 88	30	161 ± 73	9.0 ± 5.2	0.82	$a = 0.127$ $b = 0.84$	0.16
1.0	Dec., Jan. 89	37	105 ± 55	7.0 ± 3.5	0.90	$a = 0.157$ $b = 0.83$	0.18
2.0	Jan. 89	38	72 ± 41	7.0 ± 3.9	0.85	$a = 0.413$ $b = 0.66$	0.24

Table 2. Respirtion of Waldeckia obesa at different temperature.

T °C	Month year	Number of animals	Mean wet weight mg ± S.D.	Mean respiration $\mu 10_2$ ind^{-1} h^{-1} ± S.D.	Correlation coeffic $/r/$	Regression $R = aW_w^b$	$a = \dfrac{R}{W_w^{0.79}}$
-1.8	Aug., Dec. 88	31	295 ± 165	10 ± 4.9	0.82	$a = 0.357$ $b = 0.58$	0.11
-1.0	Aug., Dec. 88	33	229 ± 147	10.9 ± 7.9	0.85	$a = 0.014$ $b = 1.21$	0.14
0.0	Aug. 88	35	218 ± 123	6.0 ± 3.6	0.83	$a = 0.049$ $b = 0.88$	0.08
1.0	March., Sept. 89	63	238 ± 107	7.1 ± 3.9	0.84	$a = 0.023$ $b = 1.04$	0.09
2.0	Sept. 89	33	252 ± 145	9.5 ± 5.9	0.90	$a = 0.052$ $b = 0.94$	0.12

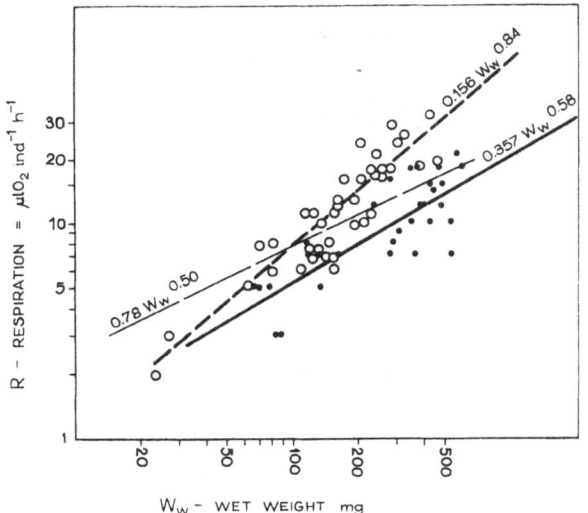

Fig. 1. Respiration of *O. plebs* [o] and *W. obesa* [●] from Admiralty Bay and *O. plebs* [– – –] from McMurdo Sound at – 1.8 °C.

(Tables 1 & 2). The respiration of *O. plebs* with the same range of body weights was similar for the animals from McMurdo Sound and Admiralty Bay (Fig. 1).

The response to temperature was similar in *O. plebs* and *W. obesa* (Tables 1 and 2). The high-est oxygen consumption was observed at – 1.8 °C; it decreased at about 0 °C, and increased again at higher temperatures. A similar response to increasing temperatures was observed in fed *O. plebs* from McMurdo Sound (Rakusa-Suszczewski, 1982, Fig. 6, lines 1 and 3). Under optimum conditions for Antarctic bottom-dwelling Amphipoda, that is, at temperatures below 0 °C and low temperature fluctuations, the difference between the resting and active metabo-lism was the highest, and the thermo-neutral zone was the narrowest. This is an adaptation to hypo-stenothermic conditions, accounting for the pos-sibility of reaching a higher metabolic rate at tem-peratures below 0 °C (Rakusa-Suszczewski, 1980) as compared with metabolism of animals living at temperatures above 0 °C when it is extra-polated to temperatures below 0 °C. The occur-rence of metabolic compensation is put in question by Maxwell & Ralph (1985).

Acknowledgements

The author gratefully acknowledges the use of a Gilson respirometer made available by the

Table 3. Correlation between the respiration and the wet weight of Antarctic Amphipods at temperature below zero.

Species	T °C	Mean wet weight ± S.D. /mg/	Mean respiration ± S.D. $\mu 10_2$ ind^{-1} h^{-1}	Authors
Paramoera walkeri	– 1.2	20.7/11.0–25.0	1.87	Rakusa-Suszczewski and Klekowski, 1973
Eusirus perdentatus	– 1.0	1640.0 ± 74.30	35.6 ± 2.43	
		1613.3 ± 83.22	43.7 ± 8.78	
		1628.0 ± 54.03	39.2 ± 4.14	
Cyphocaris richardi	– 1.4	640.0	28.9	Opaliński &
		1092.0 ± 41.17	42.7 ± 5.27	
		1062.2 ± 58.11	41.2 ± 4.80	
	– 1.0	1470.0	113.8	Jaździewski, 1978Eurythenes
gryllus		690.0	74.0	
		950.0 ± 285.3	87.3 ± 27.0	
Orchomene plebs	– 1.8	194.0 ± 101	13.0 ± 4.9	
	– 1.0	148.0 ± 86	9.9 ± 6.6	
Waldeckia obesa	– 1.8	295.0 ± 165	10.0 ± 4.9	present paper
	– 1.0	229.0 ± 147	10.9 ± 7.9	

Total: $R = 0.1742 \, W_{w}^{0.79}$ $r = 0.9$

National Science Foundation at the U.S. base
Palmer Station.

References

Arnaud, P. M., K. Jazdzewski, P. Presler & J. Sicinski, 1986.
Preliminary survey of benthic invertebrates collected by
Polish Antarctic Expeditions in Admiralty Bay (King
George Island, South Shetland Islands, Antarctica). Pol.
Polar Res. 7: 7–24.

Bellan-Santini, D., 1972. Invertebrés Marins. Des XIIeme et
XVeme expeditions Antarctiques Francaises en terre
Adélie. 10 Amphipodes Gammariens. Tethys Suppl. 4:
157–238.

Lipski, M., 1987. Variations of physical conditions, nutrients
and chlorophyll a contents in Admiralty Bay (King
George, Island, South Shetland Islands 1929). Pol. Polar
Res. 8: 307–332.

Littlepage, J. L., 1965. Oceanographic investigations in
McMurdo Sound Antarctica. In M. O. Lee (ed.), Biology
of the Antarctic Seas. 2: 1–37. Amer. Geophys. Union,
Washington, D.C.

Lowry, J. K. & S. Bullock, 1976. Cataloque of the Marine
Gammaridean Amphipoda of the Southern Ocean. The
Royal Society of New Zealand, Wellington, N.Z.
pp. 1–187.

Maxwell, I. G. & R. Ralph, 1985. Non cold adapted
metabolism in the decapod Chorismus antarcticus and other
sub-Antarctic marine crustaceans. In W. R. Siegfried,
R. P. Condy & R. M. Laws (eds.), Antarctic nutrient cycles
and food webs. Springer-Verlag, Berlin.

Opalinski, K. W. & K. Jazdzewski, 1978. Respiration of
some Antarctic amphipods. Pol. Arch. Hydrobiol. 25:
643–655.

Presler, P., 1986. Necrophagous invertebrates of the
Admiralty Bay of King George Island (South Shetland
Islands, Antarctica). Pol. Polar Res. 7: 25–61.

Rakusa-Suszczewski, S., 1975. Respiration and osmoregulation as the expression of the adaptation of the invertebrates and fishes to life under hypostenothermic conditions. Pol. Arch. Hydrobiol. 22: 521–552.

Rakusa-Suszczewski, S., 1980. Hypostenothermic organisms. Pol. Polar Res. 1: 231–241.

Rakusa-Suszczewski, S., 1982. The biology and metabolism
of Orchomene plebs (Hurley, 1965) (Amphipoda:
Gammaridea) from McMurdo Sound Ross Sea, Antarctic.
Polar Biol. 1: 47–54.

Rakusa-Suszczewski, S. & R. Z. Klekowski, 1973. Biology
and respiration of the Antarctic Amphipoda (Paramoera
walkeri Stebbing) in the summer. Pol. Arch. Hydrobiol. 20:
475–488.

Sucsenja, L. M., 1972. Intensivnost dychanija rakoobraz-
nych. Kiev-Naukov dumka, pp. 195.

Hydrobiologia **223**: 181–187, 1991.
L. Watling (ed.), VIIth International Colloquium on Amphipoda.
© 1991 *Kluwer Academic Publishers.*

Eco-physiological characteristic of some common caprellid species in the Possjet Bay (the Japan Sea)

S.V. Vassilenko
Zoological Institute Academy of Sciences, Leningrad, B-034, USSR

Key words: Caprellidae, Japan Sea, ecology, physiology

Abstract

Parameters of equations relating wet and dry body weight to length have been calculated for four species of caprellids from the Japan Sea (Possjet Bay), specifically *Caprella cristibrachium*, *C. kroyeri*, *C. penantis*, and *C. bispinosa*. A common equation was obtained for oxygen consumption rate in relation to body weight. It was shown that within the Order Amphipoda the metabolic rate of caprellids is 1.5 times lower than that of gammarids. The differences in the fecundity of gravid females of four species and females of one species in the spring and summer, respectively, are shown. At the same time the relationship between number of eggs per female and wet body weight of female is approximated by one equation. The equation obtained shows that caprellids represent an ecologically homogeneous group.

Introduction

Research in the field of production hydrobiology has become widely developed in the Soviet Union over the recent years. It has been described in large reviews (Zaika, 1972; Khmeleva, 1973, 1988; Greze, 1977; Winberg, 1979; Alimov, 1981, 1989; Khmeleva & Golubev, 1984).

To assess the condition of aquatic ecosystems one needs quantitative data on the populations of several species involved in the process of cycling of substances and energy transformation. It has become necessary to study ecologo-physiological characteristics of animals at the level of an individual population including the quantitative study of metabolic rates, feeding, growth, production, and reproductive patterns. As is well known, amphipods play an important role in the life of the seas as well as in freshwaters. In some benthic biocenoses they are predominant in numbers and biomass. For example, *Caprella cristibrachium* attains in surf areas a density of 95 000 specimens m^{-2} with a biomass of 80 g m^{-2}.

There are a large number of reports dealing with the study of the life cycle, relationship of number of eggs per clutch in females to body length and mass (Abolmasova, 1975; Steele & Steele, 1970, 1972; Jazdzewski, 1970). There are also reviews dealing with peculiarities of reproduction and production of crustaceans (in particular of some amphipod species – Khmeleva & Golubev, 1984; Dulepov *et al.*).

Material and methods

Information on the ecology of caprellids from the Pacific Seas of the U.S.S.R. is very scanty. In 1981 I studied size-weight characteristics and oxygen consumption rates in four caprellid species collected in Possjet Bay (the Japan Sea) as follows:

1. *Caprella cristibrachium* Mayer (Pacific widespread boreal species occurring from the Bay of Peter the Great to the Aleutian Islands, a common species of the rocky surf intertidal zone, inhabiting algae).
2. –3. *C. kroyeri* De Haan and *C. mutica* Schurin (West Pacific subtropical species spread from the Kyushu Island and Yellow Sea up to northern Hokkaido and the Bay of Peter the Great, inhabiting algae at a depth of 1.5 m to 12 m in semi-enclosed well warmed inlets).
4. *C. penantis* (widespread in subtropical and tropical waters of the world's oceans, inhabiting the intertidal zone of rocky capes of Possjet Bay).
5. *C. bispinosa* Mayer (West Pacific widespread boreal species occurring from Honshu Island to Paramushir Island at depths of 1 to 21 m on algae.

To evaluate the relationship between number of eggs, mass of clutch, and body mass of gravid females, we have examined also species *C. cristibrachium*, *C. penantis* and *C. bispinosa*, collected in the Possjet Bay (the Japan Sea) in May, June, July 1965–1983.

For experiments, caprellids were collected by diving. The caprellids were delivered to the laboratory and sustained in cages in running water from 3 to 20 days at a temperature of 18–20 °C, similar to that of their natural environment. Oxygen consumption rate was measured using closed bottles. Animals were held in the bottles for 5 to 8 hours. After the experiment water from respirometers was poured into pycnometers using a siphon. Oxygen content in the pycnometer was determined with the Winkler method.

After the experiment the specimens were measured and weighed then dried at a temperature of 60 °C for 10 to 14 days and then weighed again. Females with eggs were measured and weighed, clutch weight and diameter of eggs were determined.

The data were processed on an 'Iskra-125' computer using a program developed by A. A. Umnov. The experimental data were approximated by linear regression equations written according to A. A. Umnov (1976) in the form:

$$y = \bar{y} + a\,(x - \bar{x}) \tag{1}$$

Equation (1) can be transformed as follows:

$$y' = \bar{y}'\,(x'/\bar{x}')^a \tag{2}$$

where $x' = L$ or W.

In this case $\bar{y}'$ and $\bar{x}'$ are geometric means of the studied values.

Parameters of the equations relating wet body weight (in mg) and dry body weight W' (in mg) to length are given in Tables 1 and 2.

Results and discussion

For *C. kroyeri* the relationship of W and L, W' and L is represented in Fig. 1 (note logarithmic scales).

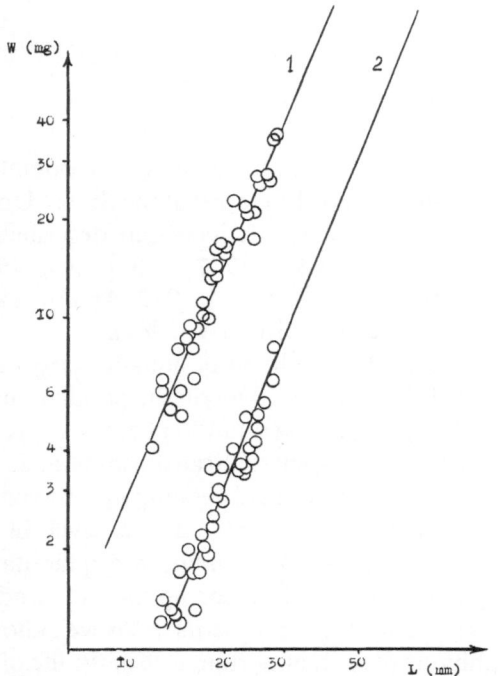

Fig. 1. Caprella kroyeri, relationship between specimen body weight and length; 1, fresh body weight; 2, dry body weight.

Table 1. Parameters of equation relating W (wet weight specimens in mg) to L (length specimens in mm)

$$W = q \left(\frac{L}{\overline{L}}\right)^a \text{ in four caprellid species.}$$

Species	n	limits of L, mm		$q \pm \sigma_q$ $(L = \overline{L})$	q_1 $(L = 1)$	$a \pm \sigma_a$	$mm^{\overline{L}}$	r
		min	max					
1. Caprella cristibrachium	35	4.0	10.5	4.8 $\pm$ 0.254	0.03	2.558 $\pm$ 0.22	7.0	0.970
2. C. krøyeri	41	9.5	28.5	12.6 $\pm$ 0.55	0.007	2.527 $\pm$ 0.19	18.9	0.974
3. C. penantis	51	3.7	8,7	3.33 $\pm$ 0.099	0.02	2.744 $\pm$ 0.12	6.4	0.988
4. C. mutica	34	5.5	15.5	5.13 $\pm$ 0.49	0.16	1.517 $\pm$ 0.32	9.5	0.859

Table 2. Parameters of equation relating W' (dry weight in mg) to L (length specimens in mm)

$$W' = q' \left(\frac{L}{\overline{L}}\right)^{a'} \text{ in four caprellid species.}$$

Species	n	limits of L, mm		$q' \pm \sigma_{q'}$ $(L = \overline{L})$	q'_1 $(L = 1)$	$a' \pm \sigma_{a'}$	r	W/W'
		min	max					
1. Caprella cristibrachium	35	4.0	10.5	1.12 $\pm$ 0.075	0.006	2.721 $\pm$ 0.282	0.959	5.0
2. C. krøyeri	41	9.5	28.5	2.69 $\pm$ 0.34	0.0016	2.523 $\pm$ 0.53	0.835	4.7
3. C. penantis	51	3.7	8.7	0.785 $\pm$ 0.03	0.004	2.833 $\pm$ 0.16	0.980	5.0
4. C. mutica	34	5.5	15.5	1.3 $\pm$ 0.086	0.021	1.822 $\pm$ 0.22	0.947	7.4

Table 3. Parameters of equation relating Q (oxygen consumption rate in mg oxygen per specimen per hour at $+20\,°C$) and W (wet weight in mg)

$$Q = Q_1 \left(\frac{W}{\overline{W}}\right)^{\kappa} \text{ in four caprellid species.}$$

Species	n	limits of W, mg		$Q_1 \pm \sigma_{Q_1}$ $(W = \overline{W})$	Q'_1 $(W = 1\text{ mg})$	$k \pm \sigma_k$	$\overline{W}$ mg	r
		min	max					
1. Caprella cristibrachium	25	0.9	16.0	0.0013 $\pm$ 0.0001	0.0005	0.891 $\pm$ 0.148	2.73	0.932
2. C. krøyeri	48	6.0	26.0	0.0043 $\pm$ 0.0003	0.0006	0.724 $\pm$ 0.124	14.1	0.866
3. C. penantis	18	0.9	7.66	0.0011 $\pm$ 0.0001	0.0006	0.672 $\pm$ 0.188	2.33	0.884
4. C. mutica	17	2.4	12.2	0.0024 $\pm$ 0.0002	0.0005	0.864 $\pm$ 0.191	5.44	0.923
5. The total for four species	108	0.9	26.0	0.0024 $\pm$ 0.0001	0.0006	0.752 $\pm$ 0.046	6.15	0.953

184

The relationship between dry and wet body weight for the four caprellid species is expressed as $W = 5.5\,W'$ where dry weight constitutes 18% of wet weight of the body (in %) (Table 1, 2). Dry weight (in %) in caprellids is equivalent to that for some other amphipods (Shapovalova, 1973).

Parameters of equations relating oxygen consumption rate and wet body weight in caprellids at 20 °C are given in Table 3. A distinct relationship between oxygen consumption rate and body weight is observed for all four caprellid species and is described by equation (Fig. 2):

$$Q = (0.0024 \pm 0.0001) \cdot (W/\overline{W})^{0.752 \pm 0.046}$$

in mg oxygen per individual per hour, where 0.0024 is oxygen consumption rate at $W = \overline{W} = 6.15$ mg.

Comparing metabolic levels in different Crus-

tacea we used equations with 'k' values close to 0.750 (Table 4). Comparison of these data has shown that metabolic level in caprellids is on the average 1.5 times lower than in Crustacea generally and 2 times lower than in gammarids. This is apparently related to the low activity of caprellids in the natural environment. They usually creep along the branches of algae, hydroids, sponges and bryozoans, sometimes swimming but the velocity of their swimming is much lower than in gammarids and hyperiids.

Analysis of the material within the genus *Caprella* shows an inter-specific dependency between the number of eggs laid and female weight. There are only scanty data for *C. kroyeri* and *C. mutica* fecundity on females collected in August with a small range of values of body weight, but they well supplement the general relationship between fecundity and female body

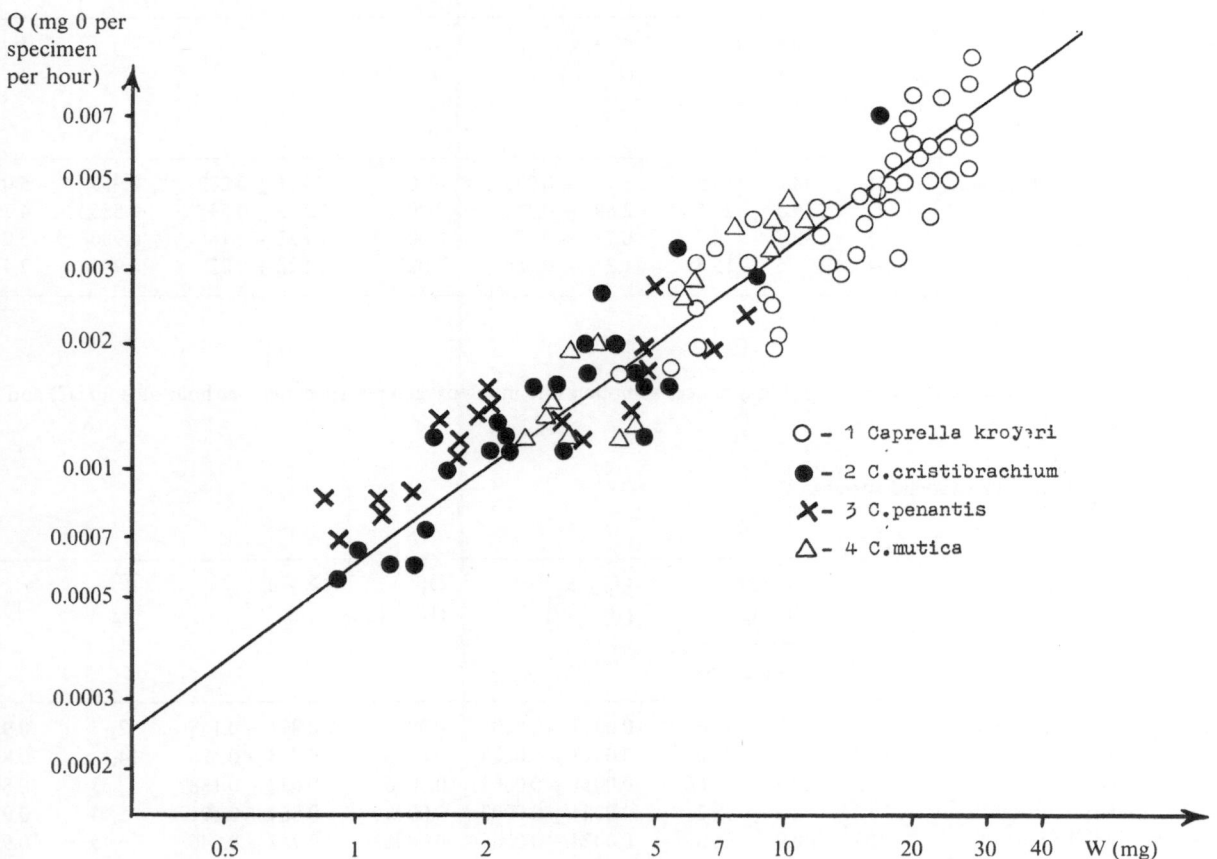

Fig. 2. Relationship between oxygen consumption rate (Q) and specimen body wet weight (W) for four caprellid species.

Table 4. Q_1 values where $k = 0.75$ in equations relating metabolic rate to weight in Crustacea and different Amphipoda groups.

Names of crustaceans groups	k (in mg oxygen per 1 gramme of wet weight)	Authors
Crustacea	0.179	Sushchenya, 1972
Gammaridea (4 species)	0.234	Ivanova, 1973 Dulepov, Dulepova, Pois, 1986
Caprellidea (4 species)	0.110	Vassilenko, 1985

weight found for caprellids. Fecundity of caprellids varies seasonally, being a function of the changing size composition of females with their individual fecundity. Thus in *C. bispinosa* and *C. cristibrachium*, females breeding in spring are much larger than females breeding in summer (from June till August). The latter are normally young females that are breeding for the first time. The number of eggs per brood in *C. cristibrachium* is 5 to 6 times higher in spring than in summer;

in *C. bispinosa* it is 18 to 20 times higher in spring than in summer (Table 5). Nevertheless, fecundity of females of all investigated species belong to a single general array and a regression line could be used to relate fecundity to female body weight (Fig. 3). The relationship between Ng (number of eggs per female) and W is approximated by one equation. The general relation between fecundity and female weight in caprellids seems to be associated with the ecological homogeneity of this group. The same may account for the very similar sizes of eggs in all caprellids (Table 6).

The data summarized by Ivanova & Vassilenko (1987) on different orders of the superclass Crustacea show that:

1. The relationship between fecundity and female body weight (size) has been calculated for all crustacean species with S-type growth from natural populations during the period of abundant breeding.
2. The relationships between overall egg weight per brood and female body weight have been calculated for all the above orders. They are

Table 5. Parameters of equation relating fecundity (N_g eggs/female) and to W (wet weight in mg)

$$N_g = y \left(\frac{W}{\overline{W}}\right)^b$$ in some caprellid species.

Species	Locality	Year of collection	n	limits of W, mg min	limits of W, mg max	limits of N_g eggs/fem min	limits of N_g eggs/fem max	$y \pm \sigma_y$ eggs/fam	$b \pm \sigma_b$	$\overline{W}$, mg	r
1. Caprella bispinosa	Japan Sea	spring 1965	28	13.2	42.8	75	322	209.1 ± 13.1	0.87 ± 0.24	29.3	0.77
	Possjet Bay	summer 1981	9	0.5	1.0	4	15	8.0 ± 1.449	1.63 ± 0.84	0.7	0.87
	all together	all together	37	0.5	42.8	4	322	94.5 ± 6.71	0.88 ± 0.04	11.85	0.99
2. C. cristibrachium	Japan Sea	summer 1965	47	1.5	5.0	4	31	16.0 ± 1.36	0.94 ± 0.26	2.5	0.68
	Possjet Bay	summer 1981	9	2.25	17.0	13	50	20.6 ± 2.02	0.63 ± 0.15	4.7	0.97
		spring 1965	46	6.0	30.0	29	160	72.4 ± 4.2	1.04 ± 0.16	15.9	0.86
		all together	102	1.5	30.0	4	160	32.3 ± 1.87	0.84 ± 0.06	6.1	0.94
3. C. penantis	Japan Sea	summer 1965	16	1.5	3.5	9	36	21.5 ± 2.93	0.86 ± 0.56	2.3	0.59
	Possjet Bay	summer 1981	12	1.5	3.5	11	30	16.1 ± 4.08	$1.15 \pm 50\%$	2.2	0.57
		all together	28	1.5	3.5	9	36	19.0 ± 2.6	1.0 ± 0.59	2.26	0.56
4. C. mutica	Japan Sea Possjet Bay	autumn 1981	14	2.2	4.3	15	41	27.3 ± 5.2	$0.4 \pm 50\%$	3.0	0.24
5. C. krøyeri	Japan Sea Possjet Bay	summer 1981	6	15.25	20.75	55	105	69.0 ± 18.2	$0.63 \pm 50\%$	17.5	0.27
6. C. advena	Iturup Island	summer 1969	9	4.5	17.0	35	85	51.5 ± 7.4	$0.41 \pm 50\%$	7.8	0.6
7. The total for Caprellidae			196	0.5	42.8	4	322	37.9 ± 1.9	0.85 ± 0.04	6.0	0.94

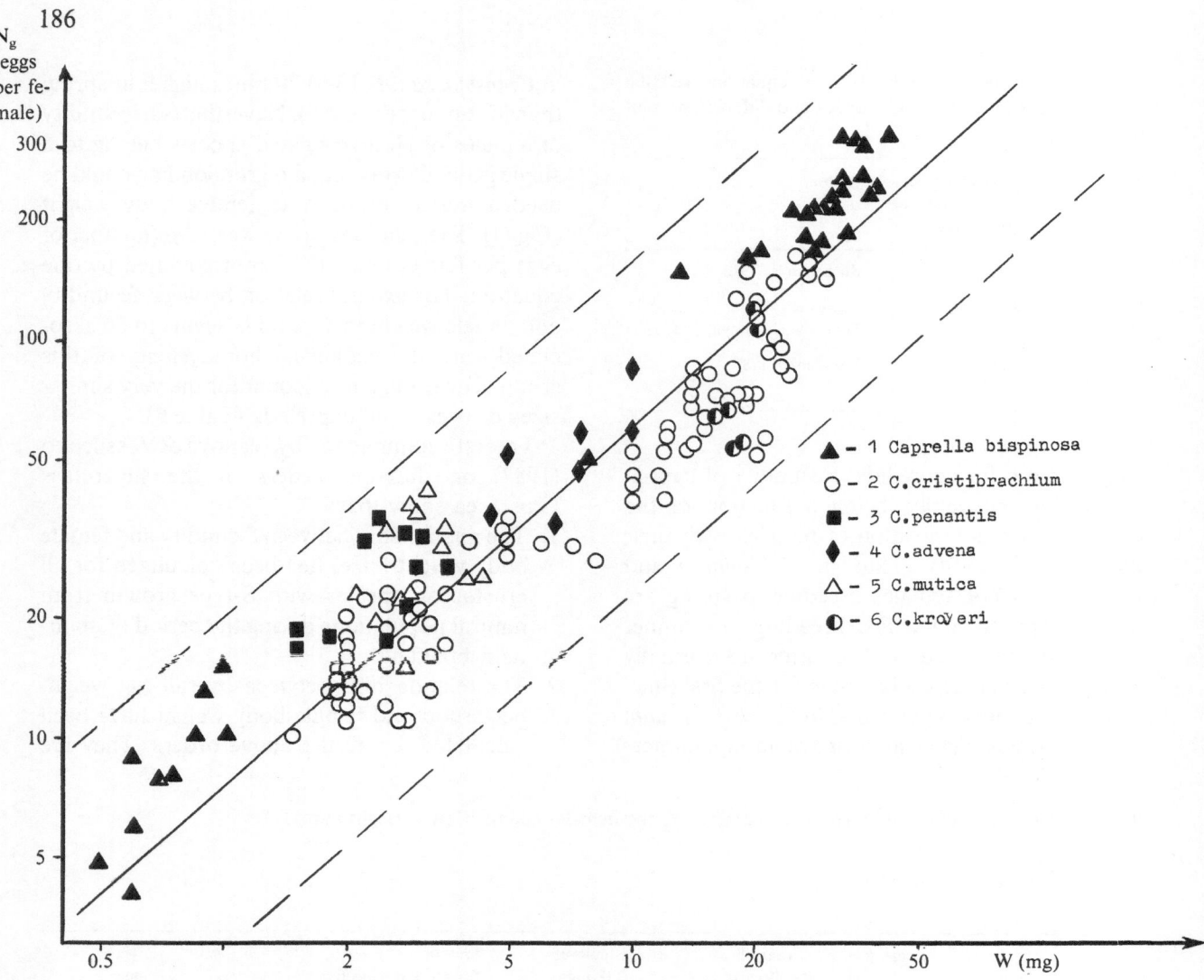

Fig. 3. Relationship between number of eggs and female body weight for several caprellid species.

Table 6. Mean egg mass at the first developmental stage and mean ratio W_g/W in some caprellid species.

Species	Egg diameter mm		Mean egg weight mg	W_g/W
	min	max		
1. Caprella bispinosa	0.30	0.37	0.021	0.15
2. C. cristibrachium	0.30	0.35	0.017	0.09
3. C. penantis	0.25	0.30	0.011	0.10
4. C. krøyeri	0.30	0.40	0.022	0.09
5. C. mutica	0.25	0.35	0.014	0.13
6. C. advena	0.40	0.45	0.039	0.24

Where W_g – the total weight of the brood (mg); W – weight (mg) respectively of the female.

approximated by a power equation with a power index from 0.77 to 1.07.

3. Comparison of the data for some orders has shown that a single equation describes the relationship between brood weight and female body weight for the whole superclass. On average for the superclass (except the order Notostraca) the W_g/W ratio appeared to be 0.16.

References

Abolmasova, G. I., 1975. Traty energiy na dykhanie i reproduksiyu yaiz u *Gammarus olivii* iz Chornogo morya.

(Energy expenditures for respiration and reproduction of eggs in *Gammarus olivii* from the Black Sea). Elementy energeticheskogo balansa morskykh bespovonochnykh. Naukova dumka, Kiev: 68–72.

Alimov, A. F., 1981. Funkcialnaya ekologiya presnovodnykh dvustvorchatykh molluskov. (The functional ecology of freshwater bivalves). Trudy Zool. Inst. AN SSSR 96: 1–248.

Alimov, A. F., 1989. Vvedenye v produktsionnuyu gidrobiolgiyu. (An introduction to production hydrobiology). Izd. Gidrometizdat, 152 pp.

Greze, I. I., 1977. Amfipody Chornogo morya i ikh biologiya. (The Amphipoda of the Black Sea and their biology). Naukova dumka, Kiev, 156 pp.

Dulepov, V. I., E. P. Dulepova & V. O. Pois, 1986. Biologiya i produktsiya rakoobraznikh Kurilskikh ostrovov. (Biology and production of the Crustacea of the Kuril Islands). Vladivostok, 356 pp.

Ivanova, L. M., 1973. Skorost' potrbleniya kisloroda donnimy bespozvonochnymi severnogo Kaspiya. (Oxygen consumption rate of benthic invertebrates on the northern Caspian). Trudy VNIRO 80 (3): 159–172.

Ivanova, M. B. & S. V. Vassilenko, 1987. Relationship between number of eggs, brood weight, and female body weight in Crustacea. Int. Revue ges. Hydrobiol. 72: 147–169.

Jazdzewski, K., 1970. Biology of Crustacea Malacostraca in the Bay of Puck, Polish Baltic Sea. Zool. pol. 20: 423–480.

Khmeleva, N. N., 1973. Biologiya i energeticheskyi balans morskikh ravnonogikh rakoobraznykh. (The biology and energetic balance of isopods). Nauka Dumka, Kiev, 184 pp.

Khmeleva, N. N., 1988. Zakonomernosti rezmnozheniya rakoobraznykh. (Reproductive patterns in Crustacea). Nauka i Technika, Minsk, 208 pp.

Khmeleva, N. N. & A. P. Golubev, 1984. Produktsiya kormovykh i promyslovykh rakoobraznykh. (Productivity of fodder and industrial crustaceans). Minsk, 216 pp.

Shapovalova, I. M., 1973. Biologiya ozernogo bokoplava *Gammarus lacustris* Sars ozera Arachlei. (The biology of *Gammarus lacustris* from Lake Arachlei). Zap. Zabaikal. filiala geograf. obshchestva SSSR 96: 121–131.

Steele, D. H. & V. J. Steele, 1970. The biology of *Gammarus* (Crustacea, Amphipoda) in the northwestern Atlantic. III. *Gammarus obtusatus* Dahl. Can. J. Zool. 48: 989–995.

Steele, D. H. & V. J. Steele, 1972. The biology of *Gammarus* (Crustacea, Amphipoda) in the northwestern Atlantic. V. *Gammarus oceanicus* Segerstrale. Can. J. Zool. 50: 801–813.

Umnov, A. A., 1976. Primenenie statistiticheskikh metodov dlya ozerki parametrov empiricheskikh uravneniy opisyvayushchikh vsiamosvyas' mezhdu energeticheskim obmenom i massoi tela zhivotnykh. (Statistical methods for the evaluation of parameters in empirical equations describing the relationships between calorific value and biomass of the animals). Zhurn. obshch. biol. 37: 71–86.

Vassilenko, S. V., 1985. Skorost' potrebleniya kisloroda i razmerno – vesovye charakteristiky chetyrekh vidov kaprellid. (The oxygen consumption rate and size-weight characteristics of four caprellids). Biol. morya 5: 40–45.

Winberg, G. G., 1979. Formirovanie predstavlenia o produktsii. (Development of theories of productivity). Obshchiye osnovy isucheniya vodnykh ekosystem. Izd. Nauka, Leningrad, pp. 114–119.

Zaika, V. E., 1972. Udelnaya produktsiya vodnykh bespozvonochnykh. (Specific productivity of the aquatic invertebrates). Kiev, 144 pp.

Hydrobiologia **223**: 189–227, 1991.
L. Watling (ed.), VIIth International Colloquium on Amphipoda.
© 1991 *Kluwer Academic Publishers.*

A review of the reproductive bionomics of aquatic gammaridean amphipods: variation of life history traits with latitude, depth, salinity and superfamily

Bernard Sainte-Marie
Direction des sciences biologiques, Institut Maurice-Lamontagne, Pêches et Océans Canada, 850 route de la Mer, C.P. 1000, Mont-Joli, Québec, G5H 3Z4 Canada

Abstract

Life history traits (mean and maximum body length of females, number of embryos per brood = brood size, embryo diameter, number of broods per female, lifespan of females) for 302 populations of aquatic gammaridean amphipods, representing 214 species in 16 superfamilies, were reviewed. The variation of these traits, of lifetime potential fecundity (i.e. the number of embryos produced per female lifespan) and of reproductive potential (i.e. the number of embryos produced per female per year), with temperature (latitude), depth, salinity and superfamily, was investigated by various univariate and multivariate methods. Gammaridean amphipods comprise semelparous and iteroparous populations and species, with semiannual, annual, biannual or perennial life cycles. However, most gammarideans studied so far are iteroparous annuals. Body length explains most of the variation in brood size and embryo diameter. The reproductive potential may be increased by increasing body size for a constant breeding frequency, by increasing brood size at the expense of smaller embryos, by increasing breeding frequency for a constant lifespan at the expense of smaller individual broods and/or embryos, and by increasing longevity for a constant breeding frequency and brood size. Combinations of these different options constitute the life history patterns of gammarideans, which vary across superfamilies, latitude and depth, and cannot simply be explained by variations in body length. High latitude species were generally characterized by biannual or perennial life histories, large body size, delayed maturity, and single or few broods with many, relatively large embryos; converse sets of traits characterized low latitude species. Deep-living species had relatively smaller broods and embryos than their shallow-living relatives, yet did not produce more broods. However, different superfamilies dominated in different habitats. The importance of natural selection relative to phylogenetic (historical) and physiological constraints in the forging of these patterns is discussed.

Introduction

Gammaridean amphipods constitute a diverse and ubiquitous suborder of Crustacea (Bousfield, 1983). They are generally dioecious, with external fertilization, and the embryos are carried by females in a ventral brood pouch, called marsupium (Schram, 1986). Detailed accounts of the reproductive and brooding behaviour of a few

gammarideans have been produced (Kinne, 1954; Borowsky, 1983, 1984, 1986; Borowsky & Borowsky, 1987; Shillaker & Moore, 1987). There exists a very large body of literature on the life history traits of gammaridean amphipods, which several workers have attempted to synthesize: reviews have either been broad (Morino, 1978; Nelson, 1980; Van Dolah & Bird, 1980; Wildish, 1982) or narrow in taxonomic scope, the latter concentrating on the superfamily Talitroidea (Wildish, 1988), the family Gammaridae (Steele & Steele, 1975c), and on the genera *Orchestia* (Wildish, 1979) and *Ampelisca* (Bellan-Santini & Dauvin, 1988).

Morino (1978) proposed a simple classification of gammaridean life histories based on breeding rhythms (seasonal or year-round breeding) and longevity, and suggested that life history types were distributed according to latitudinal (temperature) gradients. He predicted semiannual populations for tropical regions, annual populations for temperate regions, and biannual or perennial populations for polar regions. Wildish (1982) recognized six basic life histories within the Gammaridea: multivoltine (more than one generation per year) semiannual, univoltine (one generation per year) or multivoltine annual, semelparous (single-brooded) biannual, and semelparous or iteroparous (multiple-brooded) perennial. Females of semiannual species or populations grow rapidly, mature early and are very fecund; this set of traits is presumably associated with warm and tropical habitats (for confirmation in talitroids, see Wildish, 1988) or with populations subjected to high rates of predation. In contrast, females of biannual or perennial species or populations tend to grow more slowly, mature later, and are less fecund; this set of traits is presumably characteristic of populations in habitats where mortality is influenced by unpredictable physical factors. Both Morino's and Wildish's classifications of gammaridean life histories were derived from relatively few observations and thus need to be tested against larger data sets.

Nelson (1980) compared average body length of reproductive females and number of embryos per brood (called brood size hereafter) for several species of aquatic gammaridean amphipods. He concluded that females were larger and embryos more numerous per brood in 'epibenthic' versus 'endobenthic' gammarideans, in brackish- versus fresh- and saltwater gammarideans, and in the family Gammaridae versus the Ampeliscidae and Haustoriidae. Furthermore, semelparous species produced more embryos per brood than iteroparous species. Although his conclusions relating to fecundity have been widely accepted, they are questionably based on comparisons of brood size alone or of the ratio of brood size to body size; the former does not take into account the known positive relation between brood size and body size (e.g. Van Dolah & Bird, 1980) and the latter is a statistically incorrect way to account for this relation.

Van Dolah & Bird (1980) reported that 'epibenthic' species of aquatic gammarideans had more, smaller embryos per brood than 'endobenthic' species, for a constant female body size, and found positive correlations between latitude and embryo size for populations of a given species. They hypothesized that 'adult mortality risk is correlated positively with egg number and inversely with egg size'. Nelson (1980) gave this adaptive hypothesis strong support, but it was challenged by Fenwick (1984) who argued that Nelson and Van Dolah & Bird had misclassified species into the epi- and endobenthic categories and that their reviews were too narrow in taxonomic scope for meaningful comparisons.

There exists compelling evidence of habitat effects on the life history traits of individual species of gammaridean amphipods. Body size at maturity, brood size, size of embryos, number of broods per female, age at maturity, and breeding season have been reported to vary intraspecifically with temperature, latitude, depth, salinity or exposure to predators (e.g. Hynes, 1954; Segerstråle, 1967, 1970; Fish & Preece, 1970; Strong, 1972; Wiederholm, 1973; Morino, 1978; Pinkster & Broodbakker, 1980; Kolding & Fenchel, 1981; Sainte-Marie & Brunel, 1983; Sheader, 1983; Skadsheim, 1984, 1989; Clarke *et al.*, 1985; Leineweber, 1985; Siegfried, 1985; Fredette & Diaz, 1986; Bellan-Santini & Dauvin, 1988; Naylor *et al.*, 1988). In particular, for individual boreal gammaridean species, a northerly (or decreasing temperature) trend of bigger bodies and smaller broods relative to body size appears to be the rule (D.H. Steele, 1967; Steele & Steele, 1975c; Van Dolah & Bird, 1980). If such trends are common to most Gammaridea, they should be obvious at the interspecific level as well. Attributing such variation to environmental or genetic effects has in practice rarely been possible (Wildish, 1970; Strong, 1972; Skadsheim, 1989).

Considering the apparent limitations of previous general reviews and the relatively large number of recent contributions to the study of the reproduction of aquatic gammaridean amphipods, it seemed timely to undertake a new review. Herein, six life history traits (mean and maximum body length of females and males, brood size, embryo diameter, number of broods per female, life expectancy of females), lifetime potential fecundity (i.e. the number of embryos produced per female lifespan) and reproductive potential (i.e. the number of embryos produced per female per year), were considered in relation to habitat and superfamily. A total of 302 aquatic gammaridean populations, representing 214 species in 16 superfamilies, were reviewed. Relationships among life history traits were explored by simple and partial correlations, and by simple and multiple regressions. Univariate statistics, canonical discriminant analysis, and analysis of covariance were used to quantify variation of reproductive traits across habitats and superfamilies.

The general objectives of this review were twofold. The first was to determine the extent and nature of correlations among selected life history traits for aquatic gammarideans. Collectively, these traits constitute the life history pattern (I shun the words strategy and tactic), which may be characteristic of populations in specific habitats and superfamilies. The second objective was to contrast life history traits, lifetime potential fecundity and reproductive potential, across habitats and superfamilies, the former being defined following rough temperature, depth and salinity gradients. Life history patterns were then discussed in light of some recent developments in life history theory (habitat templets and r-K-A selection).

Materials and methods

Derivation of data

Published literature on the reproduction of aquatic gammaridean amphipods was reviewed.

My aim was to make the population and species list as exhaustive as possible, but inevitably some publications must have been overlooked, with no prejudice intended. Other publications could not be secured through interlibrary loans or by other means. The sole criterion for inclusion in this study was that reports provide information at least on the mean body length of reproductive females and on the mean number of embryos per brood. Additionally, where available, I gathered information on maximum body length of females and males, on the longevity of reproductive females, on the frequency of brooding during female lifespan (semelparous or iteroparous, and maximum number of broods produced per female), on the diameter of embryos, and on the habitat (temperature, salinity, depth) occupied by individual populations or species. The following remarks pertain to the derivation and presentation of raw data in Appendix 1.

Taxonomic affiliation. The status of each species was verified, especially those appearing in older publications, to ascertain that they had not been synonimized or attributed to a new genus or superfamily. Species were grouped according to superfamily, following Bousfield's (1983) classification of the gammaridean Amphipoda. All species belonging to the 'Gammarus' complex were referred to the genus *Gammarus*, because attempts to divide the latter have not generated consensus. Of course, the value of predictive regressions and conclusions derived for superfamilies depends closely on the timeliness of the adopted taxonomic groupings.

Some species were studied in more than one site, by the same or different authors. Data from different authors were always included in my review, while those from the same author, based on populations in different areas or seasons, were included only if life history traits or habitats differed markedly. Appendix 1 also presents incidental information for species which were not studied with the purpose of elucidating reproductive bionomics (taxonomy, ecology, genetics or physiology motivated the work). This body of literature was not scrutinized systematically, and was not the preferred source of data, but was included when species belonged to poorly represented superfamilies or habitats.

Body length of reproductive females. Body length of gammaridean amphipods is generally measured from the anterior end of the cephalon to the distal end of the telson. However, amphipods have occasionally been measured from the anterior end of the cephalon to the base of the telson. Since the telson generally contributes only marginally to total body length, the difference between both measures of body length was presumed to be negligible. When not provided directly in the original study, mean body length was, in decreasing order of preference, derived from raw data, calculated as the average of minimum and maximum body length, or inferred from graphs. In all cases, maximum body length was also noted.

A few workers used relative indices of female body length based on measures of the cephalon (e.g. Goedmakers, 1981; Fenwick, 1985), of an article of the second antennae (Gaylor, 1922), of the basal segment of a pereopod (Moore, 1981), or on a partial measure of body length (e.g. Dexter, 1971; Fish, 1975). While these measures are easier to obtain and less prone to error than total body length, they are obviously less suitable for interspecific comparisons of length-fecundity relationships. Still fewer workers have used female weight as a standard by which to compare number of embryos per brood (e.g. Cheng, 1942; Sameoto, 1969b; Duncan, 1969 for a terrestrial example) or age determined by counts of articles on the antennular flagella of a terrestrial gammaridean (Tamura & Koseki, 1974). Some of the workers who used non-conventional measures of body size provided equations to estimate total body length from the length index or body weight, and their results were incorporated into this review. Ultimately, body volume may prove to be the best measure of amphipod size for comparative purposes (Wildish & Frost, 1991).

Half-range of mature female body length. A statistic was developed to characterize variability in body length of females at maturity. The half-range of mature female body length, called *HMFBL* for sake of brevity, was determined by the equation

$$HMFBL = BL_{max} - BL_{mean},$$

where *BL* is body length of mature females. This index, corrected for body length, became the *HMFBL* ratio (*HMFBLr*), given by the equation:

$$HMFBLr = (BL_{max} - BL_{mean})/BL_{mean}.$$

The *HMFBL* ratio was intended to serve as an index of the number of broods produced by females, when that information was lacking in the original study. Gammaridean amphipods must moult to oviposit (Charniaux-Cotton, 1985). In iteroparous populations and species, successive brooding instars may be interspaced by resting (V.J. Steele, 1967) or preparatory stages (Ingram & Hessler, 1987). There may be no growth when females moult from preparatory to brooding stages, but moulting from brooding to preparatory stages is apparently always accompanied by an increment in body size (e.g. Bone, 1972; Ingram & Hessler, 1987). When no preparatory or resting stage exists, i.e. broods are carried by each sequential mature instar, females apparently grow at each moult (e.g. Sexton, 1928). Hence, whatever the case may be, there should exist a positive relationship between the *HMFBL* ratio and the number of broods.

Brood size, i.e. number of embryos per brood. Since brood mortality occurs in several gammaridean species, ranging over the full incubation period from 0 to 58.5% of initial oviposited eggs (see review by Moore, 1981), a standard embryo developmental stage should be used to compare brood size. Stage V embryos (*sensu* Thurston, 1968) represent effective recruits, but data for this developmental stage are rarely available and are unreliable because Stage V embryos may temporarily exit the marsupium (e.g. Embody, 1911; Nayar, 1956; Sheader & Chia, 1970; Borowsky, 1980b; Moore, 1981; Shillaker & Moore, 1987). My alternative was to consider numbers of Stage I embryos, for which most authors have provided data. However, a small number of studies have been included in which authors did not specify the embryo developmental stage or counted only Stage II or III embryos. Still fewer studies, mainly of deep-sea populations, provided only counts of oocytes in gonads. Brood size may be overestimated for some of these populations because oocyte maturation may be accompanied by a decrease in oocyte numbers (Hessler et al., 1978). Where mean brood size was not given directly in the original study, it was, in order of decreasing preference, calculated from raw data, predicted from regression lines of brood size on body length, estimated as the mean of minimum and maximum values, or inferred from graphs.

Embryo diameter was the mean of measurements of the long and short axes of Stage I embryos. Several authors gave only the measure of the long axis of embryos, so it was necessary to convert these data for comparisons. For this purpose, measurements of the small axis (SA, in mm), of the long axis (LA, in mm) and of embryo diameter (ED, in mm) of Stage I embryos of 19 species (from Ivanov, 1961; Kanneworf, 1965; Bregazzi, 1972; Thurston, 1974; Morino, 1978) were regressed and the following predictive regression equations determined:

$$SA = 0.7556\,LA + 0.0263,\ r^2 = 0.984$$
$$ED = 0.8778\,LA + 0.0131,\ r^2 = 0.996$$
$$ED = 1.1510\,SA - 0.0113,\ r^2 = 0.995$$
$$\text{range: } SA = 0.35\text{--}1.15 \text{ mm}, LA = 0.43\text{--}1.53 \text{ mm}$$

The size of Stage I embryos in deep-sea species was estimated by authors or myself with growth curves of ova (in Hessler et al., 1978; Ingram & Hessler, 1987).

Semelparity, iteroparity and maximum number of broods. In most of the reviewed literature, the semelparous or iteroparous condition of females was determined by examination of the ovaries. Alternatively, the maximum number of broods produced per female was determined from size-frequency (polymodal) analyses, growth factors and inferred moult instars, or laboratory cultures. The maximum number of broods may vary in different generations of multivoltine populations: in these cases, upper values were used herein.

Lifespan of females was recorded where available from the literature. Investigators mainly used cohort analysis or laboratory cultures to determine age. Several authors presented a range of lifespans, corresponding to life expectancies of females in different generations (summer or winter) of multivoltine populations, so upper values were retained for analyses.

Alternative indices of fecundity. *Lifetime potential fecundity* (LF, in number of embryos per female) is given by the equation:

$$LF = BS * NB,$$

where BS is the mean number of embryos in a brood and NB is the maximum number of broods produced by a female during her lifespan. Wildish (1982) proposed and discussed another index of fecundity, standardized for a 12-mo lifespan, called the reproductive potential (R, in number of embryos female^{-1} yr^{-1}):

$$R = b * n * p,$$

where b is brood size, n is the number of broods per year per female, and p is the proportion of adult females in relation to the total number of adult females and males present in the population. Maternally-biased sex ratios may increase the number of female descendants and, consequently, of offspring calculated over several generations (Wildish, 1971, 1982). Skewed adult sex ratios are frequent in the Gammaridea (Moore, 1981; Wildish, 1982; Costello & Myers, 1989). However, they may often be unrelated to direct maternal effects: apparent or real biases may result from parasitism (Bulnheim, 1978), from differential longevity (Heller, 1968; Sheader, 1978) or maturation rates (Gable & Croker, 1977) of sexes, from spatial segregation of sexes in the horizontal or vertical planes (e.g. Bregazzi, 1972; Bosworth,

1976; Smith & Baldwin, 1984), or from differential mortality of sexes in the juvenile and subadult stages (Heller, 1968; Moore, 1981). For most gammaridean populations, it may be difficult to ascribe biased sex ratios either to maternal or ecological effects, so I have elected to disregard this variable in my calculations of *reproductive potential* (*RP*, in number of embryos per female per year), which is thus simply:

$$RP = BS * NB * 12/LS,$$

where *BS* is brood size, *NB* is the number of broods per female, and *LS* is the female's lifespan in months.

Temperature. Populations were classified according to habitat temperature either as warm- or cold-living. Warm designated shallow tropical or warm temperate habitats; cold designated polar or cold-temperate habitats, as well as bathyal or abyssal environments. The northern limit of warm areas was set by a line running from the most easterly point of Cape Cod (United States) to the middle of the westerly mouth of the English Channel, to Calais (France), to the southern extremity of the Kyü Shü Island (Japan), and across to the northern border of California (United States). The Mediterranean Sea and inland seas of Europe and eastern Asia were included in the warm habitats. The southern limit of warm areas was set by a line running from the northern border of Argentina, across the Atlantic and Pacific well to the south of the African continent and Australia, through the Strait of Cook (New Zealand), and to Santiago (Chile). Thus, excepting the few deep-living gammarideans, the classifications of populations following temperature in fact reflects a latitudinal separation.

Salinity. Populations were described as inhabiting freshwater, brackish ($\leq 20‰$ average over a tidal cycle) or marine ($> 20‰$ average over a tidal cycle) habitats. Supralittoral and salt-marsh talitroids were classified as marine forms.

Depth. Deep-living populations of gammarideans were those inhabiting waters deeper than 200 m. Shelf-living populations of gammarideans were those encountered in depths less than 200 m, including supralittoral and salt-marsh talitroids.

Treatment of data

Canonical discriminant analysis was used to compare variation of life history traits of populations and species across superfamilies, and across temperature, salinity and depth gradients. In this analysis, linear combinations of independent variables, here the life history traits, are constructed to produce discriminant functions (Legendre & Legendre, 1984). The relative importance of each discriminant function in separating the groups (defined as populations in different habitats or superfamilies) is shown by the percentage of the total eigenvalue and by the strength of the canonical correlation. Group means give the average position of groups in reduced multidimensional space when there is more than one discriminant function, or along a simple axis when there is only one discriminant function (i.e. two groups). The standardized canonical coefficients indicate what independent variables contribute most to the discrimination of groups along each of the axes (discriminant functions). Only four life history traits, body length, brood size, embryo diameter and the *HMFBL* ratio, were used in canonical discriminant analyses: this was done to maximize the number of superfamilies under comparison, because data on lifespan were too often lacking. The minimum number of observations necessary for inclusion of a group in analyses was arbitrarily fixed at five. All data were $\log_{10}$-transformed prior to analyses so they met or approached a normal distribution.

Frequencies were tested for independence of classification criteria by means of the G-test with William's correction (Sokal & Rohlf, 1981). The mean and standard deviation were calculated for life history traits, lifetime potential fecundity and reproductive potential. Group means were compared with the non-parametric Kruskal-Wallis test, a single-factor analysis of variance by ranks (Sokal & Rohlf, 1981).

Relationships between life history traits were explored through the use of simple (Pearson) and partial correlations. Simple correlations measure the relation between two independent variables; partial correlations measure the relation between two independent variables while holding the effects of other variables constant (Sokal & Rohlf, 1981). Correlation matrices were derived from listwise comparisons of variables, i.e. only populations with complete sets of observations for all life history traits were used.

Predictive (Model I) simple or multiple regressions were calculated for selected relationships. When necessary, slopes (regression coefficients) and elevation of significant regressions were compared by analysis of variance (ANOVA) and covariance (ANCOVA). Note that elevation may be compared only among lines with homogeneous slopes (Sokal & Rohlf, 1981). Prior to all correlation and regression analyses, data were $\log_{10}$-transformed to satisfy conditions of normality and/or because mean and variance were positively correlated. The interpretation of coefficients for regressions of log-transformed dependent and independent variables has been discussed at length by White & Gould (1965).

Results

General observations

Information on the life history traits of aquatic gammaridean amphipods was available for a total of 302 populations, representing 214 species in 16 superfamilies (Appendix 1). The vast majority of these populations were from the northern hemisphere. Eight superfamilies were well represented: the Gammaroidea (20.2% of reviewed populations), the Pontoporeioidea (13.9%), the Corophioidea (12.6%), the Lysianassoidea (11.6%), the Eusiroidea (9.3%), the Talitroidea (8.6%) and the Ampeliscoidea (8.0%). Cold and warm water populations accounted for 71.2% and 28.8% of investigated populations, respectively. Populations were predominantly marine (75.8%), and the remaining were from brackish (15.2%) or fresh (9.0%) waters. Finally, 96.4% of reviewed populations were from waters < 200 m deep and 3.6% were from deeper waters.

Extreme values for life history traits reviewed in the literature were the following. The lilljeborgioid *Seborgia minima* (Bousfield, 1970) had the smallest females (0.9 mm mean body length) and broods (1 embryo). The greatest mean body length (≥ 23 cm) and largest mean embryo diameter (~ 9.11 mm) were found in the lysianassoid *Alicella gigantea* (Barnard & Ingram, 1986), while the lysianassoid *Anonyx nugax* produced the largest mean broods (630 embryos), with one ~ 47-mm female carrying in excess of 950 embryos (Kuznetsov, 1964). The smallest embryos (0.23 mm diameter) were found in the commensal corophioid *Gammaropsis inaequistylis* (Steele *et al.*, 1986). Several gammaridean females live only a few weeks or months (Appendix 1), in contrast to the ~ 13-yr lifespan of the lysianassoid *Eurythenes gryllus* (Ingram & Hessler, 1987). The lifespan of the abyssal giant

Table 1. Sex-related differences in maximum adult body length (*BL*), expressed as the relative frequency of total observations (N), for populations from superfamilies of gammaridean amphipods with > 10 observations. Females and males were considered equal in size when the difference in body length between both sexes was < 2.5% of the larger value.

Superfamily	♀ *BL* > ♂ *BL*	♀ *BL* = ♂ *BL*	♀ *BL* < ♂ *BL*	N
Gammaroidea	2.5%	0.0%	97.5%	40
Talitroidea	25.0%	33.3%	41.7%	12
Ampeliscoidea	31.3%	62.5%	6.2%	16
Corophioidea	47.1%	35.3%	17.6%	17
Lysianassoidea	88.2%	5.9%	5.9%	17
Pontoporeioidea	94.4%	5.6%	0.0%	18

Alicella gigantea is likely greater, but no estimate of longevity exists at present. Finally, the maximum number of broods produced per female varies from one, in a flurry of gammarideans (Appendix 1), to 26 in the gammaroid *Gammarus chevreuxi* (Sexton, 1928).

In gammaridean amphipods, maximum female body length may be greater, equal or less than maximum male body length (Table 1); the occurrence of such cases is apparently closely dependent on superfamily affiliation ($G = 61.20$, $df = 10$, $P < 0.001$). Female gammaroids were virtually always smaller than males (97.5% of cases), with the sole possible exception of *Gammarellus angulosus* (see Fig. 2 in D.H. Steele & Steele, 1972b). In contrast, body length of females was equal to or greater than that of males in pontoporeioids (100% of cases), lysianassoids (> 94%), ampeliscoids (> 93%) and corophioids (> 82%). Finally, talitroids formed a mixed group with populations or species where female maximum body length exceeded, equalled or was less than that of males (Table 1).

The stegocephaloid *Stegocephalus inflatus* (D.H. Steele, 1967), some leucothoids (Schram, 1986) and the lysianassoid genus *Acantiostoma* (Lowry & Stoddart, 1986) may have small males because they are protandrous hermaphrodites. The corophioid *Corophium bonnellii* apparently represents another departure from dioecism, since it is probably parthenogenetic (Moore, 1981; Costello & Myers, 1989).

Number of broods, HMFBL ratio and classification of life histories

The simple correlation between the maximum number of broods produced per female and the *HMFBL* ratio was highly significant ($r = 0.64$, $N = 93$, $P < 0.001$). In cold waters, the *HMFBL* ratio ranged from 0.0110 to 0.3478 in semelparous populations and from 0.1304 to 0.7846 in iteroparous populations. The *HMFBL* ratio can hence be used to separate semelparous from iteroparous populations: this separation was carried out only for the cold-water habitat, because only there

were semelparous populations adequately represented. All unclassified cold-water populations with a *HMFBL* ratio < 0.1304 were considered to be semelparous, while populations with a *HMFBL* ratio > 0.3478 were considered to be iteroparous (Appendix 1).

Iteroparity was apparently more frequent than semelparity within the suborder Gammaridea (Appendix 1 and Fig. 1A). Semelparous populations were significantly more frequent in cold

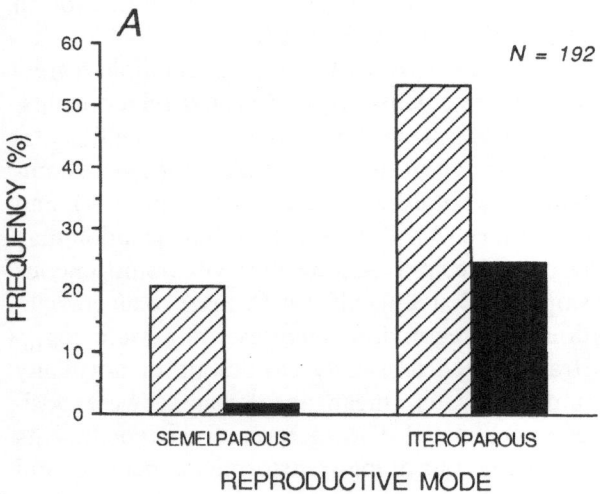

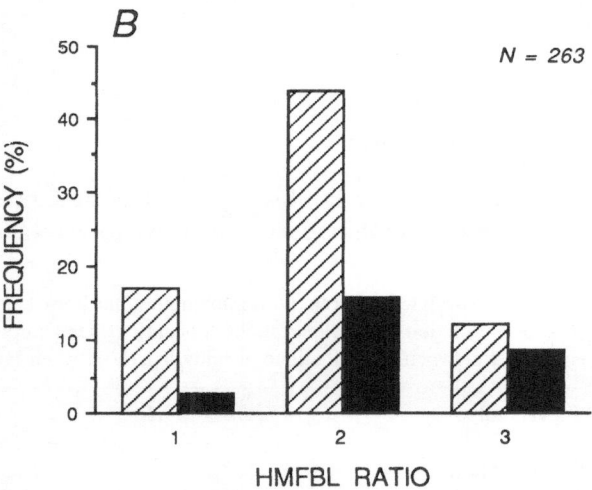

Fig. 1. Frequency distributions of (A) semelparous and iteroparous populations and of (B) *HMFBL* ratios for aquatic gammaridean amphipods in cold (hatched bars) and warm (black bars) habitats. *HMFBL* ratios: 1 is < 0.1304, 2 is ≥ 0.1304 and ≤ 0.3478, 3 is > 0.3478.

than in warm waters (Fig. 1A; $G = 12.45$, $df = 1$, $P < 0.001$). This conclusion was supported by a broader-based analysis using the *HMFBL* ratio in lieu of the number of broods per female (Fig. 1B), which indicated that distribution of *HMFBL* ratios was dependent on temperature ($G = 11.05$, $df = 2$, $P < 0.01$).

In cold waters, frequency distributions of mean female body length relative to the semelparous/-iteroparous condition or the *HMFBL* ratio suggested that small females tended to be more iteroparous than large ones (Fig. 2), but testing of this hypothesis yielded conflicting conclusions. The occurrence of semelparity and iteroparity was independent of mean female body length (Fig. 2A; $G = 2.64$, $df = 2$, $P > 0.1$), while the *HMFBL* ratio was highly dependent on mean female body length (Fig. 2B; $G = 16.18$, $df = 4$, $P < 0.01$). In warm waters, the range of body lengths was too small and females of the semelparous habit or with small *HMFBL* ratios were too few for meaningful conclusions.

All corophioids and gammaroids reviewed herein were iteroparous (Appendix 1). Nelson (1980) erroneously listed four semelparous Gammaroidea: *Gammaracanthus loricatus*, *Gammarellus angulosus*, *Gammarus setosus* and *G. wilkitzkii*. This misinterpretation may have arisen from the ambiguous wording in a review paper by Steele & Steele (1975c), which refers to these species as single-brooded arctic gammaroids. However, careful reading of papers by V.J. Steele (1967), V.J. Steele & Steele (1970) and D.H. Steele & Steele (1972b, 1975a, 1976) indicates that these species are in fact iteroparous (confirmed herein by their large *HMFBL* ratios), but that they produce only one brood per year.

Semelparity was represented to some extent in most other superfamilies with populations in cold waters (Appendix 1). The Ampeliscoidea and Phoxocephaloidea were reported to be mostly iteroparous, even in cold waters. However, females of the potentially double-brooded, cold-living ampeliscoids and phoxocephaloids, may only rarely produce a second brood (Carrasco & Arcos, 1984; Slattery, 1985; Bellan-Santini & Dauvin, 1988).

To suit all amphipod life histories reviewed herein, Wildish's (1982) proposed classification must be expanded and modified slightly to include eight categories: the (multivoltine) semelparous and iteroparous semiannuals (lifespan < 12 mo.), the semelparous annual ($12 \leq LS < 24$ mo.) and (multivoltine) iteroparous annuals, the semel-

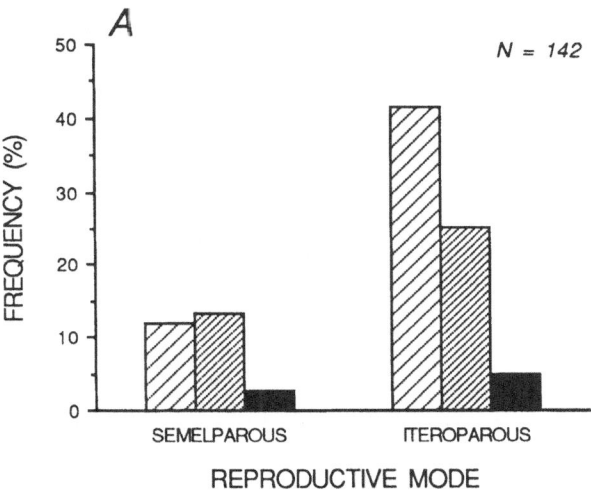

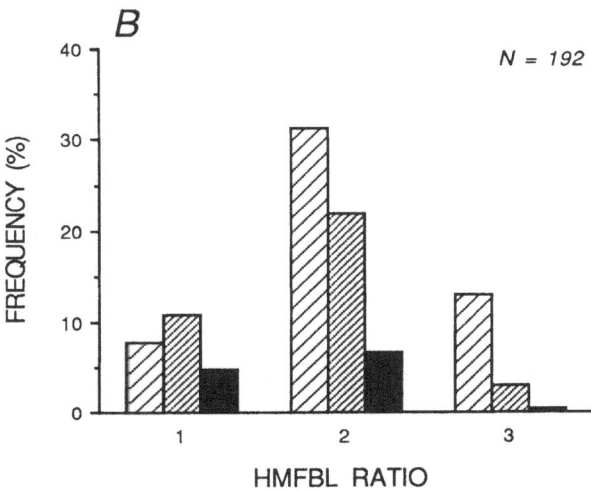

Fig. 2. Frequency distributions of (A) semelparous and iteroparous populations and of (B) *HMFBL* ratios (see legend of Fig. 1) as a function of body length (mm) for gammaridean amphipods in cold habitats. Lightly hatched bars represent small females (< 10 mm body length), heavily hatched bars represent medium females (10 to < 20 mm), dark bars represent large females (≥ 20 mm).

parous and iteroparous biannuals ($24 \leq LS$ 36 mo.), and the semelparous and iteroparous perennials ($LS \geq 36$ mo.). The distinction between semelparous and iteroparous semiannual populations is necessary, since females in each of the alternating 4- and 8-mo. generations of the ampeliscoids *Ampelisca abdita* and *A. vadorum* only breed once at the northern extremity of their distributional range (Mills, 1967; Nelson, 1978; Van Dolah & Bird, 1980). Furthermore, addition of an iteroparous biannual category is required to accommodate the ampeliscoid *A. armoricana*, the corophioid *Leptocheirus pinguis*, the gammaroids *Gammarus lacustris* and *Pallasea quadrispinosa*, and the lysianassoids *Hippomedon propinquus*, *Psammonyx nobilis* and *P. terranovae*, all of which may produce ≥ 2 broods in a 2-yr lifespan (Appendix 1).

Annual populations were by far the most common, representing 54.2% of all cases ($N = 107$); followed by semiannual (19.6%), biannual (14.0%) and perennial (12.2%). Iteroparity prevailed in semiannual (90.5%) and annual (82.8%) populations, but was significantly less frequent ($G = 6.94$, $df = 2$, $P < 0.05$) in biannual and perennial populations (60.7%).

Correlations among life history traits

Simple (Pearson) correlations showed that mean female body length, brood size, embryo diameter and female lifespan were all positively and significantly intercorrelated; the number of broods per female was negatively and significantly correlated with other life history traits (Table 2). The positive partial correlations between brood size and body length, and between embryo diameter and body length, were significant but of rather limited interest since they were obvious or have been repeatedly demonstrated. However, other partial correlations were of particular interest. The positive and significant partial correlation between body size and number of broods indicates that large females have more broods than small females when investment into individual broods is comparable (same number and size of embryos). The significant negative partial correlations between brood size and embryo diameter, between the number of broods and embryos diameter, and between the number of broods and brood size, show what comprises are possible among these life history traits for a constant body size. The positive partial correlation between lifespan and body length was marginally not significant ($r = 0.22$, $P < 0.1$). Using *HMFBL* in place of the number of broods substantially increased sample size ($N = 65$) and yielded the same significant partial correlations as above, except for the partial correlation between lifespan and body length which was significant ($r = 0.29$, $P < 0.05$).

Canonical discriminant analyses and univariate statistics for life history traits

Canonical discriminant analysis was unable to separate gammaridean populations according to salinity (Table 3). Of the four life history traits considered in discriminant analyses, only the

Table 2. Pearson (below diagonal) and partial (above diagonal) correlations between selected life history traits of gammaridean amphipods: BL = body length of females, BS = brood size, ED = embryo diameter, NB = number of broods per female, LS = lifespan. All correlations based on listwise comparisons of log-transformed data, $N = 51$. *** = $P < 0.001$, * = $P < 0.05$, ns = not significant.

	BL	BS	ED	NB	LS
BL	–	0.71***	0.58***	0.34*	0.22ns
BS	0.81***	–	−0.32*	−0.48***	0.13ns
ED	0.73***	0.52***	–	−0.29*	0.19ns
NB	−0.34*	−0.53***	−0.38**	–	−0.11ns
LS	0.69***	0.63***	0.60***	−0.39**	–

Table 3. Statistics for canonical discriminant analyses across habitats and superfamilies.

Discriminant function	% total eigenvalues	Canonical correlation	F	P
Temperature				
1	100.0	0.439	7.21	<0.001
Bathymetry				
1	100.0	0.533	12.00	<0.001
Salinity				
1	59.0	0.200	1.06	0.394
2	41.0	0.168	1.17	0.326
Superfamily				
1	64.7	0.648	3.70	<0.001
2	21.1	0.437	2.17	0.004
3	12.6	0.352	1.60	0.110
4	1.6	0.135	0.48	0.752

HMFBL ratio differed significantly with salinity, being greater in brackish than in marine populations (Table 5). Accordingly, the mean number of broods per female differed significantly with salinity and was also greatest in brackish populations.

Gammaridean populations separated neatly following temperature (Table 3). Judging from the values of the standardized canonical coefficients and group means in Table 4, cold water populations had larger embryos but smaller HMFBL ratios (i.e. fewer broods). However, the Kruskal-Wallis test (Table 5) indicated that all life history traits used in discriminant analysis differed significantly with temperature. On average, females, broods, and embryos were larger, while HMFBL ratios were smaller (i.e. fewer broods), in cold- than in warm-water gammaridean populations. The difference between the mean number of broods in cold and warm waters was marginally not significant.

The sharpest abiotic contrast between gammaridean populations was obtained in comparing deep- and shallow-living populations, as seen by the higher canonical correlation (0.533) in Table 3. Standardized canonical coefficients and group means indicated that females of deep-living populations had larger bodies and smaller HMFBL ratios (i.e. fewer broods) than females of shallow-living populations (Table 4). However, the Kruskal-Wallis test pointed to significant differences between mean body length, embryo diameter and HMFBL ratios (Table 5).

Superfamilies were neatly segregated by canonical discriminant analysis (Table 3), with embryo diameter and HMFBL ratio contributing most to separation along the first discriminant function, and body length and brood size contributing most to separation along the second discriminant function (Table 4 and Fig. 3). Superfamilies most separated by their group means were the lysianassoids, characterized by large embryos and large body size; the gammaroids, characterized by large HMFBL ratios (i.e. numerous broods) and rather large body size; the corophioids, charac-

Table 4. Standardized canonical coefficients and group means derived for significant discriminant functions presented in Table 3. Gammaridean populations are described as living in cold (C), warm (W), shallow (S) or deep (D) habitats. BL = body length of females, BS = brood size, ED = embryo diameter, HMFBLr = HMFBL ratio.

Discriminant function	BL	BS	ED	$HMFBLr$	Group means
Temperature					
1	−0.32	0.31	0.96	−0.46	C = 0.23 W = −0.99
Bathymetry					
1	−1.66	1.25	−0.16	0.17	S = 0.14 D = −2.79
Superfamily					
1	0.48	0.18	−1.02	0.94	See
2	1.46	−0.88	0.09	0.56	Fig. 3

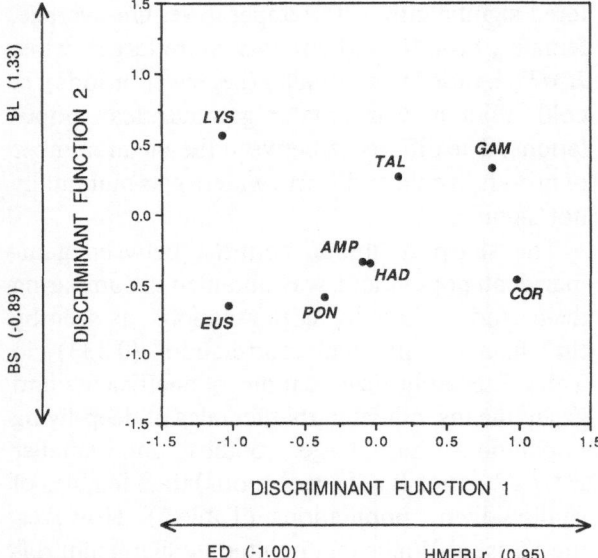

Fig. 3. Plot of group means derived from canonical discriminant analysis of the gammaridean superfamilies Ampeliscoidea (AMP), Corophioidea (COR), Eusiroidea (EUS), Gammaroidea (GAM), Hadzioidea (HAD), Lysianassoidea (LYS), Pontoporeioidea (PON) and Talitroidea (TAL). *ED* = embryo diameter, *HMFBLr* = *HMFBL* ratio, *BL* = body length, *BS* = brood size.

terized by large *HMFBL* ratios (i.e. numerous broods) and large brood size; and the eusiroids, characterized by large embryos and large brood size. Mean values and the Kruskal-Wallis test (Table 5), corroborated the marginal positions of these four superfamilies within the suborder Gammaridea. Lysianassoids had the largest mean embryo diameter (1.18 mm), gammaroids had the second largest mean *HMFBL* ratio (0.32), corophioids had the largest mean *HMFBL* ratio (0.34), and eusiroids had the second largest mean embryo diameter (0.70 mm) and largest mean brood size (66.9 embryos).

Although lifespan could not be used in canonical discriminant analyses, because data were lacking for too many gammaridean populations, mean values and the Kruskal-Wallis test pointed to some interesting variations across habitats and superfamilies (Table 6). Mean lifespan was significantly greater in cold- than in warm-living populations, but did not vary with salinity. Data

were lacking for all but one deep-living species (the lysianassoid *Eurythenes gryllus*), but one can expect the mean lifespan of all large deep-living gammarideans to be quite high because of the positive correlations between body length and lifespan (Table 2 and in text above) and of the not greater than average growth rates of deep- compared to shallow-living crustaceans (Mauchline, 1988a). Mean lifespan varied significantly across superfamilies, and was by far greatest in the Eusiroidea (31 mo.) and Lysianassoidea (30 mo.).

Embryo size, brood size, lifetime potential fecundity, and reproductive potential

Most of the variation in brood size was accounted for by body size (Table 7). However, prediction of brood size was improved significantly through the combined use of body size, embryo diameter, and *HMFBLr* or the number of broods per female, as independent variables in a multiple regression (Table 7).

Comparisons of regressions of brood size or of

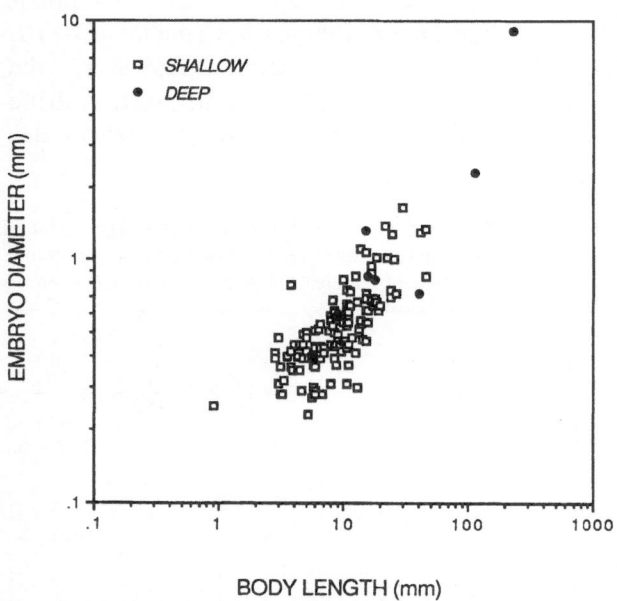

Fig. 4. Scattergram of embryo diameter as a function of body length of gammaridean amphipods.

Table 5. Mean and standard deviation of life history traits of gammaridean populations in different habitats and superfamilies. Means are presented only for groups with ≥ 5 observations for body length, brood size, embryo diameter and *HMFBL* ratio. A Kruskal-Wallis test was used to compare class levels with ≥ 5 observations; sample size is given in parentheses. AMP = Ampeliscoidea, COR = Corophioidea, EUS = Eusiroidea, GAM = Gammaroidea, HAD = Hadzioidea, LYS = Lysianassoidea, PON = Pontoporeioidea, TAL = Talitroidea, KW = Kruskal-Wallis statistic, *** = $P < 0.001$, ** = $P < 0.01$, * = $P < 0.05$, *ns* = not significant.

Groups	Body length (mm)	Brood size	Embryo diameter (mm)	*HMFBL* ratio	Number of broods
Temperature					
Cold	13.0 ± 18.3 (215)	42.0 ± 72.2 (215)	0.69 ± 0.84 (118)	0.23 ± 0.13 (190)	3.5 ± 3.8 (98)
Warm	7.0 ± 2.9 (87)	19.2 ± 14.4 (87)	0.38 ± 0.07 (34)	0.29 ± 0.15 (73)	6.5 ± 6.8 (14)
KW	25.63^{***}	9.60^{**}	39.22^{***}	7.90^{**}	3.42 ($P = 0.06$)
Bathymetry					
Deep	45.9 ± 68.4 (11)	50.6 ± 57.7 (11)	2.02 ± 2.93 (8)	0.15 ± 0.07 (9)	1.5 ± 0.7 (2)
Shallow	9.9 ± 7.0 (291)	34.9 ± 62.4 (291)	0.55 ± 0.24 (144)	0.25 ± 0.14 (254)	3.9 ± 4.4 (110)
KW	9.29^{**}	$1.35ns$	9.48^{**}	5.98^{*}	–
Salinity					
Freshwater	8.8 ± 2.8 (27)	24.7 ± 12.7 (27)	0.44 ± 0.14 (9)	0.28 ± 0.15 (24)	4.9 ± 4.6 (15)
Brackish	9.4 ± 4.0 (46)	23.2 ± 14.8 (46)	0.54 ± 0.17 (19)	0.30 ± 0.17 (40)	6.2 ± 6.2 (21)
Marine	11.9 ± 17.9 (229)	39.1 ± 70.6 (229)	0.65 ± 0.82 (124)	0.24 ± 0.13 (199)	3.0 ± 3.5 (76)
KW	$0.68ns$	$0.62ns$	$2.28ns$	6.40^{*}	15.71^{***}
Superfamily					
AMP	9.6 ± 3.4 (24)	27.8 ± 14.8 (24)	0.51 ± 0.12 (12)	0.19 ± 0.08 (21)	1.3 ± 0.5 (12)
COR	6.6 ± 3.7 (38)	28.7 ± 37.8 (38)	0.38 ± 0.08 (19)	0.34 ± 0.15 (27)	5.0 ± 3.2 (9)
EUS	17.0 ± 9.5 (28)	66.9 ± 54.6 (28)	0.70 ± 0.30 (12)	0.16 ± 0.13 (28)	1.0 ± 0.0 (6)
GAM	12.2 ± 7.3 (61)	48.5 ± 86.2 (61)	0.53 ± 0.16 (33)	0.32 ± 0.12 (59)	6.6 ± 5.8 (25)
HAD	9.0 ± 6.9 (12)	21.4 ± 18.3 (12)	0.47 ± 0.14 (9)	0.31 ± 0.30 (8)	8.0 ± 12.1 (3)
LYS	23.1 ± 40.9 (35)	59.7 ± 113.5 (35)	1.18 ± 1.78 (23)	0.19 ± 0.10 (33)	2.6 ± 1.9 (20)
PON	6.0 ± 2.5 (42)	14.0 ± 16.5 (42)	0.52 ± 0.16 (12)	0.22 ± 0.10 (39)	1.9 ± 3.0 (16)
TAL	8.6 ± 3.9 (26)	16.8 ± 9.3 (26)	0.52 ± 0.18 (13)	0.28 ± 0.12 (19)	5.1 ± 2.7 (10)
KW	89.01^{***}	63.57^{***}	37.78^{***}	58.37^{***}	55.94^{***}

embryo diameter on body length, across habitats, appear in Table 8. Slopes and elevations of regressions of brood size on body length did not differ with temperature or salinity, but the slope of the regression was significantly less in deep- than in shallow-living populations. The regression of embryo diameter on body length was significantly steeper in cold- than in warm-water populations, in marine- compared to brackish populations, and in deep- compared to shallow-living populations. However, few data points were available to characterize the deep-living populations (Fig. 4). All slopes of regressions of embryo diameter on body length were much less than unity, indicating that relative embryo size

decreased with increasing body length of females, as noted in other crustaceans (Mauchline, 1988b).

Differences between slopes and regressions of brood size on body length of females in different superfamilies were marginally not significant (Table 9). Note the very small slope for the Eusiroidea. The coefficient of determination for the Talitroidea regression was extremely small ($r^2 = 0.04$) and the slope did not differ significantly from zero, despite the fairly large number of observations. Elevation of regression lines differed significantly among superfamilies (Table 9): adjusted mean brood size, presented in Table 10, was greatest in corophioids ($\log ABS = 1.49$) and least in stegocephaloids ($\log ABS = 1.02$). For a

Table 6. Lifespan (in mo.), lifetime potential fecundity (in embryos) and reproductive potential (in embryos female^{-1} yr^{-1}) of gammaridean amphipods in different habitats and superfamilies. A Kruskal-Wallis test was used to compare groups with ≥ 5 observations; sample size is given in parentheses. See Table 5 for abbreviations.

Groups	Lifespan	Lifetime fecundity	Reproductive potential
Temperature			
Cold	19.6 ± 17.4 (102)	87.7 ± 114.0 (98)	66.9 ± 122.4 (75)
Warm	11.8 ± 6.3 (30)	119.3 ± 144.2 (14)	355.1 ± 342.5 (7)
KW	6.47*	0.03*ns*	10.10**
Bathymetry			
Deep	156.0 (1)	83.6 ± 19.2 (2)	–
Shallow	16.8 ± 10.3 (131)	91.8 ± 119.0 (110)	91.5 ± 170.1 (78)
KW	–	–	–
Salinity			
Freshwater	16.5 ± 6.8 (15)	119.1 ± 118.5 (15)	121.1 ± 253.2 (12)
Brackish	12.6 ± 4.9 (25)	145.8 ± 166.4 (21)	161.0 ± 231.3 (15)
Marine	19.5 ± 18.4 (92)	71.2 ± 95.9 (76)	66.0 ± 119.5 (55)
KW	2.93*ns*	12.97**	10.79**
Superfamily			
AMP	17.6 ± 9.9 (18)	33.6 ± 30.7 (12)	27.1 ± 12.0 (9)
COR	10.3 ± 6.8 (16)	71.6 ± 65.7 (9)	162.6 ± 222.4 (8)
EUS	31.0 ± 18.0 (6)	107.6 ± 54.7 (6)	42.3 ± 13.1 (4)
GAM	15.4 ± 9.7 (31)	167.7 ± 171.0 (25)	157.0 ± 226.1 (21)
HAD	17.5 ± 17.7 (2)	151.1 ± 183.6 (3)	445.6 ± 601.8 (2)
LYS	30.1 ± 32.6 (19)	81.8 ± 132.6 (20)	43.3 ± 48.9 (15)
PON	18.8 ± 5.9 (20)	34.2 ± 35.5 (16)	14.1 ± 14.5 (10)
TAL	11.0 ± 4.0 (10)	95.4 ± 63.0 (10)	61.2 ± 44.5 (4)
KW	32.15***	37.35***	32.81***

given body size, corophioids produced more embryos per brood than all other gammaridean superfamilies, with the exception of the Eusiroidea. The Ampeliscoidea, Eusiroidea and Gammaroidea, tended to have larger adjusted mean brood sizes than the Pontoporeioidea, Lysianassoidea and Stegocephaloidea.

Regressions of $\log_{10}$ embryo diameter and $\log_{10}$ body length were significant for the Ampeliscoidea (slope = 0.37, $N = 12$, $P < 0.05$), Corophioidea (slope = 0.25, $N = 19$, $P < 0.01$), Eusiroidea (slope = 0.57, $N = 12$, $P < 0.001$), Gammaroidea (slope = 0.47, $N = 33$, $P < 0.001$), Hadzioidea (slope = 0.30, $N = 9$, $P < 0.01$) and

Table 7. Predictive simple and multiple regressions of brood size on body length (*BL*, in mm), embryo diameter (*ED*, in mm), *HMFBL* ratio (*HMFBLr*), number of broods per female (*NB*) and lifespan (*LS*, in mo.) for all gammaridean populations. All coefficients in the multiple regressions are significant at the 1% level or less. *** = $P < 0.001$.

Equation of regression	N	r^2	F
$\log BS = -0.328 \log HMFBLr + 1.087$	261	0.04	12.02***
$\log BS = -0.343 \log NB + 1.462$	112	0.10	12.85***
$\log BS = 1.162 \log ED + 1.625$	151	0.25	48.54***
$\log BS = 1.221 \log BL + 0.152$	302	0.59	433.02***
$\log BS = 1.686 \log BL - 0.944 \log ED - 0.585$	152	0.67	152.05***
$\log BS = 1.714 \log BL - 1.087 \log ED - 0.200 \log HMFBLr - 0.772$	126	0.70	94.34***
$\log BS = 1.480 \log BL - 0.546 \log ED - 0.411 \log NB - 0.138$	71	0.71	55.03***

Table 8. Predictive regressions of brood size (*BS*) and of embryo diameter (*ED*, in mm) on body length (*BL*, in mm) for gammaridean populations in different habitats. Significant regression lines are compared by analysis of variance and of covariance. *** = $P < 0.001$, ** = $P < 0.01$, * = $P < 0.05$, *ns* = not significant.

Groups	Brood size	Embryo diameter
Temperature		
Warm	$\log BS = 1.326\log BL + 0.086$	$\log ED = 0.157\log BL - 0.547$
	$N = 87, r^2 = 0.49, F = 82.91{***}$	$N = 34, r^2 = 0.17, F = 6.72{*}$
Cold	$\log BS = 1.218\log BL + 0.147$	$\log ED = 0.539\log BL - 0.785$
	$N = 215, r^2 = 0.60, F = 313.66{***}$	$N = 118, r^2 = 0.67, F = 232.86{***}$
Slopes	$F = 0.38ns$	$F = 17.08{***}$
Elevation	$F = 0.50ns$	–
Bathymetry		
Shallow	$\log BS = 1.348\log BL + 0.050$	$\log ED = 0.449\log BL - 0.717$
	$N = 291, r^2 = 0.62, F = 464.89{***}$	$N = 144, r^2 = 0.57, F = 184.47{***}$
Deep	$\log BS = 0.937\log BL + 0.171$	$\log ED = 0.695\log BL - 0.919$
	$N = 11, r^2 = 0.73, F = 24.54{***}$	$N = 8, r^2 = 0.81, F = 25.33{**}$
Slopes	$F = 5.15{*}$	$F = 8.51{**}$
Salinity		
Freshwater	$\log BS = 0.832\log BL + 0.562$	$\log ED = 0.433\log BL - 0.783$
	$N = 27, r^2 = 0.21, F = 6.54{*}$	$N = 9, r^2 = 0.26, F = 2.52ns$
Brackish	$\log BS = 0.973\log BL + 0.364$	$\log ED = 0.326\log BL - 0.591$
	$N = 46, r^2 = 0.44, F = 34.91{***}$	$N = 19, r^2 = 0.52, F = 18.47{***}$
Marine	$\log BS = 1.261\log BL + 0.112$	$\log ED = 0.550\log BL - 0.802$
	$N = 229, r^2 = 0.62, F = 377.40{***}$	$N = 124, r^2 = 0.69, F = 277.56{***}$
Slopes	$F = 1.63ns$	$F = 5.80{*}$
Elevator	$F = 0.53ns$	–

Lysianassoidea (slope = 0.65, $N = 23$, $P < 0.001$). Slopes differed significantly ($F = 4.91$, $P < 0.001$).

Mean values of the lifetime potential fecundity and reproductive potential for gammaridean populations in different habitats and superfami-

Table 9. Predictive regressions of brood size (*BS*) on body length (*BL*, in mm) for superfamilies with significant relations comprising ≥ 5 observations. The predictive regression for the Talitroidea was not significant despite the large sample size.

	Equation of regression	N	r^2	F
Ampeliscoidea	$\log BS = 1.335\log BL + 0.089$	24	0.49	21.30***
Corophioidea	$\log BS = 1.465\log BL + 0.141$	38	0.61	55.83***
Eusiroidea	$\log BS = 0.866\log BL + 0.668$	28	0.28	9.97**
Gammaroidea	$\log BS = 1.490\log BL + 0.084$	61	0.59	83.81***
Hadzioidea	$\log BS = 1.263\log BL + 0.073$	12	0.74	29.18***
Leucothoidea	$\log BS = 1.439\log BL - 0.118$	6	0.70	9.30*
Lysianassoidea	$\log BS = 1.062\log BL + 0.204$	35	0.55	40.54***
Oedicerotoidea	$\log BS = 1.978\log BL - 0.446$	8	0.69	13.67*
Pontoporeioidea	$\log BS = 1.806\log BL - 0.375$	42	0.72	103.79***
Stegocephaloidea	$\log BS = 1.010\log BL + 0.132$	5	0.88	21.61*
Slopes	$F = 1.83$ ($P = 0.06$)			
Elevation	$F = 4.58{***}$			
Talitroidea	$\log BS = 0.288\log BL + 0.904$	26	0.04	1.08ns

Table 10. Probability levels for multiple comparisons of logarithmic estimates of brood size adjusted for body length of females (*ABS*) in gammaridean superfamilies with significant predictive regressions and ≥ 5 observations (see Table 9). The boxed area encloses pairs which differ or tend to differ significantly. AMP = Ampeliscoidea, COR = Corophioidea, EUS = Eusiroidea, GAM = Gammaroidea, HAD = Hadzioidea, LEU = Leucothoidea, LYS = Lysianassoidea, OED = Oedicerotoidea, PON = Pontoporeioidea, STE = Stegocephaloidea.

	log*ABS*	COR	EUS	AMP	GAM	OED	HAD	PON	LEU	LYS	STE
COR	1.49	–									
EUS	1.40	0.19	–								
AMP	1.35	0.05	0.58	–							
GAM	1.35	0.02	0.44	0.93	–						
OED	1.32	0.09	0.47	0.72	0.75	–					
HAD	1.27	0.01	0.19	0.38	0.37	0.72	–				
PON	1.23	0.00	0.02	0.07	0.04	0.40	0.63	–			
LEU	1.19	0.01	0.12	0.21	0.20	0.42	0.59	0.79	–		
LYS	1.17	0.00	0.00	0.01	0.00	0.20	0.30	0.43	0.86	–	
STE	1.02	0.00	0.00	0.01	0.01	0.06	0.09	0.12	0.31	0.25	–

lies are shown in Table 6. Univariate analyses indicated that reproductive potential, but not lifetime potential fecundity, was significantly greater in warm than in cold waters. Regardless of the index used, gammaridean populations were significantly more fecund in brackish and fresh waters than in marine environments. Nothing can be said about the reproductive potential of deep-living populations, because data on the lifespan and number of broods for the calculation of reproductive potential were too few. Lifetime potential fecundity and reproductive potential varied significantly with superfamily affiliation: pontoporeioids and ampeliscoids were the least fecund of gammaridean amphipods according to either index, gammaroids and eusiroids had the highest lifetime potential fecundity, and corophioids and gammaroids had the greatest reproductive potential.

Discussion

Correlations among life history traits

Simple correlations indicated that all life history traits were intercorrelated for populations of aquatic gammarideans (Table 2). A novel, although intuitively obvious finding, was the significant positive correlation between the maximum number of broods produced per female and the *HMFBL* ratio. This occurs in iteroparous females because each oviposition is preceded by one or more moults (Charniaux-Cotton, 1985), which are generally accompanied by an increment in body length. The coefficient of the correlation between the number of broods per female and the *HMFBL* ratio was rather small ($r = 0.64$), indicating scatter about the mean trend. The unexplained variance may result from several factors, other than obvious imprecisions on measurements of body length. Alternating generations of females in multivoltine populations may mature at different body lengths (e.g. Mills, 1967; Nelson, 1980; Van Dolah & Bird, 1980), yielding excessively large *HMFBL* ratios when mean and maximum body length are considered irrespectively of specific generations. Also, in a given population, semelparous or iteroparous females belonging to a same generation may mature asynchronously and in different moult instars (Nair & Anger, 1979a; Sagar, 1980; Sainte-Marie & Brunel, 1983; see Caine, 1979 for a caprellid example), iteroparous females may oviposit irregularly in the instars following the initial moult to maturity (e.g. Legueux, 1926; Sexton, 1928; Nair & Anger, 1979a), or their breeding life may be punctuated by obligate resting stages, accompanied by growth (e.g. Kinne, 1953a; V.J. Steele, 1967; Morino, 1978).

Partial correlations (Table 2) offer a different perspective into relationships between life history traits, because they consider relations between two variables while removing the effects of other variables. Brood size and embryo diameter were positively and strongly correlated with body length; this has long been known for malacostracan crustaceans (e.g. Cheng, 1942; Jensen, 1958; Steele & Steele, 1975c; Mauchline, 1988b). Brood size and embryo diameter were negatively correlated with each other: for a given female body size and number of broods, production of large broods thus entails smaller embryos, and *vice versa*. The negative relation between brood and embryo size, given a fixed body size, had previously been reported for gammarideans (e.g. Steele & Steele, 1975c; Van Dolah & Bird, 1980) and for a variety of other crustaceans (Kerfoot, 1977; Clarke, 1979; Corey, 1981; Mauchline, 1988b).

Partial correlations pointed to other important compromises which may occur between reproductive traits. *HMFBL* and the number of broods produced per female were negatively correlated with embryo diameter (Table 2) and with brood size. This means that for a given body size, females breeding more frequently produce fewer and/or smaller embryos than females breeding less frequently. Hence, considerable variation of the reproductive output (measured as the ratio of brood volume or weight to female body length or

volume) may be expected between gammaridean populations and species, as demonstrated for a few species (Wildish, 1982; Clarke *et al.*, 1985). Nelson (1980) reported that semelparous females produced more embryos per brood than iteroparous females, but the former were also larger, and the effect of body length was not accounted for by ANCOVA. Implications of repeat-breeding for embryo size were hitherto unrecognized. The strong negative partial correlations among several gammaridean life history traits imply that there are limits on directional selection for any given trait (Doyle & Hunte, 1981; Skadsheim, 1990).

The reduction in brood and/or embryo size with increasing frequency of reproduction may be due to physical and/or physiological constraints. In iteroparous species, ova of the maturing brood 'compete' for limited body cavity space and nutritional reserves with underlying germinal tissues, containing subsequent broods in a more or less developed state, and with the gut which generally remains functional (Sainte-Marie *et al.*, 1990). In contrast, the body cavity of some semelparous species may be invaded by the ovaries (Bregazzi, 1972), completely constricting the gut and causing the ventral sternites to bulge outwards (Sainte-Marie *et al.*, 1990). Additional constraints on brood and embryo size are marsupium capacity, which is somewhat expandable (Sainte-Marie *et al.*, 1990), and the probable existence of a minimum viable embryo size (Mauchline, 1988b). Ultimately, brood and embryo size are limited by the amount of energy devoted by the female to reproduction, which depends on growth and maintenance costs (Clarke, 1987).

Based on an unspecified number of observations, Ingram & Hessler (1987) proposed a general predictive equation for brood size of gammaridean amphipods, which used body length as an independent variable. Their equation diverges sharply from mine (Table 7), notably because their data were not log-transformed prior to regression analysis. At any rate, such general predictive equations are of rather limited value, considering the significant variations of regressions of brood size on body length across some habitats and superfamilies (Tables 8, 9 and 10).

Among the superfamilies which were represented by a large number of observations, only the Talitroidea yielded a non significant regression size on body length. The extreme diversity of habitats occupied by this superfamily – terrestrial, supralittoral, fully aquatic; cold and warm; fresh water, brackish and marine (see Friend & Richardson, 1986; Wildish, 1988) – may result in species having very different life history patterns.

The strong partial correlations among life history traits observed herein imply that more than body size is necessary for good predictions of brood size of gammaridean amphipods. The best general predictive regression was a multiple regression, with body length, embryo diameter and the *HMFBL* ratio or number of broods per female, as independent variables (Table 7) which, in that order, explained decreasing proportions of variation in brood size.

Comparisons of gammaridean life history patterns

Most previous reviews of peracaridan life history patterns have focused only on brood size, body length and, to a lesser extent, embryo diameter (Steele & Steele, 1975c; Nelson, 1980; Van Dolah & Bird, 1980; Corey, 1981; Luxmoore, 1982). However, these variables were not the only, and certainly not always the most important, traits characterizing life history patterns of aquatic gammaridean amphipods. For example, the *HMFBL* ratio (i.e. number of broods) contributed more than brood size to the separatory power of the first discriminant function, in two out of three canonical discriminant analyses (Table 4).

Different life history patterns are often interpreted as 'strategies' or 'tactics', resulting from natural selection on covarying life history traits, to adjust fecundity so that some measure of individual fitness is maximized (Williams, 1966). Adaptationists therefore commonly infer mortality rates from fecundity. Nelson (1980) and Van Dolah & Bird (1980) measured gammaridean fecundity as brood size, standardized or not for body length, and hypothesized that adult mortality is greater in species with larger broods.

However, as argued above, and in light of the observed differences between number of broods produced by females in different habitats or superfamilies, brood size alone is a very poor index of total offspring production and hence, in the adaptationist scheme, of mortality rates. The most desirable index of fecundity will depend on the purpose of the study (Wildish, 1982): lifetime potential fecundity is a good overall indicator, but reproductive potential, which actually is a fecundity rate, seems most suitable for the inference of mortality rates. Indices of fecundity are not interchangeable: one is led to strikingly different conclusions depending on which index is used. For instance, mean brood size was significantly greater in gammaridean females of cold- than of warm-water habitats, but the reverse was true of reproductive potential (Table 6), while regressions of brood size on body length (Table 8) and lifetime potential fecundity (Table 6) did not differ significantly.

In this review, gammaridean superfamilies were represented inconsistently across habitats. Hence, there may be confounding or interactive effects within and between habitats and superfamilies (e.g. Brown, 1983); for example, most deep-living species reviewed herein were lysianassoids. This observation is important for the discussion of habitat and superfamily effects on life history traits.

Habitat harshness and life history patterns

Patterns of geographical and ecological variation of life history traits (mostly 'clutch' size) have been recorded for a great variety of animal taxa, either at the intra- or interspecific level (e.g. Bagenal, 1966; Steele & Steele, 1975c; Clarke, 1979; Kaplan & Salthe, 1979; Ricklefs, 1980; Van Dolah & Bird, 1980; Berven, 1982; Koenig, 1984; Healey & Heard, 1984; Belk *et al.*, 1990). Various hypotheses have been put forth to account for these trends in life history patterns; some of the most recent and most widely considered hypotheses link them to some index of habitat harshness: for instance, seasonality of food resources ('Ashmole's hypothesis': Ashmole, 1963; Ricklefs, 1980), stability ('bethedging': Schaffer, 1974), or predictability ('r-K selection': MacArthur & Wilson, 1967; Pianka, 1970; redefined, re-interpreted and expanded as 'r-K-adversity selection': Greenslade, 1972; 1983). The habitat would thus provide 'the templet on which evolution forges characteristic life-history strategies' (Southwood, 1977; 1988). There is some agreement about the habitat characteristics which are of evolutionary significance to the organism's life history: the first is the frequency of disturbance and the second is productivity or environmental adversity (e.g. Southwood, 1977; 1988; Hildrew & Townsend, 1987; Greenslade, 1983).

The r-K-A selection hypothesis predicts that animals will tend to be K-selected in highly productive (favourable) and rarely disturbed (predictable) habitats, A-selected in poorly productive (adverse) and rarely disturbed habitats, and r-selected in frequently disturbed (unpredictable) habitats (Greenslade, 1983). Following the Greenslade (1983) and Southwood (1988) habitat templets, and with respect to criteria used to distinguish habitats of gammaridean amphipods in my review, the following generalizations seem possible. High latitudes and the deep sea are adverse but predictable habitats, thus A-selected (see Greenslade's 1983 re-interpretation of Clarke's 1979 data); brackish habitats are very unpredictable (Southwood, 1988), thus r-selected. The predicted species attributes are: for K-selection, intermediate longevity, intermediate maturity and intermediate fecundity; for A-selection, great longevity, late maturity and low fecundity; for r-selection, short longevity, early maturity and high fecundity (Greenslade, 1983).

Latitude

High-latitude (cold-water) gammaridean amphipods were characterized in general by univoltinism, large body size, delayed maturity, great longevity, large embryos, and few broods in a lifetime. The opposite set of traits tended to typify low-latitude (warm-water) populations. Reproductive potential was much greater in the latter only because of the greater number of broods and shorter lifespan of females (i.e. increased reproductive tempo). These general observations support and extend previous observations and predictions on the latitudinal distribution of gammaridean life history patterns (Morino, 1978; Wildish, 1982, 1988; Bellan-Santini & Dauvin, 1988).

There was no significant difference between regressions of brood size on body length for high- and low-latitude gammarideans. However, the slope of the regression of embryo diameter on body length was significantly greater in cold- than in warm-living populations (Table 6). These conclusions are robust, since restricting analyses to consider only shallow-living, marine populations gives the same results (Sainte-Marie, pers. observ.). They imply that brood volume was greater in high-latitude than in low-latitude gammarideans, for any given body size. Steele & Steele (1975c) noted that the ratio between brood volume and parent volume tended to be greater in polar than in boreal or temperate populations of the Gammaridae and of the eusiroid *Calliopius laeviusculus*. In light of my findings, the proximate reason for this trend seems be the greater number of broods produced by females of warm- versus cold-water species, which entails a reduction of relative embryo size and, ultimately, of brood volume. A reduction in the number of broods carried per female with increasing latitude occurs also at the intraspecific level, in at least one lysianassoid species (Sainte-Marie, unpubl. observ.), in the ampeliscoids *Ampelisca abdita* and *A. brevicornis*, and possibly in the pontoporeioid *Amphiporeia lawrenciana* (Appendix 1). Such latitudinal trends may be related to the poorly understood interactions between investments into maintenance, growth and reproduction (Clarke, 1987).

The greater longevity, later maturity and smaller reproductive potential of high- compared to low-latitude gammaridean species is consistent with A-selection. However, fecundity, measured as brood size, brood size adjusted for body length or lifetime potential fecundity, was significantly greater or no less in high- than in low-latitude gammarideans. These latter observations, essentially based on northern hemisphere gammarideans, contrast with reports of reduced 'clutches' of antarctic relative to temperate benthos (reviewed by Clarke, 1979). It is possible that life history patterns of the antarctic fauna differ from those of the arctic fauna. However, Thurston (1974) was unable to detect any difference between embryo size of antarctic and arctic or boreal gammarideans. Moreover, there is no significant difference between regressions of embryo diameter or brood size on body length for antarctic and arctic gammaridean populations reviewed herein (Sainte-Marie, pers. obser.), which were mostly eusiroids and lysianassoids at both poles (raw data in Schellensberg, 1926; Stephensen, 1923, 1944; MacGinitie, 1955; Kuznetsov, 1964; Bregazzi, 1972; Rakusa-Suszczewski, 1972, 1982; Thurston, 1974; Sagar, 1980).

For a given body size, the constant brood size with increasing latitude (herein) also contrasts with many reported intraspecific trends of decreasing peracarid brood size with increasing latitude (D.H. Steele, 1967; Van Dolah & Bird, 1980; Wägele, 1987) or decreasing ambient temperature (summer versus winter populations, see for example Hynes, 1955; Heller, 1968; Vlasblom, 1969; Chambers, 1977; Van Dolah & Bird, 1980; Kolding & Fenchel, 1981; Moore, 1981; Sheader, 1983; Skadsheim, 1984; Naylor et al., 1988). There are however exceptions to this rule, where brood size increases with temperature (Dexter, 1971; Fish, 1975). The reason for this and the above discrepancies may be rooted in the extreme diversity of gammaridean species, superfamilies and/or ecological types (see below), and of their unequal representation with latitude. Larger broods, for a given body size, are expected of the mainly semelparous eusiroids and lysianassoids which dominate high-latitude populations in my database. Phylogenetic constraints may therefore override latitudinal variation in life history traits, as demonstrated by Stebbins (1989) for a northern hemisphere isopod genus which shows a smaller-north trend in female body size, in contrast to the bigger-north trend apparent in other isopod genera.

Many benthic animals apparently synchronize brood release with optimum conditions for survival (e.g. Thorson, 1950; Todd & Doyle, 1981). Semelparity or reduced frequency of breeding, decreased voltinism or univoltinism, and possibly increased longevity of gammarideans at high latitudes, are almost certainly related to the extreme seasonality of resources. The slow development of large embryos (McLaren, 1966; Steele & Steele, 1973b; Wear, 1974; Todd & Doyle, 1981), particularly at cold temperatures (Wittmann, 1984), along with obligate resting stages (V.J. Steele, 1967; Bone, 1972), increase the time required to produce a brood and thus contribute to synchronize offspring release with the yearly pulse in productivity. It has also been argued that reduced growth rates, longer inter-brood periods or delayed sexual maturity may be due to severe (seasonal) food limitation (e.g. Clarke, 1980, 1983, 1987; Siegfried, 1985). Physiological constraints and/or food limitation may thus play a key rôle in the evolution of life history traits of high-latitude animals (e.g. Thorson, 1950; Vance, 1973; Luxmoore, 1982; Wildish, 1982).

Depth

Apparently, deep-living gammaridean populations were characterized mainly by their large body size (Table 4). But this conclusion hinges on data for a few extremely big deep-sea lysianassoids (*Alicella gigantea*, *Eurythenes gryllus*, *Hirondellea gigas*), which have been the focus of research because of their peculiar ecology and presumed importance in the deep sea (e.g. Hessler et al., 1978; Smith & Baldwin, 1984; Ingram & Hessler, 1987). One may hence argue that they are not representative of deep-sea gammarideans in general. Comparisons of body size of all gammarideans (Barnard, 1962) and of lysianassoids in particular (Steele, 1983), across depth and latitudinal gradients, indeed suggest that deep-sea assemblages differ from shallow, cold-water assemblages only in the presence of a few extremely large members.

Although females of deep-sea gammarideans were on average 4.5 times larger than their shallow-living counterparts, brood size was not significantly greater (Table 5). Sainte-Marie *et al.* (1990) compared regressions of brood size and embryo diameter on body length and concluded that the rate of increase of brood size relative to body length was greater in shallow- than in deep-living lysianassoids, while the rate of increase of embryo size was similar for both groups. A broader analysis of all shallow- and deep-living gammarideans indicated that the regression between brood size and body length was identical in both groups, but that the rate of increase of embryo size relative to body length was greater in deep-living gammarideans (Table 8). It is unlikely that the smaller relative size of broods of deep-living gammarideans results from a greater frequency of breeding, given the significantly smaller mean *HMFBL* ratio of deep- compared to shallow-living Gammaridea or Lysianassoidea (Table 5).

One can only speculate about the reproductive potential of deep-living lysianassoids, because information on the number of broods and/or lifespan were lacking. Reproductive potential may be small, because the *HMFBL* ratio is less in deep- than in shallow-living populations of gammarideans (and of lysianassoids in particular). However, for the abyssal lysianassoid *Eurythenes gryllus*, Ingram & Hessler (1987) inferred a lifespan of 156 mo. and 5 potential brooding instars (XV, XVII, XIX, XXI and XXIII), which would yield a maximum reproductive potential of 74.6 embryos female^{-1} yr^{-1}. This value is similar to the mean value of 91.5 embryos female^{-1} yr^{-1} for shallow-living gammarideans (Table 6) and greater than the mean value of 43.3 embryos female^{-1} yr^{-1} for cold-living lysianassoids.

Deep-sea lysianassoids seem to be A-selected. Nutrient limitation is regarded as an important factor structuring deep-sea communities (e.g. Stockton & DeLaca, 1982) and life history traits of deep-living invertebrates (Thorson, 1950; Mileikovsky, 1971; Vance, 1973). The large body size of deep-sea lysianassoids may be an adaptation to efficiently forage on unpredictable and ephemeral carrion over large areas of bottom (Sainte-Marie, 1986). At the same time, nutrient limitation may be selecting for small body size in suspension and detritus feeders (Thiel, 1979; Carney *et al.*, 1983), with yet unknown effects on the life history traits of detritivorous gammarideans.

Salinity

Univariate comparisons of life history traits, of lifetime potential fecundity and of reproductive potential across salinity gradients pointed only to differences in *HMFBL* ratios, number of broods, lifetime potential fecundity and reproductive potential, with greater values of these variables occurring in brackish water (Tables 5 and 6). Regression analysis indicated that the rate of size increase of embryos relative to body length was greater in marine than in brackish gammarideans (Table 8); this is no doubt related to the greater breeding frequency observed in the latter. These results differ markedly from those of Nelson (1980), who reported that female body and brood size were greater in brackish than in marine gammaridean species in general. The significantly greater lifetime potential fecundity and reproductive potential (but not brood size) of gammaridean populations in brackish waters, and their tendency to be shorter-lived, is consistent with r-selection.

The present comparisons may have been biased because gammaroids represented $\geq 42\%$ of observations for any given trait in fresh-water populations, $\geq 33\%$ of observations in brackish waters, but $\leq 18\%$ of observations in marine waters. Considering only the Gammaroidea, there was no significant difference between any of the life history traits, lifetime potential fecundity or reproductive potential (Sainte-Marie, pers. observ.), but the number of observations was small. At the intraspecific level, there exists empirical or experimental evidence for the effect of salinity on life history traits of the Gammaroidea (Hynes, 1954; Pinkster & Broodbakker, 1980; Skadsheim, 1989) and Talitroidea (Wildish, 1970, 1982).

Superfamily and ecological habit

Life history patterns differed sharply among gammaridean superfamilies. Lysianassoids and eusiroids had fewer broods, larger embryos and a lower reproductive potential than gammaroids and corophioids (Fig. 3, Tables 5 and 6). These two latter superfamilies were unique in retaining iteroparity even at high latitudes, a trait which enhances reproductive potential and lifetime potential fecundity. These same life history patterns may also be evidenced by a restricted analysis of only cold water, shallow-living species (Sainte-Marie, pers. obser.).

The relatively clear separation of gammaridean superfamilies should not be interpreted as a sign of extreme uniformity of life history traits for species of a given superfamily. Ecological habits are quite similar among species of the Gammaroidea or Pontoporeioidea, but other superfamilies are remarkably heterogeneous. For instance, the Lysianassoidea comprise predators, carrion-feeders, omnivorous scavengers, detritivores, herbivores, associated species (commensals and parasites), as well as typically endobenthic, epibenthic and suprabenthic forms (Besner, 1976; Bousfield, 1983: Sainte-Marie & Brunel, 1985). These different lifestyles may be correlated with very different life history traits. The below-average body size of some corophioids (e.g. *Gammaropsis inaequistylis*) and lysianassoids (e.g. *Opisa tridentata* and *Euonyx chelatus*), and of many leucothoids, coincides with their known commensal or symbiotic associations with macrofauna (Vader, 1983; Steele *et al.*, 1986; Bousfield, 1987; Comely & Ansell, 1988). Compared to free-living isopods, commensal isopods are smaller in order to fit on their host, shorter- or longer-lived depending on their host's life expectancy, and less fecund, presumably reflecting the reduced risks associated with a commensal lifestyle (Marsden, 1982; Stebbins, 1989). The small body size of most pontoporeioids, and of some lysianassoids and lilljeborgioids, may result from their endobenthic lifestyle (Nelson, 1980) and microphagous feeding habits; the large body size of some eusiroids, lysianassoids and oedicerotoids may relate to their long-range, suprabenthic foraging capabilities and to their predaceous or necrophageous feeding habits (Sainte-Marie & Brunel, 1985).

Nelson (1980) and Van Dolah & Bird (1980) compared some life history traits of so-called endo- and epibenthic gammarideans, irrespective of superfamily affiliation. Their conclusions were the following: on average, 'epibenthic' females were larger than 'endobenthic' females (Nelson, 1980); mean brood size as well as the ratio of brood size to body length (Nelson, 1980), and brood size adjusted for body length (Van Dolah & Bird, 1980), were greater in 'epi-' than in 'endobenthic' females; mean size of embryos was greater in 'endo-' than in 'epibenthic' species (Van Dolah & Bird, 1980). It was hypothesized that the numerous embryos in broods of 'epibenthic' females reflected a greater mortality risk to adults, due to predation (see also Wildish, 1982) and an inclement environment, and that the large embryos of 'endobenthic' females represented an adaptation to maximize survivorship of juveniles facing severe predation or competition for resources (see also Smith & Fretwell, 1974; Stearns, 1976; Kerfoot, 1977; Todd & Doyle, 1981).

Both Nelson's and Van Dolah & Bird's reviews contained several common flaws, some of which were pointed out by Fenwick (1984). Firstly, brood size was used as an index of fecundity. This approach skirts the problem of iteroparity and longevity, both of which must be taken into account if one attempts to relate fecundity to risks of adult mortality (Wildish, 1982). Secondly, the majority of species and populations belonged to the 'epibenthic' Gammaroidea and to the 'endobenthic' Pontoporeioidea (72% of reviewed populations in Nelson, 1980; 80% in Van Dolah & Bird, 1980); thus comparisons of 'epi-' and 'endobenthic' species were in essence comparisons between two superfamilies. But the Gammaroidea, whose classification as epibenthic was criticized by Fenwick (1984), differ from the Pontoporeioidea in more than just their supposed 'epibenthic' habit. Gammaroids are essentially littoral amphipods which live in a high-risk environment: exposure at low tide, osmotic stress and temperature shocks are common threats. Moreover, the Gammaroidea may be unique among the Gammaridea in having a lengthy precopula during which the male carries the female (Kinne, 1954; Borowsky, 1984; Borowsky & Borowsky, 1987); this requires the males to be large (robust) relative to females, a characteristic which is shared fully by no other superfamily reviewed herein (Table 1), and may entail greater vulnerability to predators. The greater fecundity of Gammaroidea may thus be simply related to phylogenetic constraints (see Wanntorp et al., 1990) or, if it is indeed adaptive, to their high-risk littoral habitat and/or precopulatory behaviour rather than to their purported 'epibenthic' habit. Finally, the classification scheme in both reviews was clearly deficient: eusiroids were grouped with corophioids and hadzioids into the 'epibenthic' category, but the former are powerful free-swimming forms (Besner, 1976; Bousfield, 1973; Sainte-Marie & Brunel, 1985), while corophioids and hadzioids are generally poor swimmers which live cryptically on hosts, in burrows or in epibenthic tubes (e.g. Enequist, 1949; Kühne & Becker, 1964; Bousfield, 1973; Frith, 1977; Atkinson et al., 1982; Sainte-Marie & Brunel, 1985; Steele et al., 1986). Lysianassoids and oedicerotoids, which were classified into the 'endobenthic' category, include some of the most natant species of gammarideans (e.g. Ingram & Hessler, 1983; Sainte-Marie & Brunel, 1985), as well as truly endobenthic species, but also some poorly mobile or cryptic associated species (Besner, 1976; Bousfield, 1987; Vader, 1983). Considering the diversity of ecological habits within some superfamilies, and the generally deficient state of our knowledge of the autoecology of gammarideans, superfamily or family are dubious criteria for classification of ecological habits.

There is little direct evidence to support the contention that adult mortality risks are greater in epibenthic than in endobenthic species: Nelson (1978, 1979a, 1979b) presented experimental evidence, based on caging experiments with a few species, that so-called epibenthic forms suffered greater mortality than so-called endobenthic forms. Other evidence also suggests that demersal predators, such as fish, may target epibenthic and suprabenthic species or developmental stages (e.g. Richards, 1963; Fincham, 1974; Stoner, 1979; Wakabara et al., 1982; Sainte-Marie & Brunel, 1985; Sudo et al., 1987). However, the differences between several epi- and endobenthic species in terms of exposure to predators are by no means clearcut and may vary during life. The swimming activities of several members of the 'endobenthic' Ampeliscoidea, Lysianassoidea and Oedicerotoidea are only seasonal in nature, as seen for instance in the ephemeral reproductive swarming of the lysianassoid Paratryphosites abyssi (Sainte-Marie & Brunel, 1985), but predators may at that time exact heavy tolls on adult populations before they breed (e.g. >80% mortality for an ampeliscoid, Klein et al., 1975).

Life history traits and size-specific or cohort mortality patterns obtained by Fenwick (1984) for some crustaceans were inconsistent with Van Dolah & Bird's (1980) predicted larger clutches/-smaller embryos and inferred greater adult mortality for 'epi-' compared to 'endobenthic' gammarideans. The data reviewed herein also show that the very epi- or suprabenthic Eusiroidea produced among the largest embryos relative to body size (in text above and see Fig. 3), in sharp contrast to predictions. All endo- and epibenthic species in Fenwick (1984) suffered moderate to high juvenile mortality and generally high adult mortality. A similar pattern has been shown for three Gammarus species (Doyle & Hunte, 1981; Steele & Steele, 1986; Skadsheim, 1990).

Interpretation of life history patterns

Habitat-specific and superfamily-specific life history patterns apparently exist within the aquatic Gammaridea. Across habitats, life history patterns were fairly consistent with r-K-A selection. Most of the variation in life history traits – and of brood size in particular – of gammaridean amphipods may be explained by body size (Tables 7 and 8), as has been demonstrated for other taxonomic groups (e.g. Stearns, 1983, 1984). Variations in the body size of gammarideans have been interpreted as adaptations to directly cope with differential, age-specific mortality rates (e.g. Nelson, 1980). However, other non-adaptive or adaptive interpretations of variations of body size across superfamilies or habitats may be more straightforward and at least as plausible (e.g. reduced body size of parasitic, commensal or deep-living detritivorous forms; gigantism of deep-living predaceous and necrophageous forms; body size constrained by temperature or food limitations). In the absence of proper life tables for the Gammaridea, interpretation of variations in body size, as of body size-independent variations of other life history traits, are thus highly conjectural; these variations may represent a 'strategy' resulting from selection or, more simply, a phylogenetic (historical) constraint (e.g. Brown, 1983; Stearns, 1983, 1984; Fenwick, 1984; Wanntorp et al., 1990) or a phenotypic expression of environmental stress (e.g. Bailey & Mackie, 1986; Clarke et al., 1985; Järvinen, 1986). Clearly, more information on age-specific mortality rates, and on their causes, is needed to reconcile ecological observation with life history theory for gammaridean amphipods.

Conclusions

Life histories of aquatic gammaridean amphipods fall into either of eight categories: semelparous or iteroparous semiannual (both multivoltine), semelparous (univoltine) or iteroparous (multivoltine) annual, semelparous or iteroparous (some multivoltine) biannual, and semelparous or iteroparous perennial (both univoltine). Most gammarideans described so far are of the iteroparous annual type. Semiannual and annual populations, with high reproductive potentials, are more characteristic of low latitude habitats, while annual and perennial gammarideans, with low reproductive potentials, are more frequent at high latitudes and in the deep sea. Exceptions exist, and they probably may be explained in terms of phylogenetic constraints or selection for particular ecological habits.

All life history traits covary, but body size explains most of the variation in brood size and embryo diameter of gammaridean amphipods. There exists four options to increase reproductive potential or lifetime fecundity of gammaridean amphipods: increase body size for a constant breeding frequency, increase brood size by reducing embryo size, increase frequency of breeding for a constant lifespan, or increase longevity for a constant breeding rate. Brood size may be predicted with simple (using body size) or preferably multiple regression equations (using body size, embryo diameter and number of broods), and these predictive functions are very specific to superfamilies and habitats.

Previous reviews of gammaridean life history patterns – excepting those of Wildish (1982, 1988) – have focused singularly on brood size, which is thought to be adaptive and directly proportional to adult mortality. However, brood size alone may be a very poor indicator of total reproductive output, because longevity and the frequency of breeding are not taken into account. The reproductive potential (from Wildish, 1982), which is actually a fecundity rate, may be a more appropriate index from which to infer mortality rates. The reproductive potential varied significantly across superfamilies and habitats, but interpretations remain highly conjectural because information on mortality rates of gammarideans was virtually lacking. Phylogenetic or physiological constraints, and not only selection, may be useful for the interpretation of gammaridean life history patterns.

Acknowledgements

I am grateful to F. Hazel for help at various stages of this work and, in particular, for drafting figures. L. Watling, D.J. Wildish and two anonymous reviewers provided useful comments on an earlier draft. Initial phases of this work were supported by a postdoctorate fellowship from the Québec Ministry of Education (FCAR).

Appendix 1. Reproductive and habitat parameters of aquatic gammaridean amphipods. Body length of females (in mm), brood size (*BS*), lifespan of females (*LS*, in mo), maximum number of broods per female (*NB*), mean embryo diameter (*ED*, in mm); information on temperature (C = cold, W = warm), salinity (B = brackish, F = freshwater, M = marine) and depth (D = deep, S = shallow) are grouped under the heading habitat. Numbers of broods in parentheses were inferred from HMFBL ratios (see text)

Species	Female length		*NE*	*LS*	*NB*	*ED*	Habitat	Authority
	Mean	Max.						
Ampeliscoidea								
Ampelisca abdita	6.6	7.7	26.0	10	1	0.43	C M S	Mills 1967
Ampelisca abdita	5.3	–	13.7	–	1	0.39	W M S	Nelson 1978
Ampelisca abdita	4.9	6.5	22.0	3	>1	–	W M S	Thoemke 1979
Ampelisca armoricana	10.8	12.4	42.4	24	>1	–	W M S	Dauvin 1988d
Ampelisca araucana	4.8	6.3	4.7	8	2	0.45	C M S	Carrasco & Arcos 1984
Ampelisca brevicornis	11.6	15.3	29.5	6	>1	0.48	W M S	Kaim-Malka 1969
Ampelisca brevicornis	13.7	15.4	23.0	15	1	–	C M S	Klein *et al.* 1975
Ampelisca brevicornis	12.8	14.7	37.1	12	1	–	C M S	Hastings 1981
Ampelisca brevicornis	11.5	14.0	32.6	18	>1	–	W M S	Dauvin 1988b
Ampelisca brevicornis	13.6	15.1	45.9	18	>1	–	W M S	Dauvin 1988b
Ampelisca diadema	9.0	11.0	12.3	–	–	0.49	W M S	Ivanov 1961
Ampelisca macrocephala	16.8	19.2	60.0	36	2	0.68	C M S	Kanneworf 1965
Ampelisca sarsi	6.7	7.8	15.6	21	>1	–	W M S	Dauvin 1989
Ampelisca spinipes	16.5	18.0	52.5	–	–	–	W M S	Bellan-Santini & Dauvin 1988
Ampelisca tenuicornis	8.8	10.5	23.6	15	2	0.37	C M S	Sheader 1977
Ampelisca tenuicornis	7.8	9.5	26.1	16	>1	–	W M S	Dauvin 1988a
Ampelisca tenuicornis	8.7	10.5	37.8	16	>1	–	W M S	Dauvin 1988a
Ampelisca typica	8.6	9.6	20.4	16	>1	–	W M S	Dauvin 1988c
Ampelisca vadorum	9.5	11.2	32.1	10	1	0.56	C M S	Mills 1967
Ampelisca vadorum	6.0	7.5	9.1	–	1	0.51	W M S	Van Dolah & Bird 1980
Ampelisca verrilli	10.6	13.7	14.6	–	>1	–	W M S	Thoemke 1979
Haploops fundiensis	7.0	–	5.0	–	1	0.43	C M S	Wildish 1982
Haploops tenuis	8.0	–	35.0	36	1	0.59	C M S	Kanneworf 1966
Haploops tubicola	10.5	11.0	45.0	36	1	0.75	C M S	Kanneworf 1966
Corophioidea								
Ampithoe lacertosa	15.1	25.5	64.0	–	>1	0.46	C M S	Heller 1968
Ampithoe longimana	5.8	–	9.4	–	>1	0.38	W M S	Nelson 1978
Ampithoe ramondi	8.0	11.0	21.0	14	–	0.31	W M S	Gilat 1962
Ampithoe rubricata	16.2	20.0	62.0	–	–	–	C M S	Skutch 1926
Ampithoe rubricata	15.3	20.0	57.5	–	–	–	C M S	Kuznetsov 1964
Ampithoe valida	9.7	13.1	22.0	–	>1	0.42	W M S	Barrett 1966
Chelura terebrans	4.0	6.0	3.7	–	8	0.45	W M S	Kühne & Becker 1964
Corophium acherusicum	4.1	6.0	18.8	–	>1	–	C M S	Onbé 1966
Corophium acherusicum	3.0	–	7.9	–	>1	0.31	W M S	Nelson 1980
Corophium arenarium	5.1	–	14.3	13	>1	–	C B S	Fish & Mills 1979
Corophium bonnellii	3.1	4.8	6.0	10	3	0.36	C M S	Moore 1978 1981
Corophium insidiosum	3.2	3.7	3.9	–	6	0.28	C M S	Sheader 1978
Corophium insidiosum	3.1	4.4	5.7	12	6	0.28	C M S	Sheader 1978
Corophium insidiosum	3.8	5.6	10.8	5	7	0.36	C M S	Nair & Anger 1979a
Coropium sextonae	4.3	5.8	16.3	12	>1	–	W M S	Hughes 1978
Corophium volutator	7.2	–	30.5	13	>1	–	C B S	Fish & Mills 1979
Corophium volutator	7.8	–	46.8	–	>1	–	C M S	Peer *et al.* 1986
Corophium volutator	6.4	–	23.8	–	>1	–	C M S	Peer *et al.* 1986
Cymadusa compta	5.8	–	13.5	–	>1	0.37	W M S	Nelson 1978

Appendix 1. (Continued)

Species	Female length						Habitat	Authority
	Mean	Max.	*NE*	*LS*	*NB*	*ED*		
Cymadusa filosa	8.5	13.0	29.5	15	–	0.39	W M S	Gilat 1962
Dulichia spinosissima	13.5	15.0	227.0	–	–	0.52	C M S	MacGinitie 1955
Erichtonius brasiliensis	5.0	6.5	9.4	12	>1	–	W M S	Hughes 1978
Gammaropsis inaequistylis	5.2	6.6	27.5	–	–	0.23	C M S	Steele *et al.* 1986
Gammaropsis longicornis	3.3	4.0	7.5	–	–	–	C M S	Thurston 1974
Gammaropsis megalops	8.8	14.0	10.5	–	–	0.45	C M S	MacGinitie 1955
Grandidierella bonnieroides	5.4	7.5	28.9	3	>1	–	W M S	Thoemke 1979
Ischyrocerus anguipes	6.8	9.0	75.0	–	–	–	C M S	Kuznetsov 1964
Ischyrocerus anguipes	4.9	6.2	34.5	–	–	–	C M S	Kuznetsov 1964
Jassa falcata	6.7	8.2	9.4	8	4	–	C M S	Nair & Anger 1979b
Jassa falcata	4.6	–	25.6	–	>1	0.29	W M S	McGovern in Nelson 1980
Jassa marmorata	4.9	–	19.2	3	–	–	W M S	Franz 1989
Jassa marmorata	6.2	–	40.3	8	–	–	W M S	Franz 1989
Lembos websteri	4.7	–	8.0	–	>1	0.39	W M S	Nelson 1978
Leptocheirus pilosus	3.5	4.0	11.0	–	–	–	C B S	Goodhart 1939
Leptocheirus pinguis	14.1	17.0	49.8	30	2	0.47	C M S	Wildish 1980
Microdeutopus gryllotalpa	6.3	8.0	21.0	4	11	–	W M S	Myers 1971
Photis reinhardi	5.0	5.5	14.0	–	–	0.50	C M S	MacGinitie 1955
Rudilemboides naglei	3.2	4.3	5.8	3	>1	–	W M S	Thoemke 1979
Crangonyctoidea								
Crangonyx gracilis	8.1	11.0	45.7	–	–	–	C F S	Embody 1911
Crangonyx gracilis	7.8	10.0	33.3	19	5	–	C F S	Hynes 1955
Crangonyx richmondensis	13.0	14.0	23.3	12	1	0.30	C F S	Judd 1963
Crangonyx richmondensis	13.8	18.6	43.0	13	1	0.54	C F S	Sprules 1967
Crangonyx rivularis	6.0	7.0	19.3	–	–	–	C F S	Judd 1963
Niphargus aquilex	4.9	6.5	2.9	8	–	–	C F S	Gledhill & Ladle 1969
Dexaminoidea								
Atylus guttatus	5.5	7.0	26.0	–	–	0.40	W M S	Ivanov 1961
Dexamine spinosa	6.0	–	25.0	–	–	–	W M S	Greze 1963
Eusiroidea								
Atyloella magellanica	18.8	19.0	114.3	–	–	0.67	C M S	Thurston 1974
Bovallia gigantea	45.2	49.0	108.0	54	1	1.34	C M S	Thurston 1968 1970 1974
Calliopius laeviusculus	9.8	13.8	85.0	–	–	–	C M S	Kuznetsov 1964
Calliopius laeviusculus	8.0	12.0	48.5	12	>1	0.45	C M S	Steele DH & Steele 1973a
Calliopius laeviusculus	10.5	12.0	65.2	12	>1	0.58	C M S	Steele DH & Steele 1973a
Calliopius laeviusculus	12.0	12.4	83.3	–	–	–	C M S	Sainte-Marie unpubl.
Djerboa furcipes	18.0	20.0	59.0	–	–	–	C M S	Thurston 1974
Epimeria loricata	25.0	30.0	20.5	–	–	–	C M S	Kuznetsov 1964
Epimeria loricata	26.3	31.0	16.0	–	–	–	C M S	Kuznetsov 1964
Epimeria monodon	24.5	25.0	55.5	–	–	1.27	C M S	Thurston 1974
Eurymera monticulosa	24.3	26.0	96.7	–	–	0.75	C M S	Thurston 1974
Liouvillea oculata	15.5	17.0	39.0	–	–	–	C M S	Thurston 1974
Metaleptamphopus pectinatus	6.0	6.5	9.3	–	–	0.43	C M S	Thurston 1974
Paramoera hurleyi	6.8	7.0	11.0	–	–	–	C M S	Thurston 1974
Paramoera walkeri	13.8	17.0	106.5	24	1	0.55	C M S	Rakusa-Suszczewski 1972
Paramoera walkeri	14.6	15.7	126.0	36	1	–	C M S	Sagar 1980
Paramoera walkeri	16.8	20.2	200.0	48	1	–	C M S	Sagar 1980

215

Appendix 1. (Continued)

| Species | Female length | | | | | | | |
	Mean	Max.	NE	LS	NB	ED	Habitat	Authority
Paramphithoe cuspidata	20.7	25.2	38.0	–	–	–	CM S	Kuznetsov 1964
Pontogeneia antarctica	17.5	19.0	49.0	–	(1)	0.65	CM S	Thurston 1974
Pontogeneia inermis	9.1	9.2	56.3	–	(1)	0.57	CM S	Sainte-Marie unpubl.
Pontogeneia simplex	15.0	16.0	23.0	–	–	–	CM S	Schellenberg 1926
Pontogeniella brevicornis	18.1	21.0	65.8	–	–	0.67	CM S	Thurston 1974
Prostebbingia gracilis	7.9	11.0	24.9	–	–	0.44	CM S	Thurston 1974
Pseudomoera gabrieli	9.5	12.0	41.0	–	–	–	W F S	Smith & Williams 1983
Rhachotropis aculeata	27.7	33.0	62.5	–	–	–	CM S	Kuznetsov 1964
Rhachotropis aculeata	38.0	43.5	236.5	–	–	–	CM S	Kuznetsov 1964
Rhachotropis oculata	12.1	15.0	24.0	–	>1	–	CM S	Desroches 1985
Stenopleura atlantica	5.8	7.5	7.0	–	–	–	CM D	Schellenberg 1926

Gammaroidea

Species	Mean	Max.	NE	LS	NB	ED	Habitat	Authority
Gammaracanthus loricatus	40.0	–	327.0	–	–	–	CM S	Barnard 1959
Gammaracanthus loricatus	46.0	60.0	599.2	–	>1	0.85	CM S	Steele DH & Steele 1976
Gammarellus angulosus	11.5	15.0	21.5	12	>1	0.65	CM S	Steele DH & Steele 1972b
Gammarellus homari	25.7	35.4	127.0	–	>1	1.00	CM S	Kuznetsov 1964, Steele 1972
Gammarus chevreuxi	7.0	9.0	30.0	18	26	–	C B S	Sexton 1928
Gammarus crinicornis	9.5	14.0	33.2	–	–	0.44	W M S	Dumay 1972
Gammarus duebeni	14.5	18.0	36.5	13	6	–	C B S	Kinne 1953a 1953b
Gammarus duebeni	10.5	13.0	17.9	15	–	–	C B S	Hynes 1954
Gammarus duebeni	11.2	15.0	24.0	15	6	0.61	C B S	Hynes 1954, 1955
Gammarus duebeni	10.9	14.0	19.8	15	5	0.65	C F S	Hynes 1954, 1955
Gammarus duebeni	14.0	18.0	25.4	12	>1	0.56	C B S	Steele DH & Steele 1969
Gammarus duebeni	13.5	15.0	40.5	–	>1	–	W B S	Jażdżewski 1973
Gammarus fasciatus	8.7	12.0	22.3	–	12	–	C F S	Embody 1911
Gammarus fasciatus	10.3	15.0	16.6	–	>1	0.46	W F S	Clemens 1950
Gammarus fasciatus	9.5	13.0	29.3	15	4	–	C F S	Hynes 1955
Gammarus finmarchicus	13.5	19.0	21.5	12	>1	0.50	CM S	Steele DH & Steele 1975b
Gammarus inaequicauda	9.5	13.0	36.0	–	–	–	W B S	Jażdżewski 1973
Gammarus lacustris	13.4	16.0	24.3	–	4	–	C F S	Embody 1911
Gammarus lacustris	11.5	14.0	21.8	15	4	–	C F S	Hynes 1955
Gammarus lacustris	10.0	13.0	17.5	24	4	–	C F S	Hynes & Harper 1972
Gammarus lawrencianus	7.0	9.0	19.2	12	>1	0.41	CM S	Steele DH & Steele 1970b
Gammarus lawrencianus	7.0	10.0	21.8	12	>1	0.41	CM S	Steele DH & Steele 1970b
Gammarus locusta	10.8	16.0	62.1	–	–	0.43	CM S	Spooner 1947
Gammarus locusta	19.6	26.0	104.5	–	–	–	CM S	Kuznetsov 1964
Gammarus locusta	23.9	29.5	130.0	–	–	–	CM S	Kuznetsov 1964
Gammarus locusta	10.0	14.0	49.0	–	–	–	W B S	Jażdżewski 1973
Gammarus marinus	14.0	18.0	16.4	11	4	–	C B S	Vlasblom 1969
Gammarus marinus	14.2	17.4	23.3	14	3	–	C B S	Skadsheim 1982
Gammarus mucronatus	6.0	8.0	14.0	–	>1	0.36	CM S	Steele DH & Steele 1975a
Gammarus mucronatus	8.1	10.0	27.4	–	–	–	W M S	Borowsky 1980a
Gammarus mucronatus	5.8	7.0	21.4	–	–	0.28	W M S	Van Dolah & Bird 1980
Gammarus mucronatus	5.5	7.5	16.8	2	>1	0.43	C B S	LaFrance & Ruber 1985
Gammarus mucronatus	8.0	11.5	35.7	8	>1	0.43	C B S	LaFrance & Ruber 1985
Gammarus mucronatus	9.6	15.5	34.6	5	>1	0.42	W B S	Fredette & Diaz 1986
Gammarus obtusatus	11.0	14.0	9.0	–	6	0.55	CM S	Sheader & Chia 1970
Gammarus obtusatus	10.5	14.0	7.0	–	4	0.61	CM S	Steele DH & Steele 1970a
Gammarus obtusatus	11.5	15.0	13.5	12	4	0.65	CM S	Steele DH & Steele 1970a

Appendix 1. (Continued)

Species	Female length						Habitat	Authority
	Mean	Max.	*NE*	*LS*	*NB*	*ED*		
Gammarus oceanicus	15.3	20.0	77.2	48	3	0.55	C M S	Steele VJ & Steele 1972
Gammarus oceanicus	15.5	20.0	37.4	48	3	0.55	C M S	Steele VJ & Steele 1972
Gammarus olivii	6.5	10.0	23.0	10	20	–	W M S	Greze 1972
Gammarus palustris	6.5	9.0	9.2	12	5	0.41	W M S	Van Dolah *et al.* 1975
Gammarus palustris	7.2	10.9	30.5	16	3	–	C B S	Gable & Croker 1977
Gammarus palustris	7.5	8.6	12.0	–	–	–	W B S	Borowsky 1980a
Gammarus pseudolimnaeus	11.5	15.0	39.0	15	3	–	C B S	Hynes & Harper 1972
Gammarus pulex	8.9	12.0	16.0	15	6	–	C F S	Hynes 1955
Gammarus salinus	12.0	14.0	76.5	–	–	–	W F S	Jażdżewski 1973
Gammarus salinus	10.5	14.5	33.0	–	7	0.31	W B S	Kolding & Fenchel 1981
Gammarus setosus	15.0	20.0	24.8	–	>1	0.69	C M S	Steele VJ & Steele 1970
Gammarus setosus	18.0	23.0	36.6	–	>1	0.69	C M S	Steele VJ & Steele 1970
Gammarus stoerensis	6.0	7.0	12.0	–	>1	0.44	C B S	Steele DH & Steele 1975a
Gammarus stoerensis	6.7	7.6	26.9	14	3	–	C B S	Skadsheim 1982
Gammarus subtypicus	11.0	13.8	21.8	–	–	0.37	W M S	Dumay 1972
Gammarus tigrinus	10.0	14.0	30.7	15	>1	0.46	C B S	Steele DH & Steele 1972a
Gammarus tigrinus	6.5	11.6	28.7	6	16	–	C F S	Chambers 1977
Gammarus troglophilus	12.8	14.6	12.0	12	>1	–	W F S	Jenio 1980
Gammarus wilkitzkii	26.5	33.0	159.1	–	>1	0.73	C M S	Barnard 1959, Steele DH & Steele 1975a
Gammarus zaddachi	10.6	13.0	27.8	–	–	0.53	C M S	Spooner 1947
Gammarus zaddachi	9.5	13.0	47.5	–	–	–	W B S	Jażdżewski 1973
Pallasea quadrispinosa	12.0	15.0	57.5	–	–	–	C F S	Samter & Weltner 1904
Pallasea quadrispinosa	11.0	14.4	28.6	24	4	0.58	C F S	Mathisen 1953
Quadrivisio lutzi	4.5	–	3.0	–	–	0.35	W M S	Stephensen 1933
Hadzioidea								
Allomelita pellucida	3.0	6.0	4.0	–	>1	0.48	W B S	Legueux 1926, Stock 1984
Casco bigelowi	15.0	–	40.0	–	1	0.62	C M S	Wildish 1982
Casco bigelowi	19.2	20.7	50.2	30	1	0.62	C M S	Wildish 1980
Elasmopus levis	7.1	9.5	22.3	–	>1	–	W M S	Borowsky 1986
Elasmopus levis	4.3	–	4.7	–	>1	0.40	W M S	Nelson 1978
Melita appendiculata	3.8	–	5.1	–	>1	0.36	W M S	Nelson 1978
Melita celericula	3.3	4.0	3.4	–	–	0.32	W M S	Croker 1971
Melita formosa	24.2	25.4	54.0	–	–	0.70	C M S	MacGinitie 1955
Melita nitida	6.9	8.7	30.0	–	>1	–	W M S	Borowsky 1980a
Melita nitida	5.8	7.0	17.7	–	–	0.30	W M S	Van Dolah & Bird 1980
Melita palmata	10.5	–	8.8	–	–	0.45	W M S	Ivanov 1961
Melita zeylanica	5.0	6.6	16.5	5	22	–	W B S	Krishnan & John 1974
Leucothoidea								
Gitanopsis squamosa	3.6	4.3	5.9	–	–	0.40	C M S	Thurston 1974
Leucothoe spinicarpa	10.7	11.0	20.7	–	–	0.66	C M S	Thurston 1974
Metopa groenlandica	5.0	–	5.0	–	–	–	C M S	Stephensen 1936
Metopa tuberculata	2.9	3.0	2.0	–	–	–	C M S	Schellenberg 1926
Metopelloides micropalpa	4.8	5.2	14.2	–	–	0.49	C M S	Sainte-Marie unpubl.
Probolisca ovata	2.9	3.8	4.5	–	–	–	C M S	Thurston 1974
Lilljeborgoidea								
Seba dubia	3.5	3.8	2.5	–	–	–	C M S	Schellenberg 1926
Seborgia minima	0.9	1.0	1.0	–	>1	0.25	W B S	Bousfield 1970

Appendix 1. (Continued)

| Species | Female length | | | | | | | |
	Mean	Max.	*NE*	*LS*	*NB*	*ED*	Habitat	Authority
Lysianassoidea								
Alicella gigantea	230.0	–	98.0	–	>1	9.11	C M D	Barnard & Ingram 1986
Anonyx nugax	42.0	46.0	630.0	40	(1)	1.30	C M S	Kuznetsov 1964, MacGinitie 1955
Anonyx sarsi	22.4	23.8	105.8	26	1	1.02	C M S	Sainte-Marie *et al.* 1990
Anonyx sarsi	18.6	19.1	84.0	26	1	1.01	C M S	Sainte-Marie unpubl.
Anonyx sp.	30.0	32.0	259.0	–	–	–	C M S	Kuznetsov 1964
Anonyx spp.	17.6	24.0	68.0	–	–	–	C M S	Kuznetsov 1964
Cheirimedon femoratus	13.0	15.0	68.9	48	1	0.67	C M S	Bregazzi 1972 1973
Euonyx chelatus	8.3	9.6	9.0	–	(1)	0.68	C M S	Comely & Ansell 1988
Eurythenes gryllus	112.8	–	194.0	156	>1	2.30	C M D	Ingram & Hessler 1987
Hippomedon kergueleni	11.5	14.5	8.8	48	4	0.74	C M S	Bregazzi 1972 1973
Hippomedon propinquus	10.0	–	10.0	–	>1	0.55	C M S	Stephensen 1923
Hippomedon propinquus	9.7	13.1	8.8	24	5	0.58	C M S	Lamarche & Brunel 1987
Hippomedon propinquus	8.8	10.9	6.5	24	5	0.58	C M S	Lamarche & Brunel 1987
Hippomedon whero	5.3	6.7	3.4	12	8	0.39	C M S	Fenwick 1985
Hirondellea gigas	40.0	43.0	97.2	–	1	0.72	C M D	Hessler *et al.* 1978
Lepidepecreum cingulatum	6.6	7.0	10.8	–	(1)	0.44	C M S	Thurston 1974
Metambasia faeroensis	5.5	6.0	2.0	–	–	0.40	C M D	Stephensen 1923
Onisimus litoralis	16.9	17.8	44.2	24	1	0.88	C B S	Sainte-Marie *et al.* 1990
Opisa tridentata	7.0	9.0	6.6	–	–	–	C M S	Bousfield 1987
Orchomene gerulicorbis	11.6	12.9	35.0	–	2	–	C M D	Thurston 1979
Orchomene plebs	17.8	21.0	28.9	–	–	0.83	C M D	Rakusa-Suszcsewski 1982
Orchomene rossi	41.0	47.0	37.1	–	>1	–	C M D	Stockton 1982
Orchomene rotundifrons	11.0	–	15.0	–	–	–	C M S	Thurston 1974
Orchomenella minuta	5.3	7.0	9.0	–	–	–	C M S	Kuznetsov 1964
Orchomenella minuta	6.2	7.0	22.0	–	–	–	C M S	Kuznetsov 1964
Orchomenella minuta	6.4	8.8	14.7	12	3	0.52	C M S	Sainte-Marie *et al.* 1990
Orchomenella minuta	5.1	5.9	16.5	12	2	0.48	C M S	Sainte-Marie unpubl.
Orchomenella pinguis	5.3	7.0	11.0	12	–	–	C M S	Kuznetsov 1964
Orchomenella pinguis	7.8	10.0	41.5	12	3	0.51	C M S	Sainte-Marie *et al.* 1990
Parawaldeckia stephenseni	4.6	5.5	12.0	–	5	–	W M S	Fincham 1974
Psammonyx nobilis	12.5	17.0	21.5	24	2	–	C M S	Scott & Croker 1976
Psammonyx terranovae	15.2	18.0	17.7	24	2	1.07	C M S	Sainte-Marie *et al.* 1990
Psammonyx terranovae	13.8	15.6	18.6	24	2	1.10	C M S	Sainte-Marie unpubl.
Pseudorchomene coatsi	14.0	16.0	62.0	–	–	–	C M S	Schellenberg 1926
Tmetonyx sp.	15.1	18.0	12.3	–	>1	1.32	C M D	Sainte-Marie unpubl.
Oedicerotoidea								
Exoediceros fossor	5.7	8.3	9.6	–	–	–	W M S	Dexter 1985
Exoediceroides maculosus	5.1	6.6	9.4	–	–	–	W M S	Dexter 1985
Monoculodes edwardsi	6.0	6.9	28.4	–	–	0.28	W M S	Van Dolah & Bird 1980
Oediceros saginatus	16.8	17.9	148.8	–	1	0.94	C M S	Sainte-Marie unpubl.
Patuki roperi	9.0	11.8	6.7	9	11	0.48	C M S	Fenwick 1985
Pontocrates altamarinus	4.0	5.0	4.8	–	–	–	C M S	Fincham 1971
Pontocrates arenarius	3.1	4.2	3.1	–	–	–	C M S	Fincham 1971
Westwoodilla caecula	6.0	–	20.0	–	–	–	C M S	Stephensen 1923
Phoxocephaloidea								
Diogidias littoralis	3.9	4.9	2.7	6	7	0.35	C M S	Fenwick 1985
Paraharpinia rotundifrons	6.5	7.5	14.3	–	(1)	0.54	C M S	Thurston 1974

Appendix 1. (Continued)

Species	Female length						Habitat	Authority
	Mean	Max.	*NE*	*LS*	*NB*	*ED*		
Protophoxus australis	8.6	11.0	19.3	14	13	0.45	C M S	Fenwick 1985
Rhepoxynius abronius	3.8	–	9.1	12	2	0.42	C M S	Slattery 1985
Rhepoxynius abronius	3.6	4.6	9.3	12	2	–	C M S	Kemp *et al.* 1985
Rhepoxynius fatigans	2.8	–	4.0	12	2	0.41	C M S	Slattery 1985
Pontoporeioidea								
Acanthohaustorius sp.	3.9	4.8	5.2	18	–	–	W M S	Croker 1967
Acanthohaustorius millsi	3.8	4.6	4.8	17	1	0.78	C M S	Sameoto 1969b, Van Dolah & Bird 1980
Acanthohaustorius millsi	5.1	6.2	7.8	17	1	–	C M S	Sameoto 1969b
Amphiporeia lawrenciana	10.0	14.0	14.6	–	>1	0.83	C M S	Downer & Steele 1979
Amphiporeia lawrenciana	8.5	9.2	25.6	–	(1)	0.61	C M S	Sainte-Marie unpubl.
Amphiporeia virginiana	5.1	6.2	8.9	12	1	–	C M S	Hager & Croker 1979
Amphiporeia virginiana	5.5	7.0	6.0	–	–	0.27	W M S	Van Dolah & Bird 1980
Bathyporeia elegans	3.5	4.5	3.5	–	–	–	C M S	Fincham 1971
Bathyporeia guilliamsoniana	7.2	–	22.0	–	–	–	W M S	Salvat 1967
Bathyporeia nana	2.4	2.9	2.4	–	–	–	C M S	Fincham 1971
Bathyporeia pelagica	5.6	6.3	7.5	–	>1	–	W M S	Salvat 1967
Bathyporeia pelagica	4.1	4.7	3.7	–	–	–	C M S	Fincham 1971
Bathyporeia pelagica	5.1	–	6.4	–	>1	0.42	C M S	Fish 1975
Bathyporeia pilosa	5.1	6.0	5.0	–	>1	–	W M S	Salvat 1967
Bathyporeia pilosa	4.4	–	6.9	–	>1	0.44	C M S	Fish 1975
Bathyporeia sarsi	5.5	6.4	9.0	–	>1	–	W M S	Salvat 1967
Bumeralius buchalius	5.0	6.0	4.8	–	–	–	W M S	Dexter 1985
Eohaustorius brevicuspus	3.9	5.0	6.9	–	13	–	C M S	Bosworth 1976
Eohaustorius estuarius	3.7	5.0	6.5	–	–	–	C B S	Bosworth 1976
Eohaustorius sencillus	2.8	3.0	6.3	12	1	0.39	C M S	Slattery 1985
Eohaustorius washingtonius	4.6	6.0	11.9	–	–	–	C M S	Bosworth 1976
Haustorius sp.	6.4	7.5	5.6	18	–	–	W M S	Croker 1967
Haustorius arenarius	9.5	13.0	11.8	–	2	–	W M S	Salvat 1967
Haustorius canadensis	6.8	9.2	7.9	–	–	–	C M S	Donn & Croker 1986
Haustorius dytiscus	4.0	4.7	4.0	18	–	–	W M S	Croker 1967
Neohaustorius biarticulatus	5.0	6.5	7.0	16	>1	–	C M S	Sameoto 1969a
Neohaustorius schmitzi	3.2	4.0	3.5	12	–	–	W M S	Croker 1967
Neohaustorius schmitzi	4.5	6.0	4.8	–	–	0.45	W M S	Van Dolah & Bird 1980
Parahaustorius longimerus	6.3	7.5	7.8	18	–	–	W M S	Croker 1967
Parahaustorius longimerus	6.6	8.4	5.3	17	1	–	C M S	Sameoto 1969b
Pontoporeia affinis	8.1	10.0	16.5	27	1	0.53	C F S	Mathisen 1953
Pontoporeia affinis	7.5	9.0	24.0	24	–	–	C F S	Moore 1979
Pontoporeia femorata	11.5	13.0	71.5	–	1	–	C M S	Marguilis 1970
Pontoporeia femorata	11.0	13.0	63.5	36	2	0.45	C M S	Steele DH & Steele 1978
Pontoporeia femorata	12.5	13.7	57.6	18	1	0.41	C M S	Wildish & Peer 1981
Pontoporeia hoyi	6.5	7.0	17.8	12	1	–	C M S	Siegfried 1985
Pontoporeia hoyi	7.0	–	14.2	24	1	–	C M S	Siegfried 1985
Pontoporeia sp.	11.0	12.1	48.3	–	(1)	0.61	C M S	Sainte-Marie unpubl.
Protohaustorius deichmannae	3.8	4.6	3.2	18	>1	–	C M S	Sameoto 1969b
Protohaustorius deichmannae	4.7	6.2	5.7	18	>1	–	C M S	Sameoto 1969b
Urohaustorius metungi	3.9	5.5	10.1	–	–	–	W M S	Dexter 1985
Urothoe brevicornis	6.4	7.9	23.0	–	1	–	W M S	Salvat 1967

Appendix 1. (Continued)

Species	Female length							
	Mean	Max.	*NE*	*LS*	*NB*	*ED*	Habitat	Authority
Stegocephaloidea								
Lafystius morhuanus	5.7	6.8	7.3	–	–	–	C M S	Bousfield 1987
Stegocephalina ingolfi	9.0	–	15.0	–	–	0.60	C M D	Stephensen 1944
Stegocephalus inflatus	31.3	40.4	65.5	–	–	–	C M S	Kuznetsov 1964
Stegocephalus inflatus	21.5	26.0	23.5	–	>1	1.37	C M S	Steele DH 1967
Stegocephalus inflatus	30.0	37.0	32.3	–	>1	1.66	C M S	Steele DH 1967
Synopioidea								
Bruzelia tuberculata	16.0	–	30.0	–	–	0.85	C M D	Stephensen 1923
Talitroidea								
Austrochiltonia australis	6.0	7.9	46.5	9	–	–	C F S	Smith & Williams 1983
Austrochiltonia subtenuis	4.0	6.0	22.1	–	>1	–	W B S	Lim & Williams 1971
Hyale barbicornis	7.5	11.5	13.5	9	–	–	C M S	Hiwatari & Kajihara 1984
Hyale nilssoni	6.2	–	20.7	12	–	–	C B S	McBane & Croker 1984
Hyale nilssoni	7.5	8.2	23.8	12	–	–	C M S	McBane & Croker 1984
Hyale prevosti	6.9	9.0	16.0	–	–	–	C M S	Kuznetsov 1964
Hyale schmidti	4.0	5.0	5.7	–	–	–	W M S	Gilat 1962
Hyalella azteca	4.3	–	8.0	–	>1	0.35	C F S	Strong 1972
Hyalella azteca	5.9	–	18.0	–	>1	0.29	C F S	Strong 1972
Hyalella azteca	6.7	–	17.0	–	>1	0.28	C F S	Strong 1972
Hyalella knickerbockeri	5.5	7.4	17.6	–	11	–	C F S	Embody 1911
Orchestia cavimana	13.5	15.5	14.6	–	4	–	C B S	Dorsman 1935
Orchestia cavimana	16.0	–	18.0	–	6	0.62	C B S	Wildish 1979
Orchestia gammarellus	15.0	18.0	19.0	–	8	0.73	C B S	Wildish 1979
Orchestia mediterranea	20.0	–	36.0	–	5	0.65	C B S	Wildish 1979
Orchestia platensis	8.4	10.7	13.5	–	>1	0.62	C B S	Nagata 1966
Orchestia platensis	9.5	12.0	18.0	12	4	–	C B S	Behbahani & Croker 1982
Orchestia platensis	9.5	12.0	24.0	10	4	0.58	C B S	Morino 1978
Orchestia roffensis	8.0	–	5.0	–	4	0.56	C B S	Wildish 1979
Parhyalella basrensis	6.3	8.5	15.3	15	4	0.39	W B S	Ali & Salman 1986
Parhyalella pietschmanni	6.5	9.0	12.9	–	>1	0.39	W M S	Steele 1973
Protoscinotus loquax	8.4	10.8	20.0	10	>1	0.50	C M S	Hughes 1982
Talitrus saltator	12.6	14.8	13.0	18	1	0.85	C B S	Williams 1978
Talorchestia margaritae	8.8	12.0	7.2	3	–	–	W B S	Venables 1981
Uhlorchestia sparinophila	7.8	9.0	5.5	–	–	–	W B S	Bousfield & Heard 1986
Uhlorchestia uhleri	9.0	10.0	6.0	–	–	–	W B S	Bousfield & Heard 1986

References

Ali, M. H. & S. D. Salman, 1986. The reproductive biology of *Parhyale basrensis* Salman (Crustacea, Amphipoda) in the Shatt al-Arab River. Estuar. coast. shelf Sci. 23: 339–351.

Ashmole, N. P., 1963. The regulation of numbers of tropical oceanic birds. Ibis 103: 458–473.

Atkinson, R. J. A., P. G. Moore & P. J. Morgan, 1982. The burrows and burrowing behaviour of *Maera loveni* (Crustacea: Amphipoda). J. Zool., Lond. 198: 399–416.

Bagenal, T. B., 1966. The ecological and geographical aspects of the fecundity of the plaice. J. mar. biol. Ass. U.K. 46: 161–186.

Bailey, R. C. & G. L. Mackie, 1986. Reproduction of a fingernail clam in contrasting habitats: life-history tactics? Can. J. Zool. 64: 1701–1704.

Barnard, J. L., 1959. Epipelagic and under ice Amphipoda of the central Arctic Basin. In V. Bushnell (ed.), Scientific studies at Fletcher's Ice Island, T-3 (1952–1955). U.S. Air Force Cambridge Research Center. Geophys. Res. Pap. 63: 115–152.

220

Barnard, J. L., 1962. South Atlantic abyssal amphipods collected by R.V. Vema. Abyssal Crustacea, Vema Res. Ser. 1: 1–78.

Barnard, J. L. & C. L. Ingram, 1986. The supergiant amphipod *Alicella gigantea* Chevreux from the North Pacific Gyre. J. Crust. Biol. 6: 825–839.

Barrett, B. E., 1966. A contribution to the knowledge of the amphipodous crustacean *Ampithoe valida* Smith 1873. Ph.D. thesis, Univ. of New Hampshire, 162 pp.

Behbahani, M. I. & R. A. Croker, 1982. Ecology of beach wrack in northern New England with special reference to *Orchestia platensis*. Estuar. coast. shelf Sci. 15: 611–620.

Belk, D., G. Anderson & S.-Y. Hsu, 1990. Additional observations on variations in egg size among populations of *Streptocephalus seali* (Anostraca). J. Crust. Biol. 10: 128–133.

Bellan-Santini, D. & J. C. Dauvin, 1988. Eléments de synthèse sur les *Ampelisca* du nord-est Atlantique. Crustaceana, Suppl. 13: 20–60.

Berven, K. A., 1982. The genetic basis of altitudinal variation in the wood frog *Rana sylvatica*. I. An experimental analysis of life history traits. Evolution 36: 962–983.

Besner, M., 1976. Ecologie et échantillonnage des populations hyperbenthiques d'amphipodes gammaridiens d'un écosystème circalittoral de l'estuaire maritime du Saint-Laurent. M.Sc. thesis, Univ. of Montréal, Montréal, 183 pp.

Bone, D. G., 1972. Aspects of the biology of the Antarctic amphipod *Bovallia gigantea* Pfeffer at Signy Island, South Orkney Islands. Br. Antarct. Surv. Bull. 27: 105–122.

Borowsky, B., 1980a. Reproductive patterns of three intertidal salt-marsh gammaridean amphipods. Mar. Biol. 55: 327–334.

Borowsky, B., 1980b. Factors that affect juvenile emergence in *Gammarus palustris* (Bousfield, 1969). J. exp. mar. Biol. Ecol. 42: 213–223.

Borowsky, B., 1983. Behaviors associated with tube-sharing in *Microdeutopus gryllotalpa* (Costa) (Crustacea: Amphipoda). J. exp. mar. Biol. Ecol. 68: 39–51.

Borowsky, B., 1984. The use of the males' gnathopods during precopulation in some gammaridean amphipods. Crustaceana 47: 245–250.

Borowsky, B., 1986. Laboratory observations of the pattern of reproduction of *Elasmopus levis* (Crustacea: Amphipoda). Mar. Behav. Physiol. 12: 245–256.

Borowsky, B. & R. Borowsky, 1987. The reproductive behaviors of the amphipod crustacean *Gammarus palustris* (Bousfield) and some insights into the nature of their stimuli. J. exp. mar. Biol. Ecol. 107: 131–144.

Bosworth, W. S., 1976. Biology of the genus *Eohaustorius* (Amphipoda: Haustoriidae) on the Oregon Coast. Ph.D. thesis, Oregon State University, Corvallis, Oregon, 194 pp.

Bousfield, E. L., 1970. Terrestrial and aquatic amphipod crustacea from Rennell Island. Nat. Hist. Rennel Isl., Br. Solomon Isl. 6: 155–168.

Bousfield, E. L., 1973. Shallow-water gammaridean Amphipoda of New England. Cornell University Press, Ithaca, N.Y., 312 pp.

Bousfield, E. L., 1983. An updated phyletic classification and palaeohistory of the Amphipoda. In F. R. Schram (ed.), Crustacean Phylogeny. A. A. Balkema, Rotterdam: 257–277.

Bousfield, E. L., 1987. Amphipod parasites of fishes of Canada. Can. Bull. Fish. Aquat. Sci. 217: 1–37.

Bousfield, E. L. & R. W. Heard, 1986. Systematics, distributional ecology, and some host-parasite relationships of *Uhlorchestia uhleri* (Shoemaker) and *U. spartinophila*, new species (Crustacea: Amphipoda), endemic to salt marshes of the Atlantic coast of North America. J. Crust. Biol. 6: 264–274.

Bregazzi, P. K., 1972. Life cycles and seasonal movements of *Cheirimedon femoratus* (Pfeffer) and *Tryphosella kergueleni* (Miers) (Crustacea: Amphipoda). Br. Antarct. Surv. Bull. 30: 1–34.

Bregazzi, P. K., 1973. Embryological development in *Tryphosella kergueleni* (Miers) and *Cheirimedon femoratus* (Pfeffer) (Crustacea: Amphipoda). Br. Antarct. Surv. Bull. 32: 63–74.

Brown, K. M., 1983. Do life history tactics exist at the intraspecific level? Data from freshwater snails. Am. Nat. 121: 871–879.

Bulnheim, H.-P., 1978. Interactions between genetic, external and parasitic factors in the sex determination of the crustacean amphipod *Gammarus duebeni*. Helgoländer wiss. Meeresunters. 31: 1–33.

Caine, E. A., 1979. Population structures of two species of caprellid amphipods (Crustacea). J. exp. mar. Biol. Ecol. 40: 103–114.

Carney, R. S., R. L. Haedrich & G. T. Rowe, 1983. Zonation of fauna in the deep sea. In G. T. Rowe (ed.), The Sea, Vol. 8. Wiley-Interscience, N.Y.: 371–398.

Carrasco, F. D. & D. F. Arcos, 1984. Life history and production of a cold-temperate population of the sublittoral amphipod *Ampelisca araucana*. Mar. Ecol. Prog. Ser. 14: 245–252.

Chambers, M. R., 1977. The population ecology of *Gammarus tigrinus* (Sexton) in the reed beds of the Tjeukemeer. Hydrobiologia 53: 155–164.

Charniaux-Cotton, H., 1985. Vitellogenesis and its control in malacostracan Crustacea. Am. Zool. 25: 197–206.

Cheng, C., 1942. On the fecundity of some gammarids. J. mar. biol. Ass. U.K. 25: 467–475.

Clarke, A., 1979. On living in cold water: K-strategies in Antarctic benthos. Mar. Biol. 55: 111–119.

Clarke, A., 1980. A reappraisal of the concept of metabolic cold adaptation in polar marine invertebrates. Biol. J. Linnean Soc. Lond. 14: 77–92.

Clarke, A., 1983. Life in cold water: the physiological ecology of marine ectotherms. Oceanogr. mar. Biol. Ann. Rev. 21: 341–453.

Clarke, A., 1987. Temperature, latitude and reproductive effort. Mar. Ecol. Prog. Ser. 38: 89–99.

Clarke, A., A. Skadsheim & L. J. Holmes, 1985. Lipid biochemistry and reproductive biology in two species of Gammaridea (Crustacea: Amphipoda). Mar. Biol. 88: 247–263.

Clemens, H. P., 1950. Life cycle and ecology of *Gammarus fasciatus* Say. Ohio State Univ., Contr. Franz Theodore Stone Inst. Hydrobiologia 12: 1–61.

Comely, C. A. & A. D. Ansell, 1988. Invertebrate associates of the sea urchin, *Echinus esculentus* L., from the scottish west coast. Ophelia 28: 111–137.

Corey, S., 1981. Comparative fecundity and reproductive strategies in seventeen species of the Cumacea (Crustacea: Peracarida). Mar. Biol. 62: 65–72.

Costello, M. R. & A. A. Myers, 1989. Breeding periodicity and sex ratios in epifaunal marine Amphipoda in Lough Hyne, Ireland. Est. Coast. Shelf Sci. 29: 409–419.

Croker, R. A., 1967. Niche diversity in five sympatric species of intertidal amphipods (Crustacea: Haustoriidae). Ecol. Monogr. 37: 173–200.

Croker, R. A., 1971. A new species of *Melita* (Amphipoda: Gammaridae) from the Marshall Islands, Micronesia. Pacif. Sci. 25: 100–108.

Dauvin, J.-C., 1988a. Biologie, dynamique, et production de populations de crustacés amphipodes de la Manche occidentale. 1. *Ampelisca tenuicornis* Liljeborg. J. exp. mar. Biol. Ecol. 118: 55–84.

Dauvin, J.-C., 1988b. Biologie, dynamique, et production de populations de crustacés amphipodes de la Manche occidentale. 2. *Ampelisca brevicornis* (Costa). J. exp. mar. Biol. Ecol. 119: 213–233.

Dauvin, J.-C., 1988c. Biologie, dynamique, et production de populations de crustacés amphipodes de la Manche occidentale. 3. *Ampelisca typica* (Bate). J. exp. mar. Biol. Ecol. 121: 1–22.

Dauvin, J.-C., 1988d. Life cycle, dynamics, and productivity of Crustacea-Amphipoda from the western English Channel. 4. *Ampelisca armoricana* Bellan-Santini et Dauvin. J. exp. mar. Biol. Ecol. 123: 235–252.

Dauvin, J.-C., 1989. Life cycle, dynamics, and productivity of Crustacea-Amphipoda from the western English Channel. 5. *Ampelisca sarsi* Chevreux. J. exp. mar. Biol. Ecol. 128: 31–56.

Desroches, M., 1985. Cycle de développement, écologie et succès de l'amphipode gammaridien planctonophage *Rhachotropis oculata* dans deux écosystèmes du golfe du Saint-Laurent. M.Sc. thesis, Univ. of Montréal, 52 pp.

Dexter, D. M., 1971. Life history of the sandy-beach amphipod *Neohaustorius schmitzi* (Crustacea: Haustoriidae). Mar. Biol. 8: 232–237.

Dexter, D. M., 1985. Distribution and life histories of abundant crustaceans of four sandy beaches of southeastern New South Wales. Aust. J. mar. Freshwat. Res. 36: 281–289.

Donn, T. E. & R. A. Croker, 1986. Life-history patterns of *Haustorius canadensis* (Crustacea: Amphipoda) in northern New England. Can. J. Zool. 64: 99–104.

Dorsman, B. A., 1935. Notes on the life history of *Orchestia bottae* Metae Edwards. Ph. D. thesis, Univ. of Leiden, 58 pp.

Downer, D. F. & D. H. Steele, 1979. Some aspects of the biology of *Amphiporeia lawrenciana* Shoemaker (Crustacea, Amphipoda) in Newfoundland waters. Can. J. Zool. 57: 257–263.

Doyle, R. W. & W. Hunte, 1981. Demography of an estuarine amphipod (*Gammarus lawrencianus*) experimentally selected for high 'r': a model of the genetic effects of environmental change. Can. J. Fish. aquat. Sci. 38: 1120–1127.

Dumay, D., (1972) 1973. Etude comparée de la fécondité de deux espèces du groupe '*Locusta*': *Gammarus crinicornis* Stock 1966 et *G. subtypicus* Stock 1966 (Amphipoda). Téthys 4: 975–980.

Duncan, K. W., 1969. The ecology of two species of terrestrial Amphipoda (Crustacea: family Talitridae) living in waste grassland. Pedobiologia 9: 323–341.

Embody, G. C., 1911. A preliminary study of the distribution, food and reproductive capacity of some freshwater amphipods. Int. Revue ges. Hydrobiol. 3: 1–33.

Enequist, P., 1949. Studies on the soft-bottom amphipods of the Skagerak. Zool. Bidr. Upps. 28: 297–492.

Fenwick, G. D., 1984. Life history tactics of brooding Crustacea. J. exp. mar. Biol. Ecol. 84: 247–264.

Fenwick, G. D., 1985. Life-histories of four co-occurring amphipods from a shallow, sand bottom at Kaikoura, New Zealand. N.Z. J. Zool. 12: 71–105.

Fincham, A. A., 1971. Ecology and population studies of some intertidal and sublittoral sand-dwelling amphipods. J. mar. biol. Ass. U.K. 51: 471–488.

Fincham, A. A., 1974. Periodic swimming behavior of amphipods in Wellington Harbour. N. Z. J. mar. Freshwat. Res. 8: 505–521.

Fish, J. D., 1975. Development, hatching and brood size in *Bathyporeia pilosa* and *B. pelagica* (Crustacea: Amphipoda). J. mar. biol. Ass. U.K. 55: 357–368.

Fish, J. D. & A. Mills, 1979. The reproductive biology of *Corophium volutator* and *C. arenarium* (Crustacea: Amphipoda). J. mar. biol. Ass. U.K. 59: 355–368.

Fish, J. D. & G. S. Preece, 1970. The annual reproductive patterns of *Bathyporeia pilosa* and *Bathyporeia pelagica* (Crustacea: Amphipoda). J. mar. biol. Ass. U.K. 50: 475–488.

Franz, D. R., 1989. Population density and demography of a fouling community amphipod. J. exp. mar. Biol. Ecol. 125: 117–136.

Fredette, T. J. & R. J. Diaz, 1986. Life history of *Gammarus mucronatus* Say (Amphipoda: Gammaridae) in warm temperate estuarine habitats, York River, Virginia. J. Crust. Biol. 6: 57–78.

Friend, J. A. & A. M. M. Richardson, 1986. Biology of terrestrial amphipods. Annu. Rev. Entomol. 31: 25–48.

Frith, D. W., 1977. A preliminary analysis of the association of amphipods *Microdeutopus damnoniensis* (Bate),

222

M. anomalus (Rathke) and *Corophium sextoni* Crawford with sponges *Halichondria panicea* (Pallas) and *Hymeniacidon perlive* (Montagu). Crustaceana 32: 113–118.

Gable, M. F. & R. A. Croker, 1977. The salt marsh amphipod, *Gammarus palustris* Bousfield, 1969 at the northern limit of its distribution. I. Ecology and life cycle. Est. coast. mar. Sci. 5: 123–134.

Gaylor, D., 1922. A study of the life history and productivity of *Hyalella knickerbokeri* Bate. Proc. Indiana Acad. Sci.: 239–250.

Gilat, E., 1962. The benthonic Amphipoda of the Mediterranean coast of Israel. II. Ecology and life history. Bull. Res. Counc. Israel 11B: 71–92.

Gledhill, T. & M. Ladle, 1969. Observations on the life-history of the subterranean amphipod *Niphargus aquilex aquilex* Schiodte. Crustaceana 16: 51–56.

Goedmakers, A., 1981. Population dynamics of three gammarid species (Crustacea, Amphipoda) in a french chalk stream. II. Standing crop. Bijdr. Dierk. 51: 31–69.

Goodhart, C. B., 1939. Notes on the bionomics of the tube-building amphipod, *Leptocheirus pilosus* Zaddach. J. mar. biol. Ass. U.K. 23: 311–325.

Greenslade, P. J. M., 1972. Evolution in the staphylinid genus *Priochirus* (Coleoptera). Evolution 26: 203–220.

Greenslade, P. J. M., 1983. Adversity selection and the habitat templet. Am. Nat. 122: 352–365.

Greze, I. I., 1963. Razmnozhenie i rost bokoplava *Dexamine spinosa* (Mont.) v Cernom more. Trudȳ sevastopol'. biol. Sta. 16: 241–255.

Greze, I. I., 1972. Main features of the life cycle of *Gammarus olivii* in the Black Sea. Zool. Zh. 51: 803–811.

Hager, R. P. & R. A. Croker, 1979. Macroinfauna of northern New England marine sand. IV. Infaunal ecology of *Amphiporeia virginiana* Shoemaker, 1933 (Crustacea: Amphipoda). Can. J. Zool. 57: 1511–1519.

Hastings, M. H., 1981. The life cycle and productivity of an intertidal population of the amphipod *Ampelisca brevicornis*. Estuar. coast. shelf Sci. 12: 665–677.

Healey, M. C. & W. R. Heard, 1984. Inter- and intra-population variation in the fecundity of chinook salmon (*Oncorhynchus tshawytscha*) and its relevance to life history theory. Can. J. Fish. aquat. Sci. 41: 476–483.

Heller, S. P., 1968. Some aspects of the biology and development of *Ampithoe lacertosa* (Crustacea: Amphipoda). Ph.D. thesis. Univ. of Washington, 132 pp.

Hessler, R. R., C. L. Ingram, A. A. Yayanos & B. R. Burnett, 1978. Scavenging amphipods from the floor of the Philippine Trench. Deep Sea Res. 25: 1029–1047.

Hildrew, A. G. & C. R. Townsend, 1987. Organization in freshwater benthic communities. In J. H. R. Gee & P. S. Giller (eds), Organisation of communities past and present. 27th Symp. British Ecological Soc., Aberystwyth, 1986. Blackwell, Oxford: 347–372.

Hiwatari, T. & T. Kajihara, 1984. Population dynamics and life cycle of *Hyale barbicornis* (Amphipoda, Crustacea) in a blue mussel zone. Mar. Ecol. Prog. Ser. 20: 177–183.

Hughes, J. E., 1982. Life history of the sandy-beach amphipod *Dogielinotus loquax* (Crustacea: Dogielinotidae) from the outer coast of Washington, USA. Mar. Biol. 71: 167–175.

Hughes, R. G., 1978. Life-histories and abundance of epizoites of the hydroid *Nemertesia antennina* (L.). J. mar. biol. Ass. U.K. 58: 313–332.

Hynes, H. B. N., 1954. The ecology of *Gammarus duebeni* Lilljeborg and its occurence in fresh water in western Britain. J. anim. Ecol. 23: 38–84.

Hynes, H. B. N., 1955. The reproductive cycle of some British freshwater Gammaridae. J. anim. Ecol. 24: 352–387.

Hynes, H. B. N. & F. Harper, 1972. The life histories of *Gammarus lacustris* and *G. pseudolimnaeus* in southern Ontario. Crustaceana, Suppl. 3: 329–341.

Ingram, C. L. & R. R. Hessler, 1983. Distribution and behavior of scavenging amphipods from the central North Pacific. Deep Sea Res. 30: 683–706.

Ingram, C. L. & R. R. Hessler, 1987. Population biology of the deep-sea amphipod *Eurythenes gryllus*: inferences from instar analyses. Deep Sea Res. 34: 1889–1910.

Ivanov, A. I., 1961. Biology of some Black Sea amphipods. Dokl. Akad. Nauk SSSR 137: 728–729. (English translation by the American Institute of Biological Sciences, Wash., D.C.)

Jażdżewski, K., 1973. Ecology of gammarids in the Bay of Puck. Oikos Suppl. 15: 121–126.

Järvinen, A., 1986. Clutch size of passerines in harsh environments. Oikos 46: 365–371.

Jenio, F., 1980. The life cycle and ecology of *Gammarus troglophilus* (Hubricht & Mackin). Crustaceana, Suppl. 6: 204–214.

Jensen, J. P., 1958. The relation between body size and number of eggs in marine malacostrakes. Meddr. Danm. Fisk-og Havunders 2: 1–25.

Judd, W. W., 1963. Studies of the Byron Bog in southern Ontario. XVI. Observations on the life cycles of two species of *Crangonyx* (Crustacea: Amphipoda). Nat. Hist. Pap. Natl Mus. Can. no. 20, 9 pp.

Kaim-Malka, A., 1969. Biologie et écologie de quelques *Ampelisca* (Crustacea-Amphipoda) de la région de Marseille. Téthys 1: 977–1022.

Kanneworf, E., 1965. Life cycle, food, and growth of the amphipod *Ampelisca macrocephala* Liljeborg from the Øresund. Ophelia 2: 305–318.

Kanneworf, E., 1966. On some amphipod species of the genus *Haploops*, with special reference to *H. tubicola* Liljeborg and *H. tenuis* sp. nov. from the Øresund. Ophelia 3: 183–207.

Kaplan, R. H. & S. N. Salthe, 1979. The allometry of reproduction: an empirical view in salamanders. Am. Nat. 113: 671–689.

Kemp, P. F., F. A. Cole & R. C. Swartz, 1985. Life history and productivity of the phoxocephalid amphipod *Rhepoxynius abronius* (Barnard). J. Crust. Biol. 5: 449–464.

Kerfoot, W. C., 1977. Competition in cladoceran communities: the cost of evolving defenses against copepod predation. Ecology 58: 303–313.

Kinne, O., 1953a. Zur biologie und physiologie von *Gammarus duebeni* Lillj., IV: Uber den ruhezustand (Reduktion des oöstegiten-borsten). Zool. Anz. 150: 41–49.

Kinne, O., 1953b. Zur biologie und physiologie von *Gammarus duebeni* Lillj., VI: Productions-biologische studie. Verh. Inst. Meeresforsch. Bremerhaven 2: 135–145.

Kinne, O., 1954. Die bedeutung der kopulation für eiablage und häutungsfrequenz bei *Gammarus*. Biol. Zentralbl. 73: 190–200.

Klein, G., E. Rachor & S. A. Gerlach, 1975. Dynamics and productivity of two populations of the benthic tube-dwelling amphipod *Ampelisca brevicornis* (Costa) in Helgoland Bight. Ophelia 14: 139–159.

Koenig, W. D., 1984. Geographic variation in clutch size in the Northern Flicker (*Colaptes auratus*): support for Ashmole's hypothesis. Auk 101: 698–706.

Kolding, S. & T. M. Fenchel, 1981. Patterns of reproduction in different populations of five species of the amphipod genus *Gammarus*. Oikos 37: 167–172.

Krishnan, L. & P. A. John, 1974. Observations on the breeding biology of *Melita zeylanica* Stebbing, a brackish water amphipod. Hydrobiologia 44: 413–430.

Kühne, H. & G. Becker, 1964. Der holz-flohkrelos *Chelura terebrans* Philippi (Amphipoda, Cheluridae): Morphologie, Verbreitung, Lebensweise, Verhalten, Entwicklung und Umweltabhangigkeit. Beihefte der Zeitschmitt für angewandte Zoologie 1: 1–141.

Kuznetsov, V. V., 1964. Biologiya massovykh i naibolee obytginykh vidov vakoobraznykh Barentseva i Belogo Morei. Akad. Nauk SSSR, Moscow, 242 pp.

LaFrance, K. & E. Ruber, 1985. The life cycle and productivity of the amphipod *Gammarus mucronatus* on a northern Massachusetts salt marsh. Limnol. Oceanogr. 30: 1067–1077.

Lamarche, G. & P. Brunel, 1987. Cycle de développement, écologie et succès de l'amphipode gammaridien *Hippomedon propinquus* dans deux écosystèmes du golfe du Saint-Laurent. Can. J. Zool. 65: 3116–3132.

Legendre, L. & P. Legendre, 1984. Ecologie numérique. Tome 2. La structure des données écologiques. Masson, Paris, 335 pp.

Legueux, M. L., 1926. Etude de la ponte chez un amphipode (*Melita pellucida* G.O. Sars). Variation du nombre et de la taille des œufs. Bull. biol. Fr. Belg. 60: 334–342.

Leineweber, P., 1985. The life cycles of four amphipod species in the Kattegat. Holarct. Ecol. 8: 165–174.

Lim, K. H. & W. D. Williams, 1971. Ecology of *Austrochiltonia subtenuis* (Sayce) (Amphipoda, Hyalellidae). Crustaceana 20: 19–24.

Lowry, J. K. & H. E. Stoddart, 1986. Protandrous hermaphrodites among the lysianassoid Amphipoda. J. Crust. Biol. 6: 742–748.

Luxmoore, R. A., 1982. The reproductive biology of some serolid isopods from the Antarctic. Polar Biol. 1: 3–11.

MacArthur, R. H. & E. O. Wilson, 1967. The theory of island biogeography. Princeton Univ. Press, Princeton, N.J., 203 pp.

McBane, C. D. & R. A. Croker, 1984. Some observations on the life history of the amphipod crustacean, *Hyale nilssoni* (Rathke), in New Hampshire. Estuaries 7: 541–545.

MacGinitie, G. E., 1955. Distribution and ecology of the marine invertebrates at Point Barrow, Alaska. Smithson. Misc. Collect. 128: 1–201.

McLaren, I. A., 1966. Predicting development rate of copepod eggs. Biol. Bull. 131: 457–469.

Marguilis, R. Y., 1970. Zhiznennyi tsikl *Pontoporeia femorata* Krøyer v Belom More (Rugozerskaya Guba). Trudỹ Belomorskogo Biologicheskii Stantsii MGU 3: 46–50.

Marsden, I. D., 1982. Population biology of the commensal asellotan *Iais pubescens* (Dana) and its sphaeromatid host *Exosphaeroma obtusum* (Dana) (Isopoda). J. exp. mar. Biol. Ecol. 58: 233–257.

Mathisen, O. A., 1953. Some investigations of the relict crustaceans in Norway with special reference to *Pontoporeia affinis* Lindstrøm and *Pallasea quadrispinosa* G.O. Sars. Nytt Mag. Zool. 1: 49–86.

Mauchline, J., 1988a. Growth and breeding of meso- and bathypelagic organisms of the Rockall Trough, northeastern Atlantic Ocean and evidence of seasonality. Mar. Biol. 98: 387–393.

Mauchline, J., 1988b. Egg and brood sizes of oceanic pelagic crustaceans. Mar. Ecol. Prog. Ser. 43: 251–258.

Mileikovsky, S. A., 1971. Types of larval development in marine bottom invertebrates, their distribution and ecological significance: a re-evaluation. Mar. Biol. 10: 193–213.

Mills, E. L., 1967. The biology of an ampeliscid amphipod crustacean sibling species pair. J. Fish. Res. Bd Can. 24: 305–355.

Moore, J. W., 1979. Ecology of a subarctic population of *Pontoporeia affinis* Lindström (Amphipoda). Crustaceana 36: 267–276.

Moore, P. G., 1978. Turbidity and kelp holdfast Amphipoda. I. Wales and S. W. England. J. exp. mar. Biol. Ecol. 32: 53–96.

Moore, P. G., 1981. The life histories of the amphipods *Lembos Websteri* Bate and *Corophium bonnellii* Milne Edwards in kelp holdfasts. J. exp. mar. Biol. Ecol. 49: 1–50.

Morino, H., 1978. Studies on the Talitridae (Amphipoda, Crustacea) in Japan. III. Life history and breeding activity of *Orchestia platensis* Krøyer. Publ. Seto Mar. Biol. Lab. 24: 245–267.

Myers, A. A., 1971. Breeding and growth in laboratory-reared *Microdeutopus gryllotalpa* Costa (Amphipoda: Gammaridea). J. nat. Hist., Lond. 5: 271–277.

Nagata, K., 1966. Studies on marine gammaridean Amphipoda of the Seto Island Sea. IV. Publ. Seto Mar. Biol. Lab. 13: 327–348.

Nair, K. K. C. & K. Anger, 1979a. Life cycle of *Corophium insidiosum* (Crustacea, Amphipoda) in laboratory culture. Helgoländer wiss. Meeresunters. 32: 279–294.

224

Nair, K. K. C. & K. Anger, 1979b. Experimental studies on the life cycle of *Jassa falcata* (Crustacea, Amphipoda). Helgoländer wiss. Meeresunters. 32: 444–452.

Nayar, K. N., 1956. The life history of a brackish water amphipod *Grandidierella bonnieri* Stebbing. Proc. Indian Acad. Sci. Sect. B 43: 178–189.

Naylor, C., J. Adams & P. Greenwood, 1988. Population dynamics and adaptive sexual strategies in a brackish water crustacean, *Gammarus duebeni*. J. anim. Ecol. 57: 493–507.

Nelson, W. G., 1978. The community ecology of seagrass amphipods: predation and community structure, life histories and biogeography. Ph.D. thesis, Duke University, Durham, North Carolina, 223 pp.

Nelson, W. G., 1979a. Experimental studies of selective predation on amphipods: consequences for amphipod distribution and abundance. J. exp. mar. Biol. Ecol. 38: 225–245.

Nelson, W. G., 1979b. An analysis of structural pattern in an eelgrass (*Zostera marina* L.) amphipod community. J. exp. mar. Biol. Ecol. 39: 231–264.

Nelson, W. G., 1980. Reproductive patterns of gammaridean amphipods. Sarsia 65: 61–71.

Onbé, T., 1966. Observations on the tubicolous amphipod, *Corophium acherusicum* Costa, in Fukuyama Harbor Area. J. Fac. Fish. Anim. Husb. Hiroshima Univ. 6: 323–338.

Peer, D. L., L. E. Linkletter & P. W. Hicklin, 1986. Life history and reproductive biology of *Corophium volutator* (Crustacea: Amphipoda) and the influence of shorebird predation on population structure in Chignecto Bay, Bay of Fundy, Canada. Neth. J. Sea Res. 20: 359–373.

Pianka, E. R., 1970. On r- and K-selection. Am. Nat. 104: 592–597.

Pinkster, S. & N. W. Broodbakker, 1980. The influence of environmental factors on distribution and reproductive success of *Eulimnogammarus obtusatus* (Dahl, 1938) and other estuarine gammarids. Crustaceana, Suppl. 6: 225–241.

Rakusa-Suszczewski, S., 1972. The biology of *Paramoera walkeri* Stebbing (Amphipoda) and the Antarctic sub-fast ice community. Pol. Arch. Hydrobiol. 19: 11–36.

Rakusa-Suszczewski, S., 1982. The biology and metabolism of *Orchomene plebs* (Hurley 1965) (Amphipoda: Gammaridea) from McMurdo Sound, Ross Sea, Antarctic. Polar. Biol. 1: 47–54.

Richards, S. W., 1963. The demersal fish population of Long Island Sound. II. Food of the juveniles from a sand-shell locality (station 1). Bull. Bingham. oceanogr. Coll. 18: 1503–1510.

Ricklefs, R. E., 1980. Geographical variation in clutch size among passerine birds: Ashmole's hypothesis. Auk 97: 38–49.

Sagar, P. M., 1980. Life cycle and growth of the Antarctic gammarid amphipod *Paramoera walkeri* (Stebbing, 1906). J. Roy. Soc. N.Z. 10: 259–270.

Sainte-Marie, B., 1986. Foraging by lysianassid amphipods. Ph.D. thesis, Dalhousie University, Halifax, 224 pp.

Sainte-Marie, B. & P. Brunel, 1983. Differences in life history and success between suprabenthic shelf populations of *Arrhis phyllonyx* (Amphipoda Gammaridea) in two ecosystems of the Gulf of St. Lawrence. J. Crust. Biol. 3: 45–69.

Sainte-Marie, B. & P. Brunel, 1985. Suprabenthic gradients of swimming activity by cold-water gammaridean amphipod crustacea over a muddy shelf in the Gulf of Saint-Lawrence. Mar. Ecol. 23: 57–69.

Sainte-Marie, B., G. Lamarche & J.-M. Gagnon, 1990. Reproductive bionomics of some shallow-water lysianassoids in the Saint Lawrence Estuary, with a review of the fecundity of the Lysianassoidea (Crustacea, Amphipoda). Can. J. Zool. 68: 1639–1644.

Salvat, B., 1967. La macrofaune carcinologique endogée des sédiments meubles intertidaux (Tanaidacés, Isopodes et Amphipodes). Ethologie, bionomie, et cycle biologique. Mém. Mus. natn. Hist. nat., Paris. 45: 1–275.

Sameoto, D. D., 1969a. Comparative ecology, life histories, and behaviour of intertidal sand-burrowing amphipods (Crustacea: Haustoriidae) at Cape Cod. J. Fish. Res. Bd Can. 26: 361–388.

Sameoto, D. D., 1969b. Some aspects of the ecology and life cycle of three species of subtidal sand-burrowing amphipods (Crustacea: Haustoriidae). J. Fish. Res. Bd Can. 26: 1321–1345.

Samter, M. & W. Weltner, 1904. Biologische eigentümlichkeiten der *Mysis relicta*, *Pallasiella quadrispinosa* und *Pontoporeia affinis* erklärt aus ihrer eiszeitlichen Entstehung. Zool. Anz. 27: 676–694.

Schaffer, W. M., 1974. Optimal reproductive effort in fluctuating environments. Am. Nat. 108: 783–790.

Schellenberg, A., 1926. Amphipoda. 3. Die gammariden der deutschen Südpolar-Expedition 1901–1903. Deutsch südpolar-Expedition, Zoologie 18: 233–414.

Schram, F. R., 1986. Crustacea. Oxford Univ. Press, New York, 606 pp.

Scott, K. J. & R. A. Croker, 1976. Macroinfauna of northern New England marine sand. III. The ecology of *Psammonyx nobilis* (Stimpson), 1853 (Crustacea: Amphipoda). Can. J. Zool. 54: 1519–1529.

Segerstråle, S. G., 1967. Observations of summer-breeding in populations of the glacial relict *Pontoporeia affinis* Lindstr. (Crustacea Amphipoda), living at greater depths in the Baltic Sea, with notes on the reproduction of *P. femorata* Krøyer. J. exp. mar. Biol. Ecol. 1: 55–64.

Segerstråle, S. G., 1970. Light control of the reproductive cycle of *Pontoporeia affinis* Lindström (Crustacea Amphipoda). J. exp. mar. Biol. Ecol. 5: 272–275.

Sexton, E. W., 1928. On the rearing and breeding of *Gammarus* in laboratory conditions. J. mar. biol. Ass. U.K. 15: 33–55.

Sheader, M., 1977. Production and population dynamics of *Ampelisca tenuicornis* (Amphipoda) with notes on the biology of its parasite *Sphaeronella longipes* (Copepoda). J. mar. biol. Ass. U.K. 57: 955–968.

Sheader, M., 1978. Distribution and reproductive biology of

Corophium insidiosum (Amphipoda) on the north-east coast of England. J. mar. biol. Ass. U.K. 58: 585–596.

Sheader, M., 1983. The reproductive biology and ecology of *Gammarus duebeni* (Crustacea: Amphipoda) in Southern England. J. mar. biol. Ass. U.K. 63: 517–540.

Sheader, M. & F.-S. Chia, 1970. Development, fecundity and brooding behaviour of the amphipod, *Marinogammarus obtusatus*. J. mar. biol. Ass. U.K. 50: 1079–1099.

Shillaker, R. O. & P. G. Moore, 1987. The biology of brooding in the amphipods *Lembos websteri* Bate and *Corophium bonnellii* Milne Edwards. J. exp. mar. Biol. Ecol. 110: 113–132.

Siegfried, C. A., 1985. Life history, population dynamics and production of *Pontoporeia hoyi* (Crustacea, Amphipoda) in relation to the trophic gradient of Lake George, New York. Hydrobiologia 122: 175–180.

Skadsheim, A., 1982. The ecology of intertidal amphipods in the Oslofjord. The life cycles of *Chaetogammarus marinus* and *C. stoerensis*. P.S.Z.N. Mar. Ecol. 3: 213–224.

Skadsheim, A., 1984. Coexistence and reproductive adaptations of amphipods: the role of environmental heterogeneity. Oikos 43: 94–103.

Skadsheim, A., 1989. Regional variation in amphipod life history: effects of temperature and salinity on breeding. J. exp. mar. Biol. Ecol. 127: 25–42.

Skadsheim, A., 1990. A cohort life table for *Gammarus salinus* (Amphipoda). Oikos 57: 207–214.

Skutch, A. F., 1926. On the habits and ecology of the tube-building amphipod *Amphitoe rubicata* Montagü. Ecology 7: 481–502.

Slattery, P. N., 1985. Life histories of infaunal amphipods from subtidal sands of Monterey Bay, California. J. Crust. Biol. 5: 635–649.

Smith, C. C. & S. D. Fretwell, 1974. The optimal balance between size and number of offspring. Am. Nat. 108: 499–506.

Smith, K. L. Jr. & R. J. Baldwin, 1984. Vertical distribution of the necrophagous amphipod, *Eurythenes gryllus*, in the North Pacific: spatial and temporal variation. Deep Sea Res. 31: 1179–1196.

Smith, M. J. & W. D. Williams, 1983. Reproductive cycles in some freshwater amphipods in southern Australia. In J. K. Lowry (ed.). Papers from the conference on the biology and evolution of crustacea. Aust. Mus. Syd. Mem. 18: 183–195.

Sokal, R. R. & F. J. Rohlf, 1981. Biometry, 2nd edition. W.H. Freeman, San Francisco, 859 pp.

Southwood, T. R. E., 1977. Habitat, the templet for ecological strategies. J. anim. Ecol. 46: 337–365.

Southwood, T. R. E., 1988. Tactics, strategies and templets. Oikos 52: 3–18.

Spooner, G. M., 1947. The distribution of *Gammarus* species in estuaries. Part I. J. mar. biol. Ass. U.K. 27: 1–52.

Sprules, W. G., 1967. The life cycle of *Crangonyx richmondensis laurentianus* Bousfield (Crustacea: Amphipoda). Can. J. Zool. 45: 877–884.

Stearns, S. C., 1976. Life history tactics: a review of the ideas. Q. Rev. Biol. 51: 3–47.

Stearns, S. C., 1983. The influence of size and phylogeny on patterns of covariation among life-history traits in the mammals. Oikos 41: 173–187.

Stearns, S. C., 1984. The effects of size and phylogeny on patterns of covariation in the life history traits of lizards and snakes. Am. Nat. 123: 56–72.

Stebbins, T. D., 1989. Population dynamics and reproductive biology of the commensal isopod *Colidotea rostrata* (Crustacea: Isopoda: Idoteidae). Mar. Biol. 101: 329–337.

Steele, D. H., 1967. The life cycle of the marine amphipod *Stegocephalus inflatus* Krøyer in the northwest Atlantic. Can. J. Zool. 45: 623–628.

Steele, D. H., 1972. Some aspects of the biology of *Gammarellus homari* (Crustacea, Amphipoda) in the northwestern Atlantic. J. Fish. Res. Bd Can. 29: 1340–1343.

Steele, D. H., 1973. The biology of *Parhyalella pietschmanni* Schellenberg, 1938 (Amphipoda, Hyalellidae) at Nosy Bé, Madagascar. Crustaceana 25: 276–280.

Steele, D. H., 1983. Size composition of lysianassid amphipods in cold and warm water habitats. In: Papers from the conference on the biology and evolution of Crustacea. Aust. Mus. Syd. Mem. 18: 113–119.

Steele, D. H. & V. J. Steele, 1969. The biology of *Gammarus* (Crustacea, Amphipoda) in the northwestern Atlantic. I. *Gammarus duebeni* Lillj. Can. J. Zool. 47: 235–244.

Steele, D. H. & V. J. Steele, 1970a. The biology of *Gammarus* (Crustacea, Amphipoda) in the northwestern Atlantic. III. *Gammarus obtusatus* Dahl. Can. J. Zool. 48: 989–995.

Steele, D. H. & V. J. Steele, 1970b. The biology of *Gammarus* (Crustacea, Amphipoda) in the northwestern Atlantic. IV. *Gammarus lawrencianus* Bousfield. Can. J. Zool. 48: 1261–1267.

Steele, D. H. & V. J. Steele, 1972a. The biology of *Gammarus* (Crustacea, Amphipoda) in the northwestern Atlantic. VI. *Gammarus tigrinus* Sexton. Can. J. Zool. 50: 1063–1068.

Steele, D. H. & V. J. Steele, 1972b. Biology of *Gammarellus angulosus* (Crustacea, Amphipoda) in the northwestern Atlantic. J. Fish. Res. Bd Can. 29: 1337–1340.

Steele, D. H. & V. J. Steele, 1973a. Some aspects of the biology of *Calliopius laeviusculus* (Krøyer) (Crustacea, Amphipoda) in the northwestern Atlantic. Can. J. Zool. 51: 723–728.

Steele, D. H. & V. J. Steele, 1973b. The biology of *Gammarus* (Crustacea, Amphipoda) in the northwestern Atlantic. VII. The duration of embryonic development in five species at various temperatures. Can. J. Zool. 51: 995–999.

Steele, D. H. & V. J. Steele, 1975a. The biology of *Gammarus* (Crustacea, Amphipoda) in the northwestern Atlantic. IX. *Gammarus wilkitzkii* Birula, *Gammarus stoerensis* Reid, and *Gammarus mucronatus* Say. Can. J. Zool. 53: 1105–1109.

Steele, D. H. & V. J. Steele, 1975b. The biology of *Gammarus* (Crustacea, Amphipoda) in the northwestern Atlantic. X. *Gammarus finmarchicus* Dahl. Can. J. Zool. 53: 1110–1115.

Steele, D. H. & V. J. Steele, 1975c. The biology of *Gammarus* (Crustacea, Amphipoda) in the northwestern Atlantic. XI. Comparison and discussion. Can. J. Zool. 53: 1116–1126.

Steele, D. H. & V. J. Steele, (1975) 1976. Some aspects of the biology of *Gammaracanthus loricatus* (Sabine) (Crustacea, Amphipoda). Astarte 8: 69–72.

Steele, D. H. & V. J. Steele, 1978. Some aspects of the biology of *Pontoporeia femorata* and *Pontoporeia affinis* (Crustacea, Amphipoda) in the northwestern Atlantic. Astarte 11: 61–66.

Steele, D. H. & V. J. Steele, 1986. The cost of reproduction in the amphipod *Gammarus lawrencianus* Bousfield, 1956. Crustaceana 51: 176–182.

Steele, D. H., R. G. Hooper & D. Keats, 1986. Two corophioid amphipods commensal on Spider crabs in Newfoundland. J. Crust. Biol. 6: 119–124.

Steele, V. J., 1967. Resting stage in the reproductive cycles of *Gammarus*. Nature 214: 1034.

Steele, V. J. & D. H. Steele, 1970. The biology of *Gammarus* (Crustacea, Amphipoda) in the northwestern Atlantic. II. *Gammarus setosus* Dementieva. Can. J. Zool. 48: 659–671.

Steele, V. J. & D. H. Steele, 1972. The biology of *Gammarus* (Crustacea, Amphipoda) in the northwestern Atlantic. V. *Gammarus oceanicus* Segerstråle. Can. J. Zool. 50: 801–813.

Stephensen, K., 1923. Crustacea Malacostraca V (Amphipoda I). Dan. Ingolf-Exped. 3: 1–100.

Stephensen, K., 1933. Fresh and brackish-water Amphipoda from Bonaire, Curaçao and Aruba. Zool. Jb. 64: 415–436.

Stephensen, K., 1936. On *Metopa groenlandica* H. J. Hansen (Crustacea, Amphipoda) Medd. Grønland 118: 1–8.

Stephensen, K., 1944. Crustacea Malacostraca VIII (Amphipoda IV). Danish Ingolf-Exp. 3C: 1–51.

Stock, J. H., 1984. Observations morphologiques et écologiques sur une population intertidale de '*Melita*' *pellucida* (Amphipoda) à Etretat (Seine-maritime, France). Cah. Biol. mar. 25: 93–106.

Stockton, W. L., 1982. Scavenging amphipods from under the Ross Ice Shelf, Antarctica. Deep Sea Res. 29: 819–835.

Stockton, W. L. & T. E. DeLaca, 1982. Food falls in the deep sea: occurrence, quality, and significance. Deep Sea Res. 29: 157–169.

Stoner, A. W., 1979. Species-specific predation on amphipod Crustacea by the pinfish *Lagodon rhomboides*: mediation by macrophyte standing crop. Mar. Biol. 55: 201–207.

Strong, D. R. Jr., 1972. Life history variation among populations of an amphipod (*Hyalella azteca*). Ecology 53: 1103–1111.

Sudo, H., M. Azuma & M. Azeta, 1987. Diel changes in predator-prey relationships between Red Sea bream and gammaridean amphipods in Shijiki bay. Nippon Suisan Gakkaishi 53: 1567–1575.

Tamura, H. & K. Koseki, 1974. Population study on a terrestrial amphipod, *Orchestia platensis japonica* (Tattersall), in a temperate forest. Jap. J. Ecol. 24: 123–139.

Thiel, H., 1979. Structural aspects of the deep-sea benthos. Ambio. spec. Rep. 6: 25–31.

Thoemke, K. W., 1979. The life histories and population dynamics of four subtidal amphipods from Tampa Bay, Florida. Ph.D. dissertation, Departement of Biology, Univ. of South Florida, 150 pp.

Thorson, G., 1950. Reproductive and larval ecology of marine bottom invertebrates. Biol. Rev. 25: 1–45.

Thurston, M. H., 1968. Notes on the life history of *Bovallia gigantea* (Pfeffer) (Crustacea, Amphipoda). Brit. Antarct. Surv. Bull. 16: 57–64.

Thurston, M. H., 1970. Growth in *Bovallia gigantea* (Pfeffer) Crustacea, Amphipoda. In: Holdgate, M. W. (ed.), Antarctic Ecology 1. Academic Press, London: 269–278.

Thurston, M. H., 1974. The Crustacea Amphipoda of Signy Island, South Orkney Islands. Brit. Antarct. Surv. Scient. Rep. 71: 1–133.

Thurston, M. H., 1979. Scavenging abyssal amphipods from the north-east Atlantic Ocean. Mar. Biol. 51: 55–68.

Todd, C. D. & R. W. Doyle, 1981. Reproductive strategies of marine benthic invertebrates: a settlement-timing hypothesis. Mar. Ecol. Prog. Ser. 4: 75–83.

Vader, W., 1983. Associations between amphipods (Crustacea: Amphipoda) and sea anemones (Anthozoa, Actinaria). Aust. Mus. Syd. Mem. 18: 141–153.

Vance, R. R., 1973. On reproductive strategies in marine benthic invertebrates. Am. Nat. 107: 339–352.

Van Dolah, R. F. & E. Bird, 1980. A comparison of reproductive patterns in epifaunal and infaunal gammaridean amphipods. Estuar. coast. mar. Sci. 11: 593–604.

Van Dolah, R. F., L. E. Shapiro & C. P. Rees, 1975. Analysis of an intertidal population of the amphipod *Gammarus palustris* using a modified version of the egg-ratio method. Mar. Biol. 33: 323–330.

Venables, B. J., 1981. Aspects of the population biology of a Venezuelan beach amphipod, *Talorchestia margaritae* (Talitridae), including estimate of biomass and daily production, and respiration rates. Crustaceana 41: 271–285.

Vlasblom, A. G., 1969. A study of a population of *Marinogammarus marinus* (Leach) in the Oosterschelde. Neth. J. Sea Res. 4: 317–338.

Wägele, J.-W., 1987. On the reproductive biology of *Ceratoserolis trilobitoides* (Crustacea: Isopoda): latitudinal variation of fecundity and embryonic development. Polar Biol. 7: 11–24.

Wakabara, Y., E. Kawakami de Rezenda & A. S. Tararam, 1982. Amphipods as one of the main food components of three pleuronectiformes from the continental shelf of South Brazil and North Uruguay. Mar. Biol. 68: 67–70.

Wanntorp, H.-E., D. R. Brooks, T. Nilsson, S. Nylin, F. Ronquist, S. C. Stearns & N. Wedell, 1990. Phylogenetic approaches in ecology. Oikos 57: 119–132.

Wear, R. G., 1974. Incubation in british decapod Crustacea and the effects of temperature on the rate and success of embryonic development. J. mar. biol. Ass. U.K. 54: 745–762.

White, J. F. & S. J. Gould, 1965. Interpretation of the coefficient in the allometric equation. Am. Nat. 99: 5–18.

Wiederholm, T., 1973. On the life cycle of *Pontoporeia affinis*

(Crustacea Amphipoda) in lake Mälaren. Zoon 1: 147–151.

Wildish, D. J., 1970. Polymorphism in *Orchestia mediterranea* A. Costa (Amphipoda, Talitridae). Crustaceana 19: 113–118.

Wildish, D. J., 1971. Adaptive significance of a biased sex ratio in *Orchestia*. Nature 233: 54–55.

Wildish, D. J., 1979. Reproductive consequences of the terrestrial habit in *Orchestia* (Crustacea, Amphipoda). Int. J. Invert. Reprod. 1: 9–20.

Wildish, D. J., 1980. Reproductive bionomics of two sublittoral amphipods in a Bay of Fundy estuary. Int. J. Invert. Reprod. 2: 311–320.

Wildish, D. J., 1982. Evolutionary ecology of reproduction in gammaridean Amphipoda. Int. J. Invert. Reprod. 5: 1–19.

Wildish, D. J., 1988. Ecology and natural history of aquatic Talitroidea. Can. J. Zool. 66: 2340–2359.

Wildish, D. J. & B. Frost, 1991. Volumetric growth in gammaridean Amphipoda. Hydrobiologia (this volume).

Wildish, D. J. & D. Peer, 1981. Methods for estimating secondary production in marine Amphipoda. Can. J. Fish. aquat. Sci. 38: 1019–1026.

Williams, G. C., 1966. Natural selection, the cost of reproduction, and a refinement of Lack's principle. Am. Nat. 100: 687–690.

Williams, G. C., 1978. The annual pattern of reproduction of *Talitrus saltator* (Crustacea: Amphipoda: Talitridae). J. Zool., Lond. 184: 231–244.

Wittmann, K. J., 1984. Ecophysiology of marsupial development and reproduction in Mysidacea (Crustacea). Oceanogr. mar. Biol. Ann. Rev. 22: 393–428.

Hydrobiologia **223**: 229–237, 1991.
L. Watling (ed.), VIIth International Colloquium on Amphipoda.
© 1991 *Kluwer Academic Publishers.*

Two types of maternal care for juveniles observed in *Caprella monoceros* Mayer, 1890 and *Caprella decipiens* Mayer, 1890 (Amphipoda: Caprellidae)*

Masakazu Aoki & Taiji Kikuchi
Amakusa Marine Biological Laboratory, Kyushu University, Tomioka, Reihoku-cho, Amakusa, Kumamoto 863-25, Japan

Key words: Amphipoda, *Caprella*, maternal care, clinging juveniles, *Sargassum* bed, reproductive strategies

Abstract

The juveniles of *Caprella monoceros* after emerging from the brood pouch cling to their mother's body for 16.0 days. The juveniles molt and grow at least four times while on their mother. The mother often grooms her juveniles with 1st gnathopods, and defends the juveniles from other caprellids. Even after moving from the mother's body to a seaweed substratum the juveniles stay near their mother for more than 10 days, and are often picked up and carried by their mother when she moves away. On the other hand, the juveniles of *Caprella decipiens* after emerging from the brood pouch do not cling to their mother, but immediately move to the seaweed substratum around the mother and stay there to molt and grow for 32.5 days. The juveniles are defended by their mother. Only at the time when their mother moves away, juveniles cling to the mother and are carried to other places. During the maternal care period, moltings of mothers seem to be suppressed in *C. monoceros*, while mothers of *C. decipiens* continue to molt. The females of *C. monoceros* mature at a smaller size and produce fewer eggs than *C. decipiens*.

Introduction

In crustaceans, maternal care of juveniles has been known and studied in crayfish of Decapoda (Little, 1975; Hazlett, 1983), but little is known in other crustaceans.

In caprellid crustaceans, the adult female lays her eggs in her brood pouch. The eggs hatched in the brood pouch emerge as juveniles and the juveniles leaving the brood pouch disperse immediately in most caprellids. On the other hand,

in *Caprella scaura* Templeton, 1836 (*C. scaura typica* Mayer, 1890) and in *Pseudoprotella phasma* (Montagu, 1804), juveniles emerged from brood pouch cling to their mother for a certain period. In particular, in *Caprella scaura*, the mother takes care of her juveniles during that period (Harrison, 1940; Lim & Alexander, 1986). However, the details and the ecological meanings of the behaviors have not been examined until now.

Among the caprellids found in a *Sargassum patens* C. Agardh bed in Amakusa, west coast of Kyushu, southern Japan (see Imada *et al.*, 1981; Aoki & Kikuchi, 1990a), another 'juvenile-cling type' maternal care was observed in *Caprella*

---

*Contributions from the Amakusa Marine Biological Laboratory, Kyushu University, No. 354.

monoceros Mayer, 1890. Moreover, a new type of maternal care was observed in *Caprella decipiens* Mayer, 1890. In this paper, we describe the details of these maternal behaviors and discuss their ecological meanings.

Materials and methods

All collections of animals for observations and experiments were carried out by SCUBA diving in a *Sargassum patens* bed on the east coast of Tomioka peninsula in Amakusa, west coast of Kyushu, Japan. The *Sargassum patens* bed extends 400 m along the coast and 100 m off shore. Water depth in the bed is 3–5 m and the bottom consists of sand and boulders.

Field observations

We observed the behavior of females with juveniles in natural condition on the calm and fine days in April, 1989. At this season, *Sargassum patens* reaches 2–3 m in height and forms a big seaweed forest. Water temperature in the habitat was 14–16 °C. More than 200 females of *Caprella monoceros* and 2 females of *C. decipiens* were observed.

Laboratory observations

Individuals for laboratory observations and rearing experiments were collected in April–May, 1989. Water temperature was 14–18 °C. Ovigerous females and females with juveniles were picked one by one from the seaweed substratum and put in separate 50 ml plastic bottles. Immediately after the collections, the caprellids in the bottles kept at field temperature were carried carefully to the tanks in our laboratory.

Clinging behavior of the Caprella monoceros juveniles

The growth stage of juveniles clinging to their mother was different between mothers even in a season. Twenty-three adult females of *C. monoceros* carrying juveniles were examined under a binocular microscope within 12 hours after the collections. We examined the clinging positions of juveniles on their mother. The females which released juveniles from brood pouches within 12 hours after the collection were recognized as 'females with the juveniles just emerged from the brood pouch'.

After the observations, the animals were fixed in buffered 5% formalin. The number of flagellar articles of antenna 1 and the body length of 6 juveniles in each cohort were examined to determine the molting time and growth of the juveniles. We measured the thickness of each part of the body and appendages of the mothers to which the juveniles were clinging, and also measured the thickness of seaweed substrata where released juveniles were clinging.

Observations by rearing experiments

Rearing experiments were carried out in the laboratory to observe the details of maternal behavior in *Caprella monoceros* and *C. decipiens*.

Animals for rearing were collected in May, 1989. Water temperature in the field was 16–18 °C. We put the females which were just ready to release juveniles in separate 140 ml glass bottles (height: 9 cm). The lid of each bottle has a hole (area: 7 cm^2) covered with 250 μm mesh. Seawater was supplied continuously through a plastic tube (diameter: 4 mm) from a tank placed 10 cm higher than the bottles. Water temperature was kept at 17–18 °C. As a light source, a 15 W fluorescent light was put 30 cm from the surface of the bottles, being controlled on either a 12L12D–14L10D time period. A piece of *Sargassum patens* branch (wet weight: about 1–2 g) was supplied as substratum for animals in each bottle. The substratum was replaced by a new one once every 3–5 days and the bottle and the lid of each bottle were cleaned every day. Natural diatoms and detritus were supplied for food. Diatoms and detritus on about 500 g (wet weight) fresh *Sargassum patens* were shaken off in the sea water tank and sieved with 250 μm mesh to eliminate large particles. The particles passed through the mesh were mixed in 6 l of seawater and about 100 ml h^{-1} of the mixed seawater was supplied continuously to each bottle at least 6 hours per day.

Reproductive characteristics of females

The ovigerous females of *C. monoceros* and *C. decipiens*, which have eggs just after ovulation, were fixed in buffered 5% formalin. The body length of each female and the number and the size of eggs were examined. The specimens of *C. monoceros* were collected in February, 1989 (water temperature: 13–15 °C) and those of *C. decipiens* were collected in April, 1989 (water temperature: 14–16 °C).

Results

Caprella monoceros

Observations of field animals

We observed many adult females carrying their juveniles on their bodies (Fig. 1a). The juveniles clinging to a mother consisted of a single cohort.

Thus, the sizes of the juveniles in their younger stages were almost the same. However, the variance of body lengths became greater in juveniles larger than 3 mm (Fig. 2). The size of juveniles on mothers differed from mother to mother even in one season. Juveniles ($n = 138$) from 23 mothers (6 juveniles from one mother) were examined to determine the molting growth of the juveniles. The relationship between body lengths and the number of antenna 1 flagellar articles is shown in Fig. 3. Juveniles change the clinging position on their mother as they grow (Fig. 4). The average thickness of eleven parts of the mother's body and the seaweed substrata were shown in Table 1.

Observations of rearing animals

Three adult females, carrying their juveniles just emerged from brood pouches, had been reared separately for 30 days. Juveniles had clung to their mother's body for 16.3 ± 3.5 days (range: 12–18 days). The juveniles molted and grew on their mother. The mother did not molt while carrying the juveniles, and she often groomed the juveniles

Fig. 1. Maternal care behavior of two caprellids; a, a female *Caprella monoceros* carrying her juveniles; b, a female *C. decipiens* staying with her juveniles. Bars: 5 mm.

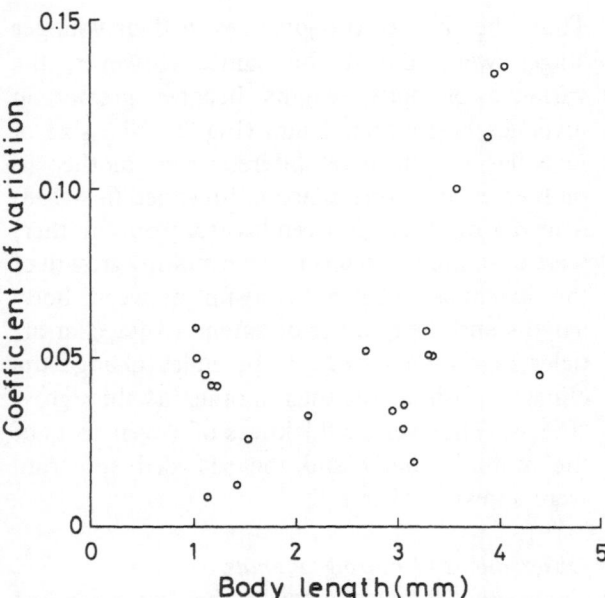

Fig. 2. Changes in coefficient of variation (S.D./Mean) of the body length of *Caprella monoceros* juveniles in each cohort of 23 adult females, with the increase of average body length of the juveniles.

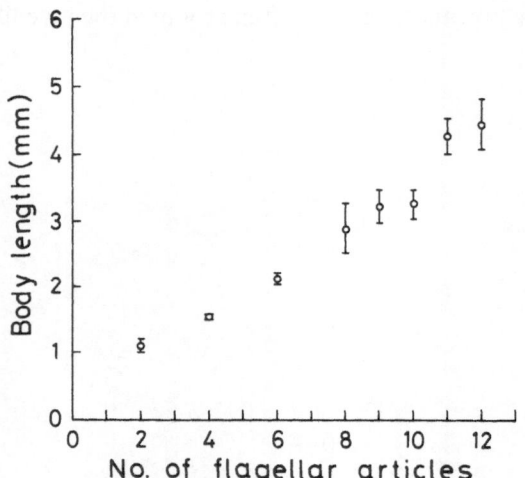

Fig. 3. Increase of body length of *Caprella monoceros* juveniles on their mother (Mean ± S.D.), with the change in the number of antenna 1 flagellar articles showing the molting growth of the juveniles.

with her first gnathopods. When other caprellids of the same species approached her, she defended juveniles from them regardless of their sex. When they fought, they faced each other and repeatedly

bent the upper half of their bodies back and forth to push each other. They continued this behavior until one of the pair turned and retreated. During this fight, they opened their second gnathopods widely but did not use them for the fight. On the other hand, if the fight continued, they sometimes used their first gnathopods for holding the opponent's body or antenna 1.

Juveniles stayed around their mother (within about 3 cm from the mother) for more than 10 days even after moving from the mother's body to the seaweed substratum. During this period, if we gave some stimuli to them, for instance, stopping the water movement or shaking the substratum they clung to, the mother moved away to another place with her juveniles. At the time, the mother swung her two pairs of antennae vigorously, and soon the juveniles gathered to cling to the mother's body. The mother often helped them with her first gnathopods and carried the juveniles to another place, where the juveniles moved to the seaweed substratum again. Three mothers with juveniles had not molted for more than 30 days. On the other hand, two females, whose juveniles were removed one week after their emergence, molted and produced eggs three days after removal of juveniles and released new juveniles 6 days after the ovulations.

Caprella decipiens

Observations of field animals
Two females staying with juveniles were observed in the *Sargassum patens* bed. In both cases, the juveniles which were about 3–4 mm in body length were staying on the branches within about 3 cm of their mother and they did not disperse further.

Observations of rearing animals
Four adult females with their juveniles which had just emerged from the brood pouches had been reared separately for 30 days. Juveniles emerged from the brood pouch immediately moved to the seaweed substratum to which their mother was clinging, and stayed within about 3 cm of their

No. of juveniles	65.0 ± 2.4	54.3 ± 9.3	44.6 ± 18.3
Body length (mm)	1.2 ± 0.2	3.0 ± 0.4	4.0 ± 0.3
No. of FA	2 – 4	6 – 10	11 – 12

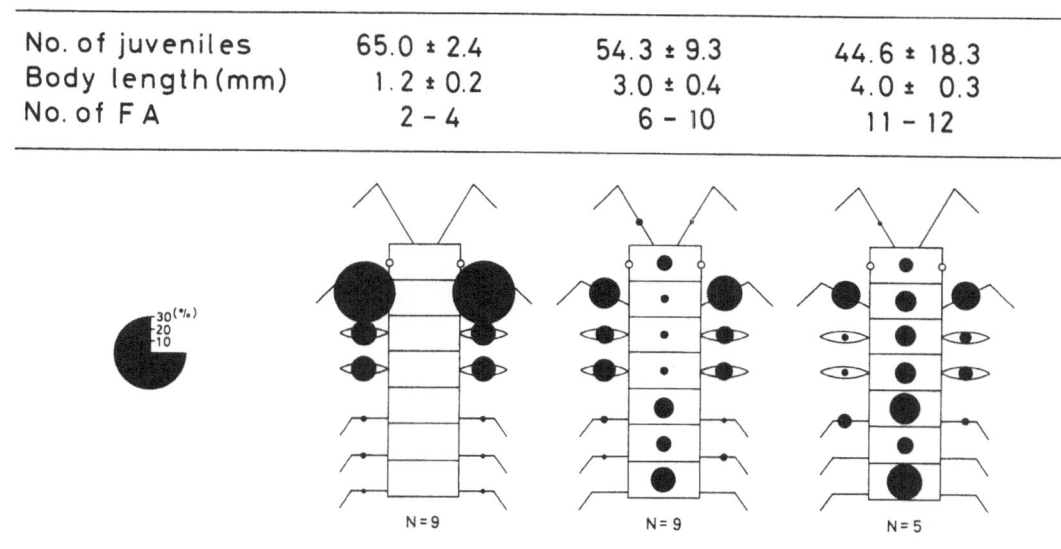

Fig. 4. Changes in the clinging positions of *Caprella monoceros* juveniles. The number of the juveniles on one mother, their body length (mean ± S.D.) and the number of antenna 1 flagellar articles (FA) are also shown.

Table 1. Size of clinging positions of juveniles; Mean ± S.D. (mm) of diameters; Gl: Gill; Gn: Gnathopod; Pd: Pereopod; Pn: Pereonite.

Bodies of mother		Appendages/Gills of mother		Branch of seaweed where detaching juvs. were clinging
Pn 2	0.55 ± 0.10	Gn 2	0.18 ± 0.02	0.93 ± 0.18
Pn 3	0.61 ± 0.08	Gl 3	0.18 ± 0.03	
Pn 4	–	Gl 4	0.18 ± 0.02	
Pn 5	0.61 ± 0.09	Pd 5	0.24 ± 0.03	
Pn 6	0.50 ± 0.05	Pd 6	0.24 ± 0.03	
Pn 7	0.44 ± 0.04	Pd 7	0.25 ± 0.03	

mother and did not disperse during the rearing period. The juveniles were defended by their mother. During this period, when we disturbed them by stopping the water flow or shaking their substrata, the juveniles were picked up by the mother and carried to another place in the same way as shown by *C. monoceros*. The mother of *C. decipiens* molted even with the juveniles staying around her. The mother molted 7.2 ± 2.8 days (*n* = 4) after the release of juveniles. One of these 4 mothers produced eggs just after molting and released new juveniles 5 days after the ovulation. In this case, we could observe that the mother and

the two different cohorts of juveniles had stayed together for 2 days, since the mother died 2 days after the releasing of the new juveniles.

Reproductive characteristics
The relationship between body size and egg number of females in both species is shown in Fig. 5. Reproductive data of both species are also shown in Table 2. In *C. monoceros*, once the females mature, the body growth suspends and the egg number increases regardless of size. On the other hand, in *C. decipiens*, maturation size of the female is larger than that of *C. monoceros* and the

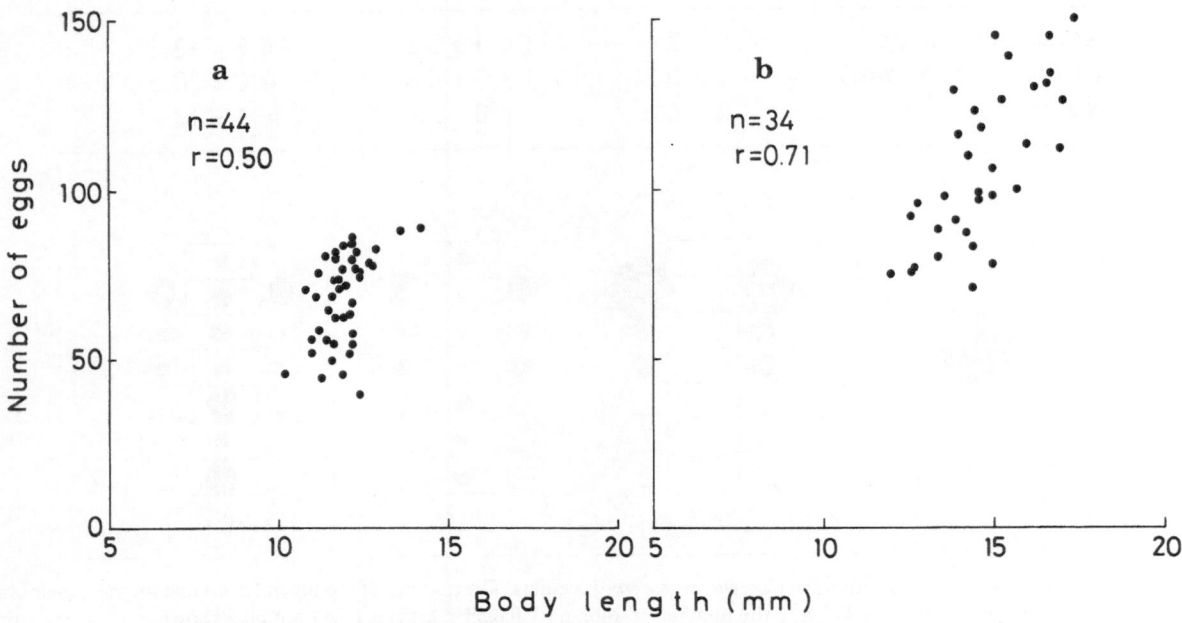

Fig. 5. Changes in the number of brooding eggs with the increase of body length. a, *Caprella monoceros* females; b, *C. decipiens* females.

Table 2. Reproductive characteristics of *Caprella monoceros* and *C. decipiens*; Mean ± S.D.

Species	Min. brooding size of females	Egg no./clutch	Egg volume (mm³)	Body size of emerged juvs.	Body size of post-cling juvs.
C. monoceros	10.2	68.6 ± 13.2	6.4 ± 0.3	1.0 ± 0.0	4.4 ± 0.4
C. decipiens	12.0	109.2 ± 25.9	11.0 ± 1.8	1.1 ± 0.6	–

egg number increases as the body length increases. The sizes of eggs and newly emerged juveniles are larger in *C. decipiens* than in *C. monoceros*.

Discussion

Maternal care in Caprella monoceros

The body sizes and the number of the first antenna flagellar articles of the juveniles were 4.0 ± 0.3 mm and 11–12 just before moving to the seaweed substrata, while 1.0 ± 0.0 mm and 2 just after emergence from the brood pouch. One molt adds at least one article in caprellids (Harrison,

1940; Caine, 1979b). Thus, according to Fig. 3, we conclude that the juveniles molt at least four times on their mother's body.

Most of the juveniles, just after emergence from the brood pouch, clung to the second gnathopods and gills which were the thinnest parts of the mother's body (Table 1). Then, the juveniles moved to thicker parts of the mother as they grew. Therefore, it appears that the positions to which juveniles cling are primarily restricted by the size of the substratum, when the juveniles are small. The juveniles, however, do not always cling to the mother's body passively. In the time when the grown up juveniles moved from the mother's body to the seaweed substratum, the thickness of the branch of seaweed they were moving to was larger than that of the thickest part of the mother's body. It means that the grown up juveniles insist on clinging to the mother's body, though they can potentially cling to thicker things.

The clinging period of juveniles in *Pseudoprotella phasma* is about 3 weeks (Harrison, 1940), in *Caprella scaura* (*C. scaura typica*) up to one week (Lim & Alexander, 1986) and in *C. monoceros* it varied between 12 and 18 days. The care observed in *P. phasma*, belonging to Aeginellidae, is similar to that in *C. scaura* and *C. monoceros*, belonging to Caprellidae. However, we believe that the clinging behavior of juveniles in *C. monoceros* and *C. scaura* are basically different from that in *P. phasma* in their nature, for the following reasons. First, in *C. monoceros* and *C. scaura*, the mothers take care of their juveniles even after the clinging period of the juveniles. On the other hand, in *P. phasma*, the juveniles fall off and sink to the bottom after their clinging period. Second, the juveniles of *C. monoceros* and *C. scaura* appear to use their mother primarily as the clinging substratum. They have dense long setae on their antenna 2 and can filter feed or scrape the substrate during their juvenile periods (Caine, 1977, 1979a; Aoki, unpublished data), which means that they need not count on their mother for food. On the other hand, the juveniles of *P. phasma* seem to count on their mother not only for substratum but also for food. Because the juveniles probably are not filter feeders, as they do not have setae on their second antennae (Caine, 1977, 1979a). Harrison (1940) quoted a report that suggests the juveniles feed on their mother's food.

The clinging behavior of the juveniles and the care for them by their mother in *C. monoceros* is quite similar to that in *C. scaura*. Moreover, the aggressive behaviors of the *C. monoceros* females carrying her juveniles look close to that between males observed by Lim and Alexander (1986). The intraspecific aggression using the second gnathopods, observed in males of *Caprella gorgonia* is often fatal and a major cause of sex ratio bias (Lewbel, 1978). On the other hand, the aggressive behavior of *C. monoceros* and *C. scaura* does not always involve the second

gnathopods and appears to be more ritualistic than that of *C. gorgonia*. The aggressive behavior observed in *C. monoceros* and *C. scaura* could be the most common in the caprellids which have 'juvenile-cling type' maternal care.

Maternal care in Caprella decipiens

The type of maternal care observed in *C. decipiens*, in which post-emergence juveniles stay and grow around their mother has not been reported in caprellids before. However, the maternal care behavior of this species shares some behavioral characteristics with *C. monoceros* and *C. scaura*. First, the juveniles stay around their mother after emergence from the brood pouch, while the juveniles of the caprellids with no maternal care disperse immediately after the emergence. Second, the mother gathers her juveniles and carrying them away to another place in case of emergency, which is the common behavior in *C. decipiens*, and in *C. monoceros* and *C. scaura* after the clinging period of juveniles. Taking account of the above two behavioral characteristics, there should be some communication between mothers and juveniles in the maternal care caprellids. In Crustacea, the chemical communication for maternal behaviors has been reported in crayfish (Little, 1975, 1976). In Amphipoda, a female sex pheromone and a male chemosensory antennal receptor for that has been reported (Dahl *et al.*, 1970). Further studies are needed to know about the possible communication for maternal care behavior in caprellids.

Ecological significance of maternal care in caprellids

Almost 100% of the *C. monoceros* juveniles of the same cohort survived until maturation when they were reared *in situ* excluding predation (Aoki & Kikuchi, 1990b). However, the survival rate of the field juveniles that is roughly estimated from Fig. 4 is 69% (44.6/65.0). Thus, the survival rate of the juveniles of *C. monoceros*, the species with maternal care, is noticeably higher than those of

Caprella leaviuscula and *Deutella californica*, species without maternal care reared in the laboratory (Caine, 1979b). Possible reasons of the high survival rate of *C. monoceros* are as follows. First, the mother defends her juveniles from attacks by other caprellids. Second, body color of mothers looks like the seaweed substratum where she is clinging; moreover, the bodies of juveniles are transparent and similar to the colonies of some hydroids growing on seaweeds such as *Plumularia setacea*. Thus, visual predators like wrasses may tend to miss them. Third, when other animals approach to them, the juveniles can move to other places with their mother, which is quicker than crawling by themselves.

The larger variations of growth observed in the late stages of clinging juveniles appears to show the presence of competition within the same cohort of juveniles. Feeding or molting could be disturbed in the dense aggregations of larger juveniles.

Reproductive strategies of Caprella monoceros and C. decipiens

Moltings of the females of *C. monoceros* seem to be repressed during the maternal care period (Fig. 5a). Thus, 'juvenile-cling type' of maternal care may be a safer but more energy consuming behavior than the care shown by *C. decipiens*. In contrast, females of *C. decipiens* can molt even if the juveniles stay with her. Moreover, the clutch size of this species is larger than *C. monoceros* and become larger as she molts and grows, and the maturation size is also larger than that of *C. monoceros*. Comparing the reproductive characteristics and the methods of maternal care between *C. monoceros* and *C. decipiens* (Fig. 5; Table 2), we conclude that the energy investment for producing eggs is larger in *C. decipiens* than in *C. monoceros*. On the other hand, the energy investment for maternal care for juveniles is larger in *C. monoceros* than in *C. decipiens*.

Phylogenetic implications of maternal care in caprellids

In Amphipoda, the behavior in which juveniles emerged from the brood pouch and return to the brood pouch has been known in *Neohaustorius schmitzi* (see Croker, 1968), in *Eulimnogammarus* (*Marinogammarus*) *obtusatus* and *E. finmarchicus* (see Sheader & Chia, 1970) and in *Gammarus palustris* (see Borowsky, 1980). Furthermore, in *Dulichia rhabdoplastis* and *Dyopedos monacanthus*, the mother constructs a mast-like structure as her territory and stays there with her juveniles for a certain period. In particular, in *D. monacanthus*, the mother on the mast defends her juveniles against other animals (McCloskey, 1970; Mattson & Cedhagen, 1989). The development of amphipods in the brood pouch can be divided into two periods: 1) from ovulation to hatching (the embryonic period), and 2) from hatching to emergence from the brood pouch (the juvenile period) (Borowsky, 1980). The former is the period of maternal care for eggs and the latter is for juveniles. Moreover, the re-entry of juveniles to the brood pouch, the defense of juveniles by their mother in her territory, and the maternal behaviors observed in caprellids are maternal care for juveniles in 'the period after the emergence from brood pouch' (the post-emergence period). It appears that some chemo- or mechano-sensory communication is present between mothers and juveniles in this period.

It is generally accepted that caprellids have evolved from podocerid-like ancestors in Gammaridea in the process of adapting to the clinging to the substratum life-style (McCain, 1970; Laubitz, 1976). It is interesting that the maternal behavior that is quite similar to the ones in caprellids was observed in the podocerids (Mattson & Cedhagen, 1989), the ancestral amphipods of caprellids. Further studies are required to know the phylogenetical relation of the mother-juvenile interaction of amphipods in the post-emergence period.

Acknowledgements

We would like to thank Drs. M. Tanaka and S. Nojima for their valuable suggestions through this study. We are indebted to Drs. T. Bowman and L. Watling for their critical reading of the draft of the manuscript. We also express our appreciation to the members of the Amakusa Marine Biological Laboratory, Kyushu University for their technical assistance and helpful suggestions for this study. This work was partially supported by the Grant-in-Aid for Scientific Research from the Ministry of Education, Science and Culture, Japan.

References

Aoki, M. & T. Kikuchi, 1990a. Habitat adaptations of caprellid amphipods and the importance of epiphytic secondary habitats in a *Sargassum patens* bed in Amakusa, southern Japan. Publ. Amakusa Mar. Biol. Lab. 10: 123–133.

Aoki, M. & T. Kikuchi, 1990b. *Caprella bidentata* Utinomi, 1947 (Amphipoda: Caprellidea), a synonym of *Caprella monoceros* Mayer, 1890 supported by experimental evidence. J. Crust. Biol., 10: 537–543.

Borowsky, B., 1980. Factors that affect juvenile emergence in *Gammarus palustris* (Bousfield, 1969). J. exp. mar. Biol. Ecol. 42: 213–223.

Croker, R. A., 1968. Return of juveniles to the marsupium in the amphipod *Neohaustorius schmitzi* Bousfield. Crustaceana 14: 215.

Caine, E. A., 1977. Feeding mechanisms and possible resource partitioning of the Caprellidae (Crustacea: Amphipoda) from Puget Sound, USA. Mar. Biol. 42: 331–336.

Caine, E. A., 1979a. Functions of swimming setae within caprellid amphipods (Crustacea). Biol. Bull. 156: 169–178.

Caine, E. A., 1979b. Population structures of two species of caprellid amphipods (Crustacea). J. exp. mar. Biol. Ecol. 40: 103–114.

Dahl, E., H. Emanuelson & C. von Mecklenburg, 1970. Pheromone transport and reception in an amphipod. Science 170: 739–740.

Harrison, R. J., 1940. On the biology of the Caprellidae. Growth and moulting of *Pseudoprotella phasma* Montagu. J. mar. biol. Ass. U.K. 24: 483–493.

Hazlett, B. A., 1983. Parental behavior in decapod Crustacea. In S. Rebach & D. W. Dunham (eds.), Studies in adaptation, the behavior in higher Crustacea. John Wiley & Sons, New York: 171–193.

Imada, K., A. Hirayama, S. Nojima & T. Kikuchi, 1981. The microdistribution of phytal amphipods on *Sargassum* seaweeds. Res. Crust. 11: 124–137.

Laubitz, D. R., 1976. On the taxonomic status of the family Caprogammaridae Kudrjaschov & Vassilenko (Amphipoda). Crustaceana. 31: 145–150.

Lewbel, G. S., 1978. Sexual dimorphism and intraspecific aggression, and their relationship to sex ratio in *Caprella gorgonia* Laubitz & Lewbel (Crustacea: Amphipoda: Caprellidae). J. exp. mar. Biol. Ecol. 33: 133–151.

Lim, S. T. A. & C. G. Alexander, 1986. Reproductive behavior of the caprellid amphipod, *Caprella scaura typica* Mayer, 1890. Mar. Behav. Physiol. 12: 217–230.

Little, E. E., 1975. Chemical communication in maternal behavior of crayfish. Nature 255: 400–401.

Little, E. E., 1976. Ontogeny of maternal behavior and brood pheromone in crayfish. J. Comp. Physiol. 112: 133–142.

Mattson, S. & T. Cedhagen, 1989. Aspects of the behavior and ecology of *Dyopedos monacanthus* (Metzger) and *D. porrectus* Bate, with comparative notes on *Dulichia tuberculata* Boeck (Crustacea: Amphipoda: Podoceridae). J. exp. mar. Biol. Ecol. 127: 253–272.

McCain, C., 1970. Familial taxa within the Caprellidea (Crustacea: Amphipoda). Proc. biol. Soc. Wash. 82: 837–842.

McCloskey, L. R., 1970. A new species of *Dulichia* (Amphipoda, Podoceridae) commensal with a sea urchin. Pac. Sci. 24: 90–98.

Sheader, M. & Chia, F.-S., 1970. Development, fecundity and brooding behavior of the amphipod, *Marinogammarus obtusatus*. J. mar. biol. Ass. U.K. 50: 1079–1099.

Hydrobiologia **223**: 239–254, 1991.
L. Watling (ed.), VIIth International Colloquium on Amphipoda.
© 1991 *Kluwer Academic Publishers.*

Tube-building behavior in *Grandidierella*, and two species of *Cerapus*

J. L. Barnard [1], K. Sandved [1] & J. D. Thomas [2]

[1] *NHB-163, Smithsonian Institution, Washington, DC 20560, USA*; [2] *Reef Foundation, Box 170, Big Pine Key, Florida 33043, USA*

Abstract

Individuals of *Grandidierella bonnieroides* and *Cerapus* sp. R commence tube formation by enrolling themselves in a blanket of detritus and gluing together clumps of the material with strands of silk in a few minutes. The initial tubes are very ragged. *G. bonnieroides* then expands the initial tube by simply dragging in more detritus. In contrast, *Cerapus* sp. R. commences building an architectured tube outward from the ragged initial tube using very fine detritus collected either by grasping nearby benthic material or by filter feeding particles from the water column. The diameter of the nearly circular architectured tube is tailored very exactly to the height of the body for which we propose a formula. The more advanced *Cerapus* sp. K skips the initial phase of blanket-tube formation and only makes short architectured tubes as long as the body, carries these about in a fashion analogous to hermit crabs, and attaches them temporarily to epfloral-faunal substrates well above the sediment surface; therefore, no detritus masses are available to *Cerapus* sp. K. Tube-building in amphipods is roughly classified into 12 kinds having many subdivisions. Advancement of, or specialization in tube building, lacks apparent correlation with morphological advancement or systematic and evolutionary deployment.

Introduction

Tube-building behavior in *Grandidierella bonnieroides*, *Cerapus* sp. K and *Cerapus* sp. R is described. These are easily studied species amenable to laboratory conditions for viewing by microtelevision. Our purpose is to build a diverse body of behavioral information on many tube building species in a search for gross systematic differences and similarities which may have value in functional morphology and evolutionary studies.

At least 12 kinds of tube-building behavior are known for gammaridean amphipods, increased from four noted by Goodhart (1938). Among these is the kind of tube formation created by *Grandidierella bonnieroides* which we term blanket-detritus-initiation. However, *G. bonnieroides* can also instigate variations of this procedure when forced into anomalous behavior in the laboratory. *Cerapus* sp. R initiates tube building to some extent in ways similar to *G. bonnieroides* but then proceeds to build a stiff parchment-like circular tube of indefinite length erectly into the seawater column from a fine-sand substrate. A morphologically similar species, *Cerapus* sp. K builds similar tubes of portable form, in length and diameter tailored to the body ('architectured'), which are carried about and temporarily attached to epifaunal-floral substrates; but *Cerapus* sp. K always remains well above sediment surfaces and lacks detritus masses to make the initial blanket-tube and begins architectured tube-construction on the epifloral substrate by constructing a silken sheath and adding particles obtained from the water column by filter feeding.

We describe the results of laboratory observations on tube building in these three species which can result in initial tubes almost identical in appearance to those collected in the field but built by species in two genera of completely distinctive systematic position. In contrast, the two species of *Cerapus* are almost identical in morphology, yet have distinctive habitat positions and strong differences in the initial mode of tube building. Various anomalous modes of behavior which are regulated by changes in the experimental apparatus and its contents are also reported.

The twelve kinds of tube-building behavior in amphipods with which we are familiar include:

(1) spinning of a simple cylindrical tube composed at first only of amphipod silk to which silt or detritus particles may eventually adhere through saltation due to water currents or passive actions of organisms in the environment, for example, *Corophium ellisi* or *C. acherusicum* as described in Barnard, Thomas, and Sandved, 1988 or in *Lembos websteri* as described by Shillaker & Moore, 1978, or as discussed by Rao & Shyamsundari, 1966; one must note that detritus or mineral grains may be mixed with the silk at the moment of tube formation but Barnard, Thomas, and Sandved (1988) induced pure silk tubes in the laboratory;

(2) the restructuring of a severely damaged tube of type 1 which has been gnarled by confrontations of amphipods within or near the tube, such as tube-theft or after confrontations which have resulted in coagulation of the original tube and its adherent particles; this results in a rebuilt tube which has large lumps of agglutinated tube material and particles, reformed into a cylindrical tube with deformities, for example in *Corophium acherusicum* or *C. ellisi* where individuals of the same species have stolen tubes of other individuals;

(3) detritus-blanket tube formation, as in *Grandidierella bonnieroides*, where the individual dives into a mass of detritus-debris, pulls it over itself like a blanket and begins to dab strands of amphipod silk to cement particles together until walls have been formed of loosely agglutinated particles around the individual; irregularities are smoothed out over time until a relatively neat tube is formed inside an externally irregular mass of debris-detritus often mixed with mineral particles of the available substrate. This method can be modified to include digging downward into un-

consolidated poorly sorted mineral sediment which can be cemented together into a tube; the insides of tubes are ultimately lined with masses of tube-silk material which are rolled into balls and then pressed onto the tube walls and mashed flat;

(4) very regular straight, often annular, perfectly cylindrical, thick, stiff tubes formed by *Cerapus* sp. R by initiating blanket-tube formation as in *Grandidierella* but then selecting one end of the mass, punching outward and gluing together selected detritus in the fashion of a bricklayer building a chimney; these tubes are anchored into the sediment and may have a sand-ball anchor attached to adjust the center of gravity; we assume *Cerapus benthophilus* Thomas & Heard (1979) belongs here but the tubes are poorly architectured;

(5) similar cementation of particles into a cylinder or shape which can be carried about like the shell of a hermit crab (for example the undescribed *Cerapus* sp. J from La Jolla, California and *Cerapus* sp. K from the Florida Keys); formation occurs on an epifloral-faunal substrate and no anchor is made, no blanket tube is initiated and the tube may be formed initially as a loose sheath or semicylinder around the organism, then later attached to the floral substrate while being constructed and ultimately detached and completed into a cylinder;

(6) cementation of a silk tube inside a gastropod shell or into or on the surface of other environmentally available substrates which are more or less portable and can be carried or moved by the amphipod (such as in *Siphonoecetes* spp.) (Gauthier, 1941; Just, 1977a, 1984; Richter, 1978);

(7) the cementation of two or more particles of organic materials which have been deliberately researched by the amphipod, such as the ostracod-like paired valves cut by *Pseudoamphithoides* from the apices of *Dictyota* blades and cemented together to form a shell around the amphipod (Just, 1977b; Lewis & Kensely, 1982) or the mixed multiple large-particle tubes of *Ampithoe* spp. (Holmes, 1901; Skutch, 1926); or the blister-like capsules formed of microscopic particles in *Leptocheirus pilosus* (Goodhart, 1938);

(8) the overrolling of an algal blade or numerous strands of filamentous algae which are then cemented into a hollow tube for the house of such species as *Cymadusa uncinata* on the blades of *Macrocystis pyrifera* in California or in anastomses of *Enteromorpha* by *Ericthonius brasiliensis* in Florida;

(9) nest-building by *Ampithoe* spp. or *Cymadusa* spp. especially noticeable in aquaria where a messy spider-web like nest of silk strands is formed in the corner of a tank and which can be draped distances of 3–5 cm from one side of the tank to another across a corner;

(10) the digging of a burrow in fine sediment which is then lined with amphipod silk (such as in *Corophium bonelli* and *C. volutator*) or the 'cemented' sandburrows of Talitridae and *Microprotopus maculatus* (Hart, 1930; Goodhart, 1938; Meadows & Reid, 1966; Ingle, 1966; Shillaker & Moore, 1978);

(11) the development of a complex tube-pocket in which the amphipod sits upside down and projects its filtering devices through a dorsal slit (Ampeliscidae) (Enequist, 1950; Mills, 1966);

(12) theft of tubes from other species or phyla and their reformation; or the simple occupation of appropriate tubes, such as tubes occupied by the non-tube dweller *Podocerus brasiliensis*, or the tubes in polychaete *Phragmatopoma* masses or in generalized fouling masses stolen by *Ericthonius* sp. and probably relined with amphipod silk; we have also observed complete replacement of individuals of the polychaete *Polydora* which built tubes on experimental plates by individuals of the amphipod *Jassa* sp. within one month in Los Angeles Harbor in 1951.

Materials and methods

Our filming equipment was described by Barnard, Thomas & Sandved (1988). Since that time we have altered our methods to employ televideo equipment and now use a film camera only for slow motion studies. Our televideo cannot record at higher than normal speed and therefore cannot be used for slow motion. Tapes of normal speed can be slowed down when replayed; this however, does not result in as good resolution as gained by film in a camera run at 64 frames per second and then projected, after development, to 24 frames per second.

Our televideo equipment is composed of a camera mounted on a tripod; a professional grade VCR to record the camera image; a CRT to monitor the input from the camera; a computer and CRT with software program to provide single-picture imaging; a stage (see Barnard, Thomas & Sandved, 1988, and below). The camera is a Panasonic WV-6000 color video with 420 lines of horizontal resolution provided with a C-mount to which is attached either a bellows with serial attachment of a Zeiss Luminar 4.5 F-4 microlens or any Nikon. The camera inputs to a Sony Trinitron PVM-127IQ color video monitor with 550 lines of resolution and to a Panasonic VHS Model AG-6010 videocassette recorder (VCR) with 3 times lapse modes and with 525 lines of horizontal resolution; the VCR can record normal continuous images or be set to record for one second every 3, 12 or 36 seconds. The cassettes are 3M professional grade. The camera is so sensitive to light and has such good resolution that light sources have been changed from powerful hot lights to small, bifurcated quartz iodine lucite cold-lights like those used with regular dissecting microscopes.

The C-mount attached to the camera is an adapter with a male C-mount screwed into the camera with an apical female-side Nikon 45 mm adapter; a miniature bellows with 45 mm male adapter with attached female Nikon 45 mm adapter; to this is attached a Nikon 45 mm adapter plate with C-mount to which is attached a 60 mm Zeiss Luminar F4.5 microlens. This microlens has been provided with a semiautomatic diaphragm and a double-release cable which when depressed closes the diaphragm to a predetermined F-stop (3 stops down from wide-open F16). As soon as the lens has been stopped down the camera is activated. This device was made from Sandved's design by Martin Forscher, Professional Camera Repair, Manhattan, New York (who will make similar items by mail-order).

With this semiautomatic diaphragm focusing device one may easily convert from horizontal to vertical movie, video or still photography of specimens in the range of 4–10 mm. Oversized 70 mm diameter plastic lids are glued by epoxy cement onto the focusing wheels for improved smoothness of operation.

Our fiber-optic lights are from Fiber Optics Specialties Inc., Datco Inc., Clearwater Florida, each split into 2 branches; optimally we use 2 sets or 4 total lights so as to illuminate the subject from two front angles, to illuminate the background and then to provide backlight to make the setae glint luminescently. Backgrounds are made from bland blue or green or black pages and propped up behind the main aquarium.

The movable stage on which the aquarium is placed is made from 3 layers of aluminum plates. The middle plate has silicon grease on both sides so that all three plates slide individually in the x and y axes; four finger knobs on the top plate allow smooth movement of the aquarium for focus; the camera is kept rigid and the aquarium moved for focusing and viewing. A z-axis motion is provided by placing the rig of aluminum plates on a large plate attached to a vertical rack and gear operated by a large knob; this was adapted from an old petroglyphic microscope stage.

Microaquarium: We modified the microaquarium used by Barnard, Sandved & Thomas (1988). The new aquarium is almost cuboidal (x, y, z axes are 105, 85 and 95 mm) composed of a box made of 5 mm thick plastic except for the front surface into which is glued optically excellent Plexiglas 1 mm thick. This is actually glued into side slots inset 3 mm from the front surface to protect this window. Additional windows behind the front are provided at depths of 6, 13, and 27 mm. These windows behind the front window are portable so that similar 1 mm thick windows can be inserted or removed from the various slots; in this way microaquaria of 105×85 mm but of varying z-axes can be assembled quickly on the single cube; for example, when the first window behind the front is inserted, the aquarium is $105 \times 85 \times 4$ mm; this is very useful for our recording of branched

algae which can simply be inserted into the 4 mm space and the branches held by the front and back windows. We made one extra replicate of this aquarium.

When we determined in our study of *Cerapus* sp. K that a current was required for optimal behavior of the organism we took the replicate aquarium, drilled a hole in the 4 mm space between the first side slots 20 mm from the top of the main aquarium and inserted a plastic nozzle with 1.5 mm opening made from a forceps cover, and glued this to plastic tubing which led by gravity from an elevated bucket of seawater. We drilled two vent holes on the opposite side of the aquarium through the back plate set at 6 mm from the front plate and drilled a large exit hole through the back of the aquarium to which was attached an overflow tube directed into a bucket below the aquarium. A small C-clamp was used to regulate inflow; this allowed us to generate horizontal currents in the microaquarium of any feasible force. Water exited through the back plate into the main aquarium and out the main aquarium through its main vent. Water from the vicinity of the laboratory on Ramrod Key was sufficiently turbid to provide numerous particles for filtering by our amphipods. The speed of current could be altered from zero to approximately 150 mm per second and was noted by the shivering effect imparted to the setae of the amphipods clinging to the branching substrates. The two species of amphipods from the high-current environment survived, indeed thrived under continuous high current motion that permitted continuous filter feeding.

To confine activities of *Grandidierella* and *Cerapus* sp. R required an aquarium of smaller proportions than shown above. To achieve this we set back-plate 3 in its slot 27 mm behind the front plate, then added a plastic block of dimensions $7 \times 27 \times 80$ mm on each side to narrow the aquarium cuboidal dimensions to $41 \times 27 \times 80$. To lower the depth we added various numbers of glass microscope slides (which fit the areal dimensions of this system) to elevate the bottom by as much as 40 mm. For example, we might then have a chamber $41 \times 27 \times 40$ mm in dimensions.

Particular methods are discussed individually under each species heading. Actions are recorded with various film cameras in regular, slow or time-lapse modes or taped with microtelevision at regular speed or in time-lapse mode. Very low light levels required by video provide no deleterious heating of the experimental chamber. This work was conducted during short trips of 10–20 days to Florida. Many events that we have observed during these short trips have been seen only once. For example, we have seen mating in only 2 species, once each, in slightly more than 1000 hours of observation during 8 trips in 7 years. Our apparatus is ideal for continuous 24 hour operation because tapes can be made during sleeping hours by setting the VCR on time-lapse.

The advantage of using motion picture film in recording the behavior of amphipods lies mainly in the greater resolution of the end product and in the fact that film can be projected onto a larger screen for a larger audience. The disadvantages are often severe; the high intensity of light needed to achieve a reasonable depth of field brings with it a high level of heat which, if not corrected, may alter the normal behavior of the animals. Price is also a consideration; 100 feet of 16 mm Ekta-chrome with development is about 10 dollars per minute, the quality of which can only be seen days later after development; cost per hour is thus 600 dollars; much of the results may be unusable.

The disadvantage of video is the difficulty of projecting images to a large audience and the lower resolution. These are outweighed by:

1. Ability to use multiple smaller light sources with a combined effective use of less than 5 percent of light output of our film lights, with much less heat.
2. Ease of positioning smaller light sources in more varied configurations with side, top and back lighting, thus achieving desired contrast, sparkle on setose structures and separation of internal organs having different refractive indexes.
3. The ability to record in higher magnification with less light and thus annulling the lower resolution to some degree.
4. Immediate rerun of tape to check quality.

5. Ease of time-lapse modes up to 72 hours per tape.
6. Operating cost reduced to about 4 dollars per hour.

Results

Grandidierella bonnieroides

This species is widely distributed in the Caribbean Sea and northward along the east coast of the United States as far as southern Georgia and along the Gulf of Mexico coast. We have observed it only in the Indian River of east Florida. Individuals were collected from suitable sites adjacent to Linkport, the Smithsonian Marine Laboratory on the grounds of Harbor Branch Foundation, 6 miles (4 km) north of Fort Pierce, Florida.

In the survey by Nelson, Cairns, and Virnstein (1982), conducted in 1978, the species was found to be widely distributed in the Indian River; however, between 1982 when we first commenced studying this species and 1988 we had considerable difficulty finding it because of the vagaries of its occurrence. Beginning in late 1987 the species was found consistently at the bottom of the sea-wall in the east channel (running north-south) and in the west part of the marina at Harbor Branch Foundation. The species occasionally 'blooms' but die-off and complete absence in areas of former bloom are more common than the presence of consistent and replicable blooms. Our positive observations on occurrence of the species are too detailed for presentation here.

Grandidierella generally lives in what is called a marginally polluted environment in canals at the foot of seawalls covered with oystershells; at the foot of boat ramps emergent into mangrove marshes; below ponded masses of dying vegetation such as turtle grass; or rarely on floating docks in masses of other kinds of fouling material such as amphipod colonies made by Corophium and Jassa amongst sponges, barnacles and various polychaete tubes. Levels of pollution are unmeasured but noted as spilled fuels in the

Indian River, some obvious but occasional domestic outfall pollution, blooms of plants fairly obviously controlled by increased levels of phosphates-nitrates in the Indian River and by the occasional recovery of unnatural human disposals. During 1989 there was a penetration of *Ericthonius brasiliensis* onto pilings, buoys and floating docks up and down the lagoon from Linkport to major bridges at Fort Pierce, with occasional subadult specimens of *Grandidierella* found on docks and pilings. This was unusual because *Grandidierella* is usually found on the benthos. The colony at Linkport marina ceased to exist but then revived in 1988–89.

In the laboratory the living individuals were elutriated into fine mesh nets or screens and placed in finger bowls. Provided with fine sediments and organic detritus so as to reform their tubes, they are kept alive for several days. Colonies can be kept for weeks or months in larger aquaria provided with foul substrate when the water is aerated through a substage filter. Exudates from damaged local sponges are fatal to most amphipods, so great care must be taken during collecting and elutriation of samples.

Tube formation

The bottom of the microaquarium is provided with a layer of fine sand formed into a steeply sloped bank 1 cm deep facing forward towards the viewer. A small patch of fine organic detritus is spread along the banked slot within the focus of the horizontally aimed camera. An individual of *Grandidierella* is introduced into the aquarium and usually finds the detritus within 60 to 240 seconds. The individual partially buries its body sideways into the upper part of the sediment, and rolls over, dragging some of the detritus across its body like a blanket. Constant activity ensues for about 10 minutes, at which time a relatively well organized, ovoid tube about twice the length of the amphipod has been constructed. We assume that the amphipod is rapidly dabbing strands of silk to baste together detritus grains; the tube may appear knobbled or ragged depending on the poor assortment of detrital grains and agglutinations. At times the amphipod will roll into a position

where one side of the potential tube is formed by the glass viewing wall so that the tube is open towards the viewer. This behavior is common; the amphipod frequently burrows into the detritus mass, or a sediment mass if detritus is not present, next to the glass viewing plate. Such behavior appears to facilitate the speed with which a tube can be formed because one side of the tube is represented by the flat glass.

This experiment was replicated successfully many times. At times the individual covered the viewing glass with tube, preventing our observation and these tests were terminated.

As in *Corophium* spp. we observed that females were quicker and more steady in making tubes than males. Delays of up to four hours occurred before males would make tubes and their progress in tube making was slow. We therefore set up a replicate series of the following experiment to see if males commonly steal tubes of females: a female was allowed to construct what appeared to be an appropriate tube two to three times body length. In replicates various sized males were introduced into the aquarium. Males as large as or larger than females were observed to capture the female tube but small males were successfully repelled by large females.

We obtained an excellent 10 hour record of the following sequence: the construction of a tube by one female against the viewing plate. This tube was lengthened to five times body length before we introduced a male larger than the female. Within one minute, the male found one of the two entrances to the tube. Thrusting his body into the tube, he clashed with the female and withdrew; he then discovered the other end of the tube and entered, causing the female to reverse in the tube and engage in a clash, the male successfully driving the female from the tube. He then began to lengthen the tube at much slower rate than had the female. We introduced more detritus and the female soon began to build a new tube on the top wall of the old tube in which the male now resided. Within two hours she had constructed a new tube four times her body length. In the process of tube-building, the female sporadically depressed the top wall of the male's tube. The male continued to bring in new detritus from outside the tube, rolling it into large balls apparently stuck together with amphipod silk and then, while inverted, plastered these thickeners against his top wall with maxillipeds and legs 1–4; he would then press the dorsal surface of his body against the new construction to mash it into place; eventually the top wall became sufficiently thick to withstand any further depression by the female on the 'second floor'.

The male and female each continued to lengthen their tubes over 5 more days before we had to discard the experiment and return the animals to the sea. The two story tube complex was dubbed an 'apartment house'. The ends of the tubes were extended but slanting upward at each end on both ends, so that the apartment house appeared to be a Chinese pagoda, with upswept eaves. We have not found similar structures in nature and undoubtedly this structure resulted from the tiny space made available to these amphipods; we have, however, found tubes in nature stuck together in random or jumbled fashion.

Continuing the experimental process we placed males in aquaria where females already had constructed tubes. The males ejected the females but in this variation we did not provide additional detritus for the female to construct a new tube. In these cases the female within 5 minutes or less began to dig into the sand next to the viewing window, to dab legs 3–4 and construct a tube out of sand; these tubes rarely included the glass wall as a side. This demonstrated the rather strong tube-forming behavior of the female regardless of detritus source and was an action which did not occur when a rejected male was left without a tube. For example, we varied the experiment by placing males alone on sand without detritus, or males that had been rejected by more powerful females, or enacting situations where a rejected female had been removed from the aquarium and replaced by a male which was then rejected by the male occupying the only tube in the aquarium.

Tubes of *Grandidierella* are made with the inside vertical diameter about 1.5 to 1.8 times as tall as the body of the animal (including the anterior legs in flexed position as a measure of the

animal height). We believe the reason for this excessive height is because the animal, when 'resting,' hangs from the ceiling of the tube by hooking pereopod 5 (we are here using the modern numbering terminology, so that this is the leg attached to segment 5) to the top or sides of the tube. The animal hangs in somewhat acrobatic fashion with the posterodorsal side of the body pressed to the ceiling but with the head and antennae hanging downward at a slight angle. We observed that this facilitates circulation through the tube because the pleopods beat rapidly, though sporadically, and create a current sufficiently strong to draw tiny particles inward from the entrance. The water flows under the individual. Activity is sporadic once the tube reaches a length about 4 times body length but even during the building process the animal occasionally eats detritus instead of forming it into tube, often reverses direction to work on the opposite end of the tube, but rarely hangs from the ceiling. Hanging time often exceeds 10 minutes once the tube is well formed. Pleopod beating is often sporadic and the animal will often cease beating pleopods for up to 60 seconds at a time. Rate of beating is also sporadic.

Antennae form the principal searching device and the principal device for dragging masses of detritus into the tube. The masses are processed into large balls often larger than the head of the organism by downward dragging of the particles into the vicinity of the maxillipeds where they are rotated by movements of antennae, legs 1–2 and maxillipeds and dabbed by legs 3–4 which apparently add strands of silk to glue together the mass. The ball or irregular mass will then be moved into the space between anterior appendages and antennae and placed into position, either at one end of the tube or plastered to the inside wall for thickening. The amphipod rotates into any possible position to make the placement and uses pereopod 5 to anchor its ultimate position; pereopod 5 is short and projects laterally; pereopods 6 and 7 are much longer, pereopod 6 projecting obliquely downward and backward, pereopod 7 bent and projecting straight backwards or slightly dorsoposteriorly. Pereopods

3–4 move very swiftly in the dabbing process while some positions seem to be assisted by or balanced by use of gnathopod 2, which is quite large.

Cerapus *species R*

Cerapus sp. R was found by Kalani Cairns of Florida Department of Natural Resources, and collected in February 1987 in the Indian River, a long tidal lagoon protected by a wide bar seaward, containing a part of the Atlantic Coast Waterway, and subject to extensive commercial and recreational development. Colonies of this species are vagarious. Kalani Cairns in monitoring the Indian River, has found patches of this species here and there but they do not endure more than a few months. The first colony we investigated occurred on the north side of an old dredged channel leading into Taylor Creek on the north boundary of Fort Pierce; this colony was dredged out 9 months later. The colony was about 1–2 meters wide and 20 meters long and was almost purely composed of amphipods in dense aggregations on fine gray sand. The aggregation occurred just at the crest of the steep and evenly molded (by previous dredging) bank marking the edge of the channel; at lowest spring tides the colony was about 5 cm above low water. This habitat was composed of very fine gray, well sorted sand densely studded with vertically erect tubes of *Cerapus* sp. R projecting a few millimeters above the sand surface. A few tanaids and polychaetes were also in the colony. The appearance of the colony was so cryptic that only by stepping on the colony could one identify it by the crunching of the tiny fine tubes. The solidity of the bottom is enhanced by these tubes because the mud bottom immediately adjacent is oozy and human feet sink slightly when implanted.

In May 1988 the species was found again by Kalani Cairns 3 miles (2 km) south of Fort Pierce at 3007 Indrio Blvd, Fort Pierce, occupying sandy masses glued to the outer edges of beach rock in 0.2 m of water depth. This colony was scattered and loosely organized, occurring in small patches of less than one quarter square meter on slight but solid elevations above the surrounding black to brown ooze.

The silty-sandy masses were composed mostly of the polychaete *Polydora* sp. and a tanaid, with scattered *Cerapus* tubes projecting above the mass. In both places the water was foul and turbid but moving (not ponded). Samples of mat, 20 × 40 cm or larger were scraped into buckets, with a slight amount of seawater covering them from the sun, returned 9 miles to the laboratory and quickly placed into clean seawater. There we broke the colonies into small fragments about 2 × 2 cm, containing 100 or fewer individuals, and placed them in large bowls with seawater covering the colonies to a depth of several cm and supplied with bubbling air. These remained viable about 5 days with aeration, daily water changes and stirring.

En masse, the activities of these colonies are incessant; many of the individuals are always in a feeding position and filtering detritus from the water, reaching out of the tube as far as possible to rake in detritus from the side of the tube or the nearby substrate, and either eating detritus or using it to construct or lengthen their tubes. We believe that there is an optimum tube length which is reached when the tube projects above the sediment more or less the same distance as the body length of the organism. Greater projection would prevent sufficient body extension for the antennae to reach the surrounding sediment surface. When the tubes are dug out from the sediment, however, they are often greatly longer than this, upwards of 5 to 10 times as long as the body, we suggest that sedimentation has occurred in these cases and the amphipods have continued upward building to escape inundation by sediment. Other species such as *Cerapus benthophilus* Thomas & Heard (1979) build extremely complex joint tube colonies of indefinite length and we assume that these tube masses are so massive that they act as a substrate on which detritus accumulates.

At first we videotaped our colonies from an upper view, making hours of tapes of activities and then rigged the microaquarium in various ways to obtain side views of colonies and individuals at substrate level.

Mature *Cerapus* colonies are composed of very densely packed erect tubes projecting above a matrix of fine gray sand or they are scattered in a matrix composed of tubes of other organisms such as polychaetes and tanaids. Sections of the mat about 2 cm square were cut apart and placed in dishes for vertical film/tape observation or placed in the smallest chamber of our side-view aquarium. Individuals were also squeezed entirely out of their tubes and introduced into either dishes or aquaria supplied with a 1 cm base of fine sand bearing a 7 mm wide × 3 mm wide × 3 mm tall patch of fine detritus. Individuals in the stock aquaria which had sawed off their tubes and were freely swimming or walking while carrying the tube were introduced into dishes or microaquaria with a 1 cm base of fine sand (median diameter approximately 62 microns).

Tube formation

Individuals free of their tube dive into the fine detritus on the fine sand base, roll, probe, and jerk backwards and forwards, on rapidly reverse themselves, somewhat in the fashion of *Grandidierella* but remaining more visible. Patches of detritus are assembled and clotted together with amphipod silk until a globular mass is formed. The amphipod then inserts its body headfirst into the middle of the mass thus creating a loose tube which completely encloses the amphipod but which has 2 open ends; the amphipod often reverses between the two ends inside the tube; after about 10 minutes of forming this rough tube one end is selected and a perfectly circular, thin-walled extension is commenced which fits the body size of the amphipod very exactly. A *Cerapus* taken from its tube and added to the colony will push its bent urosome into the soft mass of the colony and then commence assembling and clotting detritus. The tube diameter appears to be controlled by the distance between the dorsal edge of the thorax and the weakly extended legs 1–4. These legs can barely be extended and manipulated within and outside the tube sufficiently to form a new tube lip. We believe this distance is equivalent to the full extension of articles 1–3 of these legs plus the width of the bent

article 4 plus a very small shim distance not yet measured.

The tube extension is formed by the amphipod grasping and dragging any available nearby detritus with the first antennae towards the space between first and second antennae. The second antennae project laterally and then are bent obliquely downward; the first antennae arch outward, upward and anteriorly in the grasping motion; the second antennae are slightly less flexible but also are whipped about in a manner which draws detritus through the water towards the maxillipeds. Very rapid antennal whipping thrusts matter towards the head where it is processed by the mouthparts in a rolling fashion and coarse particles are rapidly ejected. A bundle of suitable material being held by the first gnathopods against the maxillipeds is passed to the second gnathopods and mashed onto the lip of the tube in an arc of about 60 degrees. We do not know if silk has already been applied to the material but the maxillipeds, gnathopods and pereopods 3–4 are used rapidly to form the material into an accretion somewhat similar to a partial row of bricks being added upward on a chimney. The final action is a clamping/rolling motion once or twice by the four members of pereopods 3–4 held together tightly in a tier; this results in a very symmetrical and precise incremental addition of material to the apex of the tube. The process then recommences and the amphipod rotates into position to add the next increment adjacent to the first.

We judge that about 4–6 rotations of 60–90 degrees each are required to finish one layer or annulus. These annuli however are somewhat spirally continuous rather than being perfect rows of 'bricks'. The amphipod works almost continuously and we judge that a length of tube as long as the body of the animal requires about half an hour of continuous and furious work. The high rate of construction is optimal because we supply a surfeit of detritus. Occasional pauses occur when the amphipod is processing and eating some of the same detritus. If detritus is not readily available in the water column the amphipod reaches out from the tube in full body extension to grasp detritus on the surrounding substrate. If the tube is well above the substrate surface the amphipod reaches downward to draw up detrital masses which are collected in the anterior space between the antennae and head by rapid whipping motions of the antennae. The tube annuli produced in experimental chambers are less distinctive than found on tubes collected from nature. We assume that this results from the uniformity of the detritus we provide the organism whereas in nature the organisms are obtaining materials both from the substrate and from filtering. The kind and color of detritus thus varies from time to time and results in annular tubes of serially distinctive composition.

Resetting of loose tubes

Loose tubes with live *Cerapus* inside are occasionally found in our laboratory stocks. They resemble the situation of the free-living *Cerapus* sp. K described below except that the tubes are generally much longer than the amphipod bodies. We have been informed by competent observers but have not as yet observed this ourselves, that *Cerapus* 'saws' off their tubes from inside by use of the serrated edge of uropods 1 or 2 (but possibly telson which has the 'best' saw-blade), then carries the tube by swimming or crawling. Both swimming and crawling are effected by grasping motions of the antennae. When tubes become stuck in unfavorable positions amongst epiflora and epifauna of the substrate the amphipod uses powerful motions to dislodge the tube and thereby continue the journey. These motions include the extreme body extension outward by antennal grasping and then a drawing of the tube outward along the body; by rolling the tube; by backwards pushing; or by reversing inside the tube, proceeding to the opposite end and employing the same motions. If the motions of the amphipod have irretrievably stranded the tube, we assume the organism must adapt; such adaptation has been observed by us for up to 2 hours as measured by restoration of normal feeding behavior at the end of the tube more suitable for feeding.

When we place one loose tube and its occupant into a microaquarium containing 1 cm of fine sand the amphipod darts to the nearest corner or edge of the aquarium leeward of the light. The tube is immediately erected and wiggled into the sand as far as possible until only a section shorter than the body length remains above the substrate. This experiment was repeated 20 + times with similar results; when a mass of tubes is already present the newly inserted individual attempts to erect and submerge its tube in the most favorable position towards the back wall of the aquarium.

An alternative method has been observed twice and the evidence for this method as a natural event has been found consistently on tubes teased out of the natural environment. Before submerging the tube, while it lies on bottom, the amphipod reaches outwards, grasps together a pile of sand and quickly cements this together into a globular or pyriform anchor which is attached to the side of the tube at one apex; after this construction the tube is then quickly elevated and submerged into the substrate, the amphipod occupying the 'anchor' end being submerged and apparently digging furiously with its antennae (digging below substrate not observable). If the substrate is much shallower than the length of the tube the amphipod ultimately ceases its struggles and emerges at the upper end projecting towards the water column. It may commence feeding by grasping rapidly for detritus in the water or lean outwards and downwards to scrape up detritus from the substrate. Feeding activity varies from (1) continuous for upwards to an hour or (2) with intermittent periods of time partially withdrawn into the tube, or (3) complete disappearance into the tube; the tube may often be suddenly jerked or slowly moved; we suspect the amphipod has proceeded to the base of its tube inside and is digging or otherwise displacing the tube.

Cerapus *species K*

This species was first found by J.D. Thomas in 1986 in Cudjoe Channel in the lower Florida Keys northwest of Big Pine Key, near Little Torch Key, at a depth of 10 meters on rocky or limestone bottom. It was collected again in October 1988 when it was studied in the laboratory for five days. In May 1989 we recollected it and with our portable microvideo system available nearby were able to record its activities for 10 days in the laboratory of the Reef Foundation on nearby Ramrod Key. *Cerapus* sp. K lives in high velocity tidal channels where tidal exchange can reach 5 knots. The amphipod is found in its portable tube clinging to various branching plants and animals such as species of the algal genera *Hypnea*, *Amphiroa*, *Codium*, *Halimedon*, *Gracilaria*, and the hydroid *Cnidoscyphus marginatus* (Allman). The branching plants and sessile animals reach heights up to 30 cm above the rocky substrate to which they are anchored. The bases of these

Fig. 1. (a) side view of *Grandidierella bonnieroides*, looking through glass plate into tubes built against plate; below, male in long tube built by and from which female was ejected, above, female constructing new tube; (b) *Cerapus* sp. K, roving male examining other individuals.

branching substrates contain large numbers of nestling, domicolous and ascidiophilous amphipods but the branches projecting upward into the water stream contain only 2 species as far as we know, *Cerapus* sp. K and *Podocerus* sp. G (a new species like *Podocerus fulanus* of California). *Cerapus* sp. K is so abundant on the plants that we estimate in some cases every millimeter of every branch is covered on at least one side by the tubes of the amphipod. During periods of slack tide, JDT has observed males swimming in the water column about 10–20 cm off the bottom. In the laboratory individuals swim by antennal movements after projecting the body halfway out of the tube. Uropod 1 bears a large ventral interramal hook which we surmise may help the individual hang on to the tube. As far as we know *Cerapus* sp. K is the only species of the genus with this feature.

This species lives in tubes of similar diameter to those of sp. R but the tubes are free and are able to be carried about by the individual. The tubes of *Cerapus* sp. K. are noticeably different in color from those of *Cerapus* sp. R. Unlike the medium-dark gray interspersed with thin brown rings, these tubes are a pinkish white interspersed with thin rings of deep caramel similar in color to the *Hypnea* stems; the overall color cast of these tubes to the human naked eye from a distance of 0.5 meter is extremely bright whitish pink or ochraceous pink. The bodies of *Cerapus* sp. K are of a brilliant pinkish purple, compared to the grayish purple of *Cerapus* sp. R.

The tubes are about as long as the length of the animal body (including antennae), and are neatly finished and slightly flared at each end. Unlike *Cerapus* sp. R, we observed no continuous construction of tubes. There is an optimal length to these tubes and once finished and neatly flared they are not altered as far as we can determine. One question we have not been able to answer is how the organism either changes an old tube or acquires a new tube of larger size to fit the enlarged body after molting. In the short periods of time we have been able to work on this species we have seen no molting, and no tube exchange. The tubes are of even diameter at both ends and no tapering extensions occur to suggest that the amphipod alters its tube after ecdysis.

We concentrated on laboratory observations of individuals collected on *Hypnea* sp. The alga was suspended in a small aquarium and observed by stereoscopic microscopy. We found tubes of *Cerapus* sp. K of all sizes, each fitted tightly to the size of the occupant, from early juvenile to full adult, either crawling about on the stems of the alga or attached by amphipod silk to the stems.

Crawling and temporary clinging are effected primarily by the use of the powerful antennae and to a slight extent by the first 4 legs. The male second legs (gnathopod 2) are powerful whereas the female gnathopod 2 is feeble. Whether or not forward positive pressure generated by currents from the pleopods is a force in movement is not known because the opaque tube hides the pleopods completely.

The attachment points of tubes and methods of attachment are variable. When first brought into the laboratory, in October 1988, some individuals had attached their tubes to the alga with a small cushion (somewhat similar to a morel mushroom) of amphipod silk on the side of the tube at various distances from either end. We touched with a probe by the observer these tubes moved very little and would spring back to primary position when released. This method of attachment was not seen again in several collections we made in May 1989 and we are puzzled at the later absence of this attachment method. Other individuals attached their tubes apparently by a single cable of amphipod silk usually at the tube lip on one end; these tubes therefore hung freely suspended by a short distance and could wave in the current we generated in the aquarium. Attachment on the alga was various; we found tubes attached so as to project horizontally at right angles to the vertical axis of the alga; tubes attached axially along the algal stems; tubes attached to other tubes; free swinging tubes attached generally on projections formed by side branches of the alga or by other *Cerapus* tubes so that the individual hung downward suspended in water; a small percentage of tubes was attached by a long narrow bead or series of beads of silk-cement along the side of tube; the entire colony of alga and tubes supporting an errant population of *Cerapus* individuals in their tubes crawling here and there; and a few individuals without tubes which appeared to be newly hatched juveniles.

Once we could observe the behavior on video in May 1989 we determined that the tube attachment cables are probably composed of at least 4 and up to 28 individual filaments. Numerous replicates showed us that when the amphipod alights on the substrate branch, the antennae grasp the branch while the amphipod extends quickly forward out of the tube, rolling legs 3 and 4 across the lip of the tube and then dabbing them on the alga. We assume that four strands are therefore emitted during each thrust. We observed varying forward pushes up to a maximum of 7. One might assume therefore that 28 strands could be laid with the two pairs of silk-legs in 7 forward pushes. The strands appear to clot together somewhat unevenly into fewer but thicker 'ropes' when seen by backlight. When we activated strong current in the aquarium the motions were undertaken rapidly so as to effect rapid attachment and prevent accidental 'blowing away' of the amphipod at the moment it alighted from a swim or after crawling into the favored position. If the tube is attached only at one end it is blown outward into the downstream direction of the current. The amphipod then reverses inside the tube and emerges from the downstream end and commences filtering with the antennae. Occasionally, the tube is also cemented by strands on the downstream side; the amphipod emerges from the downstream end and 'sails' the tube downward to a position where the antennae grasp the branch and strands are emitted to cement it into position parallel to the branch. The branch may be at any possible angle to the current or to gravity. Filter feeding in the fully attached tube can occur from either end and one supposes that the amphipod chooses the end with the most favorable purchase to the available current.

In only one of more than 10 000 + observations did we find one individual which attached one end of the tube to the substrate and then used the antennae in a swimming motion to soar at right angles directly into the current upstream; the amphipod filter-fed for about 5 minutes in this remarkable position.

We adjusted the current flow in the microaquarium in various degrees up to levels which would blow away unattached individuals.

Roving males

In medium current velocities as denoted by the rapid vibration of antennae and setae on the individuals, the largest males detach themselves with their tubes after short feeding cycles and rove the branching substrate. They investigate almost every individual larger than a small juvenile. Females generally remain attached during the high current feeding cycles. Many encounters between the roving males and 'female' individuals appear to be almost violent in human terms, the male moving toward the female and rapidly thrashing her antennae. Certain females return this thrashing but others consistently withdraw into their tubes. One large female which had mated about an hour earlier was successful in repelling every male intruder without having to withdraw. Many smaller females were forced to withdraw at which time the male might inject his antennae deeply into the female tube in jerky, though occasionally prolonged thrusts. The male would quickly investigate each end of the female tube, then depart along the branch to the next individual.

We assume the male in checking females for readiness to mate. We never observed this behavior in *Cerapus* sp. R.

Tube-building in juvenile

One juvenile estimated to be recently fledged, was followed with our video camera and the process of tube building was observed until a roving male accidentally destroyed the event 11 hours later.

The tiny juvenile was first seen clinging to one or more threads stretched along the hypotenuse position of right angled limbs of the alga *Hypnea* sp.

In the rapid motions of this individual, which we judged to be 0.4 mm long, there evolved within an hour a faint pink, thin, translucent shroud-hood attached to the main holding thread. The

amphipod remained inside the arc of this semilunar hood. We could detect the faint cast of this shroud best when it became about as long as one body segment of the individual and when the juvenile would partially move in and out of the narrow shroud. At the end of two hours the shroud had changed to incandescent white as it had apparently been thickened by addition of the white detritus (composition unknown) typical of the water in the Keys channels. The tube seemed to be composed of a granular, slightly prickly mosaic of particles glued together. The juvenile was protected from the powerful current of the aquarium by hanging in the plant axil but continuously darted its antennae into the stream to catch detritus particles which varied from almost invisible to as large as the amphipod head. The juvenile appeared to eat some of the particles but molded most of them into either end of the semilunar tube in the same fashion as in *Cerapus* sp. R. At the end of three hours a definite shell forming a partially closed U-shape now cemented on both limbs to the alga had evolved. At the end of 10 hours and 58 minutes the shell was about 6–7 body segments long but had not been detached from the alga nor had the unfilled arc of the cylinder been completed. Destruction by a roving male then occurred, the tube being split into two parts and the juvenile lost in the current.

We processed incoming collections of algae and hydroids in search of similar tubes from natural conditions but found only 2 probable examples, also partially destroyed.

Mating

We observed one apparent mating. A roving male had apparently detected a female ready for mating and we first found the pair quiescent perhaps a few minutes after their linkage. The female had begun to extend her body outward from the tube while the much larger male assumed a position at her top right side but in the same direction. Both slowly extended their bodies farther from the tube and at 15 minutes after we first found them we saw a spray of white particles drifting downstream from the vicinity of the abdomens of both individuals. This may have been sperm which escaped the transfer process. We otherwise saw no actual sperm transfer. At the end of 20 minutes the male resumed ordinary activities, but the quiescent female was knocked out of her tube and lost in the current.

Discussion

Literature on tube-building in amphipods is not diverse; bibliographies in the items given in our Literature Cited have most of the pertinent papers. A few taxonomic papers since 1758 also have drawings of tubes attributed to various species. No survey of their formation has been done.

A wide diversity of tube kinds has been discovered as outlined in the Introduction to this paper. Our studies of 6 species, 3 herein and 3 in Barnard, Thomas & Sandved (1988) plus the studies cited above, demonstrate that distinct behavioral modes are associated with the resultant tube forms. Certain similarities occur among some of the species we and others have studied whereas other species and species groups have wholly distinct and diverse behavioral modes.

The simplest form of tube building appears to be that in which the organism simply emits strands of amphipod silk which are manipulated into the form of a tube (*Corophium* spp.). Foreign particles need not be available for integration into the tube but we suspect that virtually all tubes in nature will include some foreign particles because they are constantly present in the environment and cannot be avoided by the organism. Of the species we have studied and those reported in the literature a pure silk tube has been built only by *Corophium* spp. and this has been induced in the laboratory by withdrawal of any foreign detritus. In most cases it would appear that the incorporation of foreign particles adds strength to the tube. Obviously, certain species actively collect nearby or passing particles obtained by search or filter feeding and incorporate them into the tube. The primitive tube-builders cement the tube to solid substrates but more advanced species may cement them to unstable substrates by multiple attachment points. More advanced species may then aggregate to form multiple tube colonies built on top of or at the side of previous members until a thick silty mass accumulates on which only the surficial tubes are viable. Another advancement, as in the siphonoectids, is the active selection of a portable substrate that can either be carried around or situated in such a position as to afford partial protection from predators.

A different form of tube building occurs in such taxa as *Grandidierella* and *Cerapus* where the organism does not spin a silken tube initially but finds detritus or inorganic particles and cements them together with silk strands, so that a tube that is not fully lined with silk eventuates. Whether or not this is a more primitive method is not yet known but we suspect it has great advantages to the organism because of the speed with which a tube can be constructed – at least as an initial tube affording optimal protection. The amphipod may then enlarge this tube or commence an entirely unique tube at a much slower rate because it can retreat to the initial tube when necessary. In *Grandidierella* the initial tube is simply elongated but in *Cerapus* sp. R a wholly new and well architectured tube is constructed as an extension of the primary tube. When this complex has reached optimal length the amphipod adds an anchor for what we assume is gravitational alteration and then erects the tube complex and digs it into the sand. We assume that if the amphipod has constructed the primary tube in a suitable position, or perhaps has anchored it or preburied it, then the architectured tube can be built directly into the water column and later erection is not required. This is a point requiring study. A final advancement in these amphipods which build cylindrical tubes, is the capability of *Cerapus* sp. K to build a circular tube similar to *Cerapus* sp. R but the tube is built with only filtrable particles above the inorganic particulate substratum on an epifloral-faunal substrate. The tube is constructed as a semilunar sleeve attached as an arc to the substrate initially but presumably is detached and finished into circular form at later times. We need further study to determine this course of events. This kind of tube is then closely tailored to body diameter and length and is carried about by the amphipod and attached from time to time at locations favorable to feeding and other activities (perhaps mating, for example). This advancement is signaled by the species, *Cerapus* sp. R, which lives on the bottom in architectured tubes of indeterminate length by the apparent capability to saw off its permanently attached tube to a suitable length and then crawl or swim to a new locality and reset the tube with a new anchor attached to the architectured tube. To what extent this occurs in nature is unknown but this is an event which can be induced in the laboratory.

A further, seemingly very simple tube is formed by certain species of *Corophium* which dig a burrow, often U-shaped (as in mud shrimps), in the sediment, and then line it with silk ('secretions' as stated in the literature).

We assume the more advanced tube formation methods are employed by species which use very little silk and select large fragments from the environment as the basic material for the tube. Much reduction in energy consumed for silk production would thus result. However, the question arises as to how much actual energy is consumed in producing silk as against the energy required, for example, by *Pseudamphithoides* to select and saw off pieces of algae in symmetrical pairs and then cement them together to form a bivalve shell.

The association between advancements in morphology and advancements in behavior and tube building methods is not worked out as yet. There may be no connection because we observe that *Corophium*, a very specialized kind of amphipod in terms of morphology (body form, antennae, some appendages highly derived), appears to use very simple tube building methods, whereas the much more generalized amphipod kind as represented by *Pseudoamphithoides*, constructs a unique tube using materials in the environment that have been very particularly selected. We suspect that various advanced modes of behavior and tubes building have no evolutionary continuity but are independently developed. An analogy might be the development of lek-behavior in Aves such as manakins and birds-of-paradise.

Undoubtedly, a great deal of study will be re-

quired to determine polarities in evolutionary deployment of behavior apart from estimated polarities that have been derived from conclusions about morphology and used for systematic classification. We already see illogical juxtapositions. Clearly, much more detailed and more microscopic examinations of morphology, function and behavior are necessary to build a comparative biology of tube building. Many more of the hundreds of tube-building amphipods need to be examined and our list of amphipod tube kinds needs to be organized into an evolutionary scheme, if possible.

Acknowledgements

We thank Dr. Mary E. Rice of Smithsonian's Linkport Station at Fort Pierce, Florida and her staff, Julianne Piraino, Sherry Reid, Woody D. Lee and, Hugh F. Reichardt. Dr. Kerry Clark of Florida Institute of Technology, Melbourne, Florida, advised us during the planning of our videotape apparatus and arrangements. We thank Joyce Coleman for helping K. Sandved with film-tape procedures. We appreciate the assistance of Drs. L. Watling and A. Hillyard of the University of Maine, Darling Marine Center, Walpole, in specifying and establishing our microtelevideo apparatus. Jan Clark and Elizabeth Harrison-Nelson of Smithsonian Institution, helped with field work. Kalani Cairns, of Florida Department of Natural Resources and formerly of Harbor Branch Foundation, assisted us many times in finding amphipod colonies in the Indian River.

References

Barnard, J. L., J. D. Thomas & K. B. Sandved, 1988. Behavior of gammaridean Amphipoda: *Corophium, Grandidierella, Podocerus,* and *Gibberosus* (American *Megaluropus*) in Florida. Crustaceana, Supplement 13: 234–244.

Enequist, P., 1950. Studies on the soft-bottom amphipods of the Skagerak. Zool. Bidr. Upps. 28: 297–492, 67 figs.

Gauthier, H., 1941. Sur l'éthologie d'un amphipode qui vit dans une coquille. Bull. Soc. Hist. nat. Afrique Nord 32: 245–266.

Goodhart, C. B., 1938. Notes on the bionomics of the tube-building amphipod, *Leptocheirus pilosus* Zaddach. J. mar. biol. Ass. U.K. 23: 311–325.

Hart, T. J., 1930. Preliminary notes on the bionomics of the amphipod. *Corophium volutator* Pallas. J. mar. biol. Ass. U.K. 26(3): 761–789.

Holmes, S. J., 1901. Observations on the habits and natural history of *Amphithoe longimana* Smith. Biol. Bull. Mar. Biol. Lab. Woods Hole 2: 165–193.

Ingle, R. W., 1966. An account of the burrowing behaviour of the amphipod *Corophium arenarium* Crawford (Amphipoda: Corophiidae). Ann. Mag. nat. Hist. 9: 309–317.

Just, J., 1977a. A new genus and species of corophiid Amphipoda from pteropod shells of the bathyal Western Atlantic, with notes on related genera (Crustacea). Steenstrupia 4: 131–138, 1 fig.

Just, J., 1977b. *Amphyllodomus incurvaria* gen. et sp.n. (Crustacea, Amphipoda), a remarkable leaf-cutting amphithoid from the marine shallows of Barbados. Zool. Scr. 6: 229–332, 3 figs.

Lewis, S. M. & B. Kensley, 1982. Notes on the ecology and behaviour of *Pseudamphithoides incurvaria* (Just) (Crustacea, Amphipoda, Ampithoidae). J. nat. Hist. 16: 267–274.

Meadows, P. S. & A. Reid, 1966. The behaviour of *Corophium volutator* (Crustacea: Amphipoda). J. Zool., Lond. 150: 387–399.

Mills, E. L., 1966. The biology of an ampeliscid amphipod crustacean sibling species pair. J. Fish Res. Bd Can. 24: 305–355.

Nelson, W. G., K. D. Cairns & R. W. Virnstein, 1982. Seasonality and spatial patterns of sea-grass-associated amphipods of the Indian River Lagoon, Florida. Bull. mar. Sci. 32: 121–129.

Rao, M. V. L. & K. Shyamasundari, 1966. Tube building habits of the fouling amphipod *Corophium triaenonyx* Stebbing at Visakhapatnam Harbour. J. zool. Soc. India 15 (1963): 134–140.

Richter, G., 1978. Einige Beobachtungen zur Lebensweise des Flohkrebses *Siphonoecetes della-vallei*. Nat. Mus. 108: 259–266.

Shillaker, R. D. & P. G. Moore, 1978. Tube building by the amphipods *Lembos websteri* Bate and *Corophium bonnellii* Milne Edwards. J. exp. mar. Biol. Ecol. 33: 169–185.

Skutch, A. F., 1926. On the habits and ecology of the tube-building amphipod *Amphithoe rubricata* Montagu. Ecology 77: 481–502.

Thomas, J. D. & R. W. Heard, 1979. A new species of *Cerapus* Say, 1817 (Crustacea: Amphipoda) from the northern Gulf of Mexico, with notes on its ecology. Proc. biol. Soc. Wash. 92: 98–105.

Hydrobiologia **223**: 255–282, 1991.
L. Watling (ed.), VIIth International Colloquium on Amphipoda.
© 1991 *Kluwer Academic Publishers.*

Precopulatory mating behavior and sexual dimorphism in the amphipod Crustacea

Kathleen E. Conlan
Zoology Division, Canadian Museum of Nature (formerly National Museum of Natural Sciences),
P.O. Box 3443, Station D, Ottawa, Ontario K1P 6P4, Canada

Key words: mating behavior, sexual dimorphism, amphipods

Abstract

Accounts in the literature of precopulatory mate-guarding in gammaridean amphipods are that males use one of two strategies for mating: either they mate-guard by carrying or attending their mates until they are ready to molt and be fertilized, or they do not guard, instead searching benthically or swarming pelagically at the time that females are ready to molt. Mate-guarding by carrying has been documented for species of the superfamilies Gammaroidea, Talitroidea, and Hadzioidea. Mate-guarding by attending has been found in the more sedentary Corophioidea and Caprellidea. Non-mate-guarders that search pelagically are species of Ampeliscoidea, Lysianassoidea, Phoxocephaloidea, Oedicerotoidea, and Pontoporeioidea. Non-mate-guarders that mate-search benthically are species of Eusiroidea, Crangonyctoidea, and Haustorioidea. Mate-guarding and non-mate-guarding males develop different secondary sex characters at maturity. Mate-guarding males have enhancements for fighting and signalling. These alterations are more elaborate in males that attend their mates than in males that carry their mates. Non-mate-guarders that search pelagically develop enhancements for swimming and sensing. Non-mate-guarders that remain benthic exhibit little change at maturity. Most mate-guarding males develop their secondary sexual characters over several molts and mate over more than one instar. Pelagic mate-searchers develop their secondary sexual characters at the last molt and mating is confined to the last instar. Females of most mate-guarding species are iteroparous, while fewer than half of non-mate-guarding species are so. It is hypothesized that mate-guarding arose more than once in the evolutionary history of amphipod Crustacea.

Introduction

A mature female amphipod is available for fertilization directly after molting when the cuticle is sufficiently flexible to allow release of the eggs through the genital pores into the brood pouch (Williamson, 1951; Lewbel, 1978; Sheader, 1983; Borowsky, 1986). A male has only this short time between molting and ovulation (a few minutes to a few hours in most amphipods – Birkhead and Clarkson, 1980; Borowsky, 1985, 1988) to deposit sperm into the brood pouch. Males of different amphipod taxa exhibit a variety of behaviors for mating (Borowsky, 1988), and a range of secondary sexual changes at maturity (Bousfield, 1979). The purpose of this paper is to (1) classify precopulatory (= preamplexing) behaviors of male amphipods from accounts in the literature; (2) examine the taxonomic distribution of the behaviors; (3) survey the geographic and ecological distribution of the behaviors; (4) assess sexual dimorphism in relation to precopulatory behavior; and (5) compare female reproductive characteristics in relation to precopulatory behavior.

Methods

Categories of male precopulatory behaviors were established from literature descriptions. Details such as species distribution, male guarding time, female fecundity, frequency of reproduction, and egg brooding time were also recorded. In accordance with Moore (1981) the terms 'generation' and 'brood' were held to be distinctly different reproductive characteristics, since it is possible for a single female to contribute more than one brood to what is biologically interpreted as a single generation. The information gathered herein concerned the former term, not the latter. For full-body analysis of sexual dimorphism, only species for which precopulatory behavior had been documented and for which mature specimens were obtainable were selected. For further analysis of gnathopod dimorphism, species in the collections of the Canadian Museum of Nature were selected at random from each genus available. In both cases the second largest adult male and female of a size series were chosen to exemplify the species at maturity. The right appendages were dissected from each specimen, stained in methylene blue, and mounted in

glycerin. The outline of each appendage was drawn from the image of an inverted microscope, and then digitized to calculate perimeter. Dimorphism was calculated as the difference between male and female in the perimeter of the appendage divided by respective body length (measured from tip of the rostrum to base of the telson on the uncurled specimen).

Results

Classification of male precopulatory behaviors

Published accounts of male precopulatory behavior were found for 97 species of amphipods (Appendix Table 1). The accounts suggested that 96 of the species could be described as being either mate-guarders or non-mate-guarders, depending on the males' behavior prior to copulation.

In mate-guarding species, a mature male located a mature female in advance of the time of the female's molt and guarded her against other males until she molted and could be fertilized. The mate-guarding species could be categorized further as being either carriers or attenders. In carrier species, the male grasped the female's dorsum or lateral plates and held her until she molted. The male was still capable of walking or swimming, the female either curling her abdomen and remaining passive or extending her body and

Table 1. Habitat distribution of mate-guarding and non-mate-guarding species listed in Appendix Table 1.

Mating category		High to mid intertidal	Mid to low intertidal	Low intertidal to subtidal	Strictly subtidal	High salinity	Brackish	Freshwater streams	Large rivers	Lakes
Mate-guarder	Aquatic carrier		×	×	×	×	×	×	×	×
	Semi-terrestrial carrier	×	×	×		×	×			
	Attender		×	×	×	×	×			
Non-mate-guarder	Pelagic searcher		×	×	×	×	×		×	×
	Benthic searcher		×	×	×	×		×		×

swimming with her partner. Carrying occurred in species of the superfamilies Gammaroidea, Hadzioidea (formerly Melitoidea), and Talitroidea. In attender species, the male did not carry the female, but maintained close proximity by placing his body over hers, remaining nearby, or sharing her tube. This occurred in species of the superfamily Corophioidea and suborder Caprellidea.

In non-mate-guarding species, the male did not locate a female in advance of her molt, and contact occurred only during copulation. Non-mate-guarding species could be assigned to the categories 'pelagic searcher' and 'benthic searcher'. In pelagic searchers, the male abandoned an infaunal life style at maturity and searched for a mate pelagically. This occurred in species of the superfamilies Ampeliscoidea, Lysianassoidea, Oedicerotoidea, Phoxocephaloidea, and Pontoporeioidea. In benthic searchers, the male did not leave the bottom sediment in order to mate-search. This was found in species of the superfamilies Crangonyctoidea, Haustorioidea, and Pontoporeioidea.

Phylogeny and distribution of mating types

In Fig. 1 the primary mating strategies of the species listed in Appendix Table 1 are mapped on a phylogeny of the Amphipoda developed by Bousfield (1979, 1982a, and 1983, and pers. comm.). This phylogeny suggests that mate-guarding behavior derived from non-mate-guarding, and that it originated more than once.

The distribution of each species whose pre-

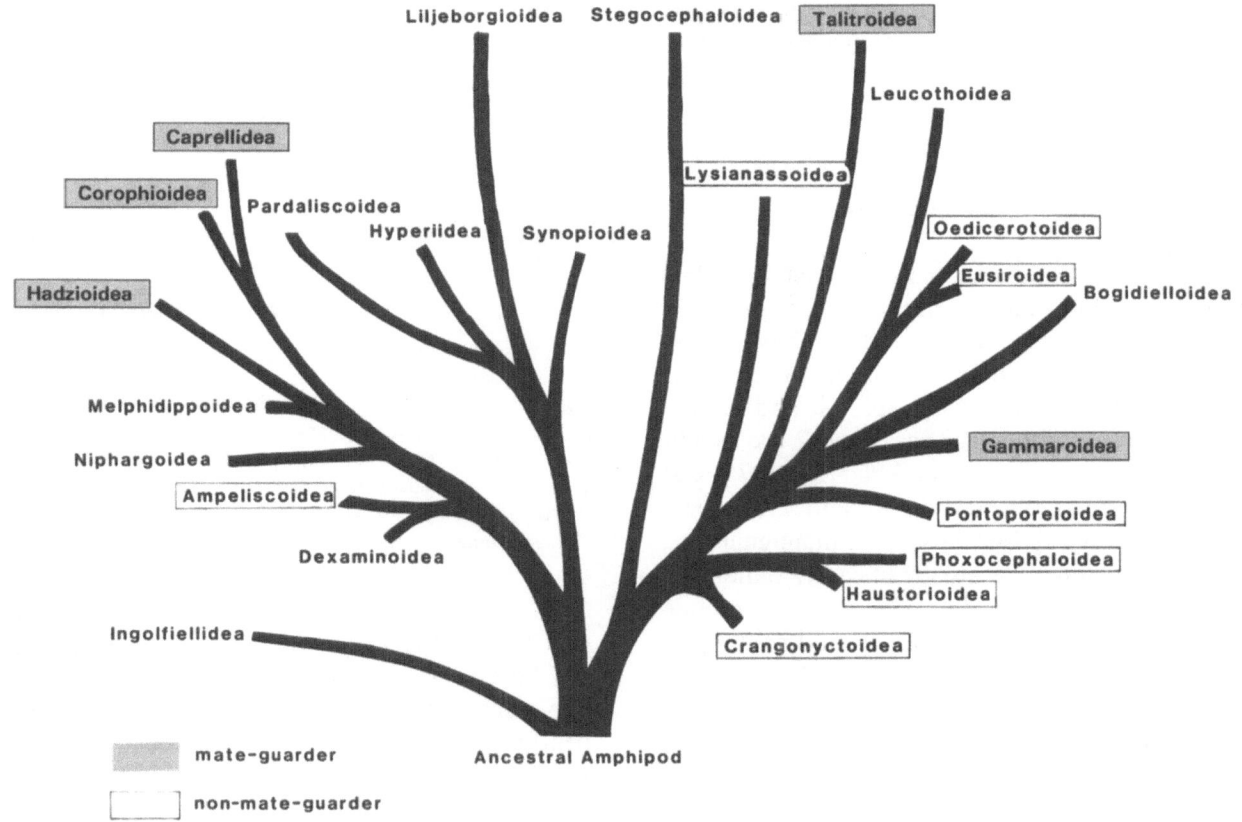

Fig. 1. Phylogeny of the Amphipoda (after Bousfield, 1979, 1982a, 1983, and pers. comm.) showing the distribution of mate-guarding and non-mate-guarding behaviors. There is no information on mating behavior for taxa that lack an outline.

258

Table 2. Geographical distribution of mate-guarding and non-mate-guarding species listed in Appendix Table 1.

Mating category		North Atlantic	North Pacific	Arctic Ocean	South Atlantic	South Pacific	Indian Ocean
Mate-guarder	Aquatic carrier	×	×	×		×	×
	Semi-terrestrial carrier	×	×		×	×	
	Attender	×	×	×	×	×	×
Non-mate-guarder	Pelagic searcher	×	×	×		×	
	Benthic searcher	×	×				

copulatory behavior was categorized is listed in Appendix Table 1 and summarized in Tables 1 and 2. Both mate-guarding and non-mate-guarding types range widely in habitat and geography. Considering the mating characteristics of each type, the absence of attenders and non-mate-guarders from the high intertidal zone may be real. So too may be the absence of pelagic searchers from freshwater streams. Other absences may be due to lack of information.

Mating type and sexual dimorphism

In Fig. 2 the sexual dimorphism of mate-guarders is compared to that of non-mate-guarders for 45 of the species listed in Appendix Table 1 (those species marked by an asterisk). For the 32 mate-guarders, the second antennae, first and second gnathopods, and peraeopods 3 and 7 are significantly dimorphic (t = 4.58, 4.96, 7.60, 3.75, and 3.78, and $d.f.$ = 29, 31, 31, 30, and 29, respectively, $P < 0.001$). For the 13 non-mate-guarders, the second antennae are significantly dimorphic at $P < 0.01$ (t = 3.37, $d.f.$ = 12). The two groups differ significantly in first and second gnathopod dimorphism (ANOVA, F = 16.34 and 20.52, $d.f.$ = 1 and 43, respectively, $P < 0.001$) and at $P < 0.025$ in the amount of second antennal dimorphism (F = 7.25, $d.f.$ = 1 and 41).

Figure 3 shows that the pelagic searchers are responsible for the second antennal dimorphism in the non-mate-guarders; there is no such altera-

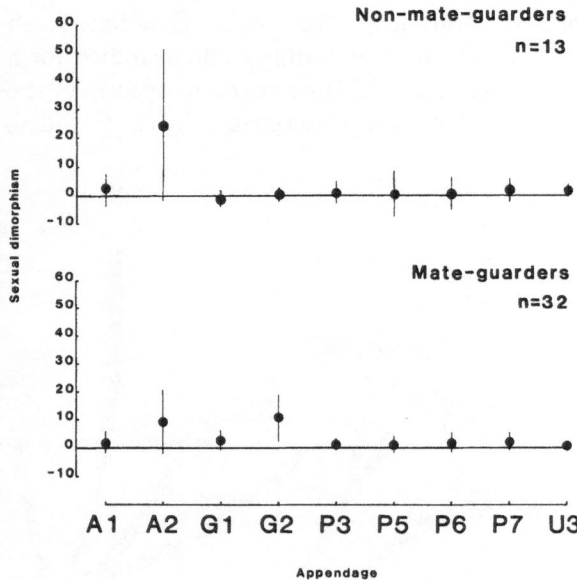

Fig. 2. Mean and one standard deviation of sexual dimorphism in mate-guarder and non-mate-guarder species marked by an asterisk in Appendix Table 1. Sexual dimorphism = (male appendage perimeter/male body length) – (female appendage perimeter/female body length).
A, antennae, G, gnathopods; P, peraeopods; U, uropod.

tion at maturity in benthic searchers (ANOVA, F = 20.68, $d.f.$ = 1 and 11, $P < 0.001$). For other appendages the differences between pelagic and benthic searchers are not significant below $P < 0.05$.

Sexual dimorphism of the aquatic carrier, semi-terrestrial carrier, and attender categories of mate-guarders is compared in Fig. 4. The first and

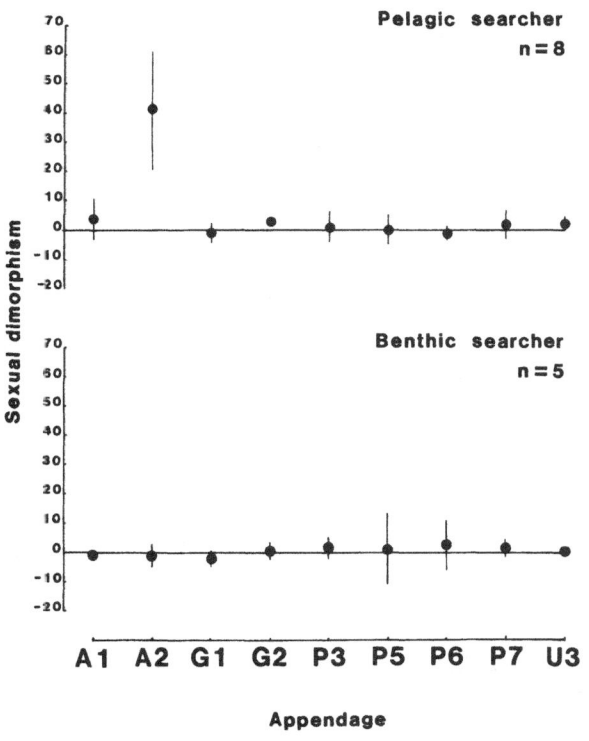

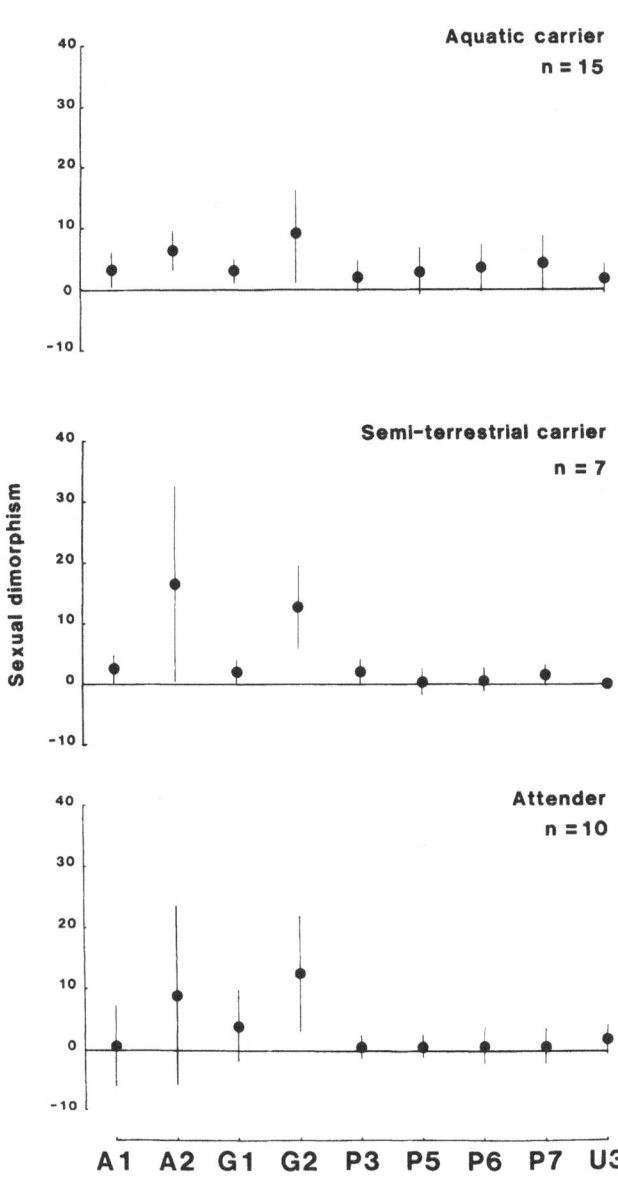

Fig. 3. Mean and one standard deviation of sexual dimorphism in pelagic and benthic categories of non-mate-guarder species marked by an asterisk in Appendix Table 1. Sexual dimorphism = (male appendage perimeter/male body length) − (female appendage perimeter/female body length).

A, antennae; G, gnathopods; P, peraeopods; U, uropod.

second antennae and first and second gnathopods are significantly dimorphic in the aquatic carriers (t = 4.37, 7.56, 6.56, and 16.86, respectively, *d.f.* = 14, P < 0.001). In the semi-terrestrial carriers and attenders, mean sexual dimorphism is significantly greater than zero only in the second gnathopods at P < 0.005 (t = 4.92 and 4.20, *d.f.* = 6 and 9, respectively). For aquatic carriers, semi-terrestrial carriers and attenders differences in sexual dimorphism are not significant at P < 0.05.

Within-taxon sexual dimorphism is compared in Fig. 5 for the enlarged gnathopod (either the first or the second) of species of Gammaroidea, Hadzioidea, aquatic Talitroidea (semi-terrestrial Talitridae excluded), Corophioidea, and Caprellidea. The species measured are listed in Appendix Table 2. In cases where precopulatory behavior has been documented, the species of Gam-

Fig. 4. Mean and one standard deviation of sexual dimorphism in carrying and attending mate-guarder species marked by an asterisk in Appendix Table 1. Sexual dimorphism = (male appendage perimeter/male body length) − (female appendage perimeter/female body length).

A, antennae; G, gnathopods; P, peraeopods; U, uropod.

maroidea, Hadzioidea and Talitroidea have been carriers (with exception of the melitid *Elasmopus levis*), and the species of Corophioidea and Caprellidea have been attenders. The aquatic

260

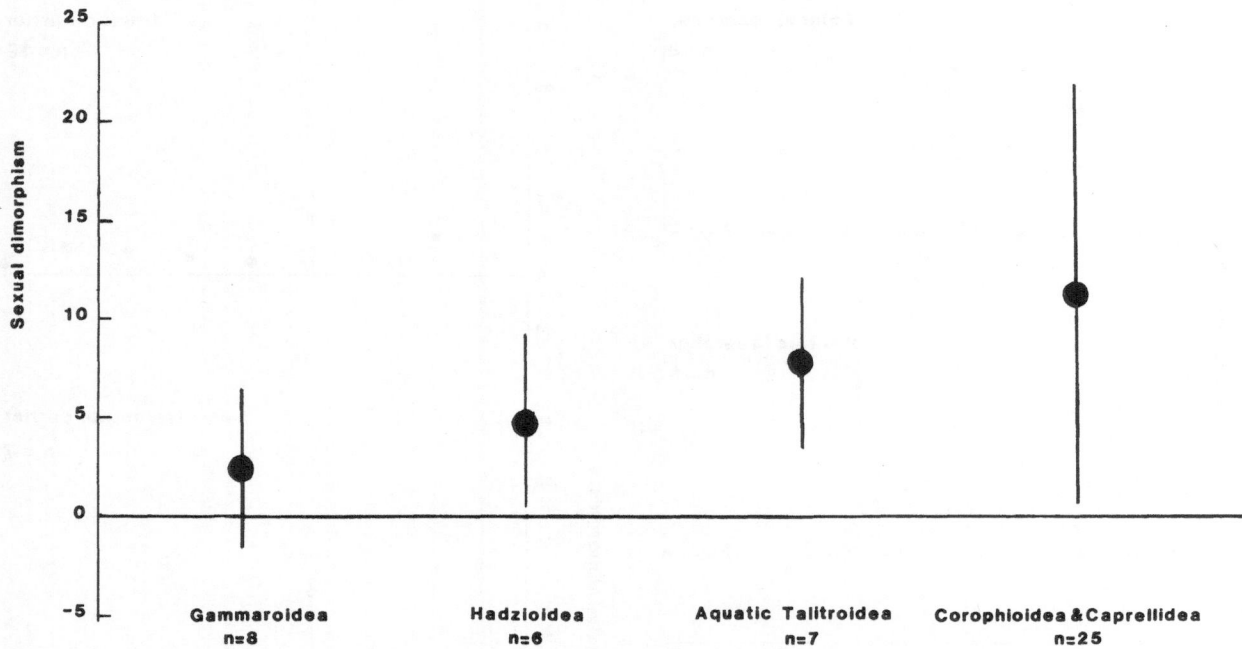

Fig. 5. Mean and one standard deviation of sexual dimorphism of the enlarged first or second gnathopod of species of Gammaroidea, Hadzioidea, aquatic Talitroidea, and Corophioidea and Caprellidea listed in Appendix Table 2. Sexual dimorphism = (male appendage perimeter/male body length) − (female appendage perimeter/female body length).

Talitroidea exhibit significantly greater sexual dimorphism of the gnathopods than do the Gammaroidea and Hadzioidea (ANOVA, $F = 27.74$ and 12.50, $d.f. = 1$ and 13 and 1 and 11, respectively, $P \leq 0.005$). The species of Corophioidea and Caprellidea have significantly greater sexual dimorphism of the gnathopods than the Gammaroidea ($F = 12.72$, $d.f. = 1$ and 31, $P < 0.005$), but differences from the Hadzioidea and aquatic Talitroidea are not significant at this probability level ($F = 6.59$ and 0.53, $d.f. = 1$ and 29 and 1 and 30, respectively). No other group comparisons are significant at this probability level, either.

Figure 6 compares sexual dimorphism of the second gnathopod among different functional groups of the semi-terrestrial Talitroidea, ordered by degree of terrestriality. The species scored are listed in Appendix Table 3. The male palustral talitrids have second gnathopods that are significantly more dimorphic than the aquatic talitrids, beach-hoppers, and landhoppers at $P < 0.05$ ($F = 6.15$, 6.25, and 7.44, $d.f. = 1$ and 10, 1 and

12, and 1 and 8, respectively), but other group comparisons do not reveal significant differences. In many of the palustral talitrids, sandhoppers, and beach-hoppers, the male's second antennae (and less frequently posterior peraeopods) are also enlarged at maturity.

Figure 7 compares the shapes of the distal segments of the enlarged second gnathopod of aquatic carriers, semi-terrestrial carriers, and attenders. Male aquatic carriers exhibit some increase in concavity or convexity of the palm of the propodus at maturity (the propodus of the juvenile male resembles that of the adult female). The shape of the males' enlarged gnathopods is relatively uniform regardless of whether the species is a member of the Gammaroidea, Hadzioidea, or aquatic Talitroidea. The enlarged gnathopod of the semi-terrestrial carriers is markedly different from that of the female (or when juvenile) and more variable in shape than is that of aquatic carriers. The male attenders exhibit the greatest and most diverse alterations in gnathopod shape, with hooks and teeth being prominent.

Table 3. Number of broods throughout a lifetime and the brooding period at defined temperatures for female amphipods of different mating types from temperate waters. Mating type is defined primarily as mate-guarder (G) or non-mate-guarder (N), and secondarily as carrier (C), attender (A), pelagic searcher (P), or benthic searcher (B). An asterisk indicates that the value combines egg maturation and hatchling brooding times.

Species	Mating type	No. broods	Egg maturation time (mean, range, or S.E.) (days)	Hatchling brood time (days)	Temperature (°C)	Reference
Echino-gammarus marinus	G : C	≤ 3				Clarke *et al.*, 1985
Eo-gammarus confervicolus	G : C		17*		10	Levings, 1980
Eulimno-gammarus obtusatus	G : C	≤ 6	13.1 12–14	3–14 4–17	7–9 standing temp.	Sheader & Chia, 1970; Steele & Steele, 1970a, 1973, 1975 Sexton & Spooner, 1940
Gammarus chevreuxi	G : C	> 1	12–14	1–2	room temp.	Sexton & Matthews, 1913
Gammarus duebeni	G : C	> 1	230 (218–236) 143 (137–148) 64 (59–71) 43 (37–49) 30 (26–33) 29 (26–33) 20 (19–22) 24	 1–5	− 1.5 0.0 5.0 8.0 10.0 10.5 15.0 12.5	Steele & Steele, 1973 1975 Hynes, 1955; Kinne, 1959
Gammarus fasciatus	G : C		7 22 8		24 15 23	Clemens, 1950 Embody, 1911
Gammarus lacustris	G : C	≤ 3	21–28 14		15 22	Hynes & Harper, 1971; Menon, 1966
Gammarus lawrencianus	G : C	> 1	82 (79–93) 17 (15–19) 13 (12–14)		0.0 10.0 13.0	Steele & Steele, 1970b, 1973, 1975
Gammarus locusta	G : C		11–12	1–4	room temp.	Blegvad, 1922
Gammarus mucronatus	G : C	> 1	8.3 (8–9) 4.6 (4–5)	0–4 0–2	17 21	Borowsky, 1980a
Gammarus oceanicus	G : C	> 1	156 (150–164) 94 (91–97) 24.8 (23–27)		− 1.5 0.0 10.0	Steele & Steele, 1972, 1973, 1975; Clarke *et al.*, 1985
Gammarus palustris	G : C	> 1	62 17 8–10 11.1 (11–12) 7.1 (7–8)	 0–6	3.5 15 20 17 21	Rees, 1975; Gable & Croker, 1977; Borowsky, 1980a; Borowsky & Borowsky, 1987

Table 3. (Continued)

Species	Mating type	No. broods	Egg maturation time (mean, range, or S.E.) (days)	Hatchling brood time (days)	Temperature (°C)	Reference
Gammarus pseudolimnaeus	G : C	>1	21 18		14.8 16.9	Embody, 1911
Jesogammarus paucisetulosus	G : C	>1	31.9 ± 2.46* 19.2 ± 1.61* 15.2 ± 1.48*		9 14 19	Hamashima & Morino, 1984
Melita nitida	G : C	>1	10.0 (0) 5.9 (5–6)	0–2 0–1	17 21	Borowsky, 1980a, 1984b
Austro-chiltonia subtenuis	G : C	>1	23.0* 16.0* 14.0* 13.3* 12.7* 12.2* 11.0* 10.3*		15 17 19 20 21 22 23.5 24	Lim & Williams, 1970
Hyale pugettensis	G : C		10–12	3–4	12	Heller, 1968
Hyalella azteca	G : C	>1	13.4 ± 0.10* 11.1 ± 0.21* 10.9 ± 0.21* 8.5 ± 0.10* 7.6 ± 0.10* 7.3 ± 0.11* 26.5* 12.7* 9.3*		20 20 20 25 25 25 15 20 25	Strong, 1972 Cooper, 1965
Parhyalella pietshmanni	G : C	>1	14–21		21	Steele, 1973
Orchestia cavimana	G : C	>1	11–15*			Wildish, 1979
Orchestia gammarellus	G : C	>1	13–15*			Williamson, 1951; Wildish, 1979
Orchestia mediterranea	G : C	>1	10–15*			Wildish, 1979
Plat-orchestia platensis	G : C	>1	15* 12* 10* 7* 7* 10* 15*		16 20 25 25 28 27 23	Morino, 1978
Talitrus saltator	G : C		approx. 30			Williamson, 1951

Table 3. (Continued)

Species	Mating type	No. broods	Egg maturation time (mean, range, or S.E.) (days)	Hatchling brood time (days)	Temperature (°C)	Reference
Ampithoe lacertosa	G : A	>1	22 (21–23) 19 (16–21)	17–19 10	8–10 11.5–15	Heller, 1968
Ampithoe rubricata	G : A	1				Skutch, 1926
Ampithoe valida	G : A	>1	5.9 ± 1.69* 10 4 (2–6)	7 (3–15) 4 (2–7)	20.6 ± 2.3 12 21	Borowsky, 1983 Barrett, 1966
Corophium bonnellii	G : A		10.4 ± 0.2 9.6 ± 0.2	≤ 1 ≤ 1	16.5–17.5 18.5–19.5	Shillaker & Moore, 1987b
Corophium insidiosum	G : A	>1	39* 11*		5 15	Sheader, 1978
Jassa marmorata	G : A	>1	16.0 (15–17) 8.8 (8–12) 10.0 (9–11)	0.4–1.25	10 15 20	Nair & Anger, 1979
Lembos websteri	G : A	>1	11.3 ± 0.2 10.1 ± 0.2	≤ 1 ≤ 1	16.5–17.5 18.5–19.5	Shillaker & Moore, 1987b
Microdeutopus gryllotalpa	G : A	>1	6–8	5	21 ± 2	Myers, 1971; Borowsky, 1980b
Caprella gorgonia	G : A	>1	15.3 ± 0.30	3	12	Lewbel, 1978
Elasmopus levis	G : A	>1	8.1 ± 0.8	1.0 ± 0.1	20	Borowsky, 1986
Ampelisca abdita	N : P	1		14		Mills, 1967
Ampelisca macrocephala	N : P	1	approx. 150		3–12	Kanneworff, 1965
Ampelisca tenuicornis	N : P	1 or 2	30–41	6–10	12–13	Sheader, 1977a
Ampelisca vadorum	N : P	1				Mills, 1967
Haploops fundiensis	N : P	1	approx. 150*		2.1–10.8	Wildish, 1984
Paramoera mohri	N : P	1	18	2	12–14	Staude, 1986
Hippomedon whero	N : P	>1				Fenwick, 1985
Parawaldeckia karaka	N : P	>1				Fincham, 1974
Patuki roperi	N : P	>1	12–19*		15	Fenwick, 1985

Table 3. (Continued)

Species	Mating type	No. broods	Egg maturation time (mean, range, or S.E.) (days)	Hatchling brood time (days)	Temperature (°C)	Reference
Arrhis phyllonyx	N : P	1	approx. 180	approx. 30	1	Sainte-Marie & Brunel, 1983
Rhepoxynius abronius	N : P	1				Slattery, 1985
Rhepoxynius fatigans	N : P	1				Slattery, 1985
Bathyporeia pelagica	N : P	>1	approx. 20*	≤4	15 ± 1	Fish, 1975
Bathyporeia pilosa	N : P	>1	approx. 15*	≤4	15 ± 1	Fish, 1975
Pontoporeia femorata	N : P	1	approx. 150*		0.5–4.0	Wildish & Peer, 1981
Eohaustorius canadensis	N : B	1	approx. 30			Donn & Croker, 1985
Eohaustorius sencillus	N : B	1				Slattery, 1985

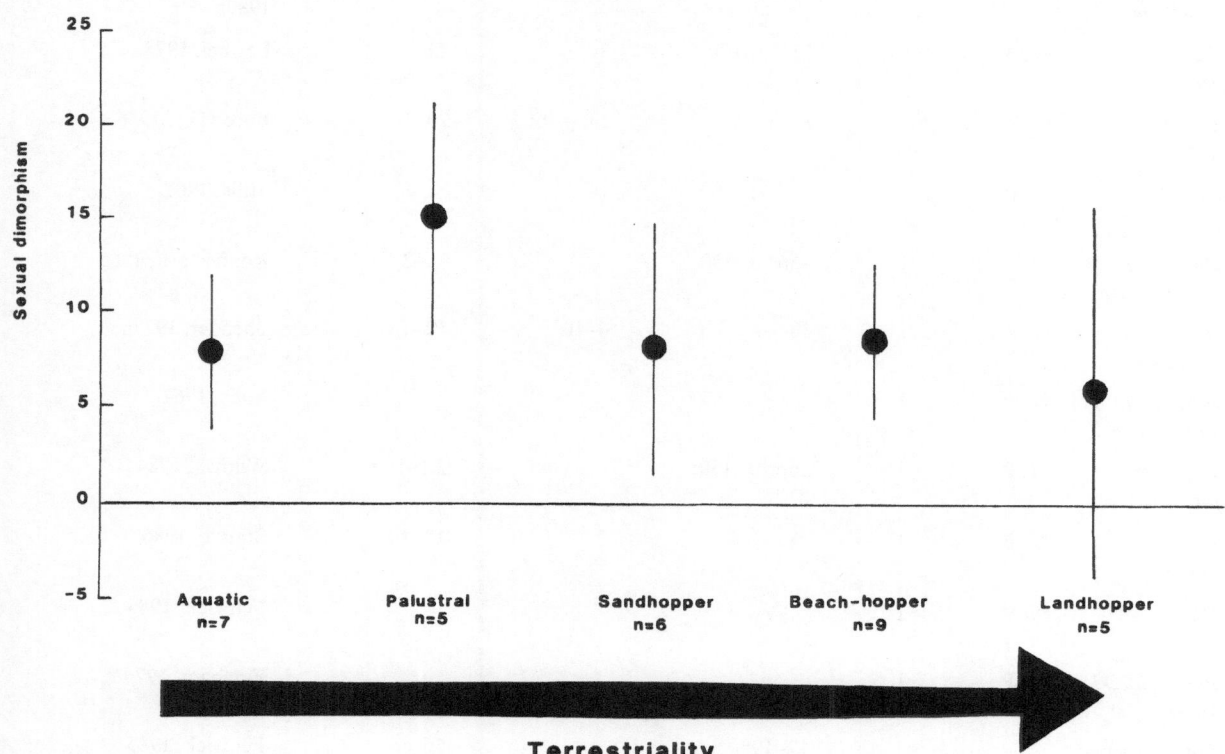

Fig. 6. Mean and one standard deviation of sexual dimorphism of the second gnathopod of aquatic, palustral, sandhopper, beach-hopper, and landhopper species listed in Appendix Table 3. The species are ordered by degree of terrestriality. Sexual dimorphism = (male appendage perimeter/male body length) − (female appendage perimeter/female body length).

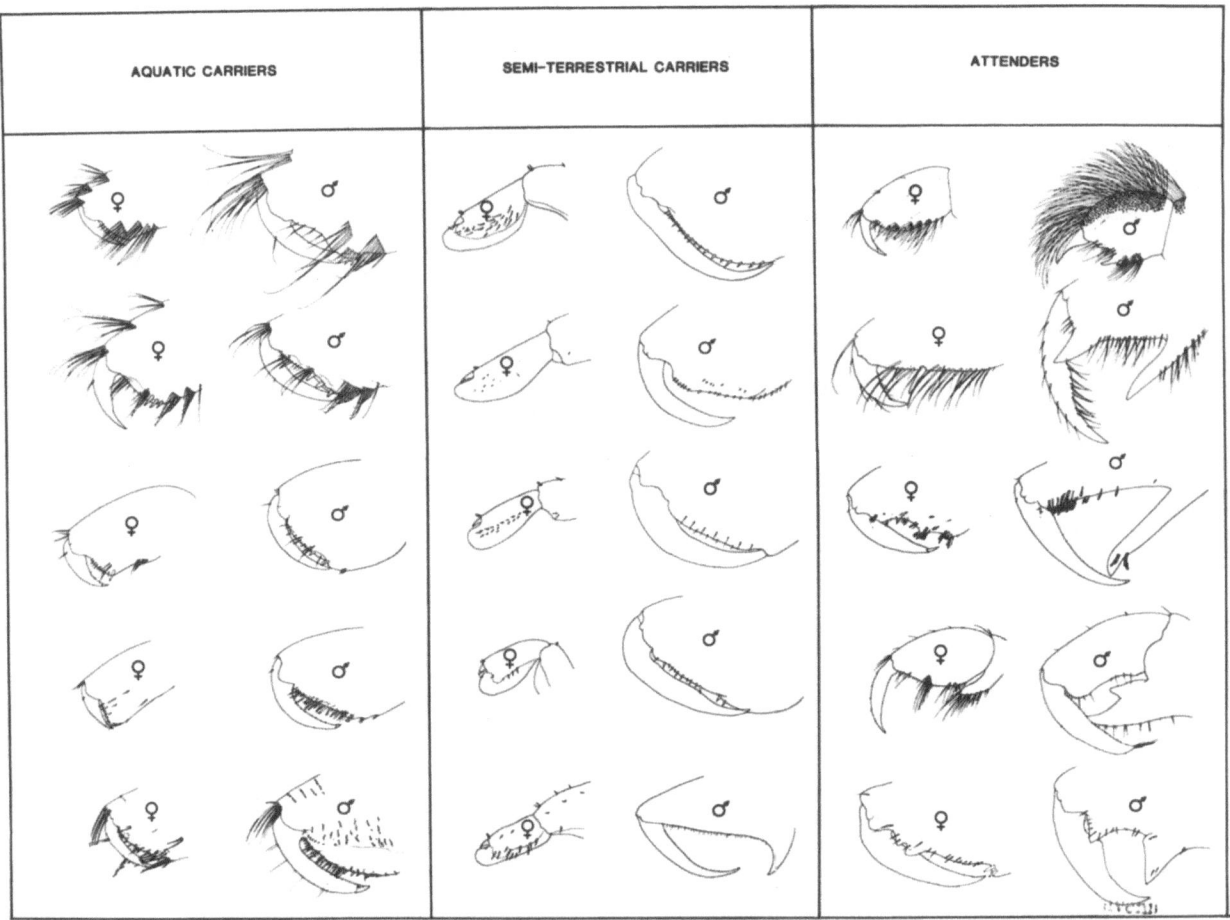

| AQUATIC CARRIERS | SEMI-TERRESTRIAL CARRIERS | ATTENDERS |

Fig. 7. Shape of the enlarged gnathopod of various mate-guarding species compared with that of the female. The juvenile male's gnathopod would resemble that of the female's. From top to bottom, aquatic carriers: *Eulimnogammarus obtusatus* (from Lincoln, 1979), *Echinogammarus marinus* (from Lincoln, 1979), *Hyale nilssoni* (from Lincoln, 1979), *Allorchestes angusta* (unpublished), and *Melita nitida* (unpublished); semi-terrestrial carriers: *Pseudorchestoidea brito* (from Lincoln, 1979); *Megalorchestia californiana* (from Bousfield, 1982b), *Traskorchestia traskiana* (from Bousfield, 1982b), *Orchestia mediterranea* (from Lincoln, 1979), and *Talorchestia deshayesii* (from Lincoln, 1979); attenders: *Lembos websteri* (from Lincoln, 1979), *Dyopedos porrectus* (from Lincoln, 1979), *Jassa falcata* (from Conlan, 1990), *Ericthonius brasiliensis* (from Lincoln, 1979), and *Caprella gorgonia* (from Laubitz and Lewbel, 1974).

Development of secondary sex characters

In Figs. 8 and 9 the development of the sexually dimorphic appendage (the second gnathopod) of a mate-guarder (the aquatic carrier *Gammarus duebeni*) is compared to the development of the sexually dimorphic appendage (the second antenna) of a non-mate-guarder (the pelagic searcher *Rhepoxynius abronius*). In *Gammarus duebeni* appendage enlargement is isometric, sug-

gesting that secondary sex change occurs over more than one molt. According to Kinne (1959) males of *Gammarus duebeni* reach maturity at about the 14th instar and mating occurs until death at about the 21st instar. In *Rhepoxynius abronius* appendage enlargement is allometric. The secondary sexual characters are formed at the last molt and mating occurs only then (Slattery, 1985).

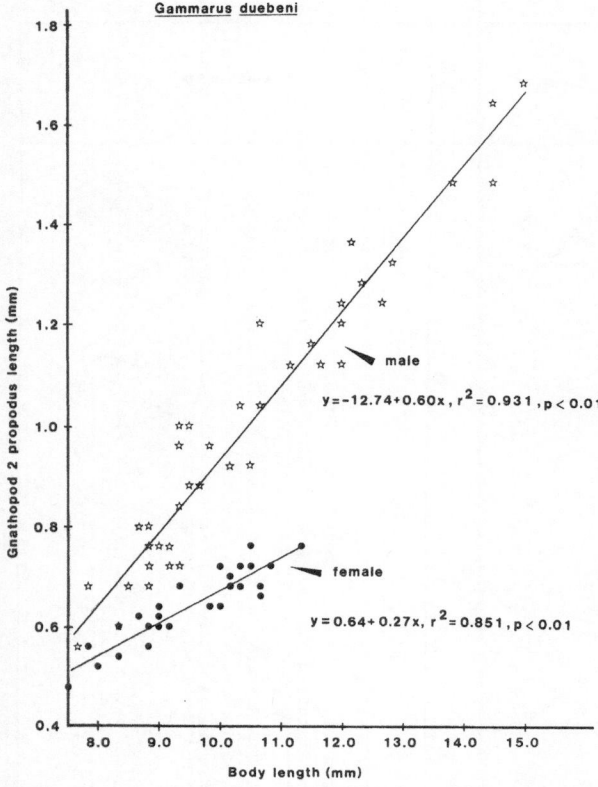

Fig. 8. Second gnathopod propodus length relative to body length compared between males and females of the mate-guarder *Gammarus duebeni*. All females are adult. The state of maturity of the males is unknown; presumably the larger males are adult.

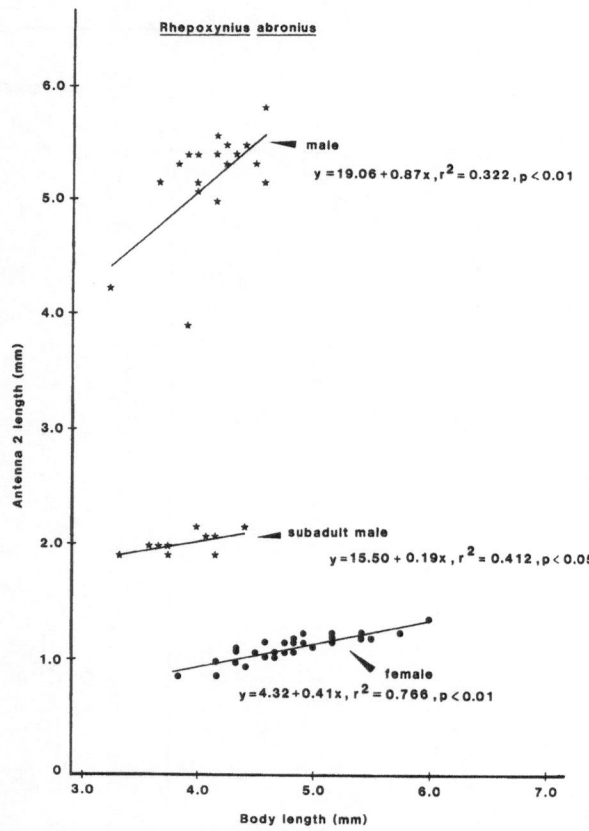

Fig. 9. Second antennal length relative to body length compared between males and females of the non-mate-guarder *Rhepoxynius abronius*. All females are adult. The presumed state of maturity of the males is indicated.

Female reproductive characteristics according to mating type

Table 3 summarizes the female reproductive characteristics of 44 species whose precopulatory behavior and reproductive characteristics have been documented. Of the 28 mate-guarders for which mating frequency is known, 27 are iteroparous (producing more than one brood in a lifetime). By comparison only 6 of the 17 non-mate-guarders are iteroparous. Interpolation of brooding time to a temperature of 15 °C suggests that female mate-guarders require 10–20 days for egg maturation and an additional 1–20 days to brood the hatchlings. Brooding time for non-mate-guarding females falls within this range at this temperature. However, for four species of

non-mate-guarders listed in Table 1 breeding occurs at temperatures colder than is usually recorded for mate-guarders, and in this case, brooding time is considerably longer.

Discussion

Function of secondary sexual characters

In mate-guarding males some anterior pair of appendages is enlarged at maturity. Usually these are the second gnathopods, and often the second antennae are enlarged as well. In non-mate-guarders, appendage enlargement is pronounced in the second antennae of pelagic searchers, but the gnathopods remain unaltered. In benthic searchers there is no enlargement.

The antennal and gnathopodal enlargement in mate-guarders may reflect enhanced sensory ability and increased aggressive interaction at maturity. Males of the carrying species *Gammarus palustris*, *G. duebeni*, *G. pulex*, *Orchestia gammarella*, *O. mediterranea*, *O. platensis*, *Orchestoidea californiana*, *O. corniculata*, *Talitrus saltator*, and *Hyalella azteca*, and the attender species *Microdeutopus gryllotalpa*, *Jassa marmorata*, *Podocerus brasiliensis*, and *Caprella gorgonia* use their enlarged gnathopods and/or antennae to defend their mates, assess their mates' reproductive state, or to displace other males (Williamson, 1951; Bowers, 1964; Strong, 1973; Behbehani, 1978; Hartnoll & Smith, 1978; Lewbel, 1978; Ward, 1983; Borowsky, 1983, 1984a, 1985; pers. obs.). The enlarged gnathopods are not used by aquatic carriers for mate-carrying, however. In these taxa carrying is accomplished by hooking the unenlarged gnathopods under the female's coxae or under an anterior peraeon segment (Kinne, 1959; Sheader & Chia, 1970; Krishnan & John, 1974; Hartnoll & Smith, 1978; Kamihira, 1981; Borowsky, 1984a; Robinson & Doyle, 1985; Fredette & Diaz, 1986; Borowsky & Borowsky, 1987). The enlarged gnathopods are held free and used for mate take-overs and defence, and to assess the female's reproductive state, possibly through contact pheromones (Strong, 1973; Hartnoll & Smith, 1978, 1980; Borowsky & Borowsky, 1987; Dick & Elwood, 1989). There is conflicting evidence concerning gnathopod use by semi-terrestrial carriers. Williamson (1951), Bowers (1964), and Van Senus (1988) reported that the beach-hopper *Orchestia gammarellus* used the enlarged gnathopod for precopulatory carrying. However, the beach-hopper *Traskorchestia traskiana* uses the unenlarged gnathopods for precopulatory carrying (pers. obs.).

The greater variety of secondary sexual enhancements in attenders than in carriers may be a result of their more sedentary and gregarious lifestyle. Attending males are not constrained by the energetic requirement of carrying a mate. However they must remain with their mates and defend them, while carriers are able to take their mates with them. Thus there would be selective pressure on attending males for large weapons, mechanisms to attract females, or means to signal territoriality and dominance. Some attender males monopolize matings (*Jassa marmorata* – Borowsky, 1985) or control mating clusters (*Siphonoecetes dellavallei* – Just, 1988). Males of the attender species *Caprella gorgonia* develop a 'poison spine' on each enlarged gnathopod at maturity with which they attempt to stab other males during agonistic encounters (Lewbel, 1978). Males of the attender *Dyopedos porrectus* threaten by snapping their enlarged gnathopods (Mattson & Cedhagen, 1989). Males of the attender *Jassa marmorata* signal dominance to juvenile males and mating intent to females with a thumb-like protuberance of the enlarged second gnathopod that is produced at maturity (Borowsky, 1985).

Some species may be able to produce sound, presumably to establish territories and to attract mates. Males of the *Floresorchestia* group of beach-hoppers have regularly spaced ridges on pleon plates 2 and 3 which appose (and are presumably rubbed by) fine, regularly spaced knobs on the posterior margin of the basis of peraeopod 7 (E.L. Bousfield, pers. comm.). Males of most species of the corophioidean genus *Photis* and some species of *Ericthonius* and *Microjassa* have a sinuous row of ridges on the outer surface of the basis of the second gnathopod and matching ridges along the inner margin of coxa 2 or 3 (Conlan, 1983 and pers. obs.). Adult males of the hadziid genus *Dulichiella* have one immensely enlarged second gnathopod propod (either left or right) which the movable dactyl apposes. Presumably the dactyl is snapped against the propodus in the manner of the large chela of snapping shrimps (E.L. Bousfield, pers. comm.).

In non-mate-guarding species, the antennal elongation demonstrated by pelagic searchers probably reflects enhanced sensory ability. Male non-mate-guarders often have enlarged and aggregated chemosensory antennal aesthetascs forming a callynophore at maturity (Lowry, 1986), and develop mechano-sensory calceoli on one or both pairs of antennae. These changes are often supplemented by eye, pleopod, uropod, and telson enlargement (e.g. in the phoxocephalids *Rhepoxynius abronius* and *R. fatigans* (Slattery, 1985), and the ampeliscids *Ampelisca abdita* and *A. vadorum* (Mills, 1967)). Males of species that search benthically show few enhancements in mate-locating ability. They may develop calceoli but the antennae do not elongate at maturity and the body is little altered. Haustoriid males may have relatively more strongly spinose burrowing appendages (E.L. Bousfield, pers. comm.), possibly to compensate for smaller body size than females.

Development of secondary sexual characters

Males of the mate-guarders *Gammarus chevreuxi*, *Microdeutopus gryllotalpa*, *Hyalella azteca*, *Caprella gorgonia*, and *Orchestia platensis* develop their secondary sexual characters over several molts and sexual maturity is reached prior to full development (Sexton, 1924; Myers, 1971; Strong, 1972; Lewbel, 1978; Morino, 1981). The attender *Jassa* is the only genus of mate-guarding amphipods known to delay development of a secondary sex character to the last instar (Borowsky, 1985; Conlan, 1989). The delay in sexual development is thought to be due to strong competition among males for females, brought on by a male-biased operational sex ratio. Delay of male secondary sexual development has been more frequently recorded for non-mate-guarders. It is known to occur in *Ampelisca tenuicornis*, *Parawaldeckia* spp., *Arrhis phyllonyx*, *Rhepoxynius abronius*, and *Rhepoxynius fatigans* (Sheader, 1977a; Lowry & Stoddart, 1983; Sainte-Marie & Brunel, 1983; Slattery, 1985). Slattery (1985) has suggested that in *Rhepoxynius abronius* and *R. fatigans* the changes in males at the last instar are disadvantageous to their normal burrowing and feeding lifestyle, thus necessitating dissociation of sexual activity from feeding and development. Terminal or near-terminal molt transformation may be characteristic of all amphipods in which the male searches for mates pelagically (Bousfield, 1979).

Prediction of mating behavior on the basis of secondary sexual development

Because mate-guarders and non-mate-guarders differ in secondary sex characteristics it may be possible to predict male precopulatory behavior on the basis of sexual dimorphism. Bousfield (1979, 1982a, and 1989) has suggested that precopulatory behavior is conservative (and Appendix Table 1 supports this), at least to the family level and in many cases to the superfamily level. For the superfamilies lacking classification in Fig. 1, Bousfield (1979, 1982a) has classified the Dexaminoidea, Melphidippoidea, Pardaliscoidea, Hyperiidea, and Synopioidea as being pelagic non-mate-guarders because the species exhibit transformations similar to those found for pelagic mate-searching species whose precopulatory behavior has been documented. Similarly, Bousfield (1982a and pers. comm.) has suggested that members of the superfamilies Leucothoidea (in the families Pleustidae and Stenothoidae), Liljeborgioidea (in the families Sebidae and Colomastigidae), Stegocephaloidea and Crangonyctoidea (in the family Paramelitidae), and Oedicerotoidea (in the families Exoedicerotidae and Paracalliopiidae) are mate-guarders because they exhibit pronounced sexual dimorphism in the gnathopods.

Reproductive characteristics

It is possible that mate-guarders and non-mate-guarders differ in life-history traits such as maturation rate, period of sexual activity, fecundity, and egg size. The majority of mate-guarders studied here were found to be iteroparous (Table 3). Fewer than half of the non-mate-guarders were found to be so. If sexual activity does not occur until the last molt in pelagic searching non-mate-guarders, then such species may devote a shorter period of their lifespan to reproduction than do mate-guarders. A non-mate-guarding male could achieve as many matings as a mate-guarding male if the source of receptive females was dependable, little time was necessary to assure successful mating, and competition for mates was low. Some pelagic non-mate-guarders appear to enhance the chance of mate location by using environmental cues to mate-search. Males and females of *Ampelisca abdita* and *Patuki roperi* rise into the plankton on night-time spring tides and mate at this time (Mills, 1967; Fenwick, 1985). Males and females of the pontoporeids *Bathyporeia pilosa* and *B. pelagica* meet pelagically with a semi-lunar frequency phased mainly with falling spring tides (Watkin, 1940; Fincham, 1970; Fish & Preece, 1970; Preece, 1971; Fish, 1975). All ages of the eusirid *Paramoera mohri* come into the plankton, and it is at this time that mates pair (Staude, 1986). Males of the lysianassid *Parawaldeckia stephenseni* search pelagically at night while females remain on the bottom (Fincham, 1974). Decreasing amount of light initiates maturation and mate-searching in *Pontoporeia affinis* (Segerstråle, 1970), and the sponge-inhabiting dexaminid *Tritaeta gibbosa* (Jones *et al.*, 1973).

Some mate-guarders also mate-search in cue to environmental rhythmicity. Adult females of *Corophium volutator*, *C. arenarium*, *C. insidiosum*, and *Lembos websteri* mate with a semi-lunar rhythmicity (Fish & Mills, 1979; Moore, 1981; Omori *et al.*, 1982). Adult males emerge from the sediment during the rising spring tide period and crawl across the mud surface in search of burrows occupied by adult females. The sexes cohabit and copulation occurs within the burrow, not within the water column (Fish & Mills, 1979). *Gammarus lacustris lacustris* initiates mating at short day-length (12 hours of light) or after several weeks of dim light, but is unresponsive to temperature (De March, 1982). Similarly, photoperiod is the key control of reproduction in *Hyalella azteca*, although temperatures are a modulating factor (De March, 1977). Sheader (1983) found that the eggs of most ovigerous females of *Gammarus duebeni* were at the same stage of development, suggesting some synchrony in breeding. However, the amount of breeding synchrony was highly variable throughout the year, being roughly controlled by mortality rates induced by changes in temperature and salinity. Many species of *Gammarus* exhibit a pulse of spring breeding timed with optimal feeding conditions (Steele & Steele, 1975; Sheader, 1978). Depending on the species and latitude, this may be followed by a summer pulse once the spring offspring mature. The summer offspring then overwinter and breed the following spring.

Evolution and radiation of mate-guarding

Mate-guarding should evolve when female receptivity is asynchronous, and the operational sex ratio is male-biased, thus causing strong competition among males for mates (Trivers, 1972). The function of mate-guarding is to ensure that the male is present during the short period that the female is receptive (Strong, 1973; Borowsky, 1984). Mate-guarding also secures the female against predatory attack by other males while her cuticle is soft after molting (Lewbel, 1978; Dick *et al.*, 1990). The evolution of mate-guarding by carrying may have allowed amphipods to radiate into areas of strong water movement and onto land (Bousfield, 1988). Many species of Gammaroidea, Hadzioidea, and Talitroidea occur in rivers and in the intertidal zone, where the risk of displacement would be presumably greater than subtidally. All terrestrial and semi-terrestrial species of amphipods are Talitroidea. Once a potential mate is located it would be advantageous to remain attached to her despite a possibly lengthy wait for mating. The proportion of the female's intermolt period during which she is carried or attended by a male may range from 4 to 53% (Strong, 1972; Hamashima & Morino, 1984; Borowsky, 1986). The duration of mate-guarding varies with temperature and salinity (Hartnoll & Smith, 1978), predation intensity (Strong, 1973), the number of competitors, and the number of available females (Ward, 1983). The duration of mate-guarding must also depend on the length of the period in which females can be mated after molting, and the ability of the female to delay ovulation. In the aquatic carrier *Gammarus pulex*, females cannot be mated after 12 hours following molting (Birkhead & Clarkson,

1980). In the aquatic carrier *Gammarus palustris*, delaying copulation for as short a time as 6 ± 6 hours after molting results in a significantly reduced fecundity. The longer the delay the greater is the reduction in fecundity (Borowsky, 1988). In semi-terrestrial and subterranean mate-guarders, the guarding time is reduced and the post-molt receptivity lengthened (Williamson, 1951; Ginet, 1967; Ridley, 1983). Mating is still possible up to 72 hours after molting in the semi-terrestrial talitrid *Orchestia gammarellus* (Campbell-Parmentier, 1963) and up to 4 days in *Talitrus saltator* (Williamson, 1951). Similarly, in the cave dweller *Niphargus orcinus virei*, in which there is no precopulatory carrying, egg laying may be delayed up to 6 to 17 days after the molt (Ginet, 1967).

Conclusions

It can be generalized from literature accounts of precopulatory mating behavior that male amphipods either mate-guard in advance of female readiness to ovulate or do not mate-guard, instead locating their mates at the time that ovulation occurs. The strategies entail different appendage alterations for the male when it matures, and possibly also different reproductive characteristics for the female. Mate-guarding may have arisen more than once from non-mate-guarding ancestors. Neither strategy seems to be geographically restricted. The two strategies seem to be taxonomically conservative to the superfamily level. Among mate-guarding taxa the male's enlarged gnathopods exhibit greater variety in shape among attenders than carriers, suggesting that a diversity of precopulatory behaviors may be found among male attenders to signal sexual intent to mates and competitors.

Acknowledgements

The impetus for this study was given by oral presentations by Dr. E.L. Bousfield to amphipod colloquia in Ottawa, Canada in 1984 and Ambleteuse, France in 1985. I am indebted to Dr. Bousfield for allowing me to cite his unpublished material. I am also grateful to Dr. Pat Weatherhead for assistance in developing the thoughts in this study and to Diana Laubitz and Les Watling for critically appraising the manuscript. Betty Borowsky, John Oliver, Peter Slattery, and Craig Staude provided institutional support for behavioral observations. Research was also conducted aboard the *R. V. Alpha Helix* (under National Science Foundation Grant DPP-8619394 to John Oliver). Many of the specimens used for the analyses of sexual dimorphism were from the collection of the Canadian Museum of Nature, Ottawa. The British Museum (Natural History), Hiroshi Morino, Peter Slattery, Craig Staude, and Don Steele also provided specimens. George Leir assisted with statistical analyses, and Ed Hendrycks and Jane Kendall with illustrations.

References

Adams, J. & P. J. Greenwood, 1983. Why are males bigger than females in pre-copula pairs of *Gammarus pulex*? Behav. Ecol. Sociobiol. 13: 239–241.

Adams, J. & P. J. Greenwood, 1985. Environmental constraints on mate choice in *Gammarus pulex* (Amphipoda). Crustaceana 50: 45–51.

Adams, J. & P. J. Greenwood, 1987. Loading constraints, sexual selection and assortative mating in peracarid Crustacea. J. Zool., Lond. 211: 35–46.

Ahsanullah, M. & A. R. Williams, 1986. Effect of uranium on growth and reproduction of the marine amphipod *Allorchestes compressa*. Mar. Biol. 93: 459–464.

Alcock, J., 1989. Animal behavior. Fourth edition. Sinauer Assoc., Mass. 596 pp.

Ali, M. H. & S. D. Salman, 1986. The reproductive biology of *Parhyale basrensis* Salman (Crustacea, Amphipoda) in the Shatt al-Arab River. Estuar. coast Shelf Sci. 23: 339–351.

Barrett, B. E., 1966. A contribution to the knowledge of the amphipodous crustacean *Ampithoe valida* Smith 1873. Ph.D. thesis, University of New Hampshire.

Behbehani, M. I., 1978. Studies on the ecology of *Orchestia platensis* Kroyer 1845 (Crustacea: Amphipoda). Ph.D. thesis, University of New Hampshire. 131 pp.

Birkhead, T. R. & K. Clarkson, 1980. Mate selection and precopulatory guarding in *Gammarus pulex*. Z. Tierpsychol. 52: 365–380.

Blegvad, H., 1922. On the biology of some Danish gammarids

and mysids (*Gammarus locusta*, *Mysis flexuosa*, *Mysis neglecta*, *Mysis inermis*). Rep. Dan. biol. Sta. 28: 1–103.

Borowsky, B., 1980a. Reproductive patterns of three intertidal salt-marsh gammaridean amphipods. Mar. Biol. 55: 327–334.

Borowsky, B., 1980b. The pattern of tube-sharing in *Microdeutopus gryllotalpa* (Crustacea: Amphipoda). Anim. Behav. 28: 790–797.

Borowsky, B., 1983. Reproductive behavior of three tube-building peracarid crustaceans: the amphipods *Jassa falcata* and *Ampithoe valida* and the tanaid *Tanais cavolinii*. Mar. Biol. 77: 257–263.

Borowsky, B., 1984a. Effects of receptive females' secretions on some male reproductive behaviors in the amphipod crustacean *Microdeutopus gryllotalpa*. Mar. Biol. 84: 183–187.

Borowsky, B., 1984b. The use of the males' gnathopods during precopulation in some gammaridean amphipods. Crustaceana 47: 245–250.

Borowsky, B., 1985. Differences in reproductive behavior between two male morphs of the amphipod crustacean *Jassa falcata* Montagu. Physiol. Zool. 58: 497–502.

Borowsky, B., 1986. Laboratory observations of the pattern of reproduction of *Elasmopus levis* (Crustacea: Amphipoda). Mar. Behav. Physiol. 12: 245–256.

Borowsky, B., 1988. Delaying copulation in the amphipod crustacean *Gammarus palustris*: effects on female fecundity and consequences for the frequency of amplexus. Mar. Behav. Physiol. 13: 359–368.

Borowsky, B., 1989. The effects of residential tubes on reproductive behaviors in *Microdeutopus gryllotalpa* (Costa) (Crustacea: Amphipoda). J. Exp. mar. Biol. Ecol. 128: 117–125.

Borowsky, B., C. A. Augelli & S. R. Wilson, 1987. Towards chemical characterization of waterborne pheromone of amphipod crustacean *Microdeutopus gryllotalpa*. J. Chem. Ecol. 13: 1673–1680.

Borowsky, B. & R. Borowsky, 1987. The reproductive behaviors of the amphipod crustacean *Gammarus palustris* (Bousfield) and some insights into the nature of their stimuli. J. Exp. mar. Biol. Ecol. 107: 131–144.

Bousfield, E. L., 1958. Fresh-water amphipod crustaceans of glaciated North America. Can. Field-Nat. 72: 55–113.

Bousfield, E. L., 1973. Shallow-water gammaridean Amphipoda of New England. Cornell University Press. 312 pp.

Bousfield, E. L., 1979. A revised classification and phylogeny of amphipod crustaceans. Trans. r. Soc. Can. 16: 343–390.

Bousfield, E. L., 1982a. Amphipoda, Gammaridea, and Ingolfiellidea. In S. B. Parker (ed), Synopsis and Classification of Living Organisms. Vol. 2. McGraw-Hill, New York: 254–294.

Bousfield, E. L., 1982b. The amphipod superfamily Talitroidea in the northeastern Pacific region. I. Family Talitridae: systematics and distributional ecology. Natl. Mus. Nat. Sci. (Ottawa) Publ. Biol. Oceanogr. 11: 1–73.

Bousfield, E. L., 1983. An updated phyletic classification and palaeohistory of the Amphipoda. In F. Schram (ed), Crustacean Phylogeny. A. A. Balkema, Rotterdam: 257–277.

Bousfield, E. L., 1988. Ordered character states as a basis for phyletic classification within the Amphipoda. Abstract. Crustaceana, Supplement 13: 279–280.

Bousfield, E. L., 1989. Revised morphological relationships within the amphipod genera *Pontoporeia* and *Gammaracanthus* and the 'glacial relict' significance of their postglacial distributions. Can. J. Fish. aquat. Sci. 46: 1714–1725.

Bowers, D. E., 1964. Natural history of two beach hoppers of the genus *Orchestoidea* (Crustacea: Amphipoda) with reference to their complemental distribution. Ecology 45: 677–696.

Bregazzi, P. K., 1972. Life cycles and seasonal movements of *Cheirimedon femoratus* (Pfeffer) and *Tryphosella kergueleni* (Miers) (Crustacea: Amphipoda). Br. Antarct. Surv. Bull. 30: 1–34.

Campbell-Parmentier, F., 1963. Vitellogenèse, maturation des ovocytes, accouplement et ponte en relation avec l'intermue chez *Orchestia gammarella* Pallas (Crustacé amphipode Talitridae). Bull. Soc. zool. Fr. 88: 474–488.

Clarke, A., A. Skadsheim & L. J. Holmes, 1985. Lipid biochemistry and reproductive biology in two species of Gammaridae (Crustacea: Amphipoda). Mar. Biol. 88: 247–263.

Clemens, H. P., 1950. Life cycle and ecology of *Gammarus fasciatus* Say. Franz Theodore Stone Institute of Hydrobiology, Ohio State University, Contribution 12. 63 pp.

Conlan, K. E., 1982. The amphipod superfamily Corophioidea in the northeastern Pacific region. Family Ampithoidea: systematics and distributional ecology. Natl. Mus. Nat. Sci. (Ottawa) Publ. Biol. Oceanogr. 10: 41–75.

Conlan, K. E., 1983. The amphipod superfamily Corophioidea in the northeastern Pacific region. 3. Family Isaeidae: systematics and distributional ecology. Natl. Mus. Nat. Sci. (Ottawa) Publ. Nat. Sci. 4: 1–75.

Conlan, K. E., 1989. Delayed reproduction and adult dimorphism in males of the amphipod genus *Jassa* (Corophioidea: Ischyroceridae): an explanation for systematic confusion. J. Crust. Biol. 9: 601–625.

Conlan, K. E., 1990. Revision of the crustacean amphipod genus *Jassa* Leach (Corophioidea: Iscyroceridae). Can. J. Zool. 10: 2031–2075.

Connell, J. H., 1963. Territorial behavior and dispersion in some marine invertebrates. Res. Popul. Ecol. 5: 87–101.

Cooper, W. E., 1965. Dynamics and production of a natural population of a fresh-water amphipod, *Hyalella azteca*. Ecol. Monogr. 35: 377–393.

Crawford, G. I., 1937. A review of the amphipod genus *Corophium*, with notes on the British species. J. mar. biol. Ass. U.K. 21: 589–630.

Crozier, W. J. & L. H. Snyder, 1923. Selective coupling of gammarids. Biol. Bull. 45: 97–104.

Dahl, E., H. Emanuelsson & C. Von Mecklenburg, 1970. Pheromone reception in the males of the amphipod *Gammarus duebeni* Lilljeborg. Oikos 21: 42–47.

De March, B. G. E., 1977. The effects of photoperiod and temperature on the induction and termination of reproductive resting stage in the freshwater amphipod *Hyalella azteca* (Saussure). Can. J. Zool. 55: 1595–1600.

De March, B. G. E., 1978. The effects of constant and variable temperatures on the size, growth, and reproduction of the freshwater amphipod *Hyalella azteca* (Saussure). Can. J. Zool. 56: 1801–1806.

De March, B. G. E., 1982. Decreased day length and light intensity as factors inducing reproduction in *Gammarus lacustris lacustris* Sars. Can. J. Zool. 60: 2962–2965.

Dick, J. T. A. & R. W. Elwood, 1989. Assessments and decisions during mate choice in *Gammarus pulex* (Amphipoda). Behaviour 109: 235–246.

Dick, J. T. A., D. E. Irvine & R. W. Elwood, 1990. Differential predation by males on moulted females may explain the competitive displacement of *Gammarus duebeni* by *G. pulex* (Amphipoda). Behav. Ecol. Sociobiol. 26: 41–45.

Dieleman, J., 1979. Swimming rhythms, migration and breeding cycles in the estuarine amphipods *Gammarus chevreuxi* and *Gammarus zaddachi*. In Naylor, E. & R. G. Hartnoll (eds), Cyclic phenomena in marine plants and animals, Proc. 13th Europ. Mar. Biol. Symp. Pergamon Press, Oxford: 415–422.

Donn, T. E., Jr. & R. A. Croker, 1985. Life-history patterns of *Haustorius canadensis* (Crustacea: Amphipoda) in northern New England. Can. J. Zool. 64: 99–104.

Downer, D. F. & D. H. Steele, 1979. Some aspects of the biology of *Amphiporeia lawrenciana* Shoemaker (Crustacea, Amphipoda) in Newfoundland waters. Can. J. Zool. 57: 257–263.

Dunham, P. J., 1986. Mate guarding in amphipods: a role for brood pouch stimuli. Biol. Bull. 170: 526–531.

Dunham, P. J., T. Alexander & A. Hurshman, 1986. Precopulatory mate guarding in an amphipod, *Gammarus lawrencianus* Bousfield. Anim. Behav. 34: 1680–1686.

Dunham, P. J. & A. Hurshman, 1988. Amphipod mate guarding decisions: deprivation versus uncertainty. Anim. Behav. 36: 609–611.

Elwood, R., J. Gibson & S. Neil, 1987. The amorous *Gammarus*: size assortative mating in *G. pulex*. Anim. Behav. 35: 1–6.

Embody, G. C., 1911. A preliminary study of the distribution, food and reproductive capacity of some fresh-water amphipods. Biol. Suppl. 3: 1–33.

Fenwick, G. D., 1985. Life-histories of four co-occurring amphipods from a shallow, sand bottom at Kaikoura, New Zealand. N.Z. J. Zool. 12: 71–105.

Fincham, A. A., 1970. Amphipods in the surf plankton. J. mar. biol. Ass. U.K. 50: 177–198.

Fincham, A. A., 1974. Periodic swimming behaviour of amphipods in Wellington Harbour. N.Z. J. mar. Freshwat. Res. 8: 505–521.

Fish, J. D., 1975. Development, hatching and brood size in *Bathyporeia pilosa* and *B. pelagica* (Crustacea: Amphipoda). J. mar. biol. Ass. U.K. 55: 357–368.

Fish, J. D. & A. Mills, 1979. The reproductive biology of *Corophium volutator* and *C. arenarium* (Crustacea: Amphipoda). J. mar. biol. Ass. U.K. 59: 355–368.

Fish, J. D. & G. S. Preece, 1970. The annual reproductive patterns of *Bathyporeia pilosa* and *Bathyporeia pelagica* (Crustacea: Amphipoda). J. mar. biol. Ass. U.K. 50: 475–488.

Fong, D. W., 1989. Morphological evolution of the amphipod *Gammarus minus* in caves: quantitative genetic analysis. Am. Midl. Nat. 121: 361–378.

Franz, D. R., 1989. Population density and demography of a fouling community amphipod. J. exp. mar. Biol. Ecol. 125: 117–136.

Fredette, T. J. & R. J. Diaz, 1986. Life history of *Gammarus mucronatus* Say (Amphipoda: Gammaridae) in warm temperate estuarine habitats, York River, Virginia. J. Crustacean Biol. 6: 57–78.

Gable, M. F. & R. A. Croker, 1977. The salt marsh amphipod, *Gammarus palustris* Bousfield, 1969 at the northern limit of its distribution. Estuar. Coast. mar. Sci. 5: 123–134.

Ginet, R., 1967. Ecologie, éthologie et biologie de *Niphargus* (Amphipodes Gammarides hypogés). Ann. Spéléol. 15: 127–376.

Goodhart, C. B., 1939. Notes on the bionomics of the tube-building amphipod, *Leptocheirus pilosus* Zaddach. J. mar. biol. Ass. U.K. 23: 311–325.

Greenwood, P. J. & J. Adams, 1984. Sexual dimorphism in *Gammarus pulex*: the effect of current flow on pre-copula pair formation. Freshwat. Biol. 14: 203–209.

Greenwood, P. J. & J. Adams, 1987. Sexual selection, size dimorphism and a fallacy. Oikos 48: 106–108.

Hamashima, W. & H. Morino, 1984. Experimental studies on the growth, survival, and breeding activities of *Jesogammarus paucisetulosus* Morino at different temperatures. Publ. Itako Hydrobiol. Stn. 1: 25–38.

Harrison, R. J., 1940. On the biology of the Caprellidae. Growth and moulting of *Pseudoprotella phasma* Montagu. J. mar. biol. Ass. U.K. 24: 483–493.

Hartnoll, R. G. & S. M. Smith, 1978. Pair formation and the reproductive cycle in *Gammarus duebenii*. J. nat. Hist. 12: 501–511.

Hartnoll, R. G. & S. M. Smith, 1980. An experimental study of sex discrimination and pair formation in *Gammarus duebenii* (Amphipoda). Crustaceana 38: 253–264.

Heller, S. P., 1968. Some aspects of the biology and development of *Ampithoe lacertosa* (Crustacea: Amphipoda). M.Sc. thesis, University of Washington, Seattle. 132 pp.

Hiwatari, T. & T. Kajihara, 1988. Experimental studies on the growth and breeding of *Hyale barbicornis* (Amphipoda, Crustacea) at different temperatures. Nippon Suisan Gakkaishi 54: 39–43.

Hughes, R. G., 1988. Dispersal by benthic invertebrates: the in situ swimming behaviour of the amphipod *Corophium volutator*. J. mar. biol. Ass. U.K. 68: 565–579.

Hunte, W. & R. A. Myers, 1988. Male investment time and

mating decisions in amphipods: uncertainty or deprivation? Anim. Behav. 36: 608–609.

Hunte, W., R. A. Myers & R. W. Doyle, 1985. Bayesian mating decisions in an amphipod, *Gammarus lawrencianus* Bousfield. Anim. Behav. 33: 366–372.

Hurley, D. E., 1968. Transition from water to land in amphipod crustaceans. Am. Zool. 8: 327–353.

Hynes, H. B. N., 1954. The ecology of *Gammarus duebeni* Lilljeborg and its occurrence in fresh water in western Britain. J. anim. Ecol. 23: 38–84.

Hynes, H. B. N., 1955. The reproductive cycle of some British freshwater Gammaridae. J. anim. Ecol. 24: 352–387.

Hynes, H. B. N. & F. Harper, 1971. The life histories of *Gammarus lacustris* and *G. pseudolimnaeus* in southern Ontario. Crustaceana Suppl. 3: 329–341.

Jenio, F., 1980. The life cycle and ecology of *Gammarus troglophilus* (Hubricht & Mackin). Crustaceana Suppl. 6: 204–215.

Jones, R. & D. C. Culver, 1989. Evidence for selection on sensory structures in a cave population of *Gammarus minus* (Amphipoda). Evolution 43: 688–693.

Jones, D. A., N. Peacock & O. F. M. Phillips, 1973. Studies on the migration of *Tritaeta gibbosa*, a subtidal benthic amphipod. Neth. J. Sea Res. 7: 135–149.

Just, J., 1988. Siphonoecetinae (Corophiidae) 6: a survey of phylogeny, distribution, and biology. Crustaceana Suppl. 13: 193–208.

Kamihira, Y., 1981. Life history of sand-burrowing amphipod *Haustorioides japonicus* (Crustacea: Dogielinotidae). Bull. Fac. Fish. Hokkaido Univ. 32: 338–348.

Kanneworff, E., 1965. Life cycle, food, and growth of the amphipod *Ampelisca macrocephala* Liljeborg from the Oresund. Ophelia 2: 305–318.

Kemp, P. F., F. A. Cole & R. C. Swartz, 1985. Life history and productivity of the phoxocephalid amphipod *Rhepoxynius abronius* (Barnard). J. Crustacean Biol. 5: 449–464.

Kinne, O., 1959. Ecological data on the amphipod *Gammarus duebeni*. A monograph. Veröff. Inst. Meeresforsch. Bremerh. 6: 177–202.

Kostalos, M. S., 1979. Life history and ecology of *Gammarus minus* Say (Amphipoda. Gammaridae). Crustaceana 37: 113–122.

Krishnan, L. & P. A. John, 1974. Observations on the breeding biology of *Melita zeylanica* Stebbing, a brackish water amphipod. Hydrobiologia 44: 413–430.

Lakshmana Rao, M. V. & K. Shyamasundari, 1963. Tube-building habits of the fouling amphipod *Corophium triaenonyx* Stebbing at Visakhapatnam Harbour. J. zool. Soc. India 15: 133–140.

Laubitz, D. R. & G. S. Lewbel, 1974. A new species of caprellid (Crustacea: Amphipoda) associated with gorgonian octocorals. Can. J. Zool. 52: 549–551.

Levings, C. D., 1980. The biology and energetics of *Eogammarus confervicolus* (Stimpson) (Amphipoda, Anisogammaridae) at the Squamish River Estuary, B.C. Can. J. Zool. 58: 1652–1663.

Lewbel, G. S., 1978. Sexual dimorphism and intraspecific aggression, and their relationship to sex ratios in *Caprella gorgonia* Laubitz & Lewbel (Crustacea: Amphipoda: Caprellidae). J. exp. mar. Biol. Ecol. 33: 133–151.

Lim, K. H. & W. D. Williams, 1971. Ecology of *Austrochiltonia subtenuis* (Sayce) (Amphipoda, Hyalellidae). Crustaceana 20: 19–24.

Lincoln, R. H., 1979. British marine Amphipoda: Gammaridea. British Museum (Natural History). 658 pp.

Lowry, J. K., 1986. The callynophore, a eucaridan/peracaridan sensory organ prevalent among the Amphipoda (Crustacea). Zool. Scr. 15: 333–349.

Lowry, J. K. & H. E. Stoddart, 1983. The amphipod genus *Parawaldeckia* in New Zealand waters (Crustacea, Lysianassoidea). J. R. Soc. N.Z. 13: 261–277.

Lyes, M. C., 1979. The reproductive behaviour of *Gammarus duebeni* (Lilljeborg), and the inhibitory effect of a surface agent. Mar. Behav. Physiol. 6: 47–55.

Mattson, S. & T. Cedhagen, 1989. Aspects of the behaviour and ecology of *Dyopedos monacanthus* (Metzger) and *D. porrectus* Bate, with comparative notes on *Dulichia tuberculata* Boeck (Crustacea: Amphipoda: Podoceridae). J. exp. mar. Biol. Ecol. 127: 253–272.

Menon, P. S., 1966. Population ecology of *Gammarus lacustris* Sars. Ph. D. thesis, University of Alberta, Canada. 117 pp.

Mills, E. L., 1963. The biology of an ampeliscid amphipod crustacean sibling species pair. Ph. D. thesis, Yale University. 160 pp.

Mills, E. L., 1967. The biology of an ampeliscid amphipod crustacean sibling species pair. J. Fish Res. Bd. Can. 24: 305–355.

Moore, P. G., 1981. The life histories of the amphipods *Lembos websteri* Bate and *Corophium bonnellii* Milne Edwards in kelp holdfasts. J. exp. mar. Biol. Ecol. 49: 1–50.

Morino, H., 1978. Studies on the Talitridae (Amphipoda, Crustacea) in Japan. III. Life history and breeding activity of *Orchestia platensis* Krøyer. Publ. Seto Mar. Biol. Lab. 24: 245–267.

Morino, H., 1981. Studies on the Talitridae (Amphipoda, Crustacea) in Japan IV. The development of gnathopod II in *Orchestia platensis* Krøyer. Publ. Seto Mar. Biol. Lab. 26: 1–13.

Morino, H., 1984. On a new freshwater species of Anisogammaridae (Gammaroidea: Amphipoda) from central Japan. Publ. Itako Hydrobiol. Stn. 1: 17–23.

Myers, A. A., 1971. Breeding and growth in laboratory-reared *Microdeutopus gryllotalpa* Costa (Amphipoda: Gammaridea). J. nat. Hist. 5: 271–277.

Nagle, J. S., 1968. Distribution of the epibiota of macroepibenthic plants. Contr. mar. Sci. 13: 105–144.

Nair, K. K. C. & K. Anger, 1979. Experimental studies on the life cycle of *Jassa falcata* (Crustacea, Amphipoda). Helgoländer wiss. Meeresunters. 32: 444–452.

Naylor, C., J. Adams & P. Greenwood, 1988a. Population dynamics and adaptive sexual strategies in a brackish water crustacean, *Gammarus duebeni*. J. anim. Ecol. 57: 493–507.

Naylor, C., J. Adams & P. Greenwood, 1988b. Variation in sex determination in natural populations of a shrimp. J. evol. Biol. 1: 355–368.

Naylor, C. & J. Adams, 1987. Sexual dimorphism, drag constraints and male performance in *Gammarus duebeni* (Amphipoda). Oikos 48: 23–27.

Omori, K., M. Tanaka & T. Kikuchi, 1982. Seasonal changes of short-term reproductive cycle in *Corophium volutator* (Crustacea: Amphipoda). – Semi-lunar or lunar cycle? Publ. Amakusa Mar. Biol. Lab. Kyushu Univ. 6: 105–117.

Pinkster, S. & N. W. Broodbakker, 1980. The influence of environmental factors on distribution and reproductive success of *Eulimnogammarus obtusatus* (Dahl, 1938) and other estuarine gammarids. Crustaceana Suppl. 6: 225–241.

Preece, G. S., 1971. The swimming rhythm of *Bathyporeia pilosa* (Crustacea: Amphipoda). J. mar. biol. Ass. U.K. 51: 777–791.

Rees, C. P., 1975. Life cycle of the amphipod *Gammarus palustris* Bousfield. Estuar. coast. mar. Sci. 3: 413–419.

Ridley, M., 1983. The explanation of organic diversity. Clarendon Press, Oxford, England. 272 pp.

Robinson, B. W. & R. W. Doyle, 1985. Trade-off between male reproduction (amplexus) and growth in the amphipod *Gammarus lawrencianus*. Biol. Bull. 168: 482–488.

Sainte-Marie, B. & P. Brunel, 1983. Differences in life history and success between suprabenthic shelf populations of *Arrhis phyllonyx* (Amphipoda Gammaridea) in two ecosystems of the Gulf of St. Lawrence. J. Crustacean Biol. 3: 45–69.

Sameoto, D. D., 1969. Comparative ecology, life histories, and behaviour of intertidal sand-burrowing amphipods (Crustacea: Haustoriidae) at Cape Cod. J. Fish. Res. Bd Can. 26: 361–388.

Segerstråle, S. G., 1970. Light control of the reproductive cycle of *Pontoporeia affinis* Lindström (Crustacea Amphipoda). J. exp. mar. Biol. Ecol. 5: 272–275.

Sexton, E. W., 1924. The moulting and growth-stages of *Gammarus* with descriptions of the normals and intersexes of *Gammarus chevreuxi*. J. mar. biol. Ass. U.K. 13: 340–401.

Sexton, E. W. & A. Matthews, 1913. Notes on the life history of *Gammarus chevreuxi*. J. mar. biol. Ass. U.K. 9: 546–556.

Sexton, E. W. & D. M. Reid, 1951. The life-history of the multiform species *Jassa falcata* (Montagu) (Crustacea Amphipoda), with a review of the bibliography of the species. J. Linn. Soc., Zool. 42: 29–91.

Sexton, E. W. & G. M. Spooner, 1940. An account of *Marinogammarus* (Schellenberg) gen. nov. (Amphipoda), with a description of a new species, *Marinogammarus pirlotti*. J. mar. biol. Ass. U.K. 24: 633–682.

Sheader, M., 1977a. Production and population dynamics of *Ampelisca tenuicornis* (Amphipoda) with notes on the biology of its parasite *Sphaeronella longipes* (Copepoda). J. mar. biol. Ass. U.K. 57: 955–968.

Sheader, M., 1977b. Breeding and marsupial development in laboratory-maintained *Parathemisto gaudichaudi* (Amphipoda). J. mar. biol. Ass. U.K. 57: 943–954.

Sheader, M., 1978. Distribution and reproductive biology of *Corophium insidiosum* (Amphipoda) on the north-east coast of England. J. mar. biol. Ass. U.K. 58: 585–596.

Sheader, M., 1983. The reproductive biology and ecology of *Gammarus duebeni* (Crustacea: Amphipoda) in southern England. J. mar. biol. Ass. U.K. 63: 517–540.

Sheader, M. & F. Chia, 1970. Development, fecundity and brooding behaviour of the amphipod, *Marinogammarus obtusatus*. J. mar. biol. Ass. U.K. 50: 1079–1099.

Shillaker, R. O. & P. G. Moore, 1987a. Tube-emergence behaviour in the amphipods *Lembos websteri* Bate and *Corophium bonnellii* Milne Edwards. J. exp. mar. Biol. Ecol. 111: 231–241.

Shillaker, R. O. & P. G. Moore, 1987b. The biology of brooding in the amphipods *Lembos websteri* Bate and *Corophium bonnellii* Milne Edwards. J. exp. mar. Biol. Ecol. 110: 113–132.

Skutch, A. F., 1926. On the habits and ecology of the tube-building amphipod *Amphithoë rubricata* Montagu. Ecology 7: 481–501.

Slattery, P. N., 1985. Life histories of infaunal amphipods from subtidal sands of Monterey Bay, California. J. Crustacean Biol. 5: 635–649.

Smallwood, M. E., 1905. The salt-marsh Amphipod: *Orchestia palustris*. Cold Spring Harbor Monographs No. 3.

Sprules, W. G., 1967. The life cycle of *Crangonyx richmondensis laurentianus* Bousfield (Crustacea: Amphipoda). Can. J. Zool. 45: 877–884.

Staude, C. P., 1986. Systematics and behavioral ecology of the amphipod genus *Paramoera* Miers (Gammaridea: Eusiroidea: Pontogeneiidae) in the eastern north Pacific. Ph. D. thesis, University of Washington, Seattle. 324 pp.

Steele, D. H., 1973. The biology of *Parhyalella pietschmanni* Schellenberg, 1938 (Amphipoda, Hyalellidae) at Nosy Bé, Madagascar. Crustaceana 25: 276–280.

Steele, D. H. & V. J. Steele, 1969. The biology of *Gammarus* (Crustacea, Amphipoda) in the northwestern Atlantic. I. *Gammarus duebeni* Lillj. Can. J. Zool. 47: 235–244.

Steele, D. H. & V. J. Steele, 1970a. The biology of *Gammarus* (Crustacea, Amphipoda) in the northwestern Atlantic. III. *Gammarus obtusatus* Dahl. Can. J. Zool. 48: 989–995.

Steele, D. H. & V. J. Steele, 1970b. The biology of *Gammarus* (Crustacea, Amphipoda) in the northwestern Atlantic. IV. *Gammarus lawrencianus* Bousfield. Can. J. Zool. 48: 1261–1267.

Steele, D. H. & V. J. Steele, 1972. The biology of *Gammarus* (Crustacea, Amphipoda) in the northwestern Atlantic. V. *Gammarus oceanicus* Segerstråle. Can. J. Zool. 50: 801–813.

Steele, D. H. & V. J. Steele, 1973. The biology of *Gammarus* (Crustacea, Amphipoda) in the northwestern Atlantic. VII. The duration of embryonic development in five species at various temperatures. Can. J. Zool. 51: 995–999.

Steele, D. H. & V. J. Steele, 1975. The biology of *Gammarus* (Crustacea, Amphipoda) in the northwestern Atlantic. XI. Comparison and discussion. Can. J. Zool. 53: 1116–1126.

Steele, D. H. & V. J. Steele, 1986. The cost of reproduction in the amphipod *Gammarus lawrencianus* Bousfield, 1956. Crustaceana 51: 176–182.

Steele, V. J., D. H. Steele & B. R. MacPherson, 1977. The effect of photoperiod on the reproductive cycle of *Gammarus setosus* Dementieva, 1931. Crustaceana, suppl. 4: 58–63.

Strong, D. R., Jr., 1972. Life history variation among populations of an amphipod (*Hyalella azteca*). Ecology 53: 1103–1111.

Strong, D. R., Jr., 1973. Amphipod amplexus, the significance of ecotypic variation. Ecology 54: 1384–1388.

Thompson, D. J. & S. J. Moule, 1983. Substrate selection and assortative mating in *Gammarus pulex* L. Hydrobiologia 99: 3–6.

Trivers, R. L., 1972. Parental investment and sexual selection. In B. Campbell (ed), Sexual selection and the descent of man. Aldine, Chicago: 136–179.

Van Dolah, R. F., 1978. Factors regulating the distribution and population dynamics of the amphipod *Gammarus palustris* in an intertidal salt marsh community. Ecol. Monogr. 48: 191–217.

Van Senus, P., 1988. Reproduction of the sandhopper, *Talorchestia capensis* (Dana) (Amphipoda, Talitridae). Crustaceana 55: 93–103.

Vlasblom, A. G., 1969. A study of a population of *Marinogammarus marinus* (Leach) in the Oosterschelde. Neth. J. Sea Res. 4: 317–338.

Ward, P. I., 1983. Advantages and a disadvantage of large size for male *Gammarus pulex* (Crustacea: Amphipoda). Behav. Ecol. Sociobiol. 14: 69–76.

Ward, P. I., 1984. The effects of size on the mating decision of *Gammarus pulex* (Crustacea, Amphipoda). Z. Tierpsychol. 64: 174–184.

Ward, P. I., 1985. The breeding behaviour of *Gammarus duebeni*. Hydrobiologia 121: 45–50.

Ward, P. I., 1986. A comparative field study of the breeding behaviour of a stream and a pond population of *Gammarus pulex* (Amphipoda). Oikos 46: 29–36.

Ward, P. I., 1987. Sexual selection and body size in *Gammarus pulex*: reply to Greenwood and Adams. Oikos 48: 108–109.

Ward, P. I., 1988. Sexual selection, natural selection, and body size in *Gammarus pulex* (Amphipoda). Am. Nat. 131: 348–359.

Ward, P. I., 1989. Mate choice in *Gammarus* (Amphipoda). J. Zool., Lond. 218: 633–635.

Watkin, E. E., 1940. The swimming and burrowing habits of the amphipod *Urothoë marina* (Bate). Proc. R. Soc. Edinb. 60: 271–280.

Watkin, E. E., 1941. The yearly life cycle of the amphipod, *Corophium volutator*. J. anim. Ecol. 10: 77–93.

Welton, J. S. & R. T. Clarke, 1980. Laboratory studies on the reproduction and growth of the amphipod, *Gammarus pulex* (L.) J. anim. Ecol. 49: 581–592.

Wilder, J., 1940. The effects of population density upon growth, reproduction, and survival of *Hyalella azteca*. Physiol. Zool. 13: 439–461.

Wildish, D. J., 1979. Reproductive consequences of the terrestrial habit in *Orchestia* (Crustacea: Amphipoda). Int. J. Invert. Reprod. 1: 9–20.

Wildish, D. J., 1984. Secondary production of four sub-littoral, soft-sediment amphipod populations in the Bay of Fundy. Can. J. Zool. 62: 1027–1033.

Wildish, D. J. & J. J. Dickinson, 1982. A new species of *Haploops* (Amphipoda, Ampeliscidae) from the Bay of Fundy. Can. J. Zool. 60: 962–967.

Wildish, D. J. & D. Peer, 1981. Methods for estimating secondary production in marine Amphipoda. Can. J. Fish. aquat. Sci. 38: 1019–1026.

Williams, J. A., 1978. The annual pattern of reproduction of *Talitrus saltator* (Crustacea: Amphipoda: Talitridae). J. Zool., Lond. 184: 231–244.

Williamson, D. I., 1951. On the mating and breeding of some semi-terrestrial amphipods. Dove Marine Laboratory Report, Third Series 12: 49–61.

Appendix Table 1.

Species of amphipods for which male precopulatory behavior has been documented in the literature. For those marked by an asterisk there were adult specimens available for analysis of sexual dimorphism of the appendages (Figs. 2 and 3).

MATE-GUARDERS: CARRIERS

Gammaroidea:

* *Anisogammarus pugettensis* (Dana): pers. obs. Northeastern Pacific Ocean. Estuaries, intertidal zone.

Echinogammarus marinus (Leach): Vlasblom, 1969 (as *Marinogammarus marinus*); Hartnoll and Smith, 1978 (as *Marinogammarus marinus*); Clarke *et al.*, 1985. Britain. Estuaries, intertidal zone.

Marinogammarus pirloti (Sexton & Spooner): Hartnoll and Smith, 1978 (as *Marinogammarus pirloti*). Britain. Estuaries, intertidal zone.

* *Eogammarus confervicolus* (Stimpson): Levings, 1980. Northeastern Pacific Ocean. Estuaries, intertidal zone, amongst algae and wood debris and under stones.

* *Eulimnogammarus obtusatus* Dahl: Sexton & Spooner, 1940 (as *Marinogammarus obtusatus*);

Sheader & Chia, 1970 (as *Marinogammarus obtusatus*); Hartnoll & Smith, 1978 (as *Marinogammarus obtusatus*); Steele & Steele, 1970a, 1973, 1975 (as *Gammarus obtusatus*); Pinkster & Broodbakker, 1980. Northeastern and northwestern Atlantic Ocean. Full salinity beaches, salt-marshes, under stones and amongst algae in the mid- to lower intertidal zone (Bousfield, 1973).

Gammarellus angulosus (Rathke): Steele & Steele, 1972a. Northwestern and northeastern Atlantic Ocean. Full salinity rocky shores in algae, low intertidal zone to – 20 m.

Gammarus chevreuxi Sexton: Sexton & Matthews, 1913; Dieleman, 1979. Northeastern Atlantic Ocean. Brackish water, in marshes, ditches, and estuaries (Lincoln, 1979).

Gammarus duebeni Liljeborg: Hynes, 1954, 1955; Kinne, 1959; Steele & Steele, 1969; Dahl *et al.*, 1970; Hartnoll & Smith, 1978, 1980; Lyes, 1979; Sheader, 1983; Borowsky, 1984b; Ward, 1985; Naylor & Adams, 1987; Naylor *et al.*, 1988a, b. Britain, Europe, and northeastern North America. Estuaries, intertidal zone, and brackish springs, lakes, and streams.

Gammarus fasciatus Say: Embody, 1911; Crozier & Snyder, 1923; Borowsky, 1984b. Northwestern Atlantic coastal estuaries and river drainages, and lakes and rivers draining into the Great Lakes and St. Lawrence River (Bousfield, 1973).

Gammarus lacustris Sars: Menon, 1966; Hynes & Harper, 1971; De March, 1982. North America and northwestern Europe. Lakes, marshes, streams, springs (Bousfield, 1958).

* *Gammarus lawrencianus* Bousfield: Borowsky, 1984b; Hunte *et al.*, 1985; Robinson & Doyle, 1985; Steele & Steele, 1970b, 1973, 1975, 1986; Dunham, 1986; Dunham *et al.*, 1986; Hunte & Myers, 1988; Dunham & Hurshman, 1988. Northwestern Atlantic Ocean. Estuaries and full salinity sandy beaches. Intertidal, under stones and amongst algae.

Gammarus locusta Linnaeus: Blegvad, 1922; Crozier & Snyder, 1923; Hartnoll & Smith, 1978. Northeastern Atlantic Ocean. Full salinity low intertidal zone to – 30 m.

Gammarus minus Say: Kostalos, 1979; Fong, 1989; Jones & Culver, 1989. Central and eastern United States. Amongst gravel, vegetation, and detritus in riffle areas of springs, spring runs, and cave streams.

Gammarus mucronatus Say: Borowsky, 1984b; Fredette & Diaz, 1986. Northwestern Atlantic Ocean. Estuaries and salt-marshes, low intertidal zone and shallowly subtidal, amongst algae. Also in rocky shore upper intertidal tide pools (Bousfield, 1973).

* *Gammarus oceanicus* Segerstråle: Steele & Steele, 1972, 1973, 1975; Clarke *et al.*, 1985 (as *Lagunogammarus oceanicus*). Northeastern and northwestern Atlantic Ocean and Arctic Ocean. Full salinity and brackish rocky shores and beaches, mid-intertidal zone to – 25 m (Lincoln, 1979).

* *Gammarus palustris* Bousfield: Rees, 1975; Gable & Croker, 1977; Van Dolah, 1978; Borowsky, 1980b, 1983, 1984b, 1985, 1988; Borowsky and Borowsky, 1987. Northwestern Atlantic Ocean. Estuaries, amongst *Spartina*.

Gammarus pseudolimnaeus Bousfield: Embody, 1911; Hynes & Harper, 1971; pers. obs. Northeastern North America. Lakes and rivers (Bousfield, 1958).

Gammarus pulex Linnaeus: Hartnoll & Smith, 1978; Birkhead & Clarkson, 1980; Welton & Clarke, 1980; Ward, 1983, 1984, 1986, 1987, 1988, 1989; Thompson & Moule, 1983; Adams & Greenwood, 1983, 1985, 1987; Greenwood & Adams, 1984, 1987; Elwood *et al.*, 1987. Britain. Freshwater streams.

Gammarus salinus Spooner: Hartnoll & Smith, 1978. Northeastern Atlantic Ocean. Estuaries, intertidal zone.

Gammarus setosus Dementieva: Steele *et al.*, 1977. Arctic Ocean, circumpolar, and Northwestern Atlantic Ocean. Mid-intertidal zone to shallowly subtidal under stones and amongst algae on muddy beaches and estuaries or in tide pools on rocky shores (Bousfield, 1973).

Gammarus troglophilus Hubricht & Mackin: Jenio, 1980. Missouri and Illinois. Cave streams, springs, and spring streams, amongst cobbles, pebbles, sand and detritus.

Gammarus zaddachi Sexton: Hartnoll & Smith, 1978; Dieleman, 1979. Northeastern Atlantic Ocean to the Arctic Ocean. Estuaries, intertidal zone and shallowly subtidal.

* *Jesogammarus paucisetulosus* Morino: Morino, 1984; Hamashima & Morino, 1984. Central Japan. Mountain streams and springs.

Locustogammarus levingsi Bousfield: pers. obs. Northeastern Pacific. Gravel beaches, in freshwater run-off, intertidal zone.

Hadzioidea:

* *Melita nitida* Smith: Borowsky, 1980a, 1984b. Northwestern Atlantic Ocean. Estuaries, mud bottoms, amongst hydroids and bryozoans (Bousfield, 1973).

Melita zeylanica Stebbing: Krishnan & John, 1974. India. In brackish waters, on tree rootlets and in tunnels of the boring isopod *Sphaeroma*.

Talitroidea:

Aquatic:

* *Allorchestes compressa* Dana: Ahsanullah & Williams, 1986. Southwestern Pacific Ocean. Full salinity, amongst seagrass.

* *Allorchestes angusta* (Dana): pers. obs. Northeastern Pacific Ocean. Full salinity rocky shores, low intertidal zone and shallowly subtidal, amongst algae.

Austrochiltonia subtenuis (Sayce): Lim & Williams, 1971. Southern Australia. Saline lakes, amongst shoreline macrophytes.

Haustorioides japonicus Kamihira: Kamihira, 1981. Northwestern Pacific Ocean. Burrowing in sandy beaches in the intertidal zone.

* *Hyale barbicornis* Hiwatari and Kajihara: Hiwatari & Kajihara, 1988. Northwestern Pacific Ocean. Rocky shores, pebble beaches, and on man-made structures in the intertidal zone, amongst algae, *Mytilus edulis*, and under stones.

* *Hyale nilssoni* (Rathke): Williamson, 1951; Moore, 1981. Northeastern and northwestern Atlantic Ocean, North Sea, Arctic Ocean. Full salinity rocky shores, intertidal zone and shallowly subtidal, amongst algae (Bousfield, 1973).

* *Hyale plumulosa* (Stimpson): Borowsky, 1984a. Northwestern Atlantic Ocean and northeastern Pacific Ocean. Full salinity rocky and cobble shores and in salt-marshes amongst *Spartina* roots, algae, and under stones. Intertidal and upper intertidal tide pools (Bousfield, 1973).

* *Hyalella azteca* Saussure: Embody, 1911; Wilder, 1940; Cooper, 1965; Strong, 1972, 1973; De March, 1978; pers. obs.; Borowsky, 1984a. Continental North America and coastal islands, from Mexico to the northern tree line, and from Atlantic to Pacific coasts (Bousfield, 1958). Amongst shoreline vegetation and detritus in streams, rivers, and lakes.

Parhyale basrensis Salman: Ali and Salman, 1986. Iraq. Estuaries, intertidal and shallow subtidal, under stones and vegetation.

* *Parhyalella pietschmanni* Schellenberg: Steele, 1973. Indo-Pacific, from Hawaii to Madagascar. Full salinity beaches, low intertidal zone at the level of low water neap tides, amongst drifting dead turtle grass.

Semi-terrestrial:

Orchestia cavimana Heller: Wildish, 1979. Mediterranean, Black, North, and Red Seas, and Atlantic coasts of North Africa, Europe, and Britain, under stones or amongst damp vegetation, usually close to fresh or brackish water (Lincoln, 1979).

* *Orchestia gammarellus* (Pallas): Williamson, 1951; Campbell-Parmentier, 1963; Wildish, 1979; Moore, 1981; Ridley, 1983. Mediterranean, Black, and North Seas, and Atlantic coast of Europe. Full salinity to brackish waters, intertidal and shallow subtidal to damp semi-terrestrial habitats well away from water (Lincoln, 1979).

* *Orchestia grillus* Bosc: Smallwood, 1905; Embody, 1911 (as *O. palustris*); Williamson, 1951 (as *O. palustris*). Northwestern Atlantic Ocean. Salt-marshes, mid-intertidal to the supratidal zone under debris and amongst *Spartina* roots and marsh grasses (Bousfield, 1973).

* *Orchestia mediterranea* Costa: Williamson, 1951; Wildish, 1979. Mediterranean, Black,

and North Seas, and Atlantic coast of Britain and Europe. Upper intertidal zone amongst decaying wrack (Lincoln, 1979).

Megalorchestia californiana Brandt: Bowers, 1964 (as *Orchestoidea californiana*). Northeastern Pacific Ocean. Exposed smooth sand beaches, upper intertidal zone (Bousfield, 1982b).

Megalorchestia corniculata (Stout): Bowers, 1964 (as *Orchestoidea corniculata*). Northeastern Pacific Ocean. Exposed coarse sand and fine gravel beaches, upper intertidal zone, burrowing under wrack (Bousfield, 1982b).

* *Platorchestia platensis* (Kröyer): Morino, 1978, 1981; Behbehani, 1978. Northeastern, northwestern, southeastern and southwestern Atlantic Ocean, northeastern, northwestern, and southeastern Pacific Ocean, and Indian Ocean on all levels of the intertidal zone and well above the high water mark (Lincoln, 1979).

* *Pseudorchestoidea brito* Stebbing: Williamson, 1951. Northeastern Atlantic Ocean. Sand or gravel beaches, usually below high water mark (Lincoln, 1979).

* *Talitrus saltator* (Montagu): Williamson, 1951; Hurley, 1968; Williams, 1978. North Sea and Atlantic coast of Europe. Sandy beaches amongst beach wrack (Lincoln, 1979).

* *Talorchestia deshayesii* (Audouin): Williamson, 1951. Mediterranean, Black, and North Seas, Atlantic coast of Europe. Burrows in fine sand on sheltered shores at about high tide level and beneath stones beyond tidal range (Lincoln, 1979).

Traskorchestia traskiana (Stimpson): pers. obs. Northeastern Pacific Ocean. Gravel, sand, and mud beaches, amongst beach wrack.

MATE-GUARDERS: ATTENDERS

Corophioidea:

* *Ampithoe lacertosa* Bate: Heller, 1968. Northeastern Pacific Ocean. Domicolous amongst algae on mud, sand, gravel, and bedrock. Full salinity low intertidal zone to −10 m (Conlan, 1982).

Ampithoe rubricata (Montagu): Skutch, 1926. Northeastern and northwestern Atlantic Ocean. Domicolous amongst algae in the intertidal zone and shallowly subtidal. Full salinity rocky shores and outer parts of estuaries (Bousfield, 1973; Lincoln, 1979).

* *Ampithoe valida* Smith: Barrett, 1966; Borowsky, 1983. Northwestern Atlantic and Northeastern Pacific coasts. Domicolous amongst algae on mud, sand, gravel, and bedrock. Brackish water and full salinity low intertidal zone to −30 m (Conlan, 1982).

Chelura terebrans Philippi: Ridley, 1983. Northeastern and northwestern Atlantic Ocean, Mediterranean, Black, and North Seas, southeastern and southwestern Pacific Ocean. Occupying burrows of the gribble *Limnoria* in submerged and waterlogged wood (Lincoln, 1979).

Corophium acutum Chevreux: Nagle, 1968. Cosmopolitan in temperate waters. Full salinity rocky shores and man-made structures in tubes amongst algae, sponges, and hydroids, shallowly subtidal (Lincoln, 1979).

Corophium arenarium Crawford: Crawford, 1937; Fish & Mills, 1979; Borowsky, 1983. Mediterranean and North Seas, Atlantic coast of Europe. Estuaries, burrowing in sandy mud in the mid- to low-intertidal zone (Lincoln, 1979).

Corophium insidiosum Crawford: Sheader, 1978; Nair & Anger, 1979. Northeastern and northwestern Atlantic Ocean, and northeastern, northwestern, and southeastern Pacific Ocean. Estuaries, low intertidal zone, building tubes on oysters, man-made structures, and amongst eelgrass (Bousfield, 1973; Lincoln, 1979).

Corophium triaenonyx Stebbing: Lakshmana Rao & Shyamasundari, 1966. Indian Ocean. Building tubes on rocky shores and man-made structures.

* *Corophium volutator* (Pallas): Watkin, 1941; Fish & Mills, 1979; Omori *et al.*, 1982; Hughes, 1988. Northeastern and northwestern Atlantic Ocean, and Mediterranean, Black, and Azov Seas. Nearly fully saline to fresh waters in inter-

tidal mud flats, salt-marsh pools, and brackish ditches (Lincoln, 1979).

Dulichia rhabdoplastis McCloskey: pers. obs. Northeastern Pacific Ocean, − 2 to − 35 m on self-constructed sediment whips attached to hydroids and tips of the spines of the red sea urchin *Strongylocentrotus franciscanus*.

Dyopedos monacanthus Metzger: Mattson & Cedhagen, 1989. Northeastern and north-western Atlantic Ocean and Arctic Ocean, on self-constructed sediment whips, mud bottom, − 15 to − 325 m (Lincoln, 1979).

Dyopedos porrectus (Bate): Mattson & Cedhagen, 1989. Northeastern and northwestern Atlantic Ocean on self-constructed sediment whips on hydroids and bryozoans, − 15 to − 700 m.

* *Ericthonius brasiliensis* (Dana): Connell, 1963; Ridley, 1983. Cosmopolitan in tropical and temperate waters. Domicolous amongst algae, hydroids, and bryozoans from the low intertidal zone to − 200 m (Lincoln, 1979).

Jassa falcata (Montagu): Sexton & Reid, 1951; Conlan, 1989. Northeastern Atlantic Ocean. High salinity rocky shores, mid- to low inter-tidal zone to − 40 m (Conlan, 1990).

* *Jassa marmorata* Holmes: Nair & Anger, 1979 (identified as *Jassa falcata*); Borowsky, 1983, 1985 (identified as *Jassa falcata*; Conlan, 1989; Franz, 1989. Northeastern and northwestern Atlantic Ocean, northeastern and northwestern Pacific Ocean, southeastern and southwestern Atlantic Ocean, southeastern and southwestern Pacific Ocean, Indian Ocean. High salinity rocky shores, strictly subtidal, to − 10 m (Conlan, 1990).

* *Lembos websteri* Bate: Shillaker & Moore, 1987a, b; Moore, 1981; Borowsky, 1983. Northeastern and northwestern Atlantic Ocean. Domicolous amongst algae and sponges in the intertidal zone to − 35 m (Bousfield, 1973; Lincoln, 1979).

Leptocheirus pilosus Zaddach: Goodhart, 1939; Ridley, 1983. Northeastern Atlantic Ocean and North Sea. Domicolous in brackish waters, rivers, creeks, and ditches (Lincoln, 1979).

Microdeutopus damnomiensis (Bate): Nagle, 1968; Borowsky, 1986. Northeastern Atlantic Ocean and Mediterranean Sea. Domicolous amongst algae on rocky shores, low intertidal zone to shallowly subtidal (Lincoln, 1979).

* *Microdeutopus gryllotalpa* Costa: Myers, 1971; Bregazzi, 1972; Borowsky, 1980a, 1983, 1984b, 1989; Borowsky *et al.*, 1987. Northeastern and northwestern Atlantic Ocean, Mediterranean and Black Seas. Domicolous amongst algae on rocky shores and man-made structures, and in eel-grass beds and salt-marshes, low intertidal zone to − 150 m (Bousfield, 1973; Lincoln, 1979).

Siphonoecetes dellavallei Stebbing: Ridley, 1983; Just, 1988. Southeastern Atlantic Ocean, Mediterranean Sea. Infralittoral fringe of the intertidal zone to − 42 m. Inhabiting self-constructed tubes or gastropod shells on mud or sand or amongst algae or eel-grass.

* *Podocerus brasiliensis* (Dana): pers. obs. Cosmopolitan in tropical and warm temperate waters. Free living, amongst algae and hydroids on rocky shores and man-made structures, or on coarse substrates, low intertidal zone to − 24 m.

Caprellidea:

* *Caprella gorgonia* Laubitz and Lewbel: Lewbel, 1978. Northeastern Pacific Ocean. On gorgonian octocorals, − 20 m to − 23 m.

Pseudoprotella phasma (Montagu): Harrison, 1940. Northeastern Atlantic Ocean. On hydroids, − 6 to − 10 m.

Hadzioidea:

* *Elasmopus levis* Smith: Borowsky, 1986, 1988 (this may be more appropriately classified as a benthic non-mate-guarder; attending is very limited compared with other species). North-western Atlantic Ocean on rocky shores, inter-tidal and shallowly subtidal, under algae and stones and amongst eelgrass (Bousfield, 1973).

NON-MATE-GUARDERS: PELAGIC

Ampeliscoidea:

* *Ampelisca abdita* Mills: Mills, 1963, 1967. Northwestern Atlantic Ocean. Domicolous, in fine sand to mud in protected bays and estuaries, low intertidal zone to − 60 m (Bousfield, 1973).

Ampelisca macrocephala Liljeborg: Kanneworff, 1965. Arctic Ocean circumpolar, northeastern and northwestern Atlantic Ocean, North and Bering Seas. Domicolous in mud, − 10 to − 280 m (Bousfield, 1973; Lincoln, 1979).

Ampelisca tenuicornis Liljeborg: Sheader, 1977a. Northeastern Atlantic Ocean, North and Mediterranean Seas. Domicolous, in fine sand or mud, 0 to − 55 m (Lincoln, 1979).

Ampelisca vadorum Mills: Mills, 1967. Northwestern Atlantic Ocean. Domicolous, in medium to coarse sand in protected bays and estuaries, low intertidal zone to > 70 m (Bousfield, 1973).

Haploops fundiensis Wildish & Dickinson: Wildish, 1984. Northwestern Atlantic Ocean. Domicolous, in mud, − 30 to − 80 m (Wildish & Dickinson, 1982).

Eusiroidea:

* *Paramoera mohri* J.L. Barnard: Staude, 1986. Northeastern Pacific Ocean. Full salinity coarse sand, gravel, and cobble sediments, midintertidal to − 10 m.

Lysianassoidea:

* *Hippomedon whero* Fenwick: Fenwick, 1985. New Zealand. Full salinity sand bottom, − 6 m.
* *Parawaldeckia karaka* Lowry and Stoddart: (identified as *P. stephenseni*) Fincham, 1974; Lowry and Stoddart, 1983. New Zealand. Full salinity sand bottom, − 1 to − 3 m.

Patuki roperi Fenwick: Fenwick, 1985. New Zealand. Full salinity sand bottom, − 6 m.

Oedicerotoidea:

* *Arrhis phyllonyx* (Sars): Sainte-Marie and Brunel, 1983. Arctic Ocean circumpolar, northeastern and northwestern Atlantic Ocean, − 10 to − 1000 m (Lincoln, 1979).
* *Pontocrates arenarius* (Bate): Fincham, 1970. Northeastern Atlantic Ocean and North Sea, − 5 to − 30 m.

Phoxocephaloidea:

* *Paraphoxus australis* K.H. Barnard: Fenwick, 1985. New Zealand. Full salinity sand bottom, − 6 m.
* *Rhepoxynius abronius* (J.L. Barnard): Slattery, 1985; Kemp *et al.*, 1985. Northeastern Pacific Ocean. High salinity sandy coasts, subtidal to at least − 37 m.

Rhepoxynius fatigans (J.L. Barnard): Slattery, 1985. Northeastern Pacific Ocean. High salinity sandy coasts, subtidal to at least − 37 m.

Pontoporeioidea:

* *Amphiporeia lawrenciana* Shoemaker: Downer & Steele, 1979. Northwestern Atlantic Ocean. Full salinity, clean sand in the very low intertidal zone to − 200 m (Bousfield, 1973).

Bathyporeia pelagica (Bate): Watkin, 1940; Fincham, 1970; Fish and Preece, 1970; Preece, 1971; Fish, 1975. Britain, intertidally just below the level of high-water neap tides (Lincoln, 1979).

* *Bathyporeia pilosa* Lindstrom: Watkin, 1940; Fincham, 1970; Fish & Preece, 1970; Preece, 1971; Fish, 1975. Northeastern Atlantic Ocean and North Sea. Intertidally on sandy beaches, above the level of high-water neap tides. May occur in brackish water (Lincoln, 1979).

Pontoporeia affinis Lindstrom: Segerstråle, 1970. Northern Europe, Asia, and North America, and the Baltic Sea. In lakes and large rivers.

* *Pontoporeia femorata* Kröyer: Wildish & Peer, 1981. Circumpolar and subarctic, burrowing in mud and sandy mud from just subtidal to > 50 m (Bousfield, 1973).

NON-MATE-GUARDERS: BENTHIC

Crangonyctoidea:

* *Crangonyx richmondensis laurentianus* Bousfield: Sprules, 1967. Northeastern North America. Bog ponds and small acidic lakes and streams (Bousfield, 1958).

Haustorioidea:

* *Eohaustorius canadensis* (Bousfield): Sameoto, 1969; Donn & Croker, 1985. Northwestern Atlantic Ocean. High salinity exposed sand beaches and marsh creeks, intertidal zone from low water to mean lower high water level (Bousfield, 1973).
Eohaustorius sawyeri Bosworth: Slattery, 1985. Northeastern Pacific Ocean. High salinity sandy coasts, shallowly subtidal to at least − 6 m.
* *Eohaustorius sencillus* Barnard: Slattery, 1985. Northeastern Pacific Ocean. High salinity sandy coasts, shallowly subtidal to at least − 9 m.
* *Neohaustorius biarticulatus* Bousfield: Sameoto, 1969. Northwestern Atlantic Ocean. High salinity exposed sand beaches, intertidal zone from low water to nearly mean high water level (Bousfield, 1973).

Eusiroidea:

* *Paramoera* n. sp.: Staude, 1986. Northeastern Pacific Ocean. High salinity and brackish water, high to low intertidal zone among coarse sediments or mussels.

UNCERTAIN STATUS

Hyperiidea:

* *Parathemisto gaudichaudi* Guerin-Meneville: Sheader, 1977b (= *Themisto compressa*). Temperate and subpolar regions in both hemispheres. Planktonic, but mating occurs on a substrate.

Appendix Table 2.

Species of aquatic carriers and attenders scored for sexual dimorphism of the enlarged gnathopod (gnathopod 1 in aorid corophioideans, gnathopod 2 in all other species).

AQUATIC CARRIERS

Gammaroidea:

Gammarus palustris
Gammarus lawrencianus
Eogammarus confervicolus
Eulimnogammarus obtusatus
Lagunogammarus oceanicus
Jesogammarus paucisetulosus
Anisogammarus pugettensis
Carinogammarus roselii

Hadzioidea:

Melita nitida
Melita sp.
Casco bigelowi
Abludomelita desdechada
Metaceradocus whakatane
Quadrivisio lutzi

Talitroidea:

Hyale nilssoni
Hyale barbicornis
Hyale plumulosa
Hyalella azteca
Parhyalella pietschmanni
Allorchestes compressus
Allorchestes angusta

ATTENDERS

Corophioidea:

Ampithoe valida
Ampithoe lacertosa
Peramphithoe plea
Podoceropsis chionoecetophila
Protomedeia fasciata
Gammaropsis thompsoni
Microdeutopus myersi

Leptocheirus plumulosus
Lembos smithi
Lembos rectangulatus
Lembos websteri
Microdeutopus gryllotalpa
Aoroides columbiae
Columbaora cyclocoxa
Ischyroceridae n. g.
Ericthonius brasiliensis
Parajassa sp.
Microjassa n. sp.
Jassa goniamera (to be transferred to a new genus (Conlan, 1989))
Jassa wandeli (to be transferred to a new genus (Conlan, 1989))
Jassa marmorata
Podocerus brasiliensis
Dyopedos arcticus
Cerapus tubularis

Caprellidea:

Caprella gorgonia

Appendix Table 3.

Species of Talitroidea scored for sexual dimorphism of the second gnathopod.

Aquatic:

Hyale nilssoni
Hyale barbicornis
Hyale plumulosa
Hyalella azteca
Parhyalella pietschmanni
Allorchestes compressus
Allorchestes angusta

Palustral:

Chelorchestia floridana
Chiltonorchestia secunda
Eorchestia nitida
Microrchestia notabilis
Uhlorchestia uhleri

Sandhopper:

Pseudorchestoidea brito
Talorchestia deshayesii
Platorchestia platensis
Megalorchestia corniculata
Platorchestia crassicornis
Parorchestia sp. (*schmitti* gp.)

Hydrobiologia **223**: 283–291, 1991.
L. Watling (ed.), VIIth International Colloquium on Amphipoda.
© 1991 *Kluwer Academic Publishers.*

Redescription of *Caprogammarus gurjanovae* Kudrjaschov & Vassilenko, 1966 (Crustacea: Amphipoda) from Hokkaido, Japan, with notes on the taxonomic status of *Caprogammarus*

Ichiro Takeuchi[1] & Shin-ichi Ishimaru[2]
[1]*Department of Fisheries, Faculty of Agriculture, The University of Tokyo, 1-1-1 Yayoi, Bunkyo-ku, Tokyo 113, Japan; Present Address: Otsuchi Marine Research Center, Ocean Research Institute, The University of Tokyo, Akahama, Otsuchi, Iwate 028-11, Japan;* [2]*Rokusei Senior High-School, Notobe, Rokusei, Ishikawa 929-16, Japan; present address: Kanazawa Women's College of Ishikawa Prefecture, 6-105 Izumihon-machi, Kanazawa, Ishikawa 921, Japan*

Abstract

Caprogammarus gurjanovae Kudrjaschov & Vassilenko, 1966 (Amphipoda: Caprellidea: Caprogammaridae) was redescribed based on the materials newly collected off Kushiro and Akkeshi, Hokkaido, Japan, which represents the southernmost record of this genus. *Caprogammarus* and *Caprella* share an identical feature, that of having the head and pereonite I 'partially fused'. Thus, *Caprogammarus* is considered to be a member of the suborder Caprellidea.

Introduction

The Caprogammaridae contains only one genus, *Caprogammarus* Kudrjaschov & Vassilenko, 1966. Kudrjaschov & Vassilenko's (1966) placement of this family under the suborder Gammaridea has not been widely accepted. McCain (1968) proposed transferring the Caprogammaridae to the suborder Caprellidea. This opinion was followed by McCain (1970), McCain & Steinberg (1970), Arimoto (1976), Laubitz (1976), Bousfield (1978, 1983) and Schram (1989), but not by Kudrjaschov & Vassilenko (1972), Vassilenko (1974, 1977) and Lowry (1976). Furthermore, Barnard (in Barnard & Karaman, 1983) placed this family under the section Corophiida of their newly erected suborder Corophiidea. So far, only Barnard & Barnard (1983) have followed the last treatment.

Only two species of *Caprogammarus* are known, namely, *C. gurjanovae* Kudrjaschov & Vassilenko, 1966 and *C. micropleopodus* Vassilenko, 1977. The genus is so far restricted to the coasts of Sakhalin Island, and Kamchatka Peninsula to Habomai Islands through Chishima Islands [= Kuril'skiye Ostrova] of the Northwest Pacific (Kudrjaschov & Vassilenko, 1966, 1972; Vassilenko, 1977). In this paper, we shall redescribe *C. gurjanovae* based on specimens newly collected off Kushiro and Akkeshi, Hokkaido, Japan.

Taxonomic description

Order Amphipoda
  Suborder Caprellidea

Family Caprogammaridae Kudrjaschov & Vassilenko, 1966
Caprogammarus Kudrjaschov & Vassilenko, 1966

Diagnosis; Flagellum of antenna II 2-segmented; mandible with molar and 3-segmented palp; gills on pereonites III and IV; vestigial coxae on gnathopod II, pereopods V, VI, and VII; pereopods III and IV reduced to one segment; abdomen composed of 3-segmented pleon and 2-segmented urosome; pleopods I to III biramous; uropods I and II uniramous; uropod III absent.

Caprogammarus gurjanovae Kudrjaschov & Vassilenko, 1966
(Figs. 1–5)

Caprogammarus gurjanovae Kudrjaschov & Vassilenko, 1966, pp. 193–198, figs. 1–4; 1972, pp. 134–148, figs. 1–5.

Material examined: 2 males, 5 females (1 mature and 4 immature), and 2 juveniles collected off Kushiro, Hokkaido, Japan (43.00 °N, 144.20 °E), July 16, 1985, by Shin-ichi Ishimaru. Arimoto's private collection No. 868, 1 immature female from Akkeshi Bay, Hokkaido, Japan (43.00 °N, 144.50 °E), August 23, 1976, by Masaki Sakaguchi.

Diagnosis: Antenna I longer than half the body length; antenna II profusely fringed with plumose setae on ventral face; pleopods I to III biramous, rami more than 10-segmented, with one pair of long plumose setae on each segment.

Description of Male (Figs. 1A, A', 2, 3): Body 18.6 mm long, head 1.1 mm, pereon 13.7 mm and abdomen 3.8 mm.

Head partially fused with pereonite I, with a forward pointing triangular projection; the articulation incomplete, leaving laterally a diagonal suture (Fig. 1A). Pereonites II and III subequal and longer than other somites. Length of pereonite I to VII: 1.5, 2.6, 2.6, 2.2, 2.2, 1.5 and 1.1 mm, respectively. Pereonite I bearing a triangular sternal process between insertions of gnathopods I. Pereonites II to VII each with one pair of anteroventral projections and another pair of lateral projections near the bases of gills or pereopods. Pereonite VII bearing one pair of posterodorsal projections and another pair of cylindrical penes on posteroventral surface. Pleonites I to III subequal, each with a posterodorsal knob (Fig. 1A'). Urosomite I very elongate. Urosomite II fased with urosomite III and telson.

Antenna I (Fig. 1A) longer than half the body. Length of peduncular segments in order of II, III, and I. Flagellum a little shorter than peduncle, composed of 27 segments.

Antenna II (Fig. 2A) shorter than peduncle of antenna I. Flagellum 2-segmented. Long plumose setae on ventral surface of peduncular segments II, III, IV and flagelar segment I.

Gnathopod I (Fig. 2B) without coxa; propodus convex, with a spine on basal part of palm.

Gnathopod II (Fig. 2C) with vestigial coxa. Basis slightly longer than pereonite II, with triangular projection on distal part. Ischium bearing a triangular projection. Merus triangular. Propodus oblong, about as long as pereonite II and twice as long as its width; a triangular palmar projection with a spine located at about one-third from proximal end.

Pereopods III (Fig. 2D) and IV (Fig. 2E) cylindrical, about seven times as long as wide.

Gills (Figs. 2D, E) oblong, longer than corresponding pereopods, and about four times as long as wide.

Pereopod V (Fig. 2F) with vestigial coxa, about two-fifths of the body length. Basis longest. Propodus concave with a knob carrying a short spine and two long setae on proximal part of palm; seven small knobs on the palm, each with one or two spines. Pereopods VI (Fig. 2G) and VII (Fig. 2H) similar to pereopod V. Length of pereopods in order of VII, VI, and V.

Pleopod I (Fig. 2I) about half as long as pereopod V; peduncle with retinacula of nine short hooked spines; rami 14-segmented, proximal segment longest, and each segment bearing two long plumose setae. Pleopods II (Fig. 2J) and III (Fig. 2K) similar to pleopod I. Length of pleopods in order of I, II, III.

Uropod I (Fig. 2L) as long as urosomite I, peduncle serrated on outer distal part; ramus cylindrical and curved upward. Uropod II (Fig. 2M) shorter than uropod I.

Mouthparts. Inner plate of maxilliped (Fig. 3A) round-pentagonal, with four short spiniform setae on inner half of distal margin and a row of several long setae; outer plate oblong, with two short spiniform and seven long setae on inner margin;

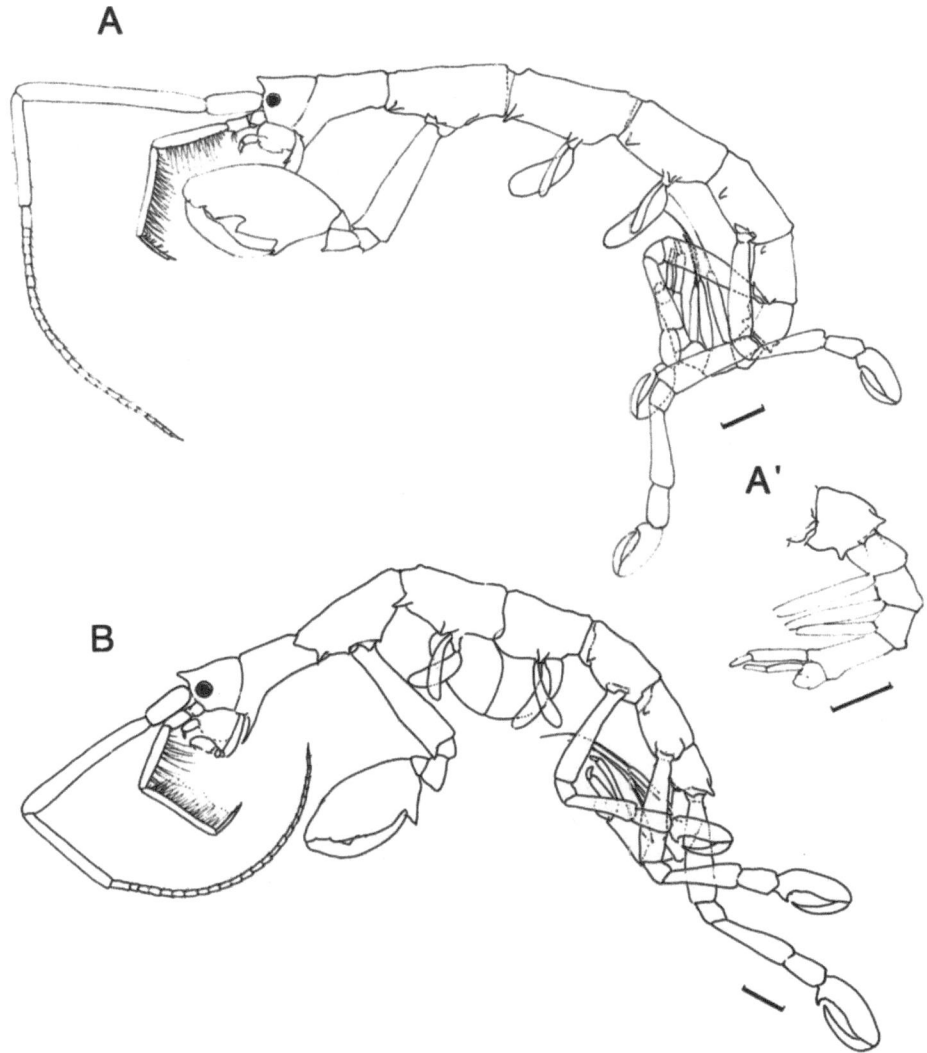

Fig. 1. Caprogammarus gurjanovae Kudrjaschov & Vassilenko, 1966. Off Kushiro, Hokkaido, Japan. Male. A, lateral view; A′, lateral view of pereonite VII and abdomen. Female. B, lateral view. Scale bars equal 1.0 mm.

second segment of palp carrying densely arranged setae on inner margin; third segment carrying dense patches of setae on inner and lateral surfaces; fourth segment claw-shaped. Outer plate of maxilla I (Fig. 3B) with six stout apical spines; distal segment of palp with five marginal spines and eight long setae arranged in an oblique row. Outer plate of maxilla II (Fig. 3C) with several setae on edge; inner plate oval, with several setae on edge. Upper lip (Fig. 3D) symmetrically notched distally. Inner lobe of lower lip (Fig. 3E)

round. Palp segments of left mandible (Fig. 3F) longer in turns of II, III and I; setal formula of segment III 1-9-1; segment II with a lateral row of several long setae; incisor divided into five teeth; lacinia mobilis carrying four teeth and two setal rows. Right mandible (Fig. 3G) similar to left one except having only one setal row.

Description of Female (Figs. 1B, 4, 5): Body length 18.8 mm long, head 1.3 mm, pereon 13.5 mm and abdomen 4.0 mm.

Flagellum of antenna I 25-segmented.

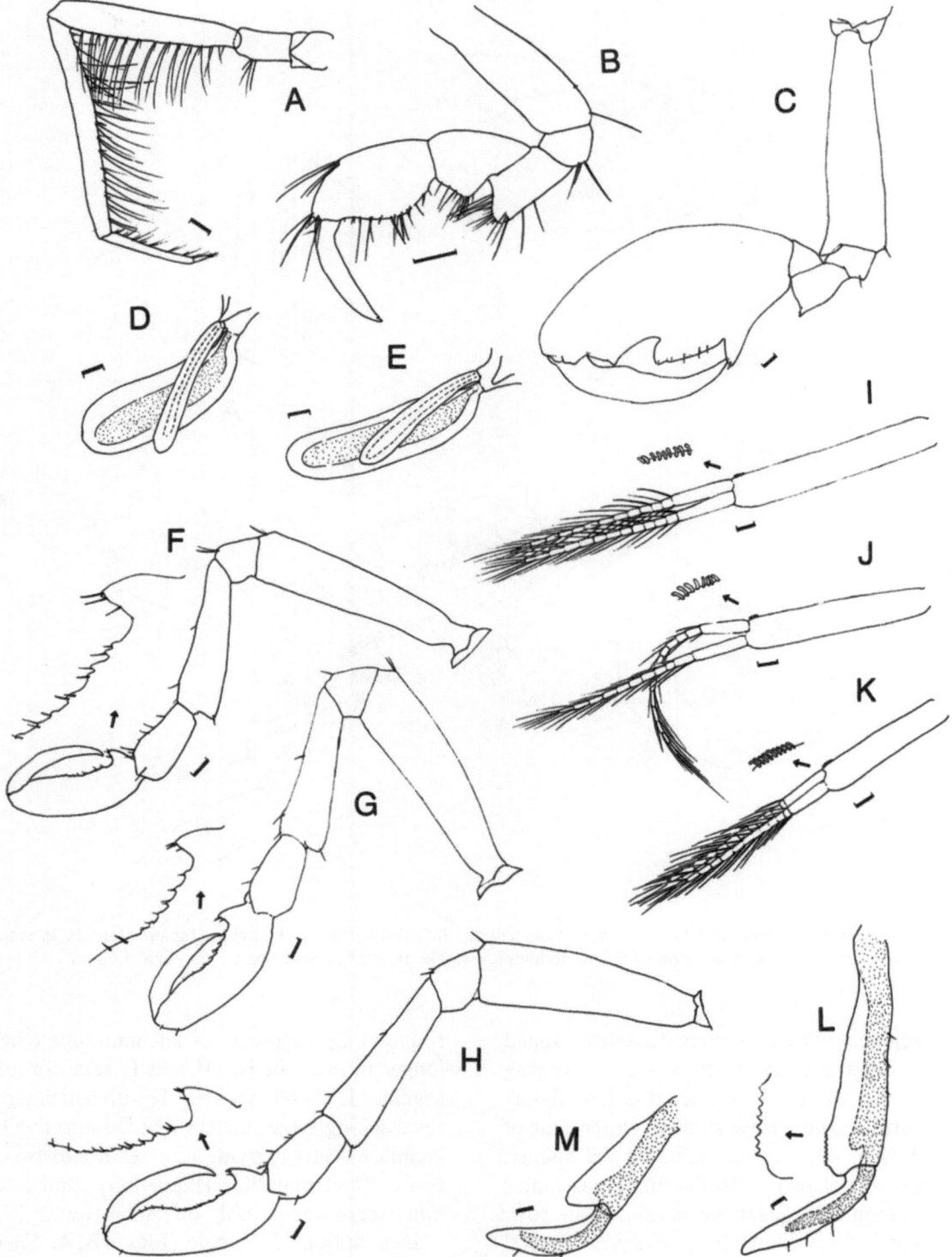

Fig. 2. Caprogammarus gurjanovae Kudrjaschov & Vassilenko, 1966. Off Kushiro, Hokkaido, Japan. Male. A, antenna II; B, gnathopod I; C, gnathopod II; D, pereopod III and gill III; E, pereopod IV and gill IV; F, pereopod V; G, pereopod VI; H, pereopod VII; I, pleopod I; J, pleopod II; K, pleopod III; L, uropod I; M, uropod II. Scale bars equal 0.2 mm.

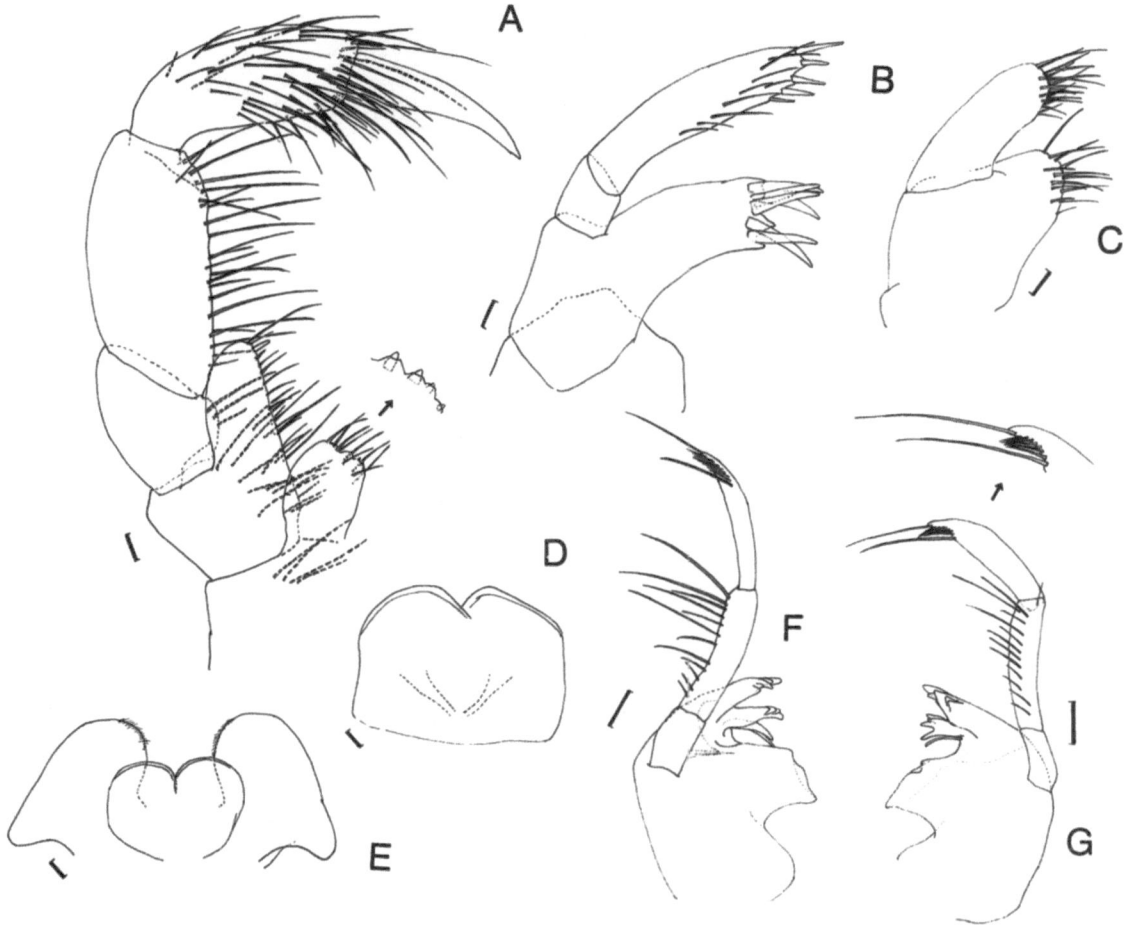

Fig. 3. Caprogammarus gurjanovae Kudrjaschov & Vassilenko, 1966. Off Kushiro, Hokkaido, Japan. Male. A, maxilliped; B, maxilla I; C, maxilla II; D, upper lip; E, lower lip; F, left mandible; G, right mandible. Scale bars equal 0.05 mm.

Oostegite on pereonite III (Fig. 5A) subtriangular, with several short setae on anterior margin and several long setae on posterior margin. Oostegite on pereonite IV with long setae on posterior margin.

Pereonite V (Fig. 5B) with genital openings located about midway along ventral surface.

Localities. Type locality: the vicinity of Paramushir Island, depth 90–210 m (Kudrjaschov & Vassilenko, 1966). Other localities: Habomai Islands (Shibotsu Island), Shikotan Island, Chishima Islands [= Kuril'skiye Ostrova] (Etorofu Island, Urup Island, Shimushir Island, Rasshua Island, Kruzenshterna Proliv, Chetvertyy Kuril'skiy Proliv), Kamchatka Peninsula (Okhotsk coast to Mys. Lopatka) and Sakhalin Island (Tatarskiy Proliv) (Kudrjaschov & Vassilenko, 1972) and off Kushiro and Akkeshi, Hokkaido (present study).

Remarks

The discovery of *Caprogammarus* Kudrjaschov & Vassilenko with peculiar characters intermediate between the Caprellidea and Gammaridea has caused confusion in the subordinal classification of amphipods. Initially, Kudrjaschov & Vassilenko (1966) placed *Caprogammarus* under the suborder Gammaridea for the following

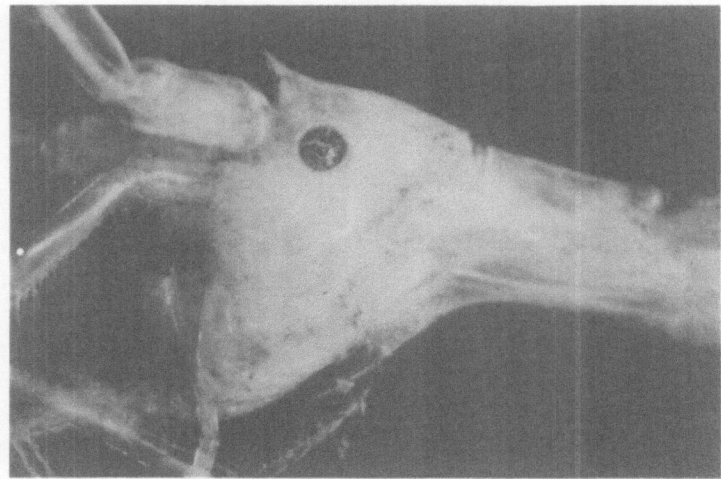

Fig. 4. Caprogammarus gurjanovae Kudrjaschov & Vassilenko, 1966. Off Kushiro, Hokkaido, Japan. Head and pereonite I of female.

reasons: 1) the head is free and not coalesced with pereonite I, 2) gnathopods I, II, pereopods V, VI, and VII have small but well-formed coxal plates, and 3) the abdomen consists of five free segments bearing three pairs of biramous pleopods and two pairs of uniramous uropods. McCain (1968), however, discussed the relationship between *Caprogammarus* and the long-bodied caprellid-like genus *Neoxenodice* Schellenberg (Gammaridea: Podoceridae), and concluded that *Caprogammarus* should be transferred to the suborder Caprellidea. He enumerated these reasons for his treatment: 1) *Caprogammarus* as well as the Caprellidea, has 2 [?1]-segmented pereopods III and

IV, while those pereopods in *Neoxenodice*, though reduced, are normally 7-segmented; 2) *Caprogammarus* bears only two pairs of gills as is common in the Caprellidea; 3) female *Caprogammarus* bears only two pairs of oostegites as is typical in the Caprellidea; and 4) most caprellids as well as *Caprogammarus* have small coxal plates on gnathopods I and II, and pereopods V to VII. His proposal was supported by McCain & Steinberg (1970), Arimoto (1976), Laubitz (1976, 1979), Bousfield (1978, 1983) and Schram (1989). In addition to McCain's (1968) opinion, Laubitz (1976) emphasized particularly that *Caprogammarus* and Caprellidea shared additional charac-

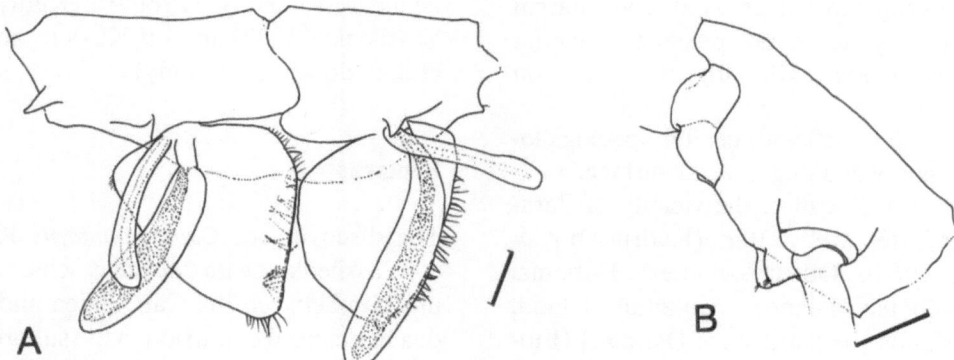

Fig. 5. Caprogammarus gurjanovae Kudrjaschov & Vassilenko, 1966. Off Kushiro, Hokkaido, Japan. Female. A, pereonites III and IV with pereopods, gills and oostegites; B, pereonite V with genital openings. Scale bars equal 1.0 mm.

teristics, i.e., the features of antennae I–II, maxilla I, maxilliped, mandibular palp and gnathopods I–II. McCain's (1968) treatment, however, has not been supported by Kudrjaschov & Vassilenko (1972), Vassilenko (1974, 1977) and Lowry (1976).

Furthermore, Barnard (1973) stated that 'the connection between the two groups [the Podoceridae and the suborder Caprellidea] is well seen in the Caprogammaridae and if any more such links are discovered, taxonomists may necessarily evaluate the Caprellidea as a superfamily and not a suborder...'. At this point in time, *Caprogammarus* was retained by Barnard in the Caprellidea and not included in the gammaridean superfamily Corophioidea, which is composed of the Podoceridae and five other families. Later, in their new classificatory arrangement of the Amphipoda, Barnard & Karaman (1983) proposed recognizing three suborders: Gammaridea, Hyperiidea, and Corophiidea. The new suborder Corophiidea was proposed on the basis of the possession of a solid fleshy telson which was thought to be plesiomorphic, and encompasses two sections then proposed: Corophiida and Caprelloida. The section Corophiida contains only the superfamily Corophioidea, and the section Caprellida includes two superfamilies: Caprelloidea and Cyamoidea. Barnard placed Caprogammaridae under the superfamily Corophioidea, while keeping other families of the old Caprellidea under the superfamily Caprelloidea (Karaman did not agree with Barnard on this point). So far only Barnard & Barnard (1983) has followed this treatment until now.

It is commonly accepted that the Caprellidea derived from podocerid-like gammarideans (McCain, 1968, 1970; Laubitz, 1976, 1979; Bousfield, 1978; Barnard & Karaman, 1983; Barnard & Barnard, 1983). Thus, it is suggested that Corophioidea is paraphyletic and the combination of the superfamily Corophioidea with the suborder Caprellidea would be monophyletic. At the present, however, the cladistics of many gammaridean families are hardly investigated. Until the cladistic relationship of gammaridean and caprellidean families can be clarified, it would

be better to keep the traditional scheme of classification that divides the order Amphipoda into four suborders, i.e., Gammaridea, Caprellidea, Hyperiidea and Ingolfiellidea.

The Hokkaido specimens examined herein agree with the original description (Kudrjaschov & Vassilenko, 1966) except for 1) the feature of articulation between the head and pereonite I, 2) 2-segmented flagellum of antenna II, 3) absence of the inner plate in maxilla I and 4) presence of the coalescent coxae in gnathopod II, as well as in pereopods V to VII.

Kudrjaschov & Vassilenko (1966; p. 197) noted 'the head which is free and not coalesced with thoracic somite I'. However, the articulation between the head and pereonite I in our *C. gurjanovae* is quite different from the other articulations between the pereon somites; it occurs as a shallow suture bent forward near the mouthparts. Consequently, the mouthparts appear conglutinated to both the head and pereonite I. Thus, it is appropriate to describe the status of the articulation in *Caprogammarus* as 'partially fused'.

The flagellum of antenna II in all specimens examined is 2-segmented. It differs from the 8-segmented situation of the original description (Kudrjaschov & Vassilenko, 1966; p. 194), but agrees with the redescription given by Kudrjaschov & Vassilenko (1972; p. 135).

Absence of the inner plate of maxilla I (Fig. 3B) does not correspond with the triangular inner plate reported in Kudrjaschov & Vassilenko (1966; p. 194), but with Kudrjaschov & Vassilenko's (1972) figure which shows no inner plate in the same appendage.

The coxae of gnathopod II, pereopods V to VII (Figs. 2C, F–H) in the Hokkaido specimens are tiny but clearly delineated from the sternite, not expanded into 'plates' as described by Kudrjaschov & Vassilenko (1966). These degenerated coxae have also been reported and/or figured in the Phtisicidae, Paracercopidae and Caprellidae of the suborder Caprellidea (e.g., McCain, 1968; Laubitz, 1970, 1972; Vassilenko, 1974; Hirayama, 1988; Takeuchi et al., 1989).

The status of the articulation between the head

290

and pereonite I in *Caprogammarus* is close to that found in the genus *Caprella* Lamarck. A detailed description of the articulation between the head and pereonite I has been neglected in the past taxonomical works on the Caprellidea. In the photograph of the head of *Caprella ungulina* Mayer, 1903 (Takeuchi *et al.*, 1989; p. 22), an extremely shallow lateral groove is recognizable. The same feature is also recognized in the figures for four species of *Caprella* by Hirayama & Kikuchi (1980; p. 185). Although the lateral suture in *Caprella* is extremely shallow, its position and direction are almost the same as those in *Caprogammarus*. On the contrary, in *Dulichia* and four related genera in the Podoceridae, the head is clearly separated from pereonite I and the mouthparts are conglutinated only to the head (Laubitz, 1977, and personal observation). Thus the similarity between the head of *Caprella* and *Caprogammarus* supports McCain's (1968) opinion that the genus *Caprogammarus* should be placed in the suborder Caprellidea.

On 25 May, 1980 at the 25th annual meeting of the Biogeographical Society of Japan held at the National Science Museum in Tokyo, Masaki Sakaguchi reported the occurrence of *Caprogammarus gurjanovae* off Akkeshi, in the north-east part of Hokkaido. His advance abstract of the report stated that the species was associated with 'tsubugai', which might refer to several species of the genus *Neptunea* of the gastropod family Buccinidae, collected from 100 m depth. The senior author (IT) found a specimen of *Caprogammarus* in Arimoto's private collection, which was labeled as collected by Masaki Sakaguchi. This specimen is probably the one that Sakaguchi reported.

Acknowledgments

We sincerely thank Prof. Ju-shey Ho, California State University, Long Beach, for his critical review of the manuscript, and the family of the late Dr. I. Arimoto for their kindness for the loan of the material. Mr. M. Sakaguchi, Nishinomiya-Higashi Senior High-School, kindly sent I.T., the senior author, the advance abstract of the 25th annual meeting of the Biogeographical Society of Japan. The localities in the Russian papers were read by Dr. S. Nagasawa, Ocean Research Institute, The University of Tokyo, to whom we must tender our cordial thanks.

References

Arimoto, I., 1976. Taxonomic studies of caprellids (Crustacea, Amphipoda, Caprellidae) found in the Japanese and adjacent waters. Spec. Publ. Seto mar. biol. Lab., Kyoto Univ., Ser. III, i–v + 229 pp.

Barnard, J. L., 1973. Revision of Corophiidae and related families (Amphipoda). Smithson. Contrib. Zool. 151: 1–27.

Barnard, J. L. & C. M. Barnard, 1983. Freshwater Amphipoda of the world. I. Evolutionary patterns. Hayfield Associates, Virginia, U.S.A., i–xvii + 358 pp.

Barnard, J. L. & G. S. Karaman, 1983. Australia as a major evolutionary centre for Amphipoda (Crustacea). Austr. Mus. Mem. 18: 45–61.

Bousfield, E. L., 1978. A revised classification and phylogeny of amphipod crustaceans. Trans. roy. Soc. Can., Ser. IV 16: 343–390.

Bousfield, E. L., 1983. An updated phyletic classification and palaeohistory of the Amphipoda. In F. R. Schram (ed.), Crustacean phylogeny. A.A. Balkema, Rotterdam, the Netherlands: 257–277.

Hirayama, A., 1988. A ghost shrimp with four-articulate fifth pereopods (Crustacea: Caprellidea: Phtisicidae) from Northwest Australia. Zool. Sci. 5: 1089–1093.

Hirayama, A. & T. Kikuchi, 1980. Caprellid fauna associated with subtidal algal beds along the coast of the Oshika Peninsula. Publ. Amakusa mar. biol. Lab., Kyushu Univ. 5: 171–188.

Kudrjaschov, V. A. & S. V. Vassilenko, 1966. A new family Caprogammaridae (Amphipoda, Gammaridea) found in the North-West Pacific. Crustaceana 10: 192–198.

Kudrjaschov, V. A. & S. V. Vassilenko, 1972. On the problem of the systematics, distribution and ecology of the amphipod *Caprogammarus gurjanovae* (Crustacea, Amphipoda, Family Caprogammaridae). Uchenye zapiski DVGU, Vladivostok 60: 134–147. (In Russian).

Laubitz, D. R., 1970. Studies on the Caprellidae (Crustacea, Amphipoda) of the American North Pacific. Nat. Mus. Can., Publ. biol. Oceanogr. 1: i–vii + 1–81.

Laubitz, D. R., 1972. The Caprellidae (Crustacea, Amphipoda) of Atlantic and Arctic Canada. Nat. Mus. Can., Publ. biol. Oceanogr. 4: 1–82.

Laubitz, D. R., 1976. On the taxonomic status of the family Caprogammaridae Kudrjaschov & Vassilenko (Amphipoda). Crustaceana 31: 145–150.

Laubitz, D. R., 1979. Phylogenetic relationships of the Podoceridae (Amphipoda, Gammaridea). Bull. biol. Soc. Washington 3: 144–152.

Lowry, J. K., 1976. *Neoxenodice cryophile*, a new podocerid from the Ross Sea, Antarctica (Amphipoda). Crustaceana 30: 98–104.

McCain, J. C., 1968. The Caprellidae (Crustacea: Amphipoda) of the Western North Atlantic. U.S. nat. Mus. Bull. 278: i-vi + 1–145.

McCain, J. C., 1970. Familial taxa within the Caprellidea (Crustacea: Amphipoda). Proc. biol. Soc. Washington 82: 837–842.

McCain, J. C. & J. E. Steinberg, 1970. Amphipoda I. Caprellidea I. Fam. Caprellidae. In H.-E. Grunter & L. B. Holthuis (eds.), Crustaceorum Catalogus 2: 1–78.

Schram, F. R., 1989. Crustacea. Oxford Univ. Press, New York, U.S.A., 606 pp.

Takeuchi, I., M. Takeda & K. Takeshita, 1989. Redescription of the bathyal caprellid, *Caprella ungulina* Mayer, 1903 (Crustacea, Amphipoda) from the North Pacific. Bull. nat. Sci. Mus., Tokyo, Ser. A (Zool.) 15: 19–28.

Vassilenko, S. V., 1974. Caprellids of the sea of USSR and adjacent waters. Opredel po Faune USSR 107: 1–287. (In Russian)

Vassilenko, S. V., 1977. A new species of amphipod *Caprogammarus micropleopodus* (Amphipoda, Caprogammaridae) inhabited on the shore of the Kurile Islands. Explorations of the Fauna of the Seas 21 (29). New Species and Genera of Marine Invertebrates: 60–66. (In Russian).

Hydrobiologia **223**: 293–299, 1991.
L. Watling (ed.), VIIth International Colloquium on Amphipoda.
© 1991 *Kluwer Academic Publishers.*

A new marine interstitial ingolfiellid (Crustacea, Amphipoda, Ingolfiellidea) from Tenerife and Hierro

Ronald Vonk[1] & Elias Sánchez[2]
[1]*Institute of Taxonomic Zoology, University of Amsterdam, P.O. Box 4766, 1009 AT Amsterdam, Netherlands*; [2]*Departamento de Zoología, Universidad de La Laguna, Tenerife, Spain*

Key words: Crustacea, Amphipoda, *Ingolfiella canariensis* n. sp., Tenerife, marine interstitial

Abstract

Ingolfiella canariensis n. sp., from coarse sand and gravel in the mediolittoral zone of Tenerife and Hierro, Canary Islands, is described. The new species shares supposedly apomorphous characters with species from comparable habitats from the Andaman Islands, Bermuda and Curaçao (Netherlands Antilles). The female of *Ingolfiella similis* Rondé-Broekhuizen & Stock, 1987, from Fuerteventura is also described.

Introduction

Along the rugged northwest coast of Tenerife many good sampling sites for interstitial stygobionts are present. The narrow volcanic sand beaches sheltered by capes and ridges of often recent lava outflows harbor *Ingolfiella, Bogidiella, Psammogammarus* and other still unidentified eyeless and microphthalmous amphipods.

The genus *Ingolfiella* now consists of 28 species, *I. canariensis* n.sp. included. Recently Ruffo & Vigna-Taglianti (1989) divided the genus into 7 subgenera. They drew the conclusion from their cladogram that *Ingolfiella* is divided into two distinct groups of subgenera characterized by the presence or absence of ocular lobes. The first group occurs in the marine environment (*Ingolfiella* s. str., *Hansenliella, Tethydiella*). The second group is associated with fresh- and anchihaline waters. The two groups seem well defined from an ecological point of view.

Biogeographically, they are less well defined for information about the male is often lacking in species from marine shallow waters. This may account for the disjunct areas in which species of these subgenera occur.

Since *I. canariensis* could not be placed in either the subgenera *Tethydiella* Ruffo & Vigna-Taglianti or *Gevgeliella* S. Karaman (sensu Stock, 1976) to which it comes closest when character states are compared, we refrain from subgeneric division.

Ingolfiella canariensis n. sp.

Material

TENERIFE (Canary Islands), 1 ♂ holotype. Station 88-594: Punta del Hidalgo, Playa de los Troches (UTM coordinates CS 37170 × 316110), 0.3 m under sediment surface, low-tide mark, Bou-Rouch (BR) biophreatical pump (see Bou, 1975); 12 Dec. 1988 (Zoölogisch Museum Amsterdam, ZMA, Amph. 108.650).

7 ♀♀ *paratypes*. Stn. 88–592: Punta del Hidalgo, beach in harbor (UTM coord. CS 36985 × 316015), 0.5 m under sediment surface, low-tide mark, BR pump; 10 Dec. 1988 (ZMA Amph. 108.651).

2 ♀♀ *paratypes*. Stn. 88–593: Punta del Hidalgo, beach left of harbor (UTM coord. CS 36975 × 316015), 0.4 m under sediment surface, low-tide mark, BR pump; 11 Dec. 1988 (ZMA Amph. 108.652).

4 ♀♀ *paratypes*. Stn. 88–596 B: boulevard of Punta del Hidalgo (UTM coord. CS 36987 × 316120), in gravel of rockpool in medio-littoral zone, 0.5 m under sediment surface, BR pump; 12 Dec. 1988 (ZMA Amph. 108.653).

3 ♀♀ *paratypes*. Stn. 88–616 A: Punta de Teno, small bay (UTM coord. CS 31175 × 313630), 0.4 m under sediment surface, low-tide mark, BR pump; 29 Dec. 1988 (ZMA Amph. 108.654).

1 ♂, 1 ♀ *paratypes*. Stn. 88–616 B: same locality as previous station, a few metres higher on the beach in coarse volcanic debris, 0.5 m under sediment surface, BR pump; 29 Dec. 1988 (ZMA Amph. 108.655).

1 ♀ *paratype*. Stn. 88–617; Punta de Teno, north beach (UTM coord. CS 31175 × 313655), strong swell, 0.2 m under sediment surface in coarse gravel between boulders, BR pump; 29 Dec. 1988 (ZMA Amph. 108.656).

1 ♂, 10 ♀♀ *paratypes*. Stn. 88–635: Mesa del Mar, beach left of hotel Sol y Mar (UTM coord. CS 36060 × 315325), tidepool at high-water mark, fine volcanic gravel, 0.3 m under sediment surface, BR pump; 16 Jan. 1989 (ZMA Amph. 108.657).

7 ♀♀ *paratypes*. Stn. 87831: boulevard of Punta del Hidalgo (UTM coord. CS 36987 × 313120), in washings of gravel and sand in rockpools in the mediolittoral zone; 23 Apr. 1987 (ZMA Amph. 108.658).

1 ♀ *paratype*. Stn. 87–20: same as previous station; 19 Apr. 1987 (ZMA Amph. 108.659).

EL HIERRO (Canary Islands)

1 ♂, 5 ♀♀ *paratypes*. Stn. 87–47: Tamaduste, harbor (UTM coord. BR 21515 × 308120), Karaman-Chappuis method (digging a hole) in muddy sand in mediolittoral zone; 29 Apr. 1987 (ZMA Amph. 108.660).

4 ♀♀ *paratypes*. Stn. 88–560: La Restinga, Jameos del Puerto (UTM coord. BR 20590 × 306075), in volcanic debris of anchiha-line cave, washings of gravel, conductivity 50.5 mS/cm; 13 Nov. 1988 (ZMA Amph. 108.661).

Description

Body length up to 1.8 mm. Body (Fig. 1) elongate; body somites about as long as high, bearing few setules. 'Ocular lobe' present, triangular.

First antenna (Fig. 3d) with 3 stout peduncle segments and 4-segmented flagellum: aesthetascs on 3 last flagellar segments and also a ribbed small aesthetasc-like structure on distal margin of terminal flagellum segment.

Second antenna (Fig. 3i) with 5-segmented peduncle (Fig. 1) and 5-segmented flagellum; terminal segment with aesthetasc-like structure on distal margin.

Right mandible (Fig. 4i) with 7 teeth on pars incisiva; lacinia mobilis with several small teeth; pars molaris sharply pointed. Left mandible (Fig. 4h) with fewer teeth on pars incisiva and lacinia mobilis; barbed spinules at base of lacinia mobilis.

First maxilla (Fig. 4a, j) with rounded inner lobe bearing 3 setae (2 in ♂); outer lobe with 6 spines of which innermost denticulate and the second trifid; 2-segmented palp with 2 apical setae.

Second maxilla (Fig. 4f) with 5 naked apical setae on outer lobe and 4 on inner lobe.

Maxilliped (Fig. 4d) with pointed endite bearing one seta; palp 5-segmented with 2 apical setae.

Coxal gills (Fig. 1, 4e) present on pereopods 3 to 5.

First gnathopod (Fig. 2d, e) with basis longer

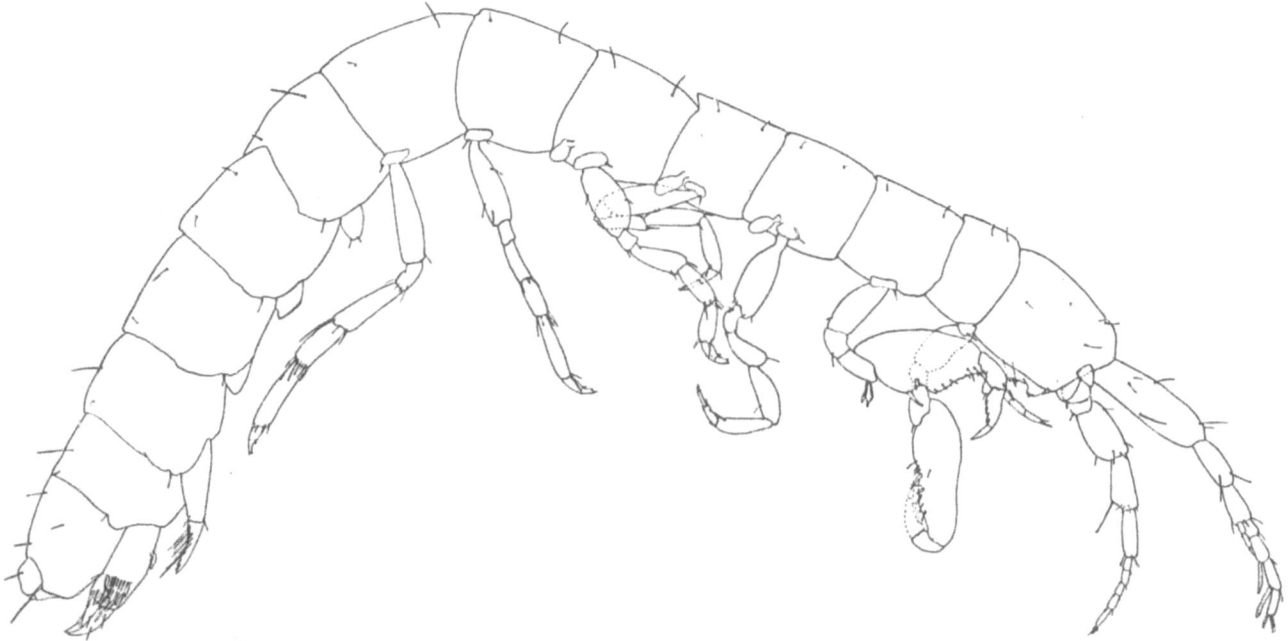

Fig. 1. *Ingolfiella canariensis* n. sp., male holotype, 1.8 mm. Playa de los Troches, Tenerife.

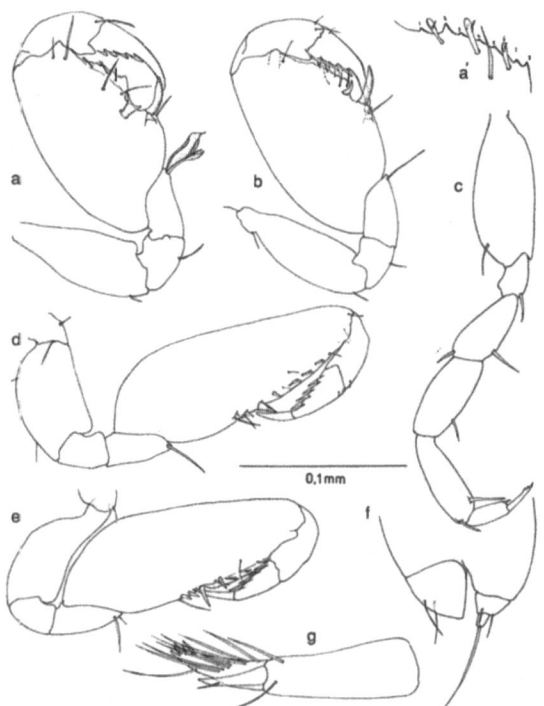

Fig. 2. *Ingolfiella canariensis* n. sp., paratypes. a, second gnathopod ♀, a′ inset drawn after a SEM photo; b, second gnathopod ♂; c, third pereopod ♀; d, first gnathopod ♂; e, first gnathopod ♂; f, telson, ♂; g, first uropod ♂.

than ischium and merus combined; merus with 1 seta anterodistally; carpus relatively elongate, palm bearing 4 spines: 1 long, 3 short and 3 Y-shaped spinules with a sensory hair coming out of the core; dactylus with 4 sharply pointed teeth.

Second gnathopod (Fig. 2a, b) carpus more oval than of first gnathopod; palm carrying 3 spines and 3 Y-shaped spinules; margin with 8 serrations (inset a′ of Fig. 2a drawn after a SEM photo). In the male a heavy forked spine near the palmar margin is present, as well as a foliaceous structure hanging from the posterior margin of the merus; dactylus with four sharp teeth.

Third pereopod (Fig. 2c) with trifid unguis (Fig. 4g). Fourth pereopod similar to third. Oostegites small and suboval (Fig. 4e), sometimes with 1 seta.

Fifth pereopod (Fig. 3a) robust, short, with broad basis; heavy spines on merus, carpus and propodus; dactylus separated from unguis by a very faint demarcation line; unguis bifid. Oostegite small, suboval. Coxal gill present.

Sixth pereopod (Fig. 3b) more slender and longer than fifth; dactylus faintly separated from unguis; unguis bifid.

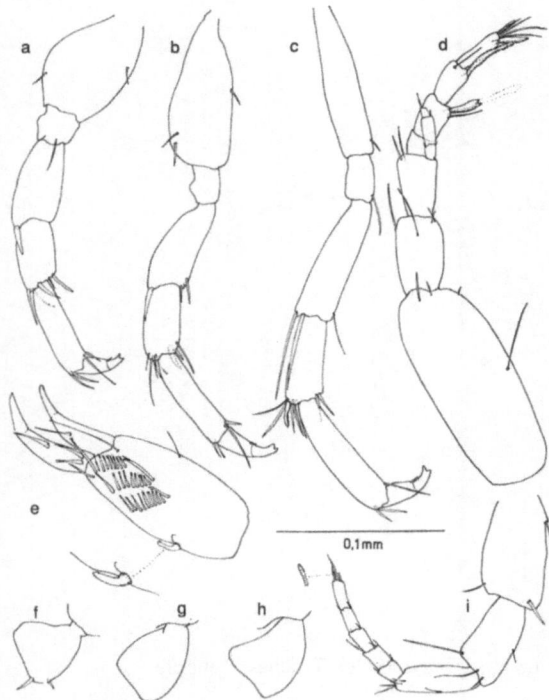

Fig. 3. *Ingolfiella canariensis* n. sp., paratypes. a, fifth pereopod ♂; b, sixth pereopod ♂; c, seventh pereopod ♂; d, first antenna ♂; e, second uropod ♀; f–h, pleopods 1–3 ♀; i, second antenna ♂.

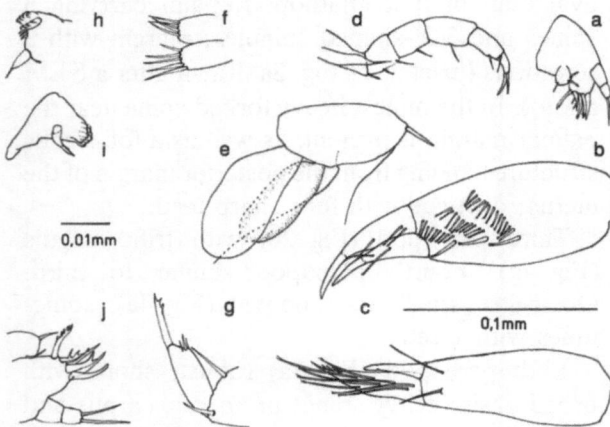

Fig. 4. *Ingolfiella canariensis* n. sp., paratypes. a, first maxilla ♀; b, second uropod ♀; c, first uropod ♀; d, maxilliped ♀; e, coxa of third pereopod with gill and oostegite; f, second maxilla ♂; g, dactylus of fourth pereopod ♂; h, left mandibular body ♂; i, right mandibular body ♂; j, first maxilla ♂.

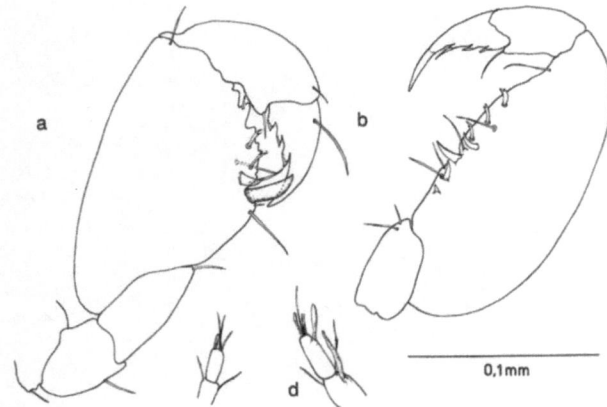

Fig. 5. *Ingolfiella similis* Rondé-Broekhuizen & Stock, 1987, female. a, second gnathopod; b, first gnathopod; c, tip of second antenna; d, tip of first antenna.

Seventh pereopod (Fig. 3c) more slender and longer than sixth; merus with 2 long spines on distoposterior corner; carpus with 6 distal spines and 1 spoon-shaped denticulate element; dactylus faintly separated from unguis; unguis bifid.

Pleopods (Fig. 3f, g, g) triangular. First pleopod in male with two distal spinules.

First uropod (Fig. 2g, 4c) biramous; peduncle with 2 setae; exopodite pointed, bearing 1 seta; endopodite longer and wider than exopodite; medial surface with 7 long setae; apex with 4 to 5 spiniform processes.

Second uropod (Fig. 3e, 4b) peduncle carrying 3 rows of spinules, bifid at tips; rami subequal, curved, pointed; exopodite with 3 setae, endopodite with 2 setae.

Third uropod (Fig. 2f) small, 2-segmented; first segment with 2 setae; second segment square at apex with 1 long distal seta.

Telson (Fig. 2f) with 2 setae.

Etymology

Named after the Canary Islands.

Remarks

The male differs from the female in the following respects:

— On the carpus of the second gnathopod a

broad, stubby, forked spine is present near the palmar margin.

– On the merus of the second gnathopod a transparent foliaceous process is present in the same position of a normal seta in the female (this process has a different structure than the 'sackartigen Gebilde' on the carpus of *I. petkovskii* S. Karaman, 1957 where it is clearly made up of two segments – rather like an aesthetasc – or than the spiniform and ciliated process as in, for instance, *I. putealis* Stock, 1977).

– There are 2 spinules at the distal and of the first pleopod (naked in ♂).

– Peduncle of the second uropod with a sub-basal, hook-shaped spine.

Only those species possessing a 4-toothed dactylus on the first and second gnathopod, and having dissimilar claws on the third to seventh pereopods, are considered. These conditions seem to be derived because they stand out against the 'normal' gammaridean condition (Dahl, 1977). Most other species in *Ingolfiella* have 0–3 teeth on the dactyli of the gnathopods and claws of similar form in the third to seventh pereopods. *I. kapuri* Coineau & Chandrasekhara, 1972 (Andaman and Nicobar Islands, mediolittoral), *I. longipes* Stock, Sket & Iliffe, 1987 (Bermuda, anchihaline), and *I. quadridentata* Stock, 1979 (Curaçao, infralittoral) share these character states with *I. canariensis*. The species are known through females only, however.

I. kapuri differs from *I. canariensis* in having: 3 palmar spines on the carpus of first gnathopod (versus 4), 7 serrations on palmar margin of carpus of second gnathopod (vs. 8); heavier unguicular spines pereopod dactyli; 1 spine on exopodite of first uropod (vs. 0 in ♀, 1 in ♂); seventh pereopod with 1 long spine on posterodistal corner of merus (vs. 2). The mouthparts of *I. kapuri* are not described.

I. longipes differs from *I. canariensis* in having: 3 spines and 2 setae on outer lobe of first maxilla (vs. 6); 1 distal spine on terminal segment of maxilliped palp (vs. 2); a spoon-shaped denticu-

lated element on the carpus of the seventh pereopod (vs. absent); 7 serrations on the palmar margin of second gnathopod carpus (vs. 8); first gnathopod without sensorial setae near edge of palmar margin; oval third pleopod (vs. triangular); bifid dactylus of third and fourth pereopods (vs. trifid).

I. quadridentata differs from *I. canariensis* in having: 1 palmar spine on carpus of first gnathopod (vs. 4); unguis of third and fourth pereopods simple with inner row of spinules (vs. trifid without spinules); carpus with 2 modified spines posterodistally (vs. absent). Of *I. quadridentata* the mouthparts are not described.

I. xarifae Ruffo, 1966, female, from shallow reefs in the Maldives, has a striking resemblance to *I. canariensis* especially with respect to gnathopod setation and spination, as well as other features of pereopodal dactyli. However, the dactylus of the first gnathopod has 3 teeth on the inner margin, while the second gnathopod has 4. This character needs to be checked in the two existing specimens, and the mouthparts are to be described.

Ingolfiella similis

Ref.: Rondé-Broekhuizen & Stock, 1987, p. 441–450

Material

FUERTEVENTURA (Canary Islands).

1 ♀. Station 87–119: end of Barranco de los Molinos (UTM coord. ES 59175 × 315760) method Karaman-Chappuis, conductivity 12.6 mS/cm; 6 May 1987 (ZMA Amph. 108.690).

1 ♂. Stn. 87–77: Las Playitas (UTM coord. ES 59990 × 312315) well about 500 m from the sea, cond. 10.7 mS/cm; 6 May 1987 (ZMA Amph. 108.689).

Description of female

Body similar to that of male; all appendages as in male (see Rondé-Broekhuizen & Stock, 1987) except for those described below: second gnathopod (Fig. 5a) without reversed element on lower margin of carpus; first pleopod without spines; second uropod without subbasal hook-shaped spine; oostegites oval.

Zoogeographic remarks

Geological data from the Canary Islands suggest that at least the western islands are separate volcanic edifices with an overall irregular and complex decrease in age from east to west. They evolved independently for at least the last 20 million years (Rondé-Broekhuizen, unpublished). Hierro, the youngest and westernmost island has a subaerial existence of 0.75 My, while Tenerife surfaced 15.7 My ago (Pitman & Talwani, 1972). They are separated by 3000 m deep water.

I. canariensis is found in marine groundwater of both islands. This fact suggests three possible explanations:
- *I. canariensis* spread, actively or passively (ships, not vicariance), from Tenerife to Hierro;
- it lives not only in the mediolittoral but also on the ocean floor and, when seabottom eruptions occur, follows the rise of a volcanic slope until it reaches shallow water.
- or *I. canariensis* invaded both islands only very recently, having evolved into a mediolittoral stygobiont from a common marine benthic ancestor.

The second explanation requires fewer speculative steps and is favored. In this case it should be possible to find *I. canariensis* also on the other western Canary Islands, i.e. La Palma, Gomera and Gran Canaria and in the seabed between those islands. For that matter; ingolfiellids have earlier been found in oceanfloor debris at great depths (*I. abyssi* Hansen, 1903; *I. atlantisi* Mills, 1967) and in dredge samples from bottoms of shallower water (*I. britannica* Spooner, 1960; *I. fuscina* Dojiri & Sieg, 1987).

Boxshall (1989) remarks that the idea of a marine crevicular fauna that is interconnected from one island to the other through deep water populations (Hart, Manning & Iliffe, 1985), can be termed a 'continuous crevicular corridor hypothesis'. He considers this hypothesis 'has considerable merit when explaining distributions between different islands within a particular archipelago, such as the Canary islands, but that it is implausible merit when explaining distributions between remote island systems surrounded by ocean floor covered by a significant depth of pelagic sediments'. As to the affinities with such distantly located species as *I. kapuri* (Andaman and Nicobar Island), *I. longipes* (Bermuda) and *I. quadridentata* (Curaçao); they all live in coastal habitats in the former realm of the circumtropical Tethys Sea (early Cretaceous, Smith *et al.*, 1981) and their phenetic similarities may stem from the time that gene flow was continuous.

Acknowledgements

In the course of the research project on the origin of insular groundwater biotas in the Atlantic financial support has come from

NATO Collaborative Research Grants Programme, contract SA. 5–2–05 (RG.0011/88); ERASMUS programme, contract ICP 88–0079 NL; NWO, Den Haag. The hospitality of the staff of the Departamento de Zoología, Universidad de La Laguna (Director: Prof. Dr. M. Ibáñez Genís) is gratefully acknowledged.

Also we wish to thank Drs. H.P. Wagner for staining part of the material, D. Platvoet for taking SEM photos and Prof. Dr. J.H. Stock for critically reading the manuscript and for providing material from Fuerteventura.

References

Bou, Cl., 1975. Les méthodes de récolte dans les eaux souterraines interstitielles. Annls. Spéléol. 29: 611–619.

Boxshall, G. A., 1989. Colonization of inland marine caves by mysophrioid copepods. J. Zool., Lond. 219: 521–526.

Coineau, N. & Rao G. Chandrasekhara, 1972. Isopodes et Amphipodes des sables intertideaux des îles Andaman et Nicobar (Golfe du Bengale). Vie Milieu (A) 22 (1A): 65–100.

Dahl, E., 1977. The amphipod functional model and its bearing upon systematics and phylogeny. Zool. Scr. 6: 221–228.

Dojiri, M. & J. Sieg, 1987. Ingolfiella fuscina, new species (Crustacea: Amphipoda) from the Gulf of Mexico and the Atlantic coast of North America, and partial redescription of I. atlantisi Mills, 1967. Proc. biol. Soc. Wash. 100: 494–505.

Hansen, H. J., 1903. The Ingolfiellidea, fam. n., a new type of Amphipoda. J. Linn. Soc. Zool 29: 117–133, pls. 14–15.

Hart, C. W., Jr., R. B. Manning & T. M. Iliffe, 1985. The fauna of Atlantic marine caves: evidence of dispersal by seafloor spreading while maintaining ties to deep waters. Proc. biol. Soc. Wash. 98: 288–292.

Karaman, S. L., 1957. Eine neue Ingolfiella aus Jugoslavien, Ingolfiella petkovskii n. sp. Folia balc. 1: 35–38.

Mills, E. L., 1967. Deep-sea Amphipoda from the western North Atlantic Ocean, 1. Ingolfiellidea and an unusual new species in the gammaridean family Pardaliscidae. Can. J. Zool. 45: 347–355.

Pitman, W. C. & M. Talwani, 1972. Sea floor spreading in the North Atlantic. geol. Soc. Am. Bull. 83: 619–646.

Rondé-Broekhuizen, B. & J. H. Stock, 1987. Stygofauna of the Canary Islands, 2: A new Ingolfiellid (Crustacea, Amphipoda) with West Indian affinities from the Canary Islands. Arch. Hydrobiol. 110: 441–450.

Ruffo, S., 1966. Ingolfiella xarifae (Crustacea Amphipoda) nuova species dell'Oceano Indiano. Mem. Mus. civ. St. Nat. Verona 14: 177–182.

Ruffo, S. & A. Vigna-Taglianti, 1989. Ricerche zoologiche della nave oceanografica 'Minerva' (C.N.R.) sulle isole circumsarde. III. Description of a new cavernicolous Ingolfiella species from Sardinia, with remarks on the systematics of the genus (Crustacea, Amphipoda, Ingolfiellidea). Ann. Mus. civ. St. Nat. Genova 87: 237–261.

Smith, A. G., A. M. Hurley & J. C. Briden, 1981. Phanerozoic paleocontinental world maps. Cambridge Earth Science Series: 1–102 (Cambridge University Press).

Spooner, G. M., 1960. The occurrence of Ingolfiella in the Eddystone shell gravel, with description of a new species. J. mar. biol. Ass. U.K. 39: 319–329.

Stock, J. H., 1977. The zoogeography of the crustacean suborder Ingolfiellidea with descriptions of new West Indian taxa. Stud. Fauna Curaçao 55: 131–146.

Stock, J. H., 1979. New data on taxonomy and zoogeography of Ingolfiellid Crustacea. Bijdr. Dierk. 49: 81–96.

Stock, J. H., B. Sket & T. M. Iliffe, 1987. Two new amphipod crustaceans from anchihaline caves in Bermuda. Crustaceana 53: 54–66.